# Representation Theory and Harmonic Analysis on Symmetric Spaces

CONTEMPORARY MATHEMATICS

714

# Representation Theory and Harmonic Analysis on Symmetric Spaces

AMS Special Session in Honor of Gestur Ólafsson's 65th Birthday
Harmonic Analysis
January 4, 2017
Atlanta, Georgia

Jens Gerlach Christensen
Susanna Dann
Matthew Dawson
Editors

AMERICAN MATHEMATICAL SOCIETY

2010 *Mathematics Subject Classification.* Primary 15A66, 17B40, 20C08, 22E46, 22E50, 32L25, 32M15, 43A90, 53C35, 81R40.

---

**Library of Congress Cataloging-in-Publication Data**

Names: Christensen, Jens Gerlach, 1975- editor. | Dann, Susanna, 1977- editor. | Dawson, Matthew, 1990- editor.

Title: Representation theory and harmonic analysis on symmetric spaces : AMS special session on harmonic analysis in honor of Gestur Ólafsson's 65th birthday, January 4, 2017, Atlanta, Georgia / Jens Gerlach Christensen, Susanna Dann, Matthew Dawson, editors.

Description: Providence, Rhode Island : American Mathematical Society, [2018] | Series: Contemporary mathematics ; volume 714 | Includes bibliographical references.

Identifiers: LCCN 2018005036 | ISBN 9781470440701 (alk. paper)

Subjects: LCSH: Ólafsson, Gestur. | Symmetric spaces–Congresses. | Harmonic analysis–Congresses. | Festschriften. | AMS: Linear and multilinear algebra; matrix theory – Basic linear algebra – Clifford algebras, spinors. msc | Nonassociative rings and algebras – Lie algebras and Lie superalgebras – Automorphisms, derivations, other operators. msc | Group theory and generalizations – Representation theory of groups – Hecke algebras and their representations. msc | Topological groups, Lie groups – Lie groups – Semisimple Lie groups and their representations. msc | Topological groups, Lie groups – Lie groups – Representations of Lie and linear algebraic groups over local fields. msc | Several complex variables and analytic spaces – Holomorphic fiber spaces – Twistor theory, double fibrations. msc | Several complex variables and analytic spaces – Complex spaces with a group of automorphisms – Hermitian symmetric spaces, bounded symmetric domains, Jordan algebras. msc | Abstract harmonic analysis – Abstract harmonic analysis – Spherical functions. msc | Differential geometry – Global differential geometry – Symmetric spaces. msc | Quantum theory – Groups and algebras in quantum theory – Symmetry breaking. msc

Classification: LCC QA649 .R47 2018 | DDC 516.3/62–dc23 LC record available at https://lccn.loc.gov/2018005036

Contemporary Mathematics ISSN: 0271-4132 (print); ISSN: 1098-3627 (online)

DOI: `http://dx.doi.org/10.1090/conm/714`

---



∞ The paper used in this book is acid-free and falls within the guidelines
established to ensure permanence and durability.
Visit the AMS home page at `http://www.ams.org/`

10 9 8 7 6 5 4 3 2 1    23 22 21 20 19 18

# Contents

# Preface

Dating back to the original work of Fourier on the solution to the heat equation using trigonometric series and the subsequent work of 19th- and early 20-century mathematicians to put it on a rigorous foundation, harmonic analysis has been an active area of research whose fruitful connections to other branches of mathematics have steadily grown stronger and more numerous over time. This volume, for instance, features articles in the areas of integral geometry, complex analysis, operator algebras, Lie algebras, special functions, and differential operators. The breadth of contributions showcases the diversity of current research in harmonic analysis and shows that it continues to be a vibrant and fruitful field of inquiry.

The connection between representation theory and harmonic analysis on symmetric spaces is, in particular, emphasized by many of the articles in this volume. This relationship has a long history, including many classical and famous results by Gelfand, Helgason, Harish-Chandra and others which have allowed, for instance, the development of a very precise description of the decomposition of the unitary, quasiregular representation associated to a Riemannian symmetric space $G/K$ as a direct integral of so-called $K$-spherical representations of $G$ with respect to a Plancherel measure given in terms of the beautiful c-function of Harish-Chandra. Several decades after these fundamental results, the role played by representation theory in analysis on symmetric spaces continues to be successfully exploited, as exhibited by many of the articles we have collected here.

This volume grew out of an AMS Special Session at the Joint Mathematics Meeting in Atlanta, Georgia in January, 2017. This session was held in honor of Gestur Ólafsson on his 65th birthday, and celebrated his many and wide-ranging contributions to the area of harmonic analysis. The session drew experts from around the globe, and several have also contributed to this volume. Gestur Ólafsson has made important contributions to harmonic analysis on symmetric spaces, unitary representations of semisimple Lie groups, reflection positivity in quantum field theory (including its connections with analysis on symmetric spaces), and wavelet theory, to name a few areas.

As students of Gestur, we are immensely grateful for the guidance and leadership he has shown us over the years. We have always found him to be exceptionally generous with his time and ideas, both as a mentor and as a collaborator. With this volume we wish to say both "thank you" and "congratulations" to our mentor.

We wish to thank the participants of the special session for making it such a success, as well as the contributors to this volume, which we hope will serve as a motivation and point of entry for young researchers, as well as provide valuable information for experts in this field. Thank you also to the many referees, without whose valued opinions this would not have been possible. Last but not least, we thank the American Mathematical Society for letting us organize the special session, as well as Chris Thivierge for her help during the preparation of this volume.

## List of Speakers

**AMS Special Session on Harmonic Analysis**
**(In Honor of Gestur Olafsson's 65th Birthday) I & II**
Joint Mathematics Meetings
Atlanta, GA
January 4, 2017

**Pierre Clare (Dartmouth College):** *Geometric pictures of intertwining operators.*

**Matthew Dawson (CIMAT, Mexico):** *Principal series representations of direct-limit groups.*

**Fulton B Gonzalez (Tufts University):** *Surjectivity of Mean Value Operators on Noncompact Symmetric Spaces.*

**Benjamin Harris (Bard College at Simon's Rock):** *Vogan-Zuckerman Characters and Semisimple Coadjoint Orbits.*

**Markus Hunziker (Baylor University):** *The Tricomi equation and complementary series representations of* $\mathrm{SL}(2,\mathbb{R})$.

**Toshiyuki Kobayashi (The University of Tokyo):** *Conformally Covariant Symmetry Breaking Operators on Differential Forms and Some Applications.*

**Sergii Myroshnychenko (Kent State University):** *On polytopes with congruent projections or sections.*

**Angela Pasquale (Université de Lorraine):** *Radial parts of differential operators and a one-parameter family of hypergeometric functions of type BC.*

**Raúl Quiroga-Barranco (CIMAT, Mexico):** *Multiplicity-free restrictions of holomorphic discrete series and Toeplitz operators.*

**Boris Rubin (Louisiana State University):** *New Inversion Formulas for the Horospherical Transform.*

**Siddhartha Sahi (Rutgers University):** *Macdonald hypergeometric functions.*

**Henrik Schlichtkrull (University of Copenhagen, Denmark):** *Real spherical spaces and their classification.*

**Robert J. Stanton (Ohio State University):** *Duality of geometries on flag manifolds associated to maximal parabolic subgroups in split G2.*

**Bent Ørsted (Aarhus University, Denmark):** *Homomorphisms between Verma modules and small representations.*

## Participants in the special session

From left to right: Raúl Quiroga-Barranco, Pierre Clare, Boris Rubin, Mark Sepanski, Markus Hunziker, Angela Pasquale, Susanna Dann, Toshiyuki Kobayashi, Matthew Dawson, Gestur Ólafsson, Sergii Myroshnychenko, Siddharta Sahi, Henrik Schlichtkrull, Ben Harris, Bent Ørsted, Robert Donley, Robert Stanton. Jens Christensen, Fulton Gonzalez

Photo courtesy of Gestur Ólafsson

# Gestur Ólafsson's contributions to Mathematics

## Books


(1) J. Hilgert and G. Ólafsson: Causal Symmetric Spaces, Geometry and Harmonic Analysis. Perspectives in Mathematics **18**, Academic Press, 1996
(2) R. Fabec and G. Ólafsson: Non-Commutative Harmonic Analysis. Drexville Publishing, 2014.
(3) K.-H. Neeb and G. Ólafsson, *Reflection Positivity: A Representation Theoretic Perspective.* To appear in Springer Briefs in Math. Physics., August 2018.


## Refereed articles and refereed chapters in books and special volumes


(1) Die Langlands-Parameter für die Flensted-Jensensche fundamentale Reihe. *Math. Scand.* **55** (1984) 229-244.
(2) Fourier and Poisson transformation associated to a semisimple symmetric space. *Invent. Math.* **90** (1987) 605-629.
(3) (with B. Ørsted) The holomorphic discrete series for affine symmetric spaces I. *Journal of Funct. Anal.* **81** (1988), 126-159.
(4) (Habilitation) Causal symmetric spaces. *Mathematica Gottingensis* **15** (1990)
(5) Symmetric Spaces of Hermitian Type. *Differential Geometry and Applications* **1** (1991), 195-233
(6) (with B. Ørsted) The holomorphic discrete series of an affine symmetric space and representations with reproducing kernels, *Trans. Amer. Math. Soc.* **326** (1991), 385-405.
(7) (with J. Hilgert and B. Ørsted) Hardy Spaces on Affine Symmetric Spaces. *J. reine und angew. Math.* **415** (1991), 189-218
(8) (with T. P. Branson) Equipartition of Energy for Waves in Symmetric Spaces. *J. Funct. Anal.* **97** (1991), 403-416.
(9) (with H. Schlichtkrull) Wave propagation on Riemannian symmetric space. *J. Funct. Anal.* **107** (1992) 270-278.
(10) (with B. Ørsted) Analytic continuation of Flensted-Jensen Representation. *Manuscripta Math.* **74** (1992), 5-23.
(11) (with J. Hilgert) Analytic extensions of representations, the solvable case. *Jap. Journ. Math.* **18** (1993) 213-290
(12) (with T. Branson and H. Schlichtkrull): A bundle valued Radon transform, with applications to invariant wave equations. *Quart. J. Math. Oxford* **45** (1994) 429-461.

(13) (with J. Faraut and J. Hilgert): Spherical functions on ordered symmetric spaces. *Ann. Inst. Fourier* **44** (1994), 927-966
(14) (with J. Faraut): Causal Semisimple Symmetric Spaces: The Geometry and Harmonic Analysis. In: Ed. Hofmann, Lawson, Vinberg: "Semigroups in Algebra, Geometry and Analysis", 3-32, 1995.
(15) (with T. Branson and H. Schlichtkrull): Huyghens' Principle in Riemannian Symmetric Spaces. *Math. Ann.* **301**, 445-462 (1995)
(16) (with T. Branson and B. Ørsted): Spectrum Generating Operators, and Intertwining Operators for Representations Induced from a maximal Parabolic Subgroup. *J. Funct. Anal.* **135** (1996) 163-205.
(17) (with B. Ørsted) Generalization of the Bargmann Transform. Proceedings of a "Workshop on Lie Theory and its Applications in Physics" Clausthal, August 1995. Ed. Dobrev, Döbner, Hilgert. World Scientific, 1996
(18) (with T. Branson): Helmholtz Operators and Symmetric Space Duality. *Invent. Math.* **129**, 63-74 (1997).
(19) (with B. Krötz and K-H. Neeb) Spherical Representations and Mixed Symmetric Spaces. *Representation Theory* **1**, 424-461 (1997).
(20) Open Problems in Harmonic Analysis on Causal Symmetric Spaces. p. 249-270. In: Positivity in Lie Theory; Open Problems. Ed. J. Hilgert, J. D. Lawson, K-H. Neeb, E. B. Vinberg, De Gruyter 1998.
(21) (with P. Jorgensen) Unitary Representations of Lie Groups with Reflection Symmetry. *J. Funct. Anal.* **158** (1998) 26-88 (1998).
(22) (With A. G. Helminck, J. Hilgert, A. Neumann) A Conjugacy Theorem for Symmetric Spaces. *Mathematische Annalen* **313** (1999), 785-791.
(23) (with T. Branson) Asymptotics of the D'Alembertian with Potential on a Pseudo-Riemannian Manifold. *Proceedings of the AMS* **127** (1999), 1339-1345.
(24) (With B. Ørsted) Causal Compactification and Hardy Spaces. *Trans. AMS* **351** (1999), 3771-3792.
(25) (with P. Jorgensen) Osterwalder-Schrader Axioms - Wightman Axioms. *Encyclopaedia of Mathematics, Supplement II* Kluwer, Jan. 2000.
(26) (with P. Jorgensen) Unitary representations and Osterwalder-Schrader Duality. Ed. R. S. Doran, V. S. Varadarajan: The Mathematical Legacy of Harish-Chandra: A Celebration of Representation Theory and Harmonic Analysis, PSPM, AM, 2000.
(27) (with A. Neumann) Minimal and Maximal Semigroups Related to Causal Symmetric Spaces. *Semigroup Forum* **61** (2000) 57–85.
(28) Analytic Continuation in Representation Theory and Harmonic Analysis. In: Global Analysis and Harmonic Analysis, ed. J. P. Bourguignon, T. Branson, and O. Hijazi. Seminares et Congr, vol 4, (2000), 201–233. Pub.: The French Math. Soc.
(29) (with N. Andersen) A Paley-Wiener Theorem for the Spherical Laplace Transform on Causal Symmetric Spaces of Rank One. *Proceedings of the AMS.* **129** (2001), 173–179.
(30) (with A. Pasquale) On the meromorphic extension of the spherical functions on noncompactly causal symmetric spaces. *J. Funct. Analysis* **181** (2001), 346–401.

(31) (with B. Krötz and K-H. Neeb) Spherical Functions on Mixed Symmetric Spaces. *Representation Theory,* **5** (2001), 43-92.

(32) (with F. Betten) Causal Compactification and Hardy Spaces for Spaces of Hermitian Type. *Pacific J. Math.* **200** (2001), 273–312.

(33) (with A. Pasquale) Regularity properties of generalized Harish-Chandra expansions. In: A. Strasburger et al. (eds.), Geometry and analysis on finite- and infinite-dimensional Lie groups , Banach Center Publications 55 (2002), 335-348. Banach Center Publications, 2002.

(34) (with M. Davidson and G. Zhang) Laguerre polynomials, restriction principle, and holomorphic representations of $\mathrm{SL}(2,\mathbb{R})$. *Acta Applicandae Mathematicae* **71** (3) (2002) 261–277.

(35) (with B. Krötz) The $c$-function for non-compactly causal symmetric spaces. *Invent. Math.* **149** (2002) 3, 647–659.

(36) (with R. Fabec) The continuous Wavelet transform and symmetric spaces. *Acta Applicandae Mathematicae* **87(1)** (2003), 261–277.

(37) (with S. Gindikin and B. Krötz) Hardy spaces for non-compactly causal symmetric spaces and the most continuous spectrum. *Math. Ann.* **327** (2003), 25–66.

(38) (with N. B. Andersen and H. Schlichtkrull) On the inversion of the Laplace and Abel Transforms for causal symmetric spaces. *Forum Math.* **15** (2003), 679–699.

(39) (with M. Davidson and G. Zhang) Laplace and Segal-Bargmann transforms on Hermitian symmetric spaces and orthogonal polynomials. *J. Funct. Anal.* **204** (2003), 157–195.

(40) (with Mark Davidson) Differential recursion relations for Laguerre functions on Hermitian matrices. *Integral Transforms and Special Functions* **14** (2003) 469–484.

(41) (with B. Krötz) The c-function for non-compactly causal symmetric spaces and its relations to harmonic analysis and representation theory Ed. S.G. Gindikin, Lie groups and symmetric spaces, In memory of F.I. Karpelevich. AMS Translations **210**, 171–194 (2003).

(42) (with M. Davidson) The Generalized Segal-Bargmann transform and Special Functions *Acta Applicandae Mathematicae,* **81** (2004), 29–50.

(43) (with A. Pasquale) Paley-Wiener theorems for the $\Theta$-spherical transform: An overview. *Acta Applicandae Mathematicae* **81** (2004), 275–309.

(44) (with D. Speegle) Wavelets, wavelet sets, and linear actions on $\mathbb{R}^n$, *Contemporary Mathematics (AMS)* **345**, *Wavelets, Frames and Operator Theory,* Eds: C. Heil, P. Jorgensen, D. Larson, 2004, 253–281.

(45) (with A. Pasquale) A Paley-Wiener Theorem for the $\Theta$-spherical Transform: The Even Multiplicity case. *Journal de mathématiques pures et appliquées* **83** (2004), 811–954.

(46) (with S. Gindikin and B. Krötz) Erratum: Holomorphic $H$-spherical distribution vectors in principal series representations. *Inventiones Mathematicae* **158** (2004), 683-684.

(47) (with S. Gindikin and B. Krötz) Holomorphic $H$-spherical distribution vectors in principal series representations. *Inventiones Mathematicae* **158** (2004), 643–682.

(48) (with B. Krötz and R. Stanton) The image of the heat kernel transform on Riemannian symmetric spaces of the noncompact type. **22** (2005), 1307–1329 *International Mathematics Research Notices.*

(49) Continuous action of Lie groups on $\mathbb{R}^n$ and frames. *International Journal of Wavelets, Multiresolution and Information Processing* **3** No. 2 (2005), 211–235.

(50) (with T. Branson and A. Pasquale) The Paley-Wiener Theorem for the Jacobi Transform and the Local Huygens' Principle for Root Systems with Even Multiplicities. *Indagationes* **16** (2005), 429–442. Special volume of *Indagationes* in honor of G. van Dijk.

(51) (with T. Branson and A. Pasquale) The Paley-Wiener Theorem and the local Huygens' principle for compact symmetric spaces. *Indagationes* **16** (2005), 393–428. Special volume of *Indagationes* in honor of G. van Dijk.

(52) (with M. Dobrescu) Wavelets without groups. *Contemp. Math.* **405** (2006), 27–40.

(53) (with A. Pasquale) Support properties and Holmgren's uniqueness theorem for differential operators with hyperplane singularities. *J. Funct. Anal.* **239** (2006), 21–43.

(54) (with S. Gindikin and B. Krötz) Horospherical model for the holomorphic discrete series and the horospherical Cauchy transform. *Compositio Mathematica* **142** (2006) 983–1008.

(55) (with R. Fabec and A. Sengupta) Fock spaces corresponding to positive definite linear transformations. *Math. Scand* **98** (2006) 262–282.

(56) (with M. Davidson and M. Aristidou) Differential Recursion Relations for Laguerre Functions on Symmetric Cones. *Bull. Sci. math.* **130** (2006), 346–263.

(57) (with E. Ournycheva, and B. Rubin) Higher-ranked wavelet transforms, ridgelets transforms, and Radon Transform on the space of matrices. *Applied and Computational Harmonic Analysis* **21** (2006) 182–203.

(58) (with S. Gindikin and B. Krötz) Holomorophic horospherical transform on non-compactly causal spaces. *IMRN* **2006** (2006), 1–47.

(59) (with S. Zheng) Function spaces associated with Schrödinger Operators: The Pöschl-Teller Potential. *J. Fourier Anal. and Appl.* **12** (2006), 653–674

(60) (with M. Aristidou and M. Davidson) Laguerre functions on symmetric cones and recursion relations in the real case. *J. Computational and Applied Mathematics.*, **199** (2007), 95–112.

(61) (with H. Schlichtkrull) The Segal-Bargmann transform for the heat equation associated with root systems. *Adv. Math.* **208** (1) (2007), 422–437.

(62) (with H. Schlichtkrull) Representation theory, Radon transform and the heat equation on a Riemannian symmetric space. *Group Representations, Ergodic Theory, and Mathematical Physics; A Tribute to George W. Mackey.* In: *Contemp. Math.*, **449** (2008), 315–344.

(63) (with H. Schlichtkrull) A local Paley-Wiener theorem for compact symmetric spaces. *Adv. Math.* **218** (2008), no. 1, 202–215.

(64) (with M. Dobrescu) Coxeter groups, wavelets, multiresolution and sampling. In *Frames and operator theory in analysis and signal processing*, 119–135, *Contemp. Math.*, **451**, Amer. Math. Soc., Providence, RI, 2008.

(65) (with S. Zheng) Harmonic Analysis Related to Schrödinger Operators. *Contemporary Math.* **464** (2008), 213–230.
(66) (with B. Rubin) Invariant functions on Grassmanians. *Contemporary math.* **464** (2008), 201–212.
(67) (with J. G. Christensen) Examples of coorbit spaces for dual pairs. *Acta Appl. Math.* **107** (2009), 25–48.
(68) (with D. Larson, and P. Massopust) Three-way tiling sets in two dimensions. *Acta Appl. Math.* **108** (2009), no. 3, 529–546.
(69) (with H. Schlichtkrull) Fourier Series on Compact Symmetric Spaces: $K$-Finite Functions of Small Support. Journal of Fourier Anal. and Appl. **16** (2010), 609–628.
(70) (with J. G. Christensen) Coorbit spaces for dual pairs. *Applied and Computational Harmonic Analysis* **31** (2011) 303–324.
(71) (with J. Wolf) Extension of Symmetric Spaces and Restriction of Weyl Groups and Invariant Polynomials. *Contemporary Math.* **544**, 85–100.
(72) (with S. Dann) Paley-Wiener Theorems with respect to the spectral parameter. *Contemporary Math.* **544** (2011), 55–84.
(73) (with H. Schlichtkrull) Fourier Transform of Spherical Distributions on Compact Symmetric Spaces. *Mathematica Scandinavica* **190** (2011), 93–113.
(74) (with A. Pasquale) The $\mathrm{Cos}^\lambda$ and $\mathrm{Sin}^\lambda$ transforms as intertwining operators between generalized principal series representations of $\mathrm{SL}(n+1,\mathbb{K})$. *Advances in Mathematics* **229** (2012), 267–293.
(75) (with A. Pasquale) Ramanujan's Master Theorem for Riemannian symmetric spaces. *Journal of Functional Analysis*, **262** (2012) 4851–4890.
(76) (with J. Christensen) and A. Mayeli) Coorbit description and atomic decomposition of Besov spaces. Numer. Funct. Anal. Optim. **33** (2012), no. 7–9, 847–871.
(77) (with M. Dawson and J. Wolf) Direct Systems of Spherical Functions and Representations, *Journal of Lie Theory* **23** (2013), 711–729.
(78) (with A. Pasquale and B. Rubin) Analytic and Group-Theoretic Aspects of the Cosine Transform.In: Eds: E. T. Quinto, F. Gonzalez, J. Christensen, Geometric Analysis and Integral Geometry, *Contemporary Math.* **598**, 167–188, Amer. Math. Soc., Providence, RI, 2013.
(79) (with R. Stanton) On the life and work of S. Helgason. In: Eds: E. T. Quinto, F. Gonzalez, J. Christensen, Geometric Analysis and Integral Geometry, *Contemporary Math.* **598**, (2013) p. 21–42,
(80) (with K. Wiboonton) The Segal-Bargmann transform on compact symmetric spaces and their direct limits. In (Eds. A. Hucleberry, I. Penkov, G. Zuckerman) Lie Groups: Structure, Actions, and Representations. In honor of Joseph A. Wolf on the Occasion of his 75th Birthday. 225–253, Progr. Math., **306**, Birkhäuser/Springer, New York, 2013.
(81) (with J. Christensen) Sampling in Spaces of Bandlimited Functions on Commutative Spaces. In: Eds: T. D.Andrews, R. Balan, J. J. Benedetto, W. Czaja, K. A. Okoudjou. Excursion in Harmonic Analysis, Volume 1, 135–69, Appl. Numer. Harmon. Anal., Birkhäuser/Springer, New York, 2013.

(82) (with B. Harris and H. He) The continuous spectrum in discrete series branching laws. *International Journal of Mathematics* **34** (2013).
(83) (with A. Pasquale) Ramanujan's master theorem for the hypergeometric Fourier transform associated with root systems. *J. Fourier Anal. Appl.*, **19** (2013), 1150–1183.
(84) (with J. Wolf) The Paley-Wiener theorem and limits of symmetric spaces. *J. Geometric Anal.*. **24** (2014), 1-31.
(85) (with K.-H. Neeb) Reflection positivity and conformal symmetry. *Journal of Functional Analysis.* **266** (2014), 2174–2224.
(86) (with J. Hilgert) The Radon transform and its dual for limits of symmetric spaces. In: Developments and Retrospectives in Lie Theory, Springer, 2014, 77–111.
(87) (with M. Dawson and R. Quiroga-Barranco) Commuting Toeplitz operators on bounded symmetric domains and multiplicity-free restrictions of holomorphic discrete series. *J. Funct. Anal.* **268** (2015), 1711–1732.
(88) (with S. Merigon and K-H. Neeb) Integrability of unitary representations on reproducing kernel spaces, *Representation Theory*, **19** (2015), 24–55.
(89) (with K-H. Neeb) Reflection positive one-parameter groups and dilations. *Complex Anal. Oper. Theory* **9** (2015), 653–721.
(90) (with K.-H. Neeb) Reflection positivity for the circle group. *J. of Physics: Conference Series* **597** (2015), 16 pages.
(91) (with V. Ho) An application of hypergeometric shift operators to the $\chi$-spherical Fourier transform. *Contemporary Math.* **650** (2015), 143–155.
(92) (with M. Dawson) A Survey of Amenability Theory for Direct-Limit Groups. *Contemporary Math.* **650** (2015), 89–109
(93) (with B. Harris and H. He) Wave Front Sets of Reductive Lie Group Representations. *Duke Math. J.* **165** (2016), 793–846.
(94) (with P. E. T. Jorgensen and K.-H. Neeb) Reflection positive stochastic processes indexed by Lie groups. SIGMA **12** (2016) 49 pages.
(95) (with J. Christensen and K. Gröchenig) New atomic decompositions for Bergman spaces on the unit ball. *Indiana Univ. Math. J* **66** (2017), 205–235.
(96) (with M. Dawson) Conical representations for direct limits of symmetric spaces. *Mathematische Zeitschrift*, **286** (2017), 1375–1419.
(97) J. Frahm, G. Ólafsson, and B. Ørated, *The Berlin form on symmetric R-spaces and reflection positivity.* Banach Center Publications **113** (2017), 134–168.
(98) K.-H. Neeb and G. Ólafsson, *Antiunitary representations and modular theory.* Banach Center Publications **113** (2017), 291–362.
(99) G. Ólafsson and R. Quiroga-Barranco, *On low-dimensional manifolds with isometric* $\mathrm{U}(p,q)$*-actions. Asian Journal of Mathematics* **21** (2017), 87-907.
(100) P. Jorgensen, K-H. Neeb, and G. Ólafsson, *Reflection positivity on real intervals. Semigroup Forum*, **96** (2018), 31–48.
(101) V. Ho and G. Ólafsson, *Paley-Wiener Theorem for Line Bundles over Compact Symmetric Spaces and New Estimates for the Heckman-Opdam Hypergeometric Functions.* Mathematische Nachrichten (2018), 25 pages. eprint = https://onlinelibrary.wiley.com/doi/pdf/10.1002/mana.201600148

## Book reviews

(1) *Lie Groups, Convex Cones and Semigroups*, by J. Hilgert, K.H. Hofmann and J.D.Lawson. Oxford Univ. Press. 1989. *Jahresbericht der DMV* **95**, (1992), 5-8
(2) *Holomorphy and Convexity in Lie Theory* by K.-H. Neeb, De Gruyter Expositions in Mathematics, 8, Berlin, New York, 2000 Jahresbericht der DMV **104** (2002)
(3) *Analysis and Probability; Wavelets, Signals, Fractals* by P. E. T. Jorgensen. Graduate Text in Mathematics, Springer, 2006. Journal of Approximation Theory, **162** (2010), 233–236.
(4) *Homogeneous spaces and equivariant embeddings* by D. A. Timashev. Encylopaedia of Mathematical Sciences **138**, Springer, 2011. In: Bulletin of the AMS **51** (2014), 349–359.

## Non refereed chapters in books, conference proceedings and other publications

(1) Several publications in *Mathematica Gottingensis*, publication of the *Sonderforschungsbereich Geometry and Analysis* at the University of Göttingen and material posted on my webpage.
(2) Ph.D. Thesis, Die Langlands-Klassifizierung, unitäre Darstellungen und die Flensted-Jensensche fundamentale Reihe. Göttingen, 1982.
(3) (with B. Ørsted) Is there an orbit method for affine symmetric spaces? In: The Orbit Method in Representation Theory, Proceedings of a Conference held in Copenhagen August to September 1988, Ed. M. Duflo, N.V. Pedersen, M. Vergne. Birkhäuser, 1990.
(4) Unitary Representations with Reflection Symmetry. In: Proceedings of an International Workshop on Lie Theory and its applications in physics. Ed. H-D. Dobner, V. K. Dobrev, J. Hilgert. World Scientific, 1999
(5) The $c$-function for symmetric spaces. A Contribution in *Proceedings of an International Workshop on Lie Theory and its applications in physics.* Ed. H-D. Dobner, V. K. Dobrev, J. Hilgert. World Scientific, 2001
(6) Groups, Wavelets, and Function Spaces. In: *Mathematisches Forschungsinstitut Oberwolfach Report No. 10/2004, Mini-Workshop: Wavelets and Frames, February 15th-February 21st, 2004*, Oberwolfach, 2004, 33–36
(7) (with H. Feichtinger, P. Jorgensen and D. Larson) Introduction to *Mathematisches Forschungsinstitut Oberwolfach Report No. 10/2004, Mini-Workshop: Wavelets and Frames, February 15th-February 21st, 2004*, Oberwolfach, 2004, 3–5
(8) (with T. Quinto) Introduction. In: Ed. G. Ólafsson and T. Quinto: The Radon Transform, Inverse Problems, and Tomography. Proceedings of Symposia in Applied Mathematics. AMS.
(9) (with E, Grinberg, P. Jorgensen, D. Larson, P. Massopust, T. Quinto, B. Rubin) Preface. In: Radon Transforms, Geometry, and Wavelets. *Contemporary Mathematics* **464**, vii–ix,AMS, 2008, Chief Editor: G. Ólafsson, editors E, Grinberg, P. Jorgensen, D. Larson, P. Massopust, T. Quinto, B. Rubin.

(10) Fourier-greining á víxlnum grúpum. *Verpill* 2009, 34–36. (Verpill is a journal for students in mathematics and physics at the University of Iceland. The English translation of the title is: Fourier analysis on Abelian groups.)

(11) (with H. Schlichtkrull) Preface to *The Selected Works of Surður Helgason*, ix–x, AMS, 2009. Editors: G. Ólafsson and H. Schlichtkrull.

(12) (with P. G. Casazza, P.E.T. Jorgensen, K.A. Kornelson, G. Kutyniok, D.R. Larson, P. Massopust, J.A. Packer, S. Silvestrov and Q. Sun) Prefase to a special volume of *Numerical Functional Analysis and Optimization* **33** (2012), 705–707.

(13) (with J. Christensen and S. D. Casey) Sampling, amenability and the Kunze-Stein phenomenon. IEEE Xplore-2015, International Conference on Sampling Theory and Applications (SampTA 2015) 68–72

## Graduate Students

(1) Co-advisor, main advisor Prof. H. Holdgrün. H. Thorleifsson (Göttingen, 1987): *Die Langlandsparameter der Leiterdarstellung von $U(p,q)$* (The Langlands parameters of the Ladder representation of $U(p,q)$).

(2) Co-advisor, main advisor Prof. H. Holdgrün. J. Kochmann (Göttingen, 1992): *Vertauschungsoperatoren zur holomorphen diskreten Reihe spezieller symmetrischer Räume* (Intertwining operators for holomorphic discrete series for special symmetric spaces).

(3) Co-advisor, main advisor Prof. H. Holdgrün. F. Betten (Göttingen, 1996): *Kausale Kompaktifizierung kompakt kausaler Räume* (Causal compactification of compactly causal spaces).

(4) T. Johansen (Louisiana State University, 2004): *Orbit Structure on the Silov Boundary of a Tube Domain and the Plancherel Decomposition of a Causally Compact Symmetric Space, with Emphasis on the Rank One Case.*

(5) I. Hossain (MS with concentration in Applications): *Query by Image Content Using Wavelets and Gibbs-Markov Random Fields.*

(6) M. Dobrescu (Louisiana State University, 2005): *Wavelet Sets with and without Groups and Multiresolution Analysis.*

(7) M. Aristidou (Louisiana State University, 2005): *Laguerre Functions Associated to Euclidean Jordan Algebras.*

(8) K. Wiboonton (Louisiana State University, 2009): *The Segal-Bargmann Transform on Inductive Limits of Compact Symmetric Spaces.*

(9) J. Christensen (Louisiana State University, 2009): *Function Spaces, Wavelets and Representation Theory.*

(10) S. Dann (Louisiana State University, 2011): *Paley-Wiener Theorems with Respect to the Spectral Parameter.*

(11) V. Ho (Louisiana State University, 2012): *Paley-Wiener Theorem for Line Bundles over Compact Symmetric Spaces.*

(12) M. Dawson (Louisiana State University, 2014): *Conical Representations for direct limits of Riemannian Symmetric Spaces.*

(13) C. A. Cross (Louisiana State University, 2015):*Partial Cosine-Funk Transforms at Poles of the Cosine-$\lambda$ Transform on Grassmann Manifolds.*

(14) A. Darweesh (Louisiana State University, 2015): *Wavelets, Coorbit Theory, and Projective Representation.* Co-advisor Jens G. Christensen.
(15) M. Hayajneh (Louisiana State University, 2016): *Twisted Reflection Positivity.*

## Postdoctoral Scholars

(1) Andreas Neumann (Lousiana State University, 1998–2000)
(2) Shijun Zheng (Lousiana State University, 2003–2005)
(3) Benjamin Harris (Lousiana State University, 2011–2014)
(4) Eli Vanney Roblero-Mendez (Lousiana State University, 2014)
(5) Stéphane Merigon (Lousiana State University, 2014)

Contemporary Mathematics
Volume **714**, 2018
https://doi.org/10.1090/conm/714/14376

# Schur $Q$-functions and the Capelli eigenvalue problem for the Lie superalgebra $\mathfrak{q}(n)$

Alexander Alldridge, Siddhartha Sahi, and Hadi Salmasian

ABSTRACT. Let $\mathfrak{l} := \mathfrak{q}(n) \times \mathfrak{q}(n)$, where $\mathfrak{q}(n)$ denotes the queer Lie superalgebra. The associative superalgebra $V$ of type $Q(n)$ has a left and right action of $\mathfrak{q}(n)$, and hence is equipped with a canonical $\mathfrak{l}$-module structure. We consider a distinguished basis $\{D_\lambda\}$ of the algebra of $\mathfrak{l}$-invariant super-polynomial differential operators on $V$, which is indexed by strict partitions of length at most $n$. We show that the spectrum of the operator $D_\lambda$, when it acts on the algebra $\mathscr{P}(V)$ of super-polynomials on $V$, is given by the factorial Schur $Q$-functions of Okounkov and Ivanov. As an application, we show that the radial projections of the spherical super-polynomials (corresponding to the diagonal symmetric pair $(\mathfrak{l}, \mathfrak{m})$, where $\mathfrak{m} := \mathfrak{q}(n)$) of irreducible $\mathfrak{l}$-submodules of $\mathscr{P}(V)$ are the classical Schur $Q$-functions. As a further application, we compute the Harish-Chandra images of the Nazarov basis $\{C_\lambda\}$ of the centre of $\mathfrak{U}(\mathfrak{q}(n))$.

## 1. Introduction

Let $G/K$ be a Hermitian symmetric space of tube type. The *Shilov boundary* of $G/K$ is of the form $G/P = K/M$, where $P = LN$ is the Siegel parabolic subgroup and $M = L \cap K$ is a symmetric subgroup of both $K$ and $L$. Let $\mathfrak{l}$, $\mathfrak{m}$, and $\mathfrak{n}$ be the complexified Lie algebras of $L$, $M$, and $N$, respectively. We set $V := \mathfrak{n}$ and regard $V$ as an $L$-module. In this setting, $V$ has the structure of a simple Jordan algebra.

The polynomial algebra $\mathscr{P}(V)$ decomposes as the multiplicity-free direct sum of simple $L$-modules $V_\lambda$, indexed naturally by partitions $\lambda$. In this situation one has canonical invariant "Capelli" differential operators of the form $\varphi^k \partial(\varphi)^k$, where $\varphi$ is the Jordan norm polynomial. The spectrum of these operators was computed by Kostant and Sahi [**11**, **12**], and a close connection with reducibility and composition factors of degenerate principal series was established by Sahi [**17**, **18**, **20**].

---

2010 *Mathematics Subject Classification.* Primary 17B10; Secondary 17B60, 58A50.

*Key words and phrases.* Capelli identity, queer Lie superalgebra, Schur $Q$-function.

Alexander Alldridge gratefully acknowledges support by the German Research Council (Deutsche Forschungsgemeinschaft DFG), grant nos. AL 698/3-1 and ZI 513/2-1, and the Institutional Strategy of the University of Cologne in the Excellence Initiative. The research of Siddhartha Sahi was partially supported by a Simons Foundation grant (509766) and of Hadi Salmasian by an NSERC Discovery Grant (RGPIN-2013-355464). This work was initiated during the Workshop on Hecke Algebras and Lie Theory, which was held at the University of Ottawa. Hadi Salmasian and Siddhartha Sahi thank the National Science Foundation (DMS-162350), the Fields Institute, and the University of Ottawa for funding this workshop.

Sahi showed [**19**] that the decomposition of $\mathscr{P}(V)$ in fact yields a distinguished basis $\{D_\lambda\}$, called the *Capelli basis*, of the subalgebra of $L$-invariant elements of the algebra $\mathscr{P}\mathscr{D}(V)$ of differential operators on $V$ with polynomial coefficients. Moreover, there is a polynomial $c_\lambda$, uniquely characterized by its degree, symmetry, and vanishing properties, such that $D_\lambda$ acts on each simple summand $V_\mu$ by the scalar $c_\lambda(\mu)$. The problem of characterizing the spectrum of the operators $D_\lambda$ is referred to as the *Capelli eigenvalue problem.*

In fact, Sahi [**19**] introduced a universal multi-parameter family of inhomogeneous polynomials that serve as a common generalization of the spectral polynomials $c_\lambda$ across all Hermitian symmetric spaces of rank $n$. Later, Knop and Sahi [**10**] studied a one-parameter subfamily of these polynomials, which already contains all the spectral polynomials. They showed that these polynomials are eigenfunctions of a class of difference operators extending the Debiard–Sekiguchi differential operators. It follows that the top degree terms of the Knop–Sahi polynomials are Jack polynomials, which for special choices of the parameter become spherical functions.

These polynomials were later studied from a different point of view by Okounkov and Olshanski, who referred to them as shifted Jack polynomials.

Subsequently, supersymmetric analogs of the Knop–Sahi shifted Jack polynomials were constructed by Sergeev and Veselov in [**28**]. More recently, two of us (Sahi and Salmasian [**22**]) have extended this circle of ideas to the setting of the triples $(\mathfrak{l}, \mathfrak{m}, V)$ of the form

$$\begin{gathered}(\mathfrak{gl}(m|n) \times \mathfrak{gl}(m|n), \mathfrak{gl}(m|n), \mathrm{Mat}_{m|n}(\mathbb{C})),\\ (\mathfrak{gl}(m|2n), \mathfrak{osp}(m|2n), S^2(\mathbb{C}^{m|2n})).\end{gathered} \tag{1.1}$$

In each of these situations one has, once again, a canonical Capelli basis of differential operators, and [**22**] establishes a precise connection to the abstract Capelli problem of Howe and Umeda [**6**]. It is further shown in Ref. [**22**] that the spectrum of the Capelli basis is given by specialisations of super analogues of Knop–Sahi polynomials, defined earlier by Sergeev and Veselov [**28**]. In the case of the triple $(\mathfrak{gl}(m|n) \times \mathfrak{gl}(m|n), \mathfrak{gl}(m|n), \mathrm{Mat}_{m|n}(\mathbb{C}))$, these results follow from earlier work of Molev [**15**], however the case $(\mathfrak{gl}(m|2n), \mathfrak{osp}(m|2n), S^2(\mathbb{C}^{m|2n}))$ is harder and requires new ideas.

The Lie superalgebras $\mathfrak{gl}(m|n)$ and $\mathfrak{osp}(m|2n)$ are examples of basic classical Lie superalgebras. Such an algebra admits an even non-degenerate invariant bilinear form and an even Cartan subalgebra, and many results for ordinary Lie algebras extend to this setting, see for instance Ref. [**2**], where spherical representations for the corresponding symmetric pairs are studied. In this paper, we show that the ideas of Ref. [**22**] can actually be extended to *non-basic* Lie superalgebras. More precisely, we consider the case of the *queer Lie superalgebra* $\mathfrak{q}(n)$, usually defined as the subalgebra of $\mathfrak{gl}(n|n)$ of matrices commuting with an odd involution [**9**]. For the present purposes, it is convenient to work with a slightly different realization of $\mathfrak{q}(n)$, which we describe below.

Let $\mathcal{E}$ be the $\mathbb{C}$-algebra generated by an odd element $\varepsilon$, with $\varepsilon^2 = 1$; thus as a superspace, $\mathcal{E} \cong \mathbb{C}^{1|1} \cong \mathbb{C} \oplus \mathbb{C}\varepsilon$. Let $\mathscr{A}$ be the associative superalgebra of $n \times n$ matrices with entries in $\mathcal{E}$. Then $\mathscr{A}$ is the associative superalgebra of type $Q(n)$, and $\mathfrak{q}(n)$ is isomorphic to $\mathscr{A}$ regarded as a Lie superalgebra via the graded commutator

$$[x, y] := xy - (-1)^{|x||y|}yx.$$

In fact $\mathscr{A}$ is also a Jordan superalgebra via the graded anticommutator, and an $\mathscr{A}$-bimodule via left and right multiplication. This bimodule structure induces a $\mathfrak{q}(n) \times \mathfrak{q}(n)$-module structure on $V := \mathscr{A}$.

In this paper, we consider the Capelli eigenvalue problem for the "diagonal" triple

$$(\mathfrak{l}, \mathfrak{m}, V) := \big(\mathfrak{q}(n) \times \mathfrak{q}(n), \mathfrak{q}(n), \mathscr{A}\big). \tag{1.2}$$

We establish a close connection with the Schur $Q$-functions $Q_\lambda$ and their inhomogeneous analogues, the *factorial Schur $Q$-functions* $Q^*_\lambda$, which were originally defined by Okounkov and studied by Ivanov [**7**]. Our main results are as follows. From Ref. [**4**], it is known that the space $\mathscr{P}(V)$ of super-polynomials on $V$ decomposes as a multiplicity-free direct sum of certain $\mathfrak{l}$-modules $V_\lambda$, which are parametrised by strict partitions $\lambda$ of length at most $n$. It follows that $\mathscr{P}(V^*)$ decomposes as a direct sum of the contragredient $\mathfrak{l}$-modules $V^*_\lambda$. In Section 4.3, we describe a certain even linear slice $\mathfrak{t}^*$ to the $M$-orbits on $V^*$. If $p$ is an $\mathfrak{m}$-invariant super-polynomial on $V^*$, then it is uniquely determined by its restriction to $\mathfrak{t}^*$. This restriction is an ordinary polynomial, and we call it the $\mathfrak{m}$*-radial part of* $p$.

THEOREM 1.1. *For every $\lambda$, the $\mathfrak{l}$-module $V^*_\lambda$ contains an $\mathfrak{m}$-spherical super-polynomial $p^*_\lambda$, which is unique up to a scalar multiple. Moreover, up to a scalar, the $\mathfrak{m}$-radial part of $p^*_\lambda$ is the Schur $Q$-function $Q_\lambda$.*

This is proved in Theorem 4.5 below. Now consider the algebra $\mathscr{P}\mathscr{D}(V)$ of polynomial coefficient differential operators on $V$. It has an $\mathfrak{l}$-module decomposition

$$\mathscr{P}\mathscr{D}(V) \cong \bigoplus_{\lambda,\mu} V_\mu \otimes V^*_\lambda \cong \bigoplus_{\lambda,\mu} \underline{\mathrm{Hom}}_{\mathfrak{l}}(V_\lambda, V_\mu),$$

and we write $D_\lambda$ for the differential operator corresponding to the identity map $\mathrm{id}_{V_\lambda} \in \underline{\mathrm{Hom}}_{\mathfrak{l}}(V_\lambda, V_\mu)$. The $D_\lambda$ are the Capelli operators, and they form a basis for the $\mathfrak{l}$-invariant differential operators acting on $\mathscr{P}(V)$. The operator $D_\lambda$ acts on each irreducible component $V_\mu$ of $\mathscr{P}(V)$ by a scalar eigenvalue $c_\lambda(\mu)$.

THEOREM 1.2. *The eigenvalues of the Capelli operator $D_\lambda$ are given by the factorial Schur $Q$-function $Q^*_\lambda$. More precisely, for all $\lambda, \mu$, we have*

$$c_\lambda(\mu) = \frac{Q^*_\lambda(\mu)}{Q^*_\lambda(\lambda)}.$$

In fact, we prove Theorem 1.2 *first* (see Theorem 3.8 below) and then use it to prove Theorem 1.1.

Compared to the cases considered in Equation (1.1), the situation in Equation (1.2) is more complicated. First, since the Cartan subalgebra of $\mathfrak{q}(n)$ is not purely even, the highest weight space of an irreducible finite dimensional $\mathfrak{q}(n)$-module is not necessarily one-dimensional. Second, unlike the basic classical cases, the tensor product of two irreducible $\mathfrak{q}(n)$-modules is not necessarily an irreducible $\mathfrak{q}(n) \times \mathfrak{q}(n)$-module, and sometimes decomposes as a direct sum of two modules which are isomorphic up to parity change. Third, the $\mathfrak{m}$-spherical vectors in $\mathscr{P}(V)$ are purely odd, whereas the $\mathfrak{m}$-spherical vectors in $\mathscr{P}(V^*)$ are purely even. These issues add to the difficulties that arise in the proofs in the case of the symmetric pair in Equation (1.2).

In [**27**, Theorem 3], Sergeev introduced the $\mathfrak{q}(n)$-analogue of the Harish-Chandra isomorphism

$$\eta : \mathscr{Z}(\mathfrak{q}(n)) \longrightarrow \mathscr{P}(\mathfrak{h}_{\bar{0}}), \tag{1.3}$$

where $\mathfrak{h}_{\bar{0}}$ denotes the even part of the Cartan subalgebra of $\mathfrak{q}(n)$. Theorem 1.2 can be reformulated in terms of the map $\eta$, as follows. The image of $\eta$ can be naturally identified with the space of $n$-variable $Q$-symmetric polynomials (see Section 3.2). We denote the actions of the first and second factors of $\mathfrak{l} = \mathfrak{q}(n) \times \mathfrak{q}(n)$ on $\mathscr{P}(V)$ by $L$ and $R$, respectively. Since the typical "$\varrho$-shift" for the Sergeev–Harish-Chandra isomorphism is equal to zero, we obtain the following reformulation of Theorem 1.2.

THEOREM 1.3. *For every Capelli operator $D_\lambda$, there exists a unique central element $z_\lambda \in \mathscr{Z}(\mathfrak{q}(n))$ such that $L(z_\lambda) = D_\lambda$. Furthermore,*

$$\eta(z_\lambda)(\mu) = \frac{Q^*_\lambda(\mu)}{Q^*_\lambda(\lambda)}.$$

The setting of the present paper was also considered by Nazarov, who constructed [**16**, Equation (4.7)] a family of invariant differential operators $\{I_\lambda\}$ using characters of the Sergeev algebra [**27**]. Nazarov also defined [**16**, Equation (4.6)] certain explicit "Capelli" elements $\{C_\lambda\}$ in $\mathscr{Z}(\mathfrak{q}(n))$, and proved [**16**, Corollary 4.6] that $I_\lambda = \gamma(C_\lambda)$, where $\gamma$ is the left action of $\mathfrak{q}(n)$ on $V$.

Although our operators $\{D_\lambda\}$ and central elements $\{z_\lambda\}$ are *different* from the $\{I_\lambda\}$ and $\{C_\lambda\}$ defined by Nazarov, one can make an *a posteriori* connection using our Proposition 3.6 below. This allows us to compute the Harish-Chandra image of Nazarov's central elements $\{C_\lambda\}$. The following result follows immediately from Theorem 4.9.

THEOREM 1.4. *The Harish-Chandra image of the operator $C_\lambda$ is given by*

$$\eta(C_\lambda)(\mu) = k_\lambda Q^*_\lambda(-\mu),$$

*where*

$$k_\lambda := (-1)^{|\lambda|}\lambda_1! \cdots \lambda_{\ell(\lambda)}! \prod_{1 \leqslant i < j \leqslant \ell(\lambda)} \frac{\lambda_i + \lambda_j}{\lambda_i - \lambda_j}.$$

We would like to mention that the polynomials $Q_\lambda$ occur in a further different scenario related to the Lie superalgebra $\mathfrak{q}(n)$. In [**26**, Theorem 1.7], Sergeev showed that the radial parts of the bispherical matrix coefficients on $\mathfrak{q}(n)\times\mathfrak{q}(n)$ with respect to the diagonal and twisted-diagonal embeddings of $\mathfrak{q}(n)$ in $\mathfrak{q}(n) \times \mathfrak{q}(n)$, are Schur $Q$-polynomials. It will be interesting to explore possible connections between our work and Sergeev's result.

We remark that it is possible to extend the results of the present paper and of Ref. [**22**] to the common setting of multiplicity-free actions on Jordan superalgebras. This will be established in a forthcoming paper [**24**]. In addition, recently Sahi and Salmasian [**23**] constructed quadratic analogues of Capelli operators on Grassmannian manifolds by lifting the Capelli basis of [**19**] via a double fibration. In the near future they plan to consider the analogous problem in the super setting.

We conclude this introduction with a brief synopsis of our paper. In Section 2, we realise the Lie superalgebras relevant to us in terms of supermatrices. In Section 3, we identify the action of $\mathfrak{l}$ on $\mathscr{P}(V)$, construct the Capelli basis, and determine the eigenvalue polynomials (Theorem 3.8). Finally, in Section 4, we study the open orbits in $V$ and $V^*$, show the existence of $\mathfrak{m}$-invariant functionals for the simple summands of $\mathscr{P}(V)$, and prove that the spherical polynomials thus defined are the classical Schur $Q$-functions (Theorem 4.5).

*Acknowledgments.* Alexander Alldridge wishes to thank the University of Ottawa, the Institute for Theoretical Physics at the University of Cologne, and the Department of Mathematics of the University of California at Berkeley for their hospitality during the preparation of this article.

The authors thank Vera Serganova and Weiqiang Wang for stimulating and fruitful conversations during the workshop, which paved the way for the present article. We also thank Alexander Sergeev for bringing Ref. [**26**] to our attention.

## 2. Lie superalgebras

The triple $(\mathfrak{l}, \mathfrak{m}, V)$ given in Equation (1.2) can be embedded inside the Lie superalgebra $\mathfrak{q}(2n)$, which can be further embedded inside $\mathfrak{gl}(2n|2n)$. This provides a concrete realisation which allows us to express the Lie superalgebras of interest as $4n \times 4n$ matrices. In order to describe it, it will be convenient to consider three commuting involutions of the algebra $\mathfrak{gl}(2n|2n)$. To this end, first we equip the space $\mathrm{Mat}_{4n\times 4n}(\mathbb{C})$ of $4n \times 4n$ complex matrices with a Lie superalgebra structure isomorphic to $\mathfrak{gl}(2n|2n)$. Instead of supermatrices in standard format, we prefer to consider those of the shape

$$x = \begin{pmatrix} A & B \\ C & D \end{pmatrix}, \quad A, B, C, D \in \mathfrak{gl}(n|n). \tag{2.1}$$

Equipped with the signed matrix commutator, the space of such matrices forms a Lie superalgebra $\mathfrak{g}$ isomorphic to $\mathfrak{gl}(2n|2n)$, the isomorphism being given by conjugation by the $4n \times 4n$ matrix

$$\widetilde{I} := \begin{pmatrix} I & 0 & 0 & 0 \\ 0 & 0 & I & 0 \\ 0 & I & 0 & 0 \\ 0 & 0 & 0 & I \end{pmatrix},$$

where $I := I_{n\times n}$ denotes the $n \times n$ identity matrix. Next let $\sigma, \varphi$, and $\theta$ be three involutions on $\mathfrak{g}$, given respectively by conjugation by the matrices

$$\Sigma := \begin{pmatrix} I & 0 & 0 & 0 \\ 0 & I & 0 & 0 \\ 0 & 0 & -I & 0 \\ 0 & 0 & 0 & -I \end{pmatrix}, \quad \Phi := \begin{pmatrix} 0 & I & 0 & 0 \\ I & 0 & 0 & 0 \\ 0 & 0 & 0 & I \\ 0 & 0 & I & 0 \end{pmatrix}, \quad \Theta := \begin{pmatrix} 0 & 0 & I & 0 \\ 0 & 0 & 0 & I \\ I & 0 & 0 & 0 \\ 0 & I & 0 & 0 \end{pmatrix}.$$

It is straightforward to verify that $\Sigma\Phi = \Phi\Sigma$, $\Phi\Theta = \Theta\Phi$, and $\Sigma\Theta = -\Theta\Sigma$. Hence, the involutions $\sigma, \varphi$, and $\theta$ commute with each other.

**2.1. The involution $\sigma$.** The subspace $\mathfrak{g}^\sigma$ of fixed points of $\sigma$ equals $\mathfrak{gl}(n|n) \times \mathfrak{gl}(n|n)$ with elements of the form

$$\begin{pmatrix} A & 0 \\ 0 & D \end{pmatrix} \text{ where } A, D \in \mathfrak{gl}(n|n).$$

The subspace $\mathfrak{g}^{-\sigma}$ of fixed points of $-\sigma$ consists of matrices of the form

$$\begin{pmatrix} 0 & B \\ C & 0 \end{pmatrix} \text{ where } B, C \in \mathfrak{gl}(n|n).$$

Thus, as a super-vector space, $\mathfrak{g}^{-\sigma} = \mathfrak{g}_+^{-\sigma} \oplus \mathfrak{g}_-^{-\sigma}$ where $\mathfrak{g}_\pm^{-\sigma}$ are respectively the spaces of $4n \times 4n$ matrices of the form

$$\begin{pmatrix} 0 & B \\ 0 & 0 \end{pmatrix} \text{ and } \begin{pmatrix} 0 & 0 \\ C & 0 \end{pmatrix}$$

with $B, C \in \mathfrak{gl}(n|n)$. In fact, $\mathfrak{g}_\pm^{-\sigma}$ are the $\pm 2$-eigenspaces of $\mathrm{ad}(\Sigma)$, where we think of $\Sigma$ as an element of $\mathfrak{g}$. Therefore, $\mathfrak{g}_\pm^{-\sigma}$ are abelian subalgebas of $\mathfrak{g}$, and together with $\mathfrak{g}^\sigma$, they form a $\mathbb{Z}$-grading of $\mathfrak{g}$. The action of $\mathfrak{g}^\sigma$ on $\mathfrak{g}_+^{-\sigma}$ is given explicitly by

$$[x, v] = \begin{pmatrix} 0 & AB - (-1)^{|B||D|}BD \\ 0 & 0 \end{pmatrix}, \tag{2.2}$$

for all homogeneous

$$x = \begin{pmatrix} A & 0 \\ 0 & D \end{pmatrix} \in \mathfrak{g}^\sigma, \quad v = \begin{pmatrix} 0 & B \\ 0 & 0 \end{pmatrix} \in \mathfrak{g}_+^{-\sigma}.$$

In what follows, we set $U := \mathfrak{g}_+^{-\sigma}$ and identify it as a super-vector space with $\mathfrak{gl}(n|n)$, *via* the map

$$\mathfrak{gl}(n|n) \longrightarrow U : B \longmapsto \begin{pmatrix} 0 & B \\ 0 & 0 \end{pmatrix}. \tag{2.3}$$

**2.2. The involution $\varphi$.** The involution $\varphi$ is induced by the parity reversing automorphism of $\mathbb{C}^{2n|2n}$, and therefore the subalgebra $\mathfrak{g}^\varphi$ of fixed points of $\varphi$ is isomorphic to $\mathfrak{q}(2n)$. It consists of all $x \in \mathfrak{g}$ as in Equation (2.1) such that the blocks $A, B, C, D \in \mathfrak{q}(n)$. Furthermore,

$$\mathfrak{l} := \mathfrak{g}^{\varphi,\sigma} \cong \mathfrak{q}(n) \times \mathfrak{q}(n).$$

We define $V := \mathfrak{g}^\varphi \cap U$. Then it is clear that

$$[\mathfrak{l}, V] \subseteq \mathfrak{g}^\varphi \cap [\mathfrak{g}^\sigma, \mathfrak{g}_+^{-\sigma}] \subseteq \mathfrak{g}^\varphi \cap \mathfrak{g}_+^{-\sigma} = V.$$

The restriction of the map defined in Equation (2.3) yields an identification of $V$ with a subspace of $\mathfrak{gl}(n|n)$ which carries the structure of $\mathfrak{q}(n)$.

**2.3. The involution $\theta$.** The algebra $\mathfrak{g}^{\sigma,\theta}$ of fixed points of both $\sigma$ and $\theta$ is isomorphic to $\mathfrak{gl}(n|n)$, and realised by supermatrices

$$\begin{pmatrix} A & 0 \\ 0 & A \end{pmatrix}, \quad A \in \mathfrak{gl}(n|n). \tag{2.4}$$

From Equation (2.2), it follows that the action of $\mathfrak{g}^{\sigma,\theta}$ on $U$ is precisely the adjoint action of $\mathfrak{gl}(n|n)$. Furthermore,

$$\mathfrak{m} := \mathfrak{l}^\theta = \mathfrak{g}^{\varphi,\sigma,\theta} \cong \mathfrak{q}(n).$$

It is realised by supermatrices of the form as in Equation (2.4) where in addition $A \in \mathfrak{q}(n)$. Moreover, the action of $\mathfrak{m}$ on $V$ is precisely the adjoint action of $\mathfrak{q}(n)$. For the following lemma, let $e \in V \subseteq U$ be the element correponding to the matrix

$$\begin{pmatrix} 0 & I_{2n\times 2n} \\ 0 & 0 \end{pmatrix},$$

and set

$$\mathfrak{s} := \mathfrak{l}^{-\theta} = \mathfrak{g}^{\varphi,\sigma,-\theta}.$$

LEMMA 2.1. *Let* $\mathsf{p} : \mathfrak{g}^\sigma \longrightarrow U$ *be the linear map defined by* $x \longmapsto x \cdot e$.

(i) *The map* $\mathsf{p}$ *is a surjection onto* $U$ *with kernel* $\mathfrak{g}^{\sigma,\theta}$. *Its restriction to* $\mathfrak{g}^{\sigma,-\theta}$ *is a linear isomorphism.*

(ii) *The restriction of* $\mathsf{p}$ *to* $\mathfrak{l}$ *is a surjection onto* $V$ *with kernel* $\mathfrak{m}$. *Its restriction to* $\mathfrak{s}$ *is a linear isomorphism onto* $V$.

(iii) *Equipped with the binary operation* $(x\cdot e)\circ(y\cdot e) := [x,[y,e]]$ *for* $x, y \in \mathfrak{g}^{\sigma,-\theta}$ *(respectively,* $x, y \in \mathfrak{s}$*) the super-vector space* $U$ *(respectively,* $V$*) becomes a Jordan superalgebra.*

PROOF. Parts (i) and (ii) follow from straightforward calculations. For (iii), it suffices to prove the statement for $U$. Note that according to Equation (2.2), we have

$$x \cdot e = \begin{pmatrix} 0 & 2A \\ 0 & 0 \end{pmatrix}, \quad \forall x = \begin{pmatrix} A & 0 \\ 0 & -A \end{pmatrix} \in \mathfrak{g}^{\sigma,-\theta}.$$

Thus, a direct calculation shows that up to normalization, $A \circ B$ coincides for $A, B \in U$ with the super-anticommutator $AB + (-1)^{|A|\cdot|B|}BA$. □

## 3. The eigenvalue polynomials

**3.1. The action of $\mathfrak{l}$ on polynomials.** Let $\mathscr{P}(V)$ denote the superalgebra of super-polynomials on $V$. Recall that $\mathscr{P}(V)$ is by definition equal to $S(V^*)$. As $\mathfrak{g}^\sigma$ acts on $V$, we obtain an induced locally finite $\mathfrak{g}^\sigma$-action on $\mathscr{P}(V^*) = S(V)$. Similar statements apply to $\mathscr{P}(V)$. We shall identify this action in terms of differential operators. To that end, we consider the complex supermanifold $\mathbb{A}(V)$, defined as the locally ringed space with underlying topological space $V_{\bar 0}$ and sheaf of superfunctions $\mathcal{O}_{\mathbb{A}(V)} := \mathcal{H}_{V_{\bar 0}} \otimes \bigwedge(V_{\bar 1})^*$, where $\mathcal{H}$ denotes the sheaf of holomorphic functions. There is a natural inclusion

$$V^* \longrightarrow \Gamma(\mathcal{O}_{\mathbb{A}(V)})$$

($\Gamma$ denoting global sections), allowing us to identify linear forms on $V$ with certain superfunctions on $\mathbb{A}(V)$. In particular, $\mathscr{P}(V)$ is a subsuperalgebra of $\Gamma(\mathcal{O}_{\mathbb{A}(V)})$.

Recall that on a supermanifold $X$, the vector fields on $X$, defined on an open set $O \subseteq X_0$ of the underlying topological space, are defined to be the superderivations of $\mathcal{O}_X|_O$ [**1**, Definition 4.1]. For any homogeneous basis $(x_a)$ of $V$, the dual basis $(x^a)$ is a coordinate system on $\mathbb{A}(V)$, and [**1**, Proposition 4.5] there are unique (and globally defined) vector fields $\frac{\partial}{\partial x^a}$ on $\mathbb{A}(V)$ of parity $|x^a|$, determined by

$$\frac{\partial}{\partial x^a}(x^b) = \delta_{ab}, \quad \forall a, b.$$

The linear action of $\mathfrak{l}$ on $V$ determines, for $v \in \mathfrak{l}$, vector fields $\mathbf{a}_v$ on $\mathbb{A}(V)$ by

$$\mathbf{a}_v(x^b) = -v \cdot x^b, \quad \forall b. \tag{3.1}$$

The sign stems from the fact that these are the fundamental vector fields for a Lie supergroup action on $\mathbb{A}(V)$, as we shall see later. By construction, for any $v \in \mathfrak{l}$, the action of $-a_v$ on $\mathscr{P}(V)$ coincides with the action of $v$ defined in the first paragraph of this subsection.

We now make this action explicit. Let $(e_k, \varepsilon_k)_{k=1,\dots,n}$ be the standard basis of $\mathbb{C}^{n|n}$. A homogeneous basis of the super-vector space $V \cong \mathfrak{q}(n)$ is determined by

$$u_{k\ell} := \begin{pmatrix} E_{k\ell} & 0 \\ 0 & E_{k\ell} \end{pmatrix}, \quad \xi_{k\ell} := \begin{pmatrix} 0 & E_{k\ell} \\ E_{k\ell} & 0 \end{pmatrix},$$

where $k,\ell = 1,\dots,n$ and $E_{k\ell}$ are the usual elementary matrices. Then

$$u_{k\ell}(e_j) = \delta_{\ell j}e_k, \qquad u_{k\ell}(\varepsilon_j) = \delta_{\ell j}\varepsilon_k,$$
$$\xi_{k\ell}(e_j) = \delta_{\ell j}\varepsilon_k, \qquad \xi_{k\ell}(\varepsilon_j) = \delta_{\ell j}e_k.$$

Let $(u^{k\ell},\xi^{k\ell})$ be the dual basis of $\mathfrak{q}(n)^* = V^*$. This determines vector fields

$$\frac{\partial}{\partial u^{k\ell}}, \quad \frac{\partial}{\partial \xi^{k\ell}}$$

on $\mathbb{A}(V)$, by the recipe given above.

Moreover, let $(a_{k\ell},\alpha_{k\ell})$ and $(b_{k\ell},\beta_{k\ell})$, respectively, be the copies of $(u_{kl},\xi_{k\ell})$ in the first and second factor of $\mathfrak{l} = \mathfrak{q}(n)\times\mathfrak{q}(n)$. By Equation (2.2), as a module over the second factor, $V$ is isomorphic to $\mathbb{C}^n\otimes(\mathbb{C}^{n|n})^*$ where $\mathfrak{q}(n)$ acts on $\mathbb{C}^{n|n}$ in the standard way. Hence, the second factor acts on $V^*$ as on $\mathbb{C}^n\otimes\mathbb{C}^{n|n}$:

$$b_{k\ell}(u^{pq}) = \delta_{\ell q}u^{pk} \qquad b_{k\ell}(\xi^{pq}) = \delta_{\ell q}\xi^{pk},$$
$$\beta_{k\ell}(u^{pq}) = \delta_{\ell q}\xi^{pk} \qquad \beta_{k\ell}(\xi^{pq}) = \delta_{\ell q}u^{pk}.$$

It follows that

$$-\mathbf{a}_{b^{k\ell}} = \sum_{p=1}^n u^{pk}\frac{\partial}{\partial u^{p\ell}} + \xi^{pk}\frac{\partial}{\partial \xi^{p\ell}}, \quad -\mathbf{a}_{\beta^{k\ell}} = \sum_{p=1}^n \xi^{pk}\frac{\partial}{\partial u^{p\ell}} + u^{pk}\frac{\partial}{\partial \xi^{p\ell}}. \tag{3.2}$$

Similarly, the first factor of $\mathfrak{l}$ acts on $V$ as on $\mathbb{C}^{n|n}\otimes\mathbb{C}^n$, and hence on $V^*$ as on $(\mathbb{C}^{n|n})^*\otimes\mathbb{C}^n$. Reasoning as above, this implies

$$-\mathbf{a}_{a^{k\ell}} = \sum_{p=1}^n u^{p\ell}\frac{\partial}{\partial u^{pk}} + \xi^{p\ell}\frac{\partial}{\partial \xi^{pk}}, \quad -\mathbf{a}_{\alpha^{k\ell}} = \sum_{p=1}^n -\xi^{p\ell}\frac{\partial}{\partial u^{pk}} + u^{p\ell}\frac{\partial}{\partial \xi^{pk}}. \tag{3.3}$$

We will presently decompose the $\mathfrak{l}$-module $\mathscr{P}(V)$. To that end, we introduce a labelling set. Let $\Lambda$ be the set of *partitions*, that is, of all finite sequences $\lambda = (\lambda_1,\dots,\lambda_m)$ of non-negative integers $\lambda_j$ such that $\lambda_1\geqslant\lambda_2\geqslant\cdots\geqslant\lambda_m$. Here, we identify $\lambda$ with any partition $(\lambda_1,\dots,\lambda_m,0,\dots,0)$ obtained from $\lambda$ by appending a finite number of zeros at its tail. If $\lambda$ can be written in the form $(\lambda_1,\dots,\lambda_\ell)$ where $\lambda_\ell>0$, then we say $\lambda$ has *length* $\ell(\lambda)=\ell$. Let $\Lambda^\ell\subseteq\Lambda$ be the set of partitions of length $\ell(\lambda)\leqslant\ell$. We also set $|\lambda| := \lambda_1+\cdots+\lambda_\ell$ if $\ell(\lambda)\leqslant\ell$. A partition $\lambda$ of length $\ell(\lambda)=\ell$ is called *strict* if $\lambda_1>\cdots>\lambda_\ell>0$. The set of all strict partitions will be denoted by $\Lambda_{>0}$, and we write $\Lambda^\ell_{>0}$ for the set of strict partitions of length at most $\ell$; that is, $\Lambda^\ell_{>0} := \Lambda^\ell\cap\Lambda_{>0}$.

For every strict partition $\lambda$ such that $\ell(\lambda)\leqslant n$, let $F_\lambda$ be the $\mathfrak{q}(n)$-highest weight module with highest weight $\lambda_1\varepsilon_1+\cdots+\lambda_n\varepsilon_n$, where the $\varepsilon_i$ are the standard characters of the even part of the Cartan subalgebra of $\mathfrak{q}(n)$. For every strict partition $\lambda$, set $\delta(\lambda):=0$ when $\ell(\lambda)$ is even, and $\delta(\lambda):=1$ otherwise. We now define an $\mathfrak{l}$-module $V_\lambda$ as follows. It is shown in [**4**, Section 2] that, as an $\mathfrak{l}$-module, the exterior tensor product $(F_\lambda)^*\boxtimes F_\lambda$ is irreducible when $\delta(\lambda)=0$, and decomposes into a direct sum of two irreducible isomorphic $\mathfrak{l}$-modules (via an odd map) if $\delta(\lambda)=1$. Following the notation of Ref. [**4**], we set

$$V_\lambda := \frac{1}{2^{\delta(\lambda)}}\big((F_\lambda)^*\boxtimes F_\lambda\big),$$

that is, we take $V_\lambda$ to be the irreducible component of $(F_\lambda)^*\boxtimes F_\lambda$ that appears in the decomposition of the super-polynomial algebra over the natural $(\mathfrak{q}(n),\mathfrak{q}(n))$-module (see Proposition 3.1). The $\mathfrak{l}$-module $V_\lambda$ is always of type $\mathsf{M}$, that is, it is irreducible as an ungraded representation. It follows that in the $\mathbb{Z}_2$-graded sense,

$$\underline{\mathrm{Hom}}_{\mathfrak{l}}(V_\lambda, V_\mu) = \delta_{\lambda\mu} \cdot \mathbb{C}, \tag{3.4}$$

where *a priori*, $\underline{\mathrm{Hom}}_{\mathfrak{l}}$ denotes the set of *all* $\mathfrak{l}$-equivariant linear maps (of any parity). In particular, all non-zero $\mathfrak{l}$-equivariant endomorphisms of $V_\lambda$ are even.

PROPOSITION 3.1. *Under the action of $\mathfrak{l}$, $\mathscr{P}^k(V)$ decomposes as the multiplicity-free direct sum of simple modules $V_\lambda$, where $\lambda$ ranges over elements of $\Lambda^n_{>0}$ which satisfy $|\lambda| = k$.*

PROOF. Recall that $\mathscr{P}(V) \cong S(V^*)$ as $\mathfrak{l}$-modules. The proposition follows from the description of the actions of the left and right copies of $\mathfrak{q}(n)$ on $V$ given above, and the results of [**4**, Section 3]. □

**3.2. Invariant polynomial differential operators.** On a complex supermanifold $X$, the *differential operators* on $X$ defined on an open set $O \subseteq X_0$ of the underlying topological space are generated as a subsuperalgebra of the $\mathbb{C}$-linear endomorphisms of $\mathcal{O}_X|_O$ by vector fields and functions. This gives a $\mathbb{C}$-algebra sheaf $\mathscr{D}_X$ that is an $\mathcal{O}_X$-bisupermodule, filtered by order.

Here, a differential operator is of *order* $\leqslant n$ if it can be expressed as a product of some functions and at most $n$ vector fields. A differential operator of order $\leqslant n$ is uniquely determined by its action on monomials of order $\leqslant n$ in some given system of coordinate functions. This follows in the usual way from the Hadamard Lemma [**13**, Lemma 2.1.8] and implies that $\mathscr{D}_X$ is locally free as a left $\mathcal{O}_X$-supermodule.

For the supermanifold $\mathbb{A}(V)$, we have a $\mathbb{C}$-superalgebra map

$$\partial : S(V) \longrightarrow \Gamma(\mathscr{D}_{\mathbb{A}(V)}).$$

It is determined by the linear map which sends any homogeneous $v \in V$ to the unique vector field $\partial(v)$ such that

$$\partial(v)(\mu) = (-1)^{|\mu||v|}\mu(v) \tag{3.5}$$

for all homogeneous $\mu \in V^*$. If $(x_a)$ is a homogeneous basis of $V$, then

$$\partial(x_a) = (-1)^{|x_a|}\frac{\partial}{\partial x^a}.$$

The image of $\partial$, denoted by $D(V)$, is the superalgebra of *constant-coefficient* differential operators on $V$. The map

$$\mathscr{P}(V) \otimes D(V) \longrightarrow \Gamma(\mathscr{D}_{\mathbb{A}(V)}) : p \otimes D \longmapsto pD \tag{3.6}$$

is an isomorphism onto a $\mathscr{P}(V)$-submodule of $\Gamma(\mathscr{D}_{\mathbb{A}(V)})$ denoted by $\mathscr{P}\mathscr{D}(V)$. Indeed, $\mathscr{P}\mathscr{D}(V)$ is a subsuperalgebra, the algebra of *polynomial differential operators*. As $\mathfrak{l}$ acts by linear vector fields, we have the bracket relation

$$[-\mathbf{a}_x, p] = -\mathbf{a}_x(p) = x \cdot p, \quad \forall x \in \mathfrak{l}, p \in \mathscr{P}(V).$$

The bilinear form

$$\mathscr{P}^k(V) \otimes D^k(V) \longrightarrow \mathbb{C} : p \otimes D \longmapsto Dp \tag{3.7}$$

is a non-degenerate pairing which is $\mathfrak{l}$-equivariant, and therefore results in a canonical $\mathfrak{l}$-module isomorphism $D^k(V) \cong \mathscr{P}^k(V)^*$. Hence the following corollary to Proposition 3.1 holds.

COROLLARY 3.2. *The space* $\mathscr{P}\mathscr{D}(V)^{\mathfrak{l}}$ *of* $\mathfrak{l}$*-invariant polynomial differential operators decomposes as follows:*

$$\mathscr{P}\mathscr{D}(V)^{\mathfrak{l}} = \bigoplus_{\lambda\in\Lambda^n_{>0}} \left(V_\lambda \otimes (V_\lambda)^*\right)^{\mathfrak{l}}.$$

There is a natural $\mathfrak{l}$-equivariant isomorphism $V_\lambda \otimes (V_\lambda)^* \cong \underline{\mathrm{End}}_{\mathbb{C}}(V_\lambda)$, so the identity element $\mathrm{id}_{V_\lambda}$ of $V_\lambda$ determines an even $\mathfrak{l}$-invariant polynomial differential operator

$$D_\lambda \in \mathscr{P}\mathscr{D}(V)^{\mathfrak{l}}.$$

That is, if $(p_j)$ is a homogeneous basis of $V_\lambda \subseteq \mathscr{P}(V)$ and $(D_j)$ is its dual basis for $(V_\lambda)^* \subseteq D(V)$, then

$$D_\lambda := \sum_j p_j D_j. \tag{3.8}$$

Schur's Lemma implies that for every $\mu \in \Lambda^n_{>0}$, there is a complex scalar $c_\lambda(\mu) \in \mathbb{C}$ such that $D_\lambda$ acts by the scalar $c_\lambda(\mu)$ on $V_\lambda \subseteq \mathscr{P}(V)$.

COROLLARY 3.3. *The operators* $D_\lambda$*, where* $\lambda$ *ranges over* $\Lambda^n_{>0}$*, form a basis of the space* $\mathscr{P}\mathscr{D}(V)^{\mathfrak{l}}$ *of* $\mathfrak{l}$*-invariant differential operators. Moreover,* $c_\lambda(\lambda) = 1$*, while* $c_\lambda(\mu) = 0$ *whenever* $|\mu| \leqslant |\lambda|$ *and* $\mu \neq \lambda$*.*

PROOF. The first statement follows from Corollary 3.2 and Equation (3.4). Since the order of the operator $D_\lambda$ is not $\leqslant |\lambda| - 1$, it vanishes on $\mathscr{P}^k(V)$ for $k < |\lambda|$. Next, assume $k = |\lambda| = |\mu|$. If $\lambda \neq \mu$ and $D_\lambda$ does not vanish on $V_\mu$, then the restriction of the bilinear form (3.7) to $V_\mu \times (V_\lambda)^*$ will be a nonzero $\mathfrak{l}$-equivariant form, hence $V_\lambda \cong V_\mu$, which is a contradiction.

It remains to compute the action of $D_\lambda$ on $V_\lambda$. Let $(p_j)$ and $D_j$ denote dual bases of $V_\lambda$ and $(V_\lambda^*)$. Then $D_\lambda p_i = \sum_j p_j D_j p_i = \sum_j p_j \delta_{ij} = p_i$ for every $i$. □

To determine $c_\lambda(\mu)$ (which will be done in the next subsection), we first need to see that it extends to a $Q$-symmetric polynomial. To that end, let $\Gamma_n$ denote the ring of *$Q$-symmetric polynomials*, that is, $n$-variable symmetric polynomials $p(x_1, \ldots, x_n)$ such that $p(t, -t, x_3, \ldots, x_n)$ does not depend on $t$. (For $n = 1$, the latter condition is vacuous.)

PROPOSITION 3.4. *For all* $\lambda \in \Lambda^n_{>0}$*, there exists a polynomial* $q^*_\lambda \in \Gamma_n$ *of degree at most* $|\lambda|$ *such that* $c_\lambda(\mu) = q^*_\lambda(\mu)$ *for all* $\mu \in \Lambda^n_{>0}$*.*

PROOF. Recall that $L$ and $R$, respectively, denote the actions of the first and second factor of $\mathfrak{l} = \mathfrak{q}(n) \times \mathfrak{q}(n)$ on $\mathscr{P}(V)$. As $\mathscr{P}(V)$ is the multiplicity-free direct sum of simple modules of the universal enveloping algebra $\mathfrak{U}(\mathfrak{l}) = \mathfrak{U}(\mathfrak{q}(n)) \otimes \mathfrak{U}(\mathfrak{q}(n))$, it follows that $L(\mathfrak{U}(\mathfrak{q}(n)))$ and $R(\mathfrak{U}(\mathfrak{q}(n)))$ are mutual commutants in $\mathscr{P}\mathscr{D}(V)$ (this double commutant property is also mentioned in [**4**]). In particular, we have

$$\mathscr{P}\mathscr{D}(V)^{\mathfrak{l}} \subseteq R(\mathfrak{U}(\mathfrak{q}(n))).$$

As $R$ is faithful, it follows that in fact

$$\mathscr{P}\mathscr{D}(V)^{\mathfrak{l}} = R(\mathscr{Z}(\mathfrak{q}(n))),$$

where $\mathscr{Z}$ denotes the centre of $\mathfrak{U}$. The latter statement also follows from the explicit construction of the Capelli operators in the work of Nazarov [**16**].

Furthermore, the simple module $V_\lambda$ occuring in the decomposition of $\mathscr{P}(V)$ is contained in the external tensor product $(F_\lambda)^* \boxtimes F_\lambda$. By Sergeev's Harish-Chandra isomorphism for $\mathfrak{q}(n)$ [**27**, Theorem 3] (see also [**5**, Theorem 2.46]), for any $u \in \mathcal{Z}(\mathfrak{q}(n))$, there exists a $Q$-symmetric polynomial $q_u^* \in \Gamma_n$ such that for all $\lambda \in \Lambda^n_{>0}$, $u$ acts on $F_\lambda$ by $q_u^*(\lambda)$.

Fix $\lambda \in \Lambda^n_{>0}$. Then the order of $D_\lambda$ is $\leqslant |\lambda|$. As $L$ is faithful, there is a unique $z_\lambda \in \mathcal{Z}(\mathfrak{q}(n))$ such that $L(z_\lambda) = D_\lambda$. Since $\mathfrak{q}(n) \subseteq \mathfrak{l}$ acts by linear vector fields, $z_\lambda$ lies in the $|\lambda|$-th part of the standard increasing filtration of $\mathfrak{U}(\mathfrak{q}(n))$. Then $q_\lambda^* := q_{z_\lambda}^*$ has degree at most $|\lambda|$, see [**5, 27**]. The assertion follows. □

DEFINITION 3.5. We call $q_\lambda^*$ the *eigenvalue polynomial* of $D_\lambda$ for $\lambda \in \Lambda^n_{>0}$.

**3.3. Schur $Q$-functions.** Our next goal is to identify the eigenvalue polynomials $q_\lambda^* \in \Gamma_n$. We first recall the definitions of certain elements $Q_\lambda$ of $\Gamma_n$, called the *$Q$-functions* of Schur, and their shifted analogues, the *factorial $Q$-functions* $Q_\lambda^*$ originally defined by Okounkov, see Ref. [**7**].

Given a sequence $(a_n)_{n\geqslant 1}$ of complex numbers, we define for any non-negative integer $k$ the *kth generalized power* of $x$ by

$$(x \mid a)^k := \prod_{i=1}^{k} (x - a_i),$$

where we set $(x|a)^0 = 1$. For every $\lambda \in \Lambda^n_{>0}$, we set

$$F_\lambda(x_1, \ldots, x_n \mid a) := \prod_{i=1}^{\ell(\lambda)} (x \mid a)^{\lambda_i} \prod_{\substack{i \leqslant \ell(\lambda) \\ i<j\leqslant n}} \frac{x_i + x_j}{x_i - x_j}.$$

We now define

$$Q_\lambda(x_1, \ldots, x_n \mid a) := \frac{2^{\ell(\lambda)}}{(n-\ell(\lambda))!} \sum_{\sigma\in S_n} F_\lambda(x_{\sigma(1)}, \ldots, x_{\sigma(n)} \mid a).$$

We remark that $Q_\lambda(x_1, \ldots, x_n \mid a)$ can be expressed as a ratio of an antisymmetric polynomial by the Vandermonde polynomial, and therefore it is also a polynomial. Two special cases of interest are

$$Q_\lambda(x_1, \ldots, x_n) := Q_\lambda(x_1, \ldots, x_n \mid \underline{0}),$$

where $\underline{0} := (0, 0, 0, \ldots)$, and

$$Q_\lambda^*(x_1, \ldots, x_n) := Q_\lambda(x_1, \ldots, x_n \mid \delta),$$

where $\delta := (0, 1, 2, 3, \ldots)$, called, respectively, the *Schur $Q$-function*, and the *factorial Schur $Q$-function.*

PROPOSITION 3.6. *Let $\lambda \in \Lambda^n_{>0}$.*

(i) *We have $Q_\lambda, Q_\lambda^* \in \Gamma_n$. Furthermore, $Q_\lambda$ is homogeneous of degree $|\lambda|$ and*

$$\deg(Q_\lambda^* - Q_\lambda) < |\lambda|.$$

(ii) *Both $(Q_\mu)_{\mu\in\Lambda^n_{>0}}$ and $(Q_\mu^*)_{\mu\in\Lambda^n_{>0}}$ are bases for the vector space $\Gamma_n$.*

(iii) *We have $Q_\lambda^*(\mu) = 0$ for every $\mu \in \Lambda^n_{>0}$ such that $|\mu| \leqslant |\lambda|$, $\lambda \neq \mu$; moreover,*

$$Q_\lambda^*(\lambda) = Q_\lambda(\lambda) = H(\lambda) := \lambda! \prod_{i<j} \frac{\lambda_i + \lambda_j}{\lambda_i - \lambda_j},$$

*where $\lambda! := \lambda_1! \cdots \lambda_n!$, and it is understood that the product extends only up to $\ell(\lambda)$.*

(iv) *The unique element of* $\Gamma_n$ *of degree at most* $|\lambda|$ *which satisfies Property* (iii) *is precisely* $Q^*_\lambda$.

PROOF. Parts (i)–(iii) follow from [**7**, § 1] (see also [**29**, § 6]). Thus, we only sketch the argument for item (iv). Set $k := |\lambda|$ and let $\Gamma_n(k) \subseteq \Gamma_n$ consist of the polynomials of degree at most $k$. From item (ii), it follows that $\dim(\Gamma_n(k)) = \#\Lambda^n_{>0}(k)$, where

$$\Lambda^n_{>0}(k) := \{\lambda \in \Lambda^n_{>0} \mid |\lambda| \leqslant k\}.$$

For every $\mu \in \Lambda^n_{>0}(k)$, we consider the linear functional $\varepsilon_\mu \in \Gamma^*_n$ defined by

$$\varepsilon_\mu(f) := f(\mu), \quad \forall f \in \Gamma_n.$$

Let the order $\subseteq$ on $\Lambda$ be defined by

$$\mu \subseteq \nu \quad \text{if and only if} \quad \forall j : \mu_j \leqslant \nu_j.$$

We choose a total order $\preceq$ on $\Lambda^n_{>0}(k)$ such that $\mu \prec \nu$ if either $|\mu| < |\nu|$ or $|\mu| = |\nu|$ and $\nu \not\subseteq \mu$. Then by [**7**, Proposition 1.16], in terms of $\preceq$, the $k \times k$ matrix

$$\left(\varepsilon_\mu(Q^*_\nu)\right)_{\mu,\nu\in\Lambda^n_{>0}(k)}$$

is triangular with no zeros on the diagonal, and therefore, invertible. It follows that the linear system

$$\varepsilon_\mu(f) = \begin{cases} H(\lambda) & \text{if } \mu = \lambda, \\ 0 & \text{if } \mu \neq \lambda. \end{cases}$$

has a unique solution in $\Gamma_n(k)$. In view of item (iii), this solution is $Q^*_\lambda$. □

REMARK 3.7. The polynomials $Q_\lambda$ also appear in Ref. [**14**]. They were initially introduced by Schur in connection with the projective representations of the symmetric group. According to [**7**, Proposition 1.16], the vanishing property of Proposition 3.6 (iii) holds in a stronger form: $Q^*_\lambda(\mu) = 0$ whenever $\lambda \not\subseteq \mu$.

THEOREM 3.8. *Let* $q^*_\lambda$ *for* $\lambda \in \Lambda^n_{>0}$ *be the eigenvalue polynomial of Proposition* 3.4. *Then*

$$q^*_\lambda = \frac{1}{\lambda!} \prod_{i<j} \frac{\lambda_i - \lambda_j}{\lambda_i + \lambda_j} Q^*_\lambda.$$

PROOF. By Proposition 3.4, $q^*_\lambda \in \Gamma_n$ satisfies the same degree, symmetry, and vanishing properties as $Q^*_\lambda$, see Proposition 3.6 (iii). The statement now follows from Proposition 3.6 (iv). □

REMARK 3.9. Nazarov [**16**, Proposition 4.8] constructs certain Capelli elements $C_\lambda \in \mathcal{Z}(\mathfrak{q}(n))$, and proves that the top degree part of their Harish-Chandra image is the classical Schur $Q$-function. Even though he refers to Ivanov's work [**7**] for the eigenvalue polynomials (see [**16**, p. 871]), his results do not include Theorem 3.8 explicitly.

## 4. The spherical polynomials

In this final section, we define $\mathfrak{m}$-spherical polynomials associated with the representations $V_\lambda$. We will show that up to a scalar multiple, these spherical polynomials are indeed the Schur $Q$-functions. As the Schur $Q$-function $Q_\lambda$ is the top-degree homogeneous part of $Q^*_\lambda$, our strategy is to prove that the spherical polynomials are equal to the top-degree homogeneous parts of the eigenvalue polynomials $q^*_\lambda$.

**4.1. Open orbits.** We begin by globalising the action of $\mathfrak{l}$ on $V$. Let $G$ be the complex Lie supergroup corresponding *via* [**3**, Theorem 7.4.5] to the supergroup pair $(\mathfrak{g}, G_0)$, where $G_0$ consists of the matrices of the form

$$\begin{pmatrix} a & 0 & b & 0 \\ 0 & a' & 0 & b' \\ c & 0 & d & 0 \\ 0 & c' & 0 & d' \end{pmatrix}, \text{ such that } \begin{pmatrix} a & b \\ c & d \end{pmatrix}, \begin{pmatrix} a' & b' \\ c' & d' \end{pmatrix} \in \mathrm{GL}(2n, \mathbb{C}),$$

and acts on $\mathfrak{g}$ by conjugation. Thus, $G_0 \cong \mathrm{GL}(2n, \mathbb{C}) \times \mathrm{GL}(2n, \mathbb{C})$. The automorphisms $\varphi$, $\sigma$, and $\theta$ of $\mathfrak{g}$ integrate to $G$, as their restrictions to $\mathfrak{g}_{\bar{0}}$ integrate to $G_0$ and

$$\mathrm{Ad}(\phi(g))(\phi(x)) = \phi(\mathrm{Ad}(g))x, \quad \forall x \in \mathfrak{g}, g \in G_0,$$

for every choice of $\phi \in \{\varphi, \sigma, \theta\}$ [**3**, Theorem 7.4.5]. In particular, there are fixed subsupergroups $G^\sigma$, $L := G^{\sigma,\varphi}$, and $M := G^{\sigma,\varphi,\theta}$ of $G$. One sees directly that

$$G_0^\sigma = \mathrm{GL}(n, \mathbb{C}) \times \mathrm{GL}(n, \mathbb{C}) \times \mathrm{GL}(n, \mathbb{C}) \times \mathrm{GL}(n, \mathbb{C}),$$
$$L_0 = \mathrm{GL}(n, \mathbb{C}) \times \mathrm{GL}(n, \mathbb{C}), \quad M_0 = \mathrm{GL}(n, \mathbb{C})$$

are connected.

Write the elements of $G_0^\sigma$ in the form

$$(A, D), \quad A = \begin{pmatrix} a & 0 \\ 0 & a' \end{pmatrix}, \quad D = \begin{pmatrix} d & 0 \\ 0 & d' \end{pmatrix}, \quad a, a', d, d' \in \mathrm{GL}(n, \mathbb{C}).$$

Then the action

$$G_0^\sigma \times U \longrightarrow U : (A, D, x) \longmapsto AxD^{-1}$$

integrates the action of $\mathfrak{g}_{\bar{0}}^\sigma$ on $U = \mathfrak{gl}(n|n)$ in such a way that

$$g \cdot (x \cdot v) = \mathrm{Ad}(g)(x) \cdot (g \cdot v), \quad \forall g \in G_0^\sigma, x \in \mathfrak{g}^\sigma, u \in U,$$

see Equation (2.2). This gives rise to a Lie supergroup action

$$G^\sigma \times \mathbb{A}(U) \longrightarrow \mathbb{A}(U),$$

where $\mathbb{A}(U)$ is the complex supermanifold corresponding to $U$, as in Section 3.1. Passing to $\varphi$-fixed points, this implies that there is an action

$$\mathbf{a} : L \times \mathbb{A}(V) \longrightarrow \mathbb{A}(V)$$

such that the fundamental vector fields

$$\mathbf{a}_x := (x \otimes 1) \circ \mathbf{a}^\sharp, \quad x \in \mathfrak{l},$$

are those determined by Equation (3.1), see [**3**, Proposition 8.3.3]. Here, $\mathbf{a}^\sharp$ is the morphism of sheaves of superalgebras that corresponds to $\mathbf{a}$. We equally obtain an $L$-action on $\mathbb{A}(V^*)$ integrating the given $\mathfrak{l}$-action.

Recall from Section 3.1 that $\mathbb{A}(V) = (V_{\bar{0}}, \mathcal{O}_{\mathbb{A}(V)})$ as a locally ringed space.

PROPOSITION 4.1. *The $L$-orbit $L \cdot e$ through the point $e \in V_{\bar{0}}$ is the homogeneous supermanifold $L/M$. The canonical morphism $j_V : L \cdot e \longrightarrow \mathbb{A}(V)$ is an $L$-equivariant open embedding.*

PROOF. As $M$ is connected, and by Lemma 2.1, the isotropy of $L$ at $e$ is $M$, see [**3**, Proposition 8.4.7]. Thus, $L \cdot e = L/M$ by definition, and the orbit morphism $\mathbf{a}_e : L \longrightarrow \mathbb{A}(V)$ factors into $\pi : L \longrightarrow L/M$ and $j_V : L \cdot e = L/M \longrightarrow \mathbb{A}(V)$ where

$\pi$ is a surjective submersion and $j_V$ is an injective immersion, both $L$-equivariant, by [**1**, Theorem 4.24]. By Lemma 2.1, the kernel of

$$T_1(\mathbf{a}_e) : T_1(L) = \mathfrak{l} \longrightarrow T_e(\mathbb{A}(V)) = V$$

is $\mathfrak{m}$ and its restriction to $\mathfrak{s}$ an isomorphism $\mathfrak{s} \longrightarrow V$. Thus, the inverse function theorem [**13**, Theorem 2.3.1] applies, and $j_V$ is an open embedding. □

REMARK 4.2. It is not hard to identify the open orbit from Proposition 4.1. Indeed, identifying $V$ with $\mathfrak{q}(n)$ *via* the isomorphism from Equation (2.3), $L \cdot e$ equals $Q(n, \mathbb{C}) \subseteq \mathbb{A}(\mathfrak{q}(n))$, defined to be the open subspace corresponding to the open subset $\mathrm{GL}(n, \mathbb{C}) \subseteq \mathfrak{gl}(n) = \mathfrak{q}(n)_{\bar{0}}$. Observe that $Q(n, \mathbb{C})$ has the structure of a complex Lie supergroup with Lie superalgebra $\mathfrak{q}(n)$.

The orbit morphism $L \longrightarrow L \cdot e$ is given by

$$(A, D) \longmapsto AD^{-1}$$

on $T$-valued points, where $T$ is any complex supermanifold.

**4.2. Spherical vectors and polynomials.** Let $\lambda \in \Lambda^n_{>0}$. Define $p^*_\lambda \in (V_\lambda)^*$ by

$$\langle p^*_\lambda, \ell \rangle := (-1)^k k!\, j^\sharp_V(\ell)(e), \quad \forall \ell \in V_\lambda \subseteq \mathscr{P}(V), \tag{4.1}$$

where $\langle \cdot, \cdot \rangle$ is the standard pairing of $(V_\lambda)^*$ with $V_\lambda$, and $j^\sharp_V$ is pullback along the morphism $j_V : L \cdot e \longrightarrow \mathbb{A}(V)$.

PROPOSITION 4.3. *For every $\lambda \in \Lambda^n_{>0}$, $p^*_\lambda$ is the up to scalars unique non-zero $\mathfrak{m}$-invariant vector of $(V_\lambda)^*$.*

PROOF. Uniqueness of the $\mathfrak{m}$-invariant in $V_\lambda$ follows from the fact that by the hom-tensor adjunction formula

$$((V_\lambda)^*)^{\mathfrak{m}} = \underline{\mathrm{Hom}}_{\mathfrak{m}}(V_\lambda, \mathbb{C}) \cong \frac{1}{2^{\delta(\lambda)}} \underline{\mathrm{Hom}}_{\mathfrak{m}}(F_\lambda, F_\lambda),$$

combined with the fact that $\dim \underline{\mathrm{End}}_{\mathfrak{m}}(F_\lambda) = 2^{\delta(\lambda)}$. (Recall that $F_\lambda$ is of type Q if and only if $\delta(\lambda) = 1$.)

Next we prove that $p^*_\lambda$ is the desired $\mathfrak{m}$-invariant. Since $j_V$ is $L$-equivariant and $M$ fixes $e$, it follows that $p^*_\lambda$ is $\mathfrak{m}$-invariant. We need to see that $p^*_\lambda \neq 0$. Above, we have noted the commutative diagram

$$\begin{array}{ccc} L & & \\ {\scriptstyle\pi}\downarrow & \searrow^{\mathbf{a}_e} & \\ L \cdot e & \xrightarrow[j_V]{} & \mathbb{A}(V) \end{array}$$

where $j_V$ is an open embedding, $\pi$ is a surjective submersion, and

$$\mathbf{a}_e = \mathbf{a} \circ (\mathrm{id} \times j_e),$$

where $j_e$ is the embedding $* \longrightarrow \mathbb{A}(V)$ of the singleton space defined by $e$. The action of $L$ is linear, *i.e.*

$$\mathbf{a}^\sharp(V^*) \subseteq \Gamma(\mathscr{O}_L) \otimes V^*,$$

and hence, because $\mathbf{a}^\sharp$ is a superalgebra morphism,

$$\mathbf{a}^\sharp(\mathscr{P}(V)) \subseteq \Gamma(\mathscr{O}_L) \otimes \mathscr{P}(V).$$

If $W \subseteq \mathscr{P}(V)$ is a graded subspace invariant under $L_0$ and $\mathfrak{l}$, then

$$\mathbf{a}^\sharp(W) \subseteq \Gamma(\mathcal{O}_L) \otimes W.$$

Indeed, we may identify

$$\Gamma(\mathcal{O}_L) = \underline{\mathrm{Hom}}_{\mathfrak{l}_{\bar{0}}}(\mathfrak{U}(\mathfrak{l}), \Gamma(\mathcal{O}_{L_0})),$$

so we may consider superfunctions $f$ on $L$ as functions $f(u;g)$ of $u \in \mathfrak{U}(\mathfrak{l})$ and $g \in L_0$. Consider the canonical extension $\mathbf{a}_u$ of the anti-homomorphism $x \longmapsto \mathbf{a}_x$ to $\mathfrak{U}(\mathfrak{l})$ and $\mathbf{a}_g := \mathbf{a} \circ (j_g \times \mathrm{id})$ for $g \in L_0$. Then by [**3**, Proposition 8.3.3], we have for $w \in W$, $u \in \mathfrak{U}(\mathfrak{l})$, and $g \in L_0$:

$$\mathbf{a}^\sharp(w)(u;g) = \mathbf{a}_u \mathbf{a}_g^\sharp(w) = \mathsf{S}_{\mathfrak{l}}(u) \cdot g^{-1} \cdot w \in W,$$

with $\mathsf{S}_{\mathfrak{l}}$ denoting the antipodal anti-automorphism of $\mathfrak{U}(\mathfrak{l})$.

Now, seeking a contradiction, assume that $p_\lambda^* = 0$, so that

$$j_e^\sharp(\ell) = j_V^\sharp(\ell)(e) = 0, \quad \forall \ell \in V_\lambda.$$

Let $\ell \in L_V^\sharp$ be arbitrary. Then

$$\pi^\sharp j_V^\sharp(\ell) = \mathbf{a}_e^\sharp(\ell) = (\mathrm{id} \otimes j_e)^\sharp \mathbf{a}^\sharp(\ell) \subseteq (\mathrm{id} \otimes j_e)^\sharp(\Gamma(\mathcal{O}_L) \otimes V_\lambda) = 0.$$

Since $\pi$ is a surjective submersion, $\pi^\sharp$ is injective, so that $j_V^\sharp(\ell) = 0$. But $j_V$ is an open embedding and $\ell$ is a superpolynomial, so $\ell = 0$, contradiction! This proves the assertion. □

**4.3. Determination of the spherical polynomials.** The Lie superalgebra $\mathfrak{g} \cong \mathfrak{q}(2n)$ carries a non-degenerate invariant odd supersymmetric bilinear form $b : \mathfrak{g} \times \mathfrak{g} \longrightarrow \mathbb{C}$, the *odd trace form.* In terms of the matrix realisation we have chosen in Equation (2.1), it is given by

$$b(x,x') := \frac{1}{2}\,\mathrm{tr}(x\Phi x'), \quad x, x' \in \mathfrak{g}$$

or explicitly,

$$b(x,x') = \mathrm{tr}\big(a\alpha' + \alpha a' + c\beta' + \beta c' + b\gamma' + \gamma b' + d\delta' + \delta d'\big),$$

$$x = \begin{pmatrix} a & \alpha & b & \beta \\ \alpha & a & \beta & b \\ \gamma & c & \delta & d \\ c & \gamma & d & \delta \end{pmatrix}, \ x' = \begin{pmatrix} a' & \alpha' & b' & \beta' \\ \alpha' & a' & \beta' & b' \\ \gamma' & c' & \delta' & d \\ c' & \gamma' & d' & \delta' \end{pmatrix}.$$

The odd form restricts to an $\mathfrak{l}$-invariant odd non-degenerate pairing of $V$ and $V^-$, allowing us to identify the $\mathfrak{l}$-modules $V^*$ and $\Pi(V^-)$.

We let $\mathfrak{t}^* := \mathbb{C}^n$. Then $\Pi(\mathfrak{t}^*)$ may be identified with a subspace of $V^-$ via the odd map

$$\mathfrak{t}^* \longrightarrow V^- : x \longmapsto \begin{pmatrix} 0 & 0 & 0 & 0 \\ 0 & 0 & 0 & 0 \\ 0 & \mathrm{diag}(x) & 0 & 0 \\ \mathrm{diag}(x) & 0 & 0 & 0 \end{pmatrix}$$

This determines an even injective linear map $\mathfrak{t}^* \longrightarrow V^*$ and hence, a morphism $\mathbb{A}(\mathfrak{t}^*) \longrightarrow \mathbb{A}(V^*)$. Thereby, we confer a meaning to $p|_{\mathfrak{t}^*}$ for polynomials $p$ in $\mathscr{P}(V^*) = S(V)$, as the pullback along this morphism.

Identifying $\Pi(V^-)$ with $V^*$, $p^*_\lambda|_{\mathfrak{t}^*}$ is the ordinary $n$-variable polynomial given by

$$p^*_\lambda|_{\mathfrak{t}^*}(x_1,\ldots,x_n) = p^*_\lambda\begin{pmatrix} 0 & 0 & 0 & 0\\ 0 & 0 & 0 & 0\\ 0 & \operatorname{diag}(x) & 0 & 0\\ \operatorname{diag}(x) & 0 & 0 & 0\end{pmatrix}, \quad \forall x_j\in\mathbb{C}. \tag{4.2}$$

By Proposition 3.1, $p^*_\lambda|_{\mathfrak{t}^*}$ is homogeneous of degree $|\lambda|$ in $\mathscr{P}(V^*)$, so $p^*_\lambda|_{\mathfrak{t}^*}$ is likewise homogeneous of degree $|\lambda|$.

DEFINITION 4.4. We call $p^*_\lambda\big|_{\mathfrak{t}^*}$ the *spherical polynomial* for $\lambda\in\Lambda^n_{>0}$.

The remainder of this subsection is devoted to the proof of the following result.

THEOREM 4.5. *Let $\lambda\in\Lambda^n_{>0}$ and $k:=|\lambda|$. Then the spherical polynomial $p^*_\lambda|_{\mathfrak{t}^*}$ is the homogeneous part of degree $k$ of the eigenvalue polynomials $q^*_\lambda$, so*

$$p^*_\lambda|_{\mathfrak{t}^*} = \frac{1}{\lambda!}\prod_{i<j}\frac{\lambda_i-\lambda_j}{\lambda_i+\lambda_j}Q_\lambda.$$

Let $\mathfrak{n}^+$ be the subalgebra of $\mathfrak{l}$ formed by the matrices

$$\begin{pmatrix} A & 0\\ 0 & D\end{pmatrix},\quad A=\begin{pmatrix} a & \alpha\\ \alpha & a\end{pmatrix},\quad D=\begin{pmatrix} d & \delta\\ \delta & d\end{pmatrix}$$

where $a$ and $\alpha$ are strictly upper triangular and $d$ and $\delta$ are strictly lower triangular. Similarly, let $\mathfrak{a}$ be the subspace of $\mathfrak{l}$ formed by the matrices

$$h_{a,\alpha} := \frac{1}{2}\begin{pmatrix} A & 0\\ 0 & -A\end{pmatrix},\quad A=\begin{pmatrix} a & \alpha\\ \alpha & a\end{pmatrix},$$

where $a$ and $\alpha$ are diagonal. Then we have the vector space decomposition

$$\mathfrak{l} = \mathfrak{m}\oplus\mathfrak{a}\oplus\mathfrak{n}^+.$$

Although $\mathfrak{a}_{\bar 0}$ is a subalgebra, $\mathfrak{a}$ is not. We still consider the basis of $\mathfrak{a}$ given by

$$h_i := h_{e_i,0} = \frac{1}{2}\begin{bmatrix} E_{ii} & 0 & 0 & 0\\ 0 & E_{ii} & 0 & 0\\ 0 & 0 & -E_{ii} & 0\\ 0 & 0 & 0 & -E_{ii}\end{bmatrix},\quad \eta_i := h_{0,e_i} = \frac{1}{2}\begin{bmatrix} 0 & E_{ii} & 0 & 0\\ E_{ii} & 0 & 0 & 0\\ 0 & 0 & 0 & -E_{ii}\\ 0 & 0 & -E_{ii} & 0\end{bmatrix},$$

where $i=1,\ldots,n$.

Let $\omega: S(\mathfrak{l})\longrightarrow\mathfrak{U}(\mathfrak{l})$ denote the canonical supersymmetrisation map, given by

$$\omega(x_1\cdots x_m) := \frac{1}{m!}\sum_{\sigma\in S_m}(-1)^{\#\{(i,j)\mid i<j,\sigma(i)>\sigma(j),|x_i|=|x_j|=\bar 1\}}x_{\sigma(1)}\cdots x_{\sigma(m)}$$

for all homogeneous $x_1,\ldots,x_m\in\mathfrak{l}$. Then, by the Poincaré–Birkhoff–Witt theorem [**25**, Chapter I, § 2.3, Corollary 1 to Theorem 1], every $u\in\mathfrak{U}(\mathfrak{l})$ can be written in a unique way as

$$u = u' + u_{\mathfrak{a}}, \tag{4.3}$$

where

$$u'\in\mathfrak{m}\mathfrak{U}(\mathfrak{l})+\mathfrak{U}(\mathfrak{l})\mathfrak{n}^+ = \mathfrak{m}\mathfrak{U}(\mathfrak{l})\oplus\omega(S(\mathfrak{a}\oplus\mathfrak{n}^+))\mathfrak{n}^+,$$

and $u_{\mathfrak{a}}$ can be written as a linear combination of monomials in the basis of $\mathfrak{a}$, that is,

$$u_{\mathfrak{a}} = \sum_{J\in\mathcal{J}} c_J h^J,$$

where

$$\mathcal{J} := \{(j_1, \ldots, j_n, j_1', \ldots, j_n') \mid \forall 1 \leqslant i \leqslant n : j_i \in \mathbb{N}, j_i' = 0, 1\},$$

the constants $c_J \in \mathbb{C}$, and

$$h^J := \prod_{i=1}^{n} h_i^{j_i} \prod_{i=1}^{n} \eta_i^{j_i'}.$$

Recall that $z_\lambda$ is the unique element of $\mathscr{Z}(\mathfrak{q}(n))$ such that $L(z_\lambda) = D_\lambda$.

LEMMA 4.6. *In the decomposition* $z_\lambda = z_\lambda' + z_{\lambda,\mathfrak{a}}$ *of Equation* (4.3), *we have* $z_\lambda \in \mathfrak{U}(\mathfrak{a}_{\bar{0}})$. *In particular, there exists a unique polynomial* $\gamma(z_\lambda) \in \mathbb{C}[x_1, \ldots, x_n]$ *such that*

$$z_{\lambda,\mathfrak{a}} = \gamma(z_\lambda)(h_1, \ldots, h_n). \tag{4.4}$$

PROOF. Let $\mathfrak{q}(n) = \mathfrak{u}^- \oplus \mathfrak{h} \oplus \mathfrak{u}^+$ be the standard triangular decomposition of $\mathfrak{q}(n)$. Furthermore, let $\mathfrak{U}(\mathfrak{q}(n))^0$ denote the centralizer of $\mathfrak{h}$ in $\mathfrak{U}(\mathfrak{q}(n))$. As usual, the Harish-Chandra projection gives rise to an $\mathrm{ad}_\mathfrak{h}$-invariant direct sum decomposition

$$\mathfrak{U}(\mathfrak{q}(n))^0 = \left(\mathfrak{u}^- \mathfrak{U}(\mathfrak{q}(n)) \cap \mathfrak{U}(\mathfrak{q}(n))\mathfrak{u}^+\right) \oplus \mathfrak{U}(\mathfrak{h}).$$

Now write $z_\lambda := z_\lambda^0 + z_{\lambda,\mathfrak{h}}$, according to the latter decomposition. From $\mathrm{ad}_\mathfrak{h}$-invariance of the decomposition, it follows that $[\mathfrak{h}_{\bar{1}}, z_{\lambda,\mathfrak{h}}] = 0$, and therefore that indeed $z_{\lambda,\mathfrak{h}} \in \mathfrak{U}(\mathfrak{h}_{\bar{0}})$. We can write $z_\lambda^0$ as a sum of monomials of the form $x_1 \cdots x_r$, where $x_1, \ldots, x_r \in \mathfrak{q}(n)$, such that $x_1 \in \mathfrak{u}^-$ and $x_r \in \mathfrak{u}^+$. Next, by the natural embedding of $\mathfrak{U}(\mathfrak{q}(n))$ into $\mathfrak{U}(\mathfrak{l})$ (as the left factor) we can consider the decomposition $z_\lambda := z_\lambda^0 + z_{\lambda,\mathfrak{h}}$ as one in $\mathfrak{U}(\mathfrak{l})$, and clearly under this embedding the monomial $x_1 \cdots x_r$ is mapped to

$$(x_1, 0) \cdots (x_r, 0) \in \mathfrak{U}(\mathfrak{l})\mathfrak{n}^+.$$

Similarly, since $z_{\lambda,\mathfrak{h}} \in \mathfrak{U}(\mathfrak{h}_{\bar{0}})$, every monomial $x_1 \cdots x_r$ of $z_{\lambda,\mathfrak{h}}$ can be written as

$$x_1 \cdots x_r = x_1' \cdots x_r',$$

where $x_i' := \frac{1}{2}\big((x_i, x_i) + (x_i, -x_i)\big)$, with $(x_i, x_i) \in \mathfrak{m}$ and $(x_i, -x_i) \in \mathfrak{a}_{\bar{0}}$. Since $\mathfrak{h}_{\bar{0}}$ is commutative, we obtain $z_{\lambda,\mathfrak{h}} \equiv \frac{1}{2^r} \prod_{i=1}^{r} (x_i, -x_i)$ modulo $\mathfrak{m}\,\mathfrak{U}(\mathfrak{l}) + \mathfrak{U}(\mathfrak{l})\mathfrak{n}^+$. The claim of the lemma now follows from uniqueness of the decomposition $z_\lambda = z_\lambda' + z_{\lambda,\mathfrak{a}}$. $\square$

LEMMA 4.7. *Consider* $e \in V$ *as a linear polynomial in* $\mathscr{P}(V^*)$. *Then, for every non-negative integer* $k \geqslant 0$, *we have* $\mathfrak{m} \cdot e^k = 0$ *and*

$$\left(\mathfrak{n}^+ \omega(S(\mathfrak{a} \oplus \mathfrak{n}^+)) \cdot e^k\right)|_{\mathfrak{t}^*} = 0.$$

PROOF. We may assume that $k \geqslant 1$. For $x \in \mathfrak{m}$, we have

$$x \cdot e^k = k(x \cdot e)e^{k-1} = 0,$$

as $x \cdot e = 0$. Next, Equations (3.2) and (3.3) imply that $\mathfrak{n}^+$ acts by vector fields that lie in the span of

$$u_{pk} \frac{\partial}{\partial u_{p\ell}}, \quad \xi_{pk} \frac{\partial}{\partial \xi_{p\ell}}, \quad u_{pk} \frac{\partial}{\partial \xi_{p\ell}}, \quad \xi_{pk} \frac{\partial}{\partial u_{p\ell}},$$

where $k > \ell$, whereas the action of $\mathfrak{a}$ is given by vector fields of the same shape where now $k = \ell$. Let $\mathcal{A}^+$ be the subalgebra of $\mathcal{P}(V^*)$ generated by

$$u_{\ell k}, \quad \xi_{\ell k}, \quad \forall k \geqslant \ell,$$

and $\mathcal{I}^+$ the ideal of $\mathcal{A}^+$ generated by

$$u_{\ell k}, \quad \xi_{\ell k}, \quad \forall k > \ell.$$

Then $\omega(S(\mathfrak{a}\oplus\mathfrak{n}^+))$ leaves $\mathcal{A}^+$ invariant and $\mathfrak{n}^+$ maps $\mathcal{A}^+$ into $\mathcal{I}^+$.

On the other hand, $e$ is the sum of all the $u_{kk}$ and $\xi_{\ell\ell}$, and hence $e^k$ is contained in $\mathcal{A}^+$. Thus, $\mathfrak{n}^+\,\omega(S(\mathfrak{a}\oplus\mathfrak{n}^+))\cdot e^k$ is contained in $\mathcal{I}^+$. As the restriction of $\mathcal{I}^+$ to $\mathfrak{t}^*$ vanishes, this proves the claim. □

For $\lambda\in\Lambda^n_{>0}$, let $D^*_\lambda\in\mathscr{P}\mathscr{D}(V^*)$ be the operator dual to $D_\lambda$, *i.e.*

$$\langle D^*_\lambda(p),q\rangle := \langle p, D_\lambda(q)\rangle,\quad \forall p\in\mathscr{P}(V^*), q\in\mathscr{P}(V).$$

PROPOSITION 4.8. *For all* $\lambda\in\Lambda^n_{>0}$ *and* $k:=|\lambda|$, *we have*

$$p^*_\lambda = (-1)^k D^*_\lambda(e^k). \tag{4.5}$$

PROOF. The polynomial $e^k\in\mathscr{P}(V^*)$ is homogeneous of degree $k$. Thus, by Proposition 3.1 and Corollary 3.3, we see that $D^*_\lambda(e^k)\in(V_\lambda)^*$. Furthermore, for every $\ell\in V_\lambda\subseteq\mathscr{P}(V)$, we have

$$\big\langle D^*_\lambda(e^k),\ell\big\rangle = c_\lambda(\lambda)\langle e^k,\ell\rangle = k!\,j^\sharp_V(\ell)(e) = (-1)^k\langle p^*_\lambda,\ell\rangle,$$

in view of Equation (4.1), and because $\ell$ is homogeneous of degree $k$. □

PROOF OF THEOREM 4.5. From the proof of Proposition 3.4, recall that there is an element $z_\lambda\in\mathcal{Z}(\mathfrak{q}(n))\subseteq\mathcal{Z}(\mathfrak{l})$ (necessarily even) of order $\leqslant k$ such that $L(z_\lambda)=D_\lambda$. Because of Lemma 4.6, we have

$$z_{\lambda,\mathfrak{a}} = \gamma(z_\lambda)(h_1,\dots,h_n)\in\mathfrak{U}(\mathfrak{a}_{\bar 0}) = S(\mathfrak{a}_{\bar 0})$$

for a unique polynomial $\gamma(z_\lambda)\in\mathbb{C}[x_1,\dots,x_n]$. By Equation (2.2), we have

$$h_{a,0}\cdot e = a,\quad \forall a=\mathrm{diag}(a_1,\dots,a_n),$$

so that

$$(z_{\lambda,\mathfrak{a}}\cdot e^k)|_{\mathfrak{t}^*}(x_1,\dots,x_n) = \gamma(z_\lambda)(x_1,\dots,x_n). \tag{4.6}$$

Let $\mathsf{S}_\mathfrak{l}:\mathfrak{U}(\mathfrak{l})\longrightarrow\mathfrak{U}(\mathfrak{l})$ denote the antipodal anti-automorphism of $\mathfrak{U}(\mathfrak{l})$, defined by $\mathsf{S}_\mathfrak{l}(x):=-x$ for $x\in\mathfrak{l}$. Then

$$\mathsf{S}_\mathfrak{l}(z_\lambda)\equiv(-1)^k z_{\lambda,\mathfrak{a}}\pmod{\mathfrak{U}(\mathfrak{l})\mathfrak{m}\oplus\mathfrak{n}^+\omega(S(\mathfrak{a}\oplus\mathfrak{n}^+))},$$

as $\mathfrak{a}_{\bar 0}$ is Abelian. Recall that we consider $\mathfrak{t}^*=\mathbb{C}^n$ as a subspace of $V^*$. Thus, for any $\mu\in\mathfrak{t}^*$, we may form $\mu^k\in\mathscr{P}^k(V)=S^k(V^*)$. In these terms, we compute by the use of Proposition 4.8 and Lemma 4.7:

$$\begin{aligned}p^*_\lambda|_{\mathfrak{t}^*}(\mu) &= \frac{1}{k!}\langle p^*_\lambda,\mu^k\rangle = \frac{(-1)^k}{k!}\langle D^*_\lambda(e^k),\mu^k\rangle\\ &= \frac{(-1)^k}{k!}\langle \mathsf{S}_\mathfrak{l}(z_\lambda)(e^k),\mu^k\rangle = \frac{1}{k!}\langle z_{\lambda,\mathfrak{a}}\cdot e^k,\mu^k\rangle.\end{aligned}$$

By Equation (4.6), this is the value of the $k$-homogeneous part of $(z_{\lambda,\mathfrak{a}}\cdot e^k)|_{\mathfrak{t}^*}$ at the point $\mu\in\mathfrak{t}^*$.

Let $\mu\in\Lambda^n_{>0}$. By the highest weight theory for $\mathfrak{q}(n)$ [**5**, § 2.3], there is a non-zero vector $v_\mu\in V_\mu$ such that $\mathfrak{n}^+v_\mu=0$ and

$$h_{a,0}\cdot v_\mu = \sum_{j=1}^n \mu_j a_j\cdot v_\mu,\quad\forall a=\mathrm{diag}(a_1,\dots,a_n).$$

In view of Equation (4.4), and because $\eta_i^2=\frac12[\eta_i,\eta_i]\in\mathfrak{m}$, we find that $V_\mu$ is generated as an $\mathfrak{m}$-module by the vectors $v^i_\mu:=\eta_i v_\mu$, $i=1,\dots,n$. Thus, there must be some $i$ such that $\langle p^*_\mu, v^i_\mu\rangle\neq 0$.

As $\mathfrak{n}^+ \cdot v_\mu^i = 0$ and $\mathfrak{m} \cdot p_\mu^* = 0$, we see that

$$c_\lambda(\mu)\,\langle p_\mu^*, v_\mu^i\rangle = \langle p_\mu^*, z_\lambda \cdot v_\mu^i\rangle = \langle p_\mu^*, z_{\lambda,\mathfrak{a}} \cdot v_\mu^i\rangle = \gamma(z_\lambda)(\mu_1,\dots,\mu_n)\,\langle p_\mu^*, v_\mu^i\rangle.$$

We conclude

$$q_\lambda^*(\mu) = c_\lambda(\mu) = \gamma(z_{\lambda,\mathfrak{a}})(\mu_1,\dots,\mu_n) = (z_{\lambda,\mathfrak{a}} \cdot e^k)|_{\mathfrak{t}^*}(\mu_1,\dots,\mu_n),$$

by Equation (4.6), and with the convention from Equation (4.2). As $\Lambda_{>0}^n$ is Zariski dense in $\mathbb{C}^n$, $p_\lambda^*|_{\mathfrak{t}^*}$ is the $k$-homogeneous part of $q_\lambda^*$. The assertion now follows directly from Theorem 3.8. □

In the next theorem, we use the notation introduced in Section 1. In particular, recall that $\eta : \mathscr{Z}(\mathfrak{q}(n)) \longrightarrow \mathscr{P}(\mathfrak{h}_{\bar{0}})$ denotes the Harish-Chandra homomorphism of Equation (1.3), and we identify the image of $\eta$ with the algebra of $Q$-symmetric polynomials in $n$ variables. Let $\mathsf{S} : \mathfrak{U}(\mathfrak{q}(n)) \longrightarrow \mathfrak{U}(\mathfrak{q}(n))$ denote the canonical anti-automorphism of the enveloping algebra $\mathfrak{U}(\mathfrak{q}(n))$ obtained by extending the map $\mathfrak{q}(n) \longrightarrow \mathfrak{q}(n) : x \longmapsto -x$.

THEOREM 4.9. *For every $\lambda \in \Lambda_{>0}$, there exists a unique element $z_\lambda \in \mathscr{Z}(\mathfrak{q}(n))$ such that $L(z_\lambda) = D_\lambda$. Furthermore,*

$$z_\lambda = r_\lambda \mathsf{S}(C_\lambda), \text{ where } r_\lambda := \frac{(-1)^{|\lambda|}}{2^{\ell(\lambda)}(\lambda_1!\cdots\lambda_{\ell(\lambda)}!)^2} \prod_{1\leqslant i<j\leqslant \ell(\lambda)} \left(\frac{\lambda_i-\lambda_j}{\lambda_i+\lambda_j}\right)^2. \tag{4.7}$$

PROOF. Existence and uniqueness of $z_\lambda$ follow as in the proof of Proposition 3.4. Note that Nazarov considers a slightly different action of $\mathfrak{l}$ on a polynomial space which decomposes as a direct sum of $\mathfrak{l}$-modules of the form $2^{-\delta(\lambda)}F_\lambda \otimes F_\lambda$. Therefore, it follows from [**16**, Proposition 4.3] that $(-1)^{|\lambda|}L(\mathsf{S}(C_\lambda))$ acts on $V_\lambda$ with the same scalar that $C_\lambda$ acts on Nazarov's $W_\lambda$ (see [**16**, Section 4]).

For the proof of Equation (4.7), note first that the "$\varrho$-shift" for Sergeev's Harish-Chandra homomorphism is zero, and that $L(z_\lambda)$ acts on the modules $V_\mu \subseteq \mathscr{P}(V)$ by a scalar given by a $Q$-symmetric polynomial in $\mu_1,\dots,\mu_n$ of degree at most $|\mu|$. From Proposition 3.6 and Corollary 3.3, it follows that $\eta(z_\lambda) = q_\lambda^*$. In particular, the action of $L(z_\lambda)$ on $V_\mu$ is by given a polynomial in $\mu$ whose leading coefficient is $(Q_\lambda^*(\lambda))^{-1}$. In addition, it follows from [**16**, Proposition 4.8] that the action of $(-1)^{|\lambda|}L(\mathsf{S}(C_\lambda))$ is by a polynomial in $\mu$ with leading coefficient

$$\lambda_1!\cdots\lambda_{\ell(\lambda)}! \prod_{1\leqslant i<j\leqslant \ell(\lambda)} \frac{\lambda_i+\lambda_j}{\lambda_i-\lambda_j}.$$

Equation (4.7) now follows from a comparison of leading coefficients and the formula for $Q_\lambda^*(\lambda)$ given in Ref. [**8**]. □

## References

[1] A. Alldridge, J. Hilgert, and T. Wurzbacher, *Superorbits*, J. Math. Jussieu, posted on 2016, 1–56, DOI 10.1017/S147474801600030X.

[2] A. Alldridge and S. Schmittner, *Spherical representations of Lie supergroups*, J. Funct. Anal. **268** (2015), no. 6, 1403–1453. MR3306354

[3] C. Carmeli, L. Caston, and R. Fioresi, *Mathematical foundations of supersymmetry*, EMS Series of Lectures in Mathematics, European Mathematical Society (EMS), Zürich, 2011. MR2840967

[4] S.-J. Cheng and W. Wang, *Remarks on the Schur–Howe–Sergeev duality*, Lett. Math. Phys. **52** (2000), no. 2, 143–153, DOI 10.1023/A:1007668930652.

[5] S.-J. Cheng and W. Wang, *Dualities and representations of Lie superalgebras*, Graduate Studies in Mathematics, vol. 144, American Mathematical Society, Providence, RI, 2012. MR3012224

[6] R. Howe and T. Umeda, *The Capelli identity, the double commutant theorem, and multiplicity-free actions*, Math. Ann. **290** (1991), no. 3, 565–619. MR1116239

[7] V. N. Ivanov, *The dimension of skew shifted Young diagrams, and projective characters of the infinite symmetric group* (Russian, with English and Russian summaries), Zap. Nauchn. Sem. S.-Peterburg. Otdel. Mat. Inst. Steklov. (POMI) **240** (1997), no. Teor. Predst. Din. Sist. Komb. i Algoritm. Metody. 2, 115–135, 292–293; English transl., J. Math. Sci. (New York) **96** (1999), no. 5, 3517–3530. MR1691642

[8] V. N. Ivanov, *A combinatorial formula for factorial Schur Q-functions* (Russian, with English and Russian summaries), Zap. Nauchn. Sem. S.-Peterburg. Otdel. Mat. Inst. Steklov. (POMI) **256** (1999), no. Teor. Predst. Din. Sist. Komb. i Algoritm. Metody. 3, 73–94, 265; English transl., J. Math. Sci. (New York) **107** (2001), no. 5, 4195–4211. MR1708560

[9] V.G. Kac, *Lie superalgebras*, Adv. Math. **26** (1977), no. 1, 8–96, DOI 10.1016/0001-8708(77)90017-2.

[10] F. Knop and S. Sahi, *Difference equations and symmetric polynomials defined by their zeros*, Int. Math. Res. Not. IMRN **10** (1996), 473–486, DOI 10.1155/S1073792896000311.

[11] B. Kostant and S. Sahi, *The Capelli identity, tube domains, and the generalized Laplace transform*, Adv. Math. **87** (1991), no. 1, 71–92, DOI 10.1016/0001-8708(91)90062-C.

[12] B. Kostant and S. Sahi, *Jordan algebras and Capelli identities*, Invent. Math. **112** (1993), no. 3, 657–664, DOI 10.1007/BF01232451.

[13] D. A. Leĭtes, *Introduction to the theory of supermanifolds* (Russian), Uspekhi Mat. Nauk **1** (1980), 3–57; English transl., Russian Math. Surveys **35** (1980), no. 1, 1–64.

[14] I. G. Macdonald, *Symmetric functions and Hall polynomials*, 2nd ed., Oxford Classic Texts in the Physical Sciences, The Clarendon Press, Oxford University Press, New York, 2015. With contribution by A. V. Zelevinsky and a foreword by Richard Stanley; Reprint of the 2008 paperback edition [ MR1354144]. MR3443860

[15] A. Molev, *Factorial supersymmetric Schur functions and super Capelli identities*, Kirillov's seminar on representation theory, Amer. Math. Soc. Transl. Ser. 2, vol. 181, Amer. Math. Soc., Providence, RI, 1998, pp. 109–137. MR1618747

[16] M. Nazarov, *Capelli identities for Lie superalgebras* (English, with English and French summaries), Ann. Sci. École Norm. Sup. (4) **30** (1997), no. 6, 847–872. MR1476298

[17] S. Sahi, *The Capelli identity and unitary representations*, Compositio Math. **81** (1992), no. 3, 247–260. MR1149169

[18] S. Sahi, *Unitary representations on the Shilov boundary of a symmetric tube domain*, Representation theory of groups and algebras, Contemp. Math., vol. 145, Amer. Math. Soc., Providence, RI, 1993, pp. 275–286. MR1216195

[19] S. Sahi, *The spectrum of certain invariant differential operators associated to a Hermitian symmetric space*, Lie theory and geometry, Progr. Math., vol. 123, Birkhäuser Boston, Boston, MA, 1994, pp. 569–576. MR1327549

[20] S. Sahi, *Jordan algebras and degenerate principal series*, J. Reine Angew. Math. **462** (1995), 1–18. MR1329899

[21] S. Sahi, *Interpolation, integrality, and a generalization of Macdonald's polynomials*, Internat. Math. Res. Notices **10** (1996), 457–471. MR1399411

[22] S. Sahi and Hadi Salmasian, *The Capelli problem for $\mathfrak{gl}(m|n)$ and the spectrum of invariant differential operators*, Adv. Math. **303** (2016), 1–38. MR3552519

[23] S. Sahi and H. Salmasian, *Quadratic Capelli operators and Okounkov polynomials*, Ann. Sci. Éc. Norm. Supér. (2018), to appear.

[24] S. Sahi, H. Salmasian, and V. Serganova, *The Capelli problem for basic classical Lie superalgebras*, In preparation.

[25] M. Scheunert, *The theory of Lie superalgebras*, Lecture Notes in Mathematics, vol. 716, Springer, Berlin, 1979.

[26] A. Sergeev, *Projective Schur functions as bispherical functions on certain homogeneous superspaces*, The orbit method in geometry and physics (Marseille, 2000), Progr. Math., vol. 213, Birkhäuser Boston, Boston, MA, 2003, pp. 421–443. MR1995387

[27] A. N. Sergeev, *The centre of enveloping algebra for Lie superalgebra $Q(n, \mathbf{C})$*, Lett. Math. Phys. **7** (1983), no. 3, 177–179. MR706205

[28] A. N. Sergeev and A. P. Veselov, *Generalised discriminants, deformed Calogero-Moser-Sutherland operators and super-Jack polynomials*, Adv. Math. **192** (2005), no. 2, 341–375. MR2128703

[29] J. R. Stembridge, *Shifted tableaux and the projective representations of symmetric groups*, Adv. Math. **74** (1989), no. 1, 87–134. MR991411

Mathematisches Institut, Mathematisch-Naturwissenschaftliche Fakultät, Universität zu Köln, Weyertal 86–90, 50931 Köln, Germany

*Current address*: Department of Mathematics, University of California at Berkeley, 969 Evans Hall, Berkeley, California 94720

*URL*: `https://aalldridge.github.io`

*Email address*: `alldridg@math.uni-koeln.de, aalldridge@berkeley.edu`

Department of Mathematics, Rutgers University, 110 Frelinghuysen Road, Piscataway, New Jersey 08854-8019

*Email address*: `sahi@math.rutgers.edu`

Department of Mathematics and Statistics, University of Ottawa, 585 King Edward Avenue, Ottawa, Ontario, Canada K1N 6N5

*Email address*: `hadi.salmasian@uottawa.ca`

Contemporary Mathematics
Volume **714**, 2018
https://doi.org/10.1090/conm/714/14331

# Analysis of free products of the general linear groups $GL_2(\mathbb{Q}_p)$ and Hecke algebras $\mathcal{H}(GL_2(\mathbb{Q}_p))$ over primes $p$

Ilwoo Cho and Palle E. T. Jorgensen

ABSTRACT. The main purposes of this paper are (i) to study a certain embedded sub-structure $\mathfrak{S}$ of the free product $C^*$-algebra $\mathbb{H}_{\mathcal{P}}$ generated by Hecke algebras $\mathcal{H}(G_p)$ over primes $p$, (ii) to obtain free-probabilisitc characterizations of the generating operators of $\mathfrak{S}$, implying number-theoretic data, and (iii) to apply the results from (i) and (ii) to $C^*$-dynamical systems.

## 1. Introduction

Our paper is interdisciplinary, and lies at the cross-roads of free probability theory, $C^*$-algebra theory, and the study of algebraic structures from number theory. In more detail, we study free product $C^*$-algebras of certain Hecke algebras $\mathcal{H}(G_p)$, indexed by primes. A main point in this paper is our free product $C^*$-probability space formed over the set $\mathcal{P}$ of all primes. In our construction, we shall make use of the Adele ring over the rational numbers, and associated $L$-functions (e.g., [18] and [19]).

In this paper, we establish free-probabilistic models for *Hecke algebras* $\mathcal{H}(G_p)$ of the *general linear groups* $G_p = GL_2(\mathbb{Q}_p)$ *over p-adic number fields* $\mathbb{Q}_p$, and corresponding *representations*. Representations of Hecke algebras have been studied in other texts (e.g., see [3] and [9]), however, we concentrate on representations obtained from our free-probabilistic structures on Hecke algebras (e.g., [12], [13], [14] and [15]). Based on our representation, one can study not only operator-algebraic structures of Hecke algebras, but also operator-theoretic properties of elements of the algebras by investigating free-distributional data (and hence, spectral information) of them.

Relations between *number theory* and *functional analysis* have been studied earlier, both pure and applied mathematical interests (e.g., [8], [9], [20], [23] and [26]). Here, we are in particular interested in Hecke algebras, and operator theory on them via *representation theory* and *free probability* (e.g., see [6], [23], [25], [27] and [28]).

---

2010 *Mathematics Subject Classification.* Primary 05E15, 11G15, 11R47, 11R56, 46L40, 46L53, 46L54, 47L30, 47L55.

*Key words and phrases.* Free probability, free product, primes, Hecke algebras, representations, $C^*$-algebras.

**1.1. Motivation and Background.** Relations between *primes* and *operators* have been studied in various earlier approaches. For instance, in [10] and [11], we studied how primes act "on" certain *von Neumann algebras* generated by *p -adic* and *Adelic measure spaces*. Independently, in [16], we studied primes as *linear functionals* acting on *arithmetic functions*. In such a case, one can understand arithmetic functions as *Krein-space operators* (See [16]). These studies are motivated by number-theoretic results (e.g., [9], [18], [19], [21], [23] and [26]) under free-probabilistic techniques (e.g., [25] and [27]).

The *p-adic number fields* $\mathbb{Q}_p$, and the *Adele ring* $\mathbb{A}_{\mathbb{Q}}$ play key roles in modern *number theory*; *analytic number theory*, *L-function theory*, and *algebraic geometry* (e.g., [18], [19] and [26]). Also, analysis on such Adelic structures gives ways for understanding vector analysis under *non-Archimedean metric* (e.g., [10], [11], [15], [16] and [26]).

In [15], the first-named author and Gillespie considered free-probabilistic models on certain embedded sub-structures of a *Hecke algebra* $\mathcal{H}(G_p)$, for each prime $p$. They characterized the "inner" freeness of $\mathcal{H}(G_p)$, determined by "*normal*" *compact-open subgroups* of $G_p$.

In [14], by extending the "inner" free-probabilistic structures of [15] to those "fully" on $\mathcal{H}(G_p)$, the first-named author established universal free-probabilistic models on Hecke algebras, and corresponding *Hilbert-space representations*. And he studied operator-theoretic properties of operators generated by *generating elements of* $\mathcal{H}(G_p)$.

In [12] and [13], from the free-probabilistic settings of [14], we studied free-probabilistic sub-structures of $\mathcal{H}(G_p)$, and corresponding *dynamical systems* of $\mathcal{H}(G_p)$. The main results of [12] and [13] illustrate again that the number-theoretic information on $\mathcal{H}(G_p)$ affects operator-theoretic properties on $\mathcal{H}(G_p)$.

In this paper, we investigate the *free product* $C^*$*-probability space* $\mathbb{H}_{\mathcal{P}}$ induced by the *noncommutative probability spaces*

$$\{(\mathcal{H}(G_p),\ \psi_p) : p \in \mathcal{P}\},$$

where $\mathcal{P}$ is the set of all primes in the set $\mathbb{N}$ of all *natural numbers* (or *positive integers*). Especially, we concentrate on the free-probabilistic sub-structure $\mathfrak{S}$ of $\mathbb{H}_{\mathcal{P}}$, generated by certain *idempotent operators* in $\mathbb{H}_{\mathcal{P}}$.

**1.2. Overview.** As corollary of our free-product constructions, we arrive at a new and canonical $C^*$-algebra which we show is built naturally from primes, and is realized in an explicit fashion in concrete Hilbert space; See main results Theorems 7.2, 8.6, 8.7, 8.8, 9.2, 9.3 (from free blocks), 9.4, 9.5 and 10.3 (from free product); and the corresponding Corollaries 8.2, 8.4, 8.5 and 8.9. But, as preparation, before we got to this, we first need a number of tools, which we develop in the material in the first half of our paper, around Theorems 3.2, 4.2, 6.1 and 6.2, and Propositions 3.4, 3.5, 4.1, 4.3, 5.1 and 5.3, dealing with $p$-adic free random variables and their moments. The latter results, we feel, will be of independent interest. And they will be used in the proofs of our main results in the second half of the paper.

In Section 2, we briefly introduce backgrounds and basic theories used in our proceeding works.

In Sections 3 and 4, we establish our free-probabilistic models on Hecke algebras $\mathcal{H}(G_p)$, for primes $p$, and consider free-distributional data of generating elements of $\mathcal{H}(G_p)$, implying the number-theoretic data.

Depending on our free probability on $\mathcal{H}(G_p)$, the Hilbert-space representations $(\mathfrak{H}_p, \alpha^p)$ are constructed and the corresponding $C^*$-algebras $\mathbb{H}(G_p)$ of $\mathcal{H}(G_p)$ are obtained in Section 5, for primes $p$. One can get suitable $C^*$-probabilistic settings on $\mathbb{H}(G_p)$, implying the free-distributional data of $\mathcal{H}(G_p)$ (obtained in Sections 3 and 4).

In Section 6, we study $C^*$-probability spaces of the $C^*$-algebras $\mathbb{H}(G_p)$ of Section 5, in detail.

In Section 7, based on the main results of Sections 5 and 6, we construct-and-study the *free product $C^*$-probability space* $\mathbb{H}_{\mathcal{P}}$ generated by the family $\{\mathbb{H}(G_p), \tau_p\}_p$ over primes $p$. Free-distributional data of generating operators of $\mathbb{H}_{\mathcal{P}}$ are characterized under *freeness*.

In Section 8, we focus on a certain embedded free-probabilistic sub-structure $\mathfrak{S}$ of the free product $C^*$-probability space $\mathbb{H}_{\mathcal{P}}$ of Section 7. We study not only free-distributional data on $\mathfrak{S}$, but also the estimations of such data on $\mathfrak{S}$.

In Section 9, we act $\mathfrak{S}$ on an arbitrarily fixed $C^*$-probability space $(A, \varphi)$, under *tensor product.* From the tensor product $C^*$-probability space $\mathfrak{S}_A = A \otimes_{\mathbb{C}} \mathfrak{S}$, free-probabilistic information affected both by $A$ and $\mathfrak{S}$ is considered. By the main results of Sections 7 and 8, one can obtain interesting Hecke-analytic-data-depending free distributions of operators in $\mathfrak{S}_A$, and corresponding estimations.

Finally, we apply our main results of Section 9 to certain concrete cases in Section 10. In particular, we act $\mathfrak{S}$ to the *group $C^*$-probability spaces*, and to the *measure-theoretic $C^*$-probability spaces.*

## 2. Preliminaries

In this section, we review concepts and backgrounds of our proceeding works.

### 2.1. Free Probability.

Readers can check Voiculescu's original analytic free probability theory in [27] (and cited papers therein), and Speicher's combinatorial free probability theory in [25] and cited papers therein. *Free probability* is a noncommutative operator-algebraic version of (classical) *probability* and *statistics* (and hence, covering commutative operator-algebraic cases). The classical *independence* of random variables from the conventional commutative stochastic analysis is replaced by so-called the *freeness*; *measures* for *measure spaces* are replaced by *linear functionals* on operator algebras, called *free probability spaces*, or *noncommutative probability spaces* (e.g., see [4], [5] and [28]).

Throughout this paper, we follow Speicher's combinatorial approach. Without detailed introduction, we use *free moments* and *free cumulants* of *free random variables* (or operators) for studying *free distributions* of the operators (e.g., [25]), and consider *free product* of free probability spaces (e.g., [23], [25] and [27]). While we include technical results, as needed, on free probability, with citations [4], [17], [23], [25] and [27], suffice it to say that free probability theory is a fast growing area of applied noncommutative analysis.

### 2.2. Hecke Algebras over $GL_2(\mathbb{Q}_p)$.

Let $p$ be a fixed *prime*, and let $\mathbb{Q}_p$ be the *p-adic number field* for $p$. This set $\mathbb{Q}_p$ is by definition the completion of the *rational numbers* $\mathbb{Q}$ with respect to the *p-adic norm*

$$|q|_p = \left|p^k \tfrac{a}{b}\right| = \left(\tfrac{1}{p}\right)^k,$$

for $q = p^k \frac{a}{b} \in \mathbb{Q}$, for some $k,\ a,\ b \in \mathbb{Z}$.

Under topology, $\mathbb{Q}_p$ is locally compact, and totally disconnected as a *Banach space*, and it contains a maximal compact subring,

$$\mathbb{Z}_p = \{x \in \mathbb{Q}_p : |x|_p \leq 1\}.$$

We call $\mathbb{Z}_p$, the *unit disk of* $\mathbb{Q}_p$, and all elements of $\mathbb{Z}_p$ are said to be the *p-adic integers* in $\mathbb{Q}_p$.

As a topological space, $\mathbb{Q}_p$ has *neighborhood bases of zero* consisting of the *subsets*

$$p^k\mathbb{Z}_p = \{p^k x : x \in \mathbb{Z}_p\}, \text{ for } k \in \mathbb{Z}.$$

In fact, set-theoretically, one has

$$\mathbb{Q}_p = \underset{k\in\mathbb{Z}}{\cup}\ p^k\mathbb{Z}_p$$

(e.g., see [26]).

Recall that the *general linear group* $GL_2(\mathbb{Q}_p)$ consists of all invertible $(2 \times 2)$-matrices over $\mathbb{Q}_p$, i.e.,

$$GL_2(\mathbb{Q}_p) = \left\{ \begin{pmatrix} a & b \\ c & d \end{pmatrix} \in M_2(\mathbb{Q}_p) \;\middle|\; \begin{matrix} a,\ b,\ c,\ d \in \mathbb{Q}_p, \\ ad - bc \neq 0 \end{matrix} \right\},$$

where $M_2(\mathbb{Q}_p)$ means the set of all $(2 \times 2)$-matrices over $\mathbb{Q}_p$.

**Notation.** Denote $GL_2(\mathbb{Q}_p)$ simply by $G_p$, if there is no confusion. □

The group $G_p$ is also locally profinite coming from the (product) topology of (those of) $\mathbb{Q}_p$, i.e., it has a neighborhood bases of the identity consisting of the compact-open subgroups

$$K_k = u_p + (p^k)\ GL_2(\mathbb{Z}_p), \text{ for all } k \in \mathbb{N},$$

where $GL_2(\mathbb{Z}_p)$ means the subset of $G_p$, whose elements are matrices in $M_2(\mathbb{Q}_p)$, having their entries in $\mathbb{Z}_p$, and where

$$u_p = \begin{pmatrix} 1 & 0 \\ 0 & 1 \end{pmatrix}, \text{ the identity matrix of } M_2(\mathbb{Q}_p).$$

Then the subgroup $K_0 = GL_2(\mathbb{Z}_p)$ forms the maximal compact subgroup of $G_p$.

DEFINITION 2.1. *Define now the Hecke algebra* $\mathcal{H}(G_p)$ *of* $G_p$ *by*
*(2.2.1)*

$$\mathcal{H}(G_p) = \left\{ f : G_p \to \mathbb{C} \;\middle|\; \begin{matrix} f \text{ has compact support,} \\ \text{and} \\ \text{it is } \rho\text{-smooth} \end{matrix} \right\}.$$

*The* $\rho$*-smoothness means that* $\mathcal{H}(G_p)$ *is a smooth representation of* $G_p$ *under right translation. In other words, for any element* $f \in \mathcal{H}(G_p)$*, there is a compact open subgroup* $K$ *of* $G_p$ *such that*

$$\rho(y)f(g) = f(gy) = f(g)$$

*for all* $g \in G_p$.

*We make* $\mathcal{H}(G_p)$ *into an associative algebra by taking* $f_1$, $f_2 \in \mathcal{H}(G_p)$ *and defining* convolution *(as a vector multiplication),*

*(2.2.2)*

$$(f_1 * f_2)(g) = \int_{G_p} f_1(x) f_2(x^{-1}g)\, d\mu_p(x)$$

*where* $\mu_p$ *denotes a* left Haar measure *on the locally compact group* $G_p$, *satisfying*

$$\mu_p\left(GL_2(\mathbb{Z}_p)\right) = 1.$$

Note also that $\mathcal{H}(G_p)$ is an algebra "without identity," but has a *system of local units*. In this vein for any compact open subgroup of $G_p$, let

$$e_K = \frac{1}{\mu_p(K)} \chi_K$$

where $\chi_K$ are the *characteristic function of the compact-open subgroup* $K$ of $G_p$.

For more about number-theoretic Hecke algebra theory, see e.g., [1], [7] and [24]. Also, readers can check the applied Hecke algebra theory in e.g., [2] and [22].

## 3. Canonical Free-Probabilistic Models in $\mathcal{H}(G_p)$

In this section, we review free-probabilistic structures obtained in [15]. Applications of these structures can be found in [12] and [13].

**3.1. The Hecke Algebra $\mathcal{H}(G_p)$.** Notice, first that, by the formal definition (2.2.1), the Hecke algebra $\mathcal{H}(G_p)$ can be re-defined algebraically by

(3.1.1)

$$\mathcal{H}(G_p) = \mathbb{C}_* \left[ \left\{ f = \sum_{j=1}^{N} t_j\ \chi_{x_j K} \;\middle|\; \begin{array}{c} N \in \mathbb{N}, \text{ and } t_j \in \mathbb{C}, \text{ and} \\ K \text{ is a compact-open subgroup of } G_p, \\ \text{depending on } f \\ \text{where } x_j \in G_p, \text{ for all } j = 1, \ldots, N \end{array} \right\} \right],$$

where $\mathbb{C}_*[X]$ mean the subalgebras of $\mathcal{M}(X)$ generated by subsets $X$ of $\mathcal{M}(G_p)$, where $\mathcal{M}(G_p)$ is the algebra consisting of all $\mu_p$-measurable functions on $G_p$, under the usual functional addition, and convolution of (2.2.2), and where $\chi_Y$ mean the usual *characteristic functions* of $\mu_p$-measurable subsets $Y$ of $G_p$, where $\mu_p$ is the *multiplicative invariant Haar measure* on $G_p$, satisfying (2.2.3) and (2.2.4). The set

(3.1.2)

$$X_p = \left\{ f = \sum_{j=1}^{N} t_j\ \chi_{x_j K} \;\middle|\; \begin{array}{c} N \in \mathbb{N}, \text{ and } t_j \in \mathbb{C}, \text{ and} \\ K \text{ is a compact subgroup of } G_p, \\ \text{depending on } f \\ \text{where } x_j \in G_p, \text{ for all } j = 1, \ldots, N \end{array} \right\},$$

generating the Hecke algebra $\mathcal{H}(G_p)$ in (3.1.1), is said to be the *generating set of* $\mathcal{H}(G_p)$, and we call elements of $X_p$, *generating elements of* $\mathcal{H}(G_p)$. i.e.,

(3.1.3)

$$\mathcal{H}(G_p) = \mathbb{C}_*[X_p],$$

where $X_p$ is the generating set (3.1.2).

By (3.1.1) and (3.1.3), one may write

(3.1.4)

$$\mathcal{H}(G_p) = \left\{ \sum_{j=1}^{N} t_j\ \chi_{x_j K_j} \;\middle|\; \begin{array}{c} N \in \mathbb{N}, \text{ and } t_j \in \mathbb{C}, \text{ and} \\ K_j \text{ are compact subgroups of } G_p, \\ \text{where } x_j \in G_p, \text{ for all } j = 1, \ldots, N \end{array} \right\},$$

set-theoretically.

PROPOSITION 3.1. *Let $\chi_{x_1K_1}$, $\chi_{x_2K_2}$ be generating elements of $\mathcal{H}(G_p)$ in $X_p$, for $x_j \in G_p$, and compact subgroups $K_j$ of $G_p$, for $j = 1, 2$. Then*
*(3.1.5)*

$$\left(\chi_{x_1K_1} * \chi_{x_2K_2}\right)(g) = \mu_p\left(x_1K_1 \cap gK_2x_2^{-1}\right),$$

*for all $g \in G_p$.*

PROOF. The proof of (3.1.5) is straightforward by (3.1.4). See [15] for more details. □

Without loss of generality,
(3.1.6)

$$\chi_K(g) = \frac{\mu_p(K \cap gK)}{\mu_p(K)}, \text{ and } e_K(g) = \frac{\mu_p(K \cap gK)}{\mu_p(K)^2},$$

for all $g \in G_p$, and all compact-open subgroups $K$ of $G_p$.

Similar to (3.1.6), for any $x \in G_p$, and compact-open subgroups $K$ of $G_p$,
(3.1.7)

$$\chi_{xK}(g) = \frac{\mu_p(xK \cap gK)}{\mu_p(xK)} = \frac{\mu_p(xK \cap gK)}{\mu_p(K)},$$

and

$$e_{xK}(g) = \frac{\mu_p(xK \cap gK)}{\mu_p(xK)^2} = \frac{\mu_p(xK \cap gK)}{\mu_p(K)^2}.$$

Notice that, by (3.1.5) and (3.1.7), if $x_1, x_2 \in G_p$, and $K_1, K_2$ are compact-open subgroups of $G_p$, then

$$\chi_{x_1K_1} * \chi_{x_2K_2}(g) = \mu_p\left(x_1K_1 \cap gK_2x_2^{-1}\right),$$

for all $g \in G_p$, and hence,
(3.1.8)

$$e_{x_1K_1} * e_{x_2K_2}(g) = \frac{\mu_p(x_1K_1 \cap gK_2x_2^{-1})}{\mu_p(K_1)\mu_p(K_2)},$$

where $e_{xK} = \frac{1}{\mu_p(xK)} \chi_{xK}$ in $\mathcal{H}(G_p)$.

We now consider specific generating elements $\chi_{xK}$ and $e_{xK}$ in $X_p$, where $K$ are "normal" compact-open subgroups of $G_p$. It is reasonable to restrict our interests to normal subgroups by the locally profiniteness of $G_p$. Recall that a subgroup $K$ is *normal* in an arbitrary group $\Gamma$, if $gK = Kg$, for all $g \in \Gamma$. As usual, we denote the normal-subgroup-inclusion by $K \lhd \Gamma$.

Define a subset $Y_p$ of the generating set $X_p$ of $\mathcal{H}(G_p)$ by
(3.1.9)

$$Y_p \overset{def}{=} \left\{\sum_{j=1}^{N} t_j \chi_{x_jK} \in X_p \,|K \lhd G_p\right\}.$$

Then one can have a subalgebra $\mathcal{H}_{Y_p}$ of $\mathcal{H}(G_p)$;
(3.1.10)

$$\mathcal{H}_{Y_p} \overset{def}{=} \mathbb{C}_*[Y_p] \text{ of } \mathcal{H}(G_p).$$

where $Y_p$ is in the sense of (3.1.9) in $X_p$.

THEOREM 3.2. *(See [15]) Let* $\chi_{x_j K_j}$, $e_{x_j K_j} \in \mathcal{H}_{Y_p}$, *where* $x_j \in G_p$, *and* $K_j \triangleleft G_p$ *compact-open, for* $j = 1, 2$, *where* $\mathcal{H}_{Y_p}$ *is in the sense of (3.1.12) in* $\mathcal{H}(G_p)$. *Then*

$$\text{(3.1.11)} \qquad \chi_{x_1 K_1} * \chi_{x_2 K_2} = \mu_p(K_1 \cap K_2)\, \chi_{x_1 x_2 K_1 K_2}, \text{ and}$$

$$\text{(3.1.12)} \qquad e_{x_1 K_1} * e_{x_2 K_2} = \frac{\mu_p(K_1 \cap K_2)\mu_p(K_1 K_2)}{\mu_p(K_1)\mu_p(K_2)}\, e_{x_1 x_2 K_1 K_2},$$

*where* $K_1 K_2 \triangleleft G_p$ *is the product group of* $K_1$ *and* $K_2$. □

The above theorem is a main result of [15]. In particular, the formulas (3.1.11) and (3.1.12) provide the fundamental tools and techniques to establish our free-probabilistic structures on $\mathcal{H}_{Y_p}$ (and hence, those on $\mathcal{H}(G_p)$ in [12] and [13]).

DEFINITION 3.1. *Let* $Y_p$ *be the subset (3.1.9) of the generating set* $X_p$, *and let* $\mathcal{H}_{Y_p} = \mathbb{C}_*[Y_p]$ *be the subalgebra (3.1.10) of the Hecke algebra* $\mathcal{H}(G_p)$, *for a fixed prime* $p$. *Then we call* $Y_p$ *and* $\mathcal{H}_{Y_p}$, *the normal sub-generating set of* $X_p$, *and the normal Hecke subalgebra of* $\mathcal{H}(G_p)$, *respectively.*

**3.2. The Normal Hecke Subalgebra $\mathcal{H}_{Y_p}$ of $\mathcal{H}(G_p)$.** In this section, we focus on the normal Hecke subalgebra $\mathcal{H}_{Y_p}$ of (3.1.10) in the Hecke algebra $\mathcal{H}(G_p)$, where $Y_p$ is the normal sub-generating set (3.1.9) of the generating set $X_p$ of $\mathcal{H}(G_p)$, for a fixed prime $p$.

**Notation.** For convenience, denote group elements $\prod_{j=1}^{N} x_j$ of $G_p$, and product subgroups $\times_{j=1}^{N} K_j$ of $G_p$ by

$$x_{1,\ldots,N}, \text{ respectively, } K_{1,\ldots,N}$$

for all $N \in \mathbb{N}$, where $x_1, \ldots, x_N \in G_p$, and $K_1, \ldots, K_N$, compact-open subgroups of $G_p$. Also, we will write

$$K_{1,\ldots,N}^{o} = K_{1,\ldots,(N-1)} \cap K_N,$$

for all $N \in \mathbb{N} - \{1\}$. We will use these notations from below. □

We obtain the following computations.

PROPOSITION 3.3. *Let* $\chi_{x_j K_j}$, $e_{x_j K_j}$ *be generating elements of the normal Hecke subalgebra* $\mathcal{H}_{Y_p}$, *for* $j \in \mathbb{N}$. *Then*

*(3.2.1)*

$$\mathop{*}_{j=1}^{N} \chi_{x_j K_j} = \left( \prod_{l=2}^{N} \mu_p(K_{1,\ldots,l}^{o}) \right) \chi_{x_{1,\ldots,N} K_{1,\ldots,N}},$$

*and*

*(3.2.2)*

$$\mathop{*}_{j=1}^{N} e_{x_j K_j} = \left( \frac{\prod_{l=2}^{N} \mu_p(K_{1,\ldots,l}^{o})}{\prod_{l=1}^{N} \mu_p(K_l)} \right) e_{x_{1,\ldots,N} K_{1,\ldots,N}},$$

*for all* $N \in \mathbb{N} - \{1\}$.

PROOF. The formulas (3.2.1) and (3.2.2) are proven by induction on (3.1.11), respectively (3.1.12). See [15], for more details. Here, the normality condition of $K_1, \ldots, K_N$ is crucial. □

**Notation.** Denote the convolution $f * \ldots * f$ of $n$-copies of $f$ simply by $f^{(n)}$, for all $n \in \mathbb{N}$, for all $f \in \mathcal{H}(G_p)$. □

For $x \in G_p$, and a normal compact-open subgroup $K$ of $G_p$,
(3.2.3)

$$(\chi_{xK})^{(n)} = \mu_p(K)^{n-1} \; \chi_{x^n K},$$

and

$$(e_{xK})^{(n)} = \left(\frac{1}{\mu_p(K)}\right)^n e_{x^n K},$$

by (3.2.1), respectively, by (3.2.2), for all $n \in \mathbb{N}$.

**3.3. Free Probability on $\mathcal{H}_{Y_p}$.** Let $\mathcal{H}(G_p)$ be the Hecke algebra generated by the general linear group $G_p = GL_2(\mathbb{Q}_p)$ over the $p$-adic number field $\mathbb{Q}_p$, for a fixed prime $p$. In this section, we restrict our interests to the normal Hecke subalgebra $\mathcal{H}_{Y_p}$ of $\mathcal{H}(G_p)$.

Let $u_p$ be the *group-identity of* $G_p$, i.e.,

$$u_p = \begin{pmatrix} 1 & 0 \\ 0 & 1 \end{pmatrix} \in G_p = GL_2(\mathbb{Q}_p).$$

Define now a linear functional $\varphi_p$ on $\mathcal{H}_{Y_p}$ by
(3.3.1)

$$\varphi_p(f) \overset{def}{=} f(u_p), \text{ for all } f \in \mathcal{H}_{Y_p}.$$

Clearly, this morphism $\varphi_p$ of (3.3.1) is a well-defined linear functional on $\mathcal{H}_{Y_p}$, and hence, the pair $(\mathcal{H}_{Y_p}, \varphi_p)$ forms a noncommutative probability space in the sense of [25] and [27].

DEFINITION 3.2. *The noncommutative probability space $(\mathcal{H}_{Y_p}, \varphi_p)$ is said to be the normal Hecke probability space (for a prime $p$), where $\varphi_p$ is in the sense of (3.3.1).*

On the normal Hecke probability space $(\mathcal{H}_{Y_p}, \varphi_p)$, one obtains the following fundamental free-moment computations.

PROPOSITION 3.4. *Let $\chi_{x_j K_j}, e_{x_j K_j}$ be generating free random variables in the normal Hecke probability space $(\mathcal{H}_{Y_p}, \varphi_p)$, for all $j \in \mathbb{N}$. Then*

$$\text{(3.3.2)} \qquad \varphi_p\left(\underset{j=1}{\overset{N}{*}} \chi_{x_j K_j}\right) = \frac{\mu_p(K^o_{1,2})...\mu_p(K^o_{1,...,N})\mu_p(x_{1,...,N}K_{1,...,N} \cap K_{1,...,N})}{\mu_p(K_{1,...,N})},$$

$$\text{(3.3.3)} \qquad \varphi_p\left(\underset{j=1}{\overset{N}{*}} e_{x_j K_j}\right) = \frac{\mu_p(K^o_{1,2})...\mu_p(K^o_{1,...,N})\mu_p(x_{1,...,N}K_{1,...,N} \cap K_{1,...,N})}{\mu_p(K_1)\mu_p(K_2)...\mu_p(K_N)\mu_p(K_{1,...,N})},$$

*for all $N \in \mathbb{N} - \{1\}$. If $N = 1$, then one has*

$$\varphi_p(\chi_{xK}) = \frac{\mu_p(xK \cap K)}{\mu_p(K)}, \text{ and } \varphi_p(e_{xK}) = \frac{\mu_p(xK \cap K)}{\mu_p(K)^2}.$$

PROOF. The free-moment formulas (3.3.2) and (3.3.3) are obtained by (3.1.7), (3.2.1), (3.2.2) and (3.3.1). See [15] for details. ☐

Let $\chi_{x_1K_1}, \ldots, \chi_{x_NK_N} \in (\mathcal{H}_{Y_p}, \varphi_p)$, for $N \in \mathbb{N}$. If $k_n^p(\ldots)$ is the free cumulant, in the sense of [25], on $\mathcal{H}_{Y_p}$ with respect to the linear functional $\varphi_p$ of (3.3.1), then

$$k_N^p\left(\chi_{x_1K_1}, \ldots, \chi_{x_NK_N}\right) = \sum_{\pi \in NC(N)} \left( \prod_{V \in \pi} \varphi_p \left( \underset{j \in V}{*} \chi_{x_{i_j}K_{i_j}} \right) \right) \mu(\pi, 1_N)$$

by the *Möbius inversion of* [25]

(3.3.4)

$$= \sum_{\pi \in NC(N)} \left( \prod_{V=(i_1,\ldots,i_{|V|}) \in \pi} (\mu_p(V)) \right) \mu(\pi, 1_N),$$

with

$$\mu_p(V) = \begin{cases} \dfrac{\left( \prod_{l=2}^{|V|} \mu_p(K^o_{i_1,\ldots,i_l}) \right) \mu_p(x_{i_1,\ldots,i_{|V|}} K_{i_1,\ldots,i_{|V|}} \cap K_{i_1,\ldots,i_{|V|}})}{\mu_p(K_{i_1,\ldots,i_{|V|}})} & \text{if } |V| > 1 \\[2ex] \dfrac{\mu_p(x_{i_1}K_{i_1} \cap K_{i_1})}{\mu_p(K_{i_1})} & \text{if } |V| = 1, \end{cases}$$

by (3.3.2), where $\mu_p(V)$ are the *block-$V$-depending free moments* for all $V \in \pi$, for all noncrossing partitions $\pi$ in the lattice $NC(N)$ of all noncrossing partitions over $\{1, \ldots, N\}$, for all $N \in \mathbb{N}$ (e.g., [25]).

PROPOSITION 3.5. *(See [15]) Let $f_j = \chi_{K_j}$, $h_j = e_{K_j}$ be free random variables in the normal Hecke free probability space $(\mathcal{H}_{Y_p}, \varphi_p)$, for $j = 1, 2$. Then*

*(3.3.5) $f_1$ and $f_2$ are free in $(\mathcal{H}_{Y_p}, \varphi_p) \Leftrightarrow \mu_p(K^o_{1,2}) = \mu_p(K_1)\mu_p(K_2)$.*

*(3.3.6) $h_1$ and $h_2$ are free in $(\mathcal{H}_{Y_p}, \varphi_p) \Leftrightarrow \mu_p(K^o_{1,2}) = \mu_p(K_1)\, \mu_p(K_2)$.* □

The proofs of (3.3.5) and (3.3.6) are based on the joint free-cumulant formula (3.3.4) (e.g., see [12], [13] and [15]).

## 4. Free Probability on $\mathcal{H}(G_p)$

In this section, we introduce an extension of our free-probabilistic model on the normal Hecke subalgebra $\mathcal{H}_{Y_p}$ of Section 3.3 to that on the Hecke algebra $\mathcal{H}(G_p)$, originally introduced in [14]. The main tools would be certain normalizing process of compact-open subgroups of $G_p$.

Let $G$ be an arbitrary group, and let $K$ be a subgroup (not necessarily a normal subgroup) of $G$. Let $K_G$ be the maximal subgroup of $K$, which is normal in $G$. i.e., it is the maximal normal subgroup of $G$ inside $K$, i.e.,

(4.1)

$$K_G \lhd G, \text{ and } K_G \leq K \leq G.$$

Remark that $K_G$ is not necessarily a normal subgroup of $K$, it may simply be a subgroup of $K$, but normal in $G$.

Such a normal subgroup $K_G$ exists for any given subgroup $K$ in $G$, because a subgroup $K$ of $G$ contains its *normal core $Core_G(K)$ of $K$ in $G$*, defined by

(4.2)

$$Core_G(K) \overset{def}{=} \bigcap_{g \in G} (g^{-1}Kg).$$

It is indeed the maximal normal subgroup of $G$ contained in $K$ (e.g., [3]). So, the normal subgroup $K_G$ in the sense of (4.1) is identified with the normal core $Core_G(K)$ of (4.2) in $G$, i.e.,

$$K_G = Core_G(K), \text{ in } G.$$

**Notation.** Let $G$ be an arbitrary group, and let $K$ be a subgroup of $G$. Then the normal core $Core_G(K)$ of $K$ in $G$ is simply denoted by $K_G$ throughout this paper.

(4.3)

$$K_G \overset{denote}{=} Core_G(K),$$

for all subgroups $K$ of $G$. □

Define now a morphism $E_p$ on the Hecke algebra $\mathcal{H}(G_p)$ by a linear transformation satisfying the condition (4.4);

(4.4)

$$E_p\left(\chi_{xK}\right) = \begin{cases} \chi_{xK_{G_p}} = \chi_{K_{G_p}x} & \text{if } xK = Kx \\ 0_{\mathcal{H}(G_p)} & \text{otherwise,} \end{cases}$$

and

$$E_p\left(\chi_{x_1K_1} * \chi_{x_2K_2}\right) = \begin{cases} \mu_p(K_{1,2}^o)\chi_{x_{1,2}K_{1,2:G_p}} & \begin{array}{l}\text{if } K_i x_j = x_j K_i, \\ \text{for all } i,\ j = 1, 2\end{array} \\ & \\ 0_{\mathcal{H}(G_p)} & \text{otherwise,} \end{cases}$$

for all $x$, $x_1$, $x_2 \in G_p$, and compact-open subgroups $K$, $K_1$, $K_2$ of $G_p$, where $0_{\mathcal{H}(G_p)}$ means the *zero-element* of the Hecke algebra $\mathcal{H}(G_p)$, and

$$K_{G_p} = Core_{G_p}(K),$$

are the normal cores (4.2) of $K$ in $G_p$, as in (4.3).

The second condition of (4.4) allows us to have

(4.5)

$$E_p\left(\underset{j=1}{\overset{N}{*}} \chi_{x_jK_j}\right) = \begin{cases} \left(\prod_{j=2}^{N} \mu_p(K_{1,\dots,j}^o)\right) \chi_{x_{1,\dots,N}K_{1,\dots,N:G_p}} & \text{or} \\ & \\ 0_{\mathcal{H}(G_p)}, & \end{cases}$$

inductively, for all $N \in \mathbb{N} - \{1\}$, by (3.2.1). In particular, if

$$x_iK_j = K_jx_i, \ \forall i,\ j = 1, \dots, N,$$

then

(4.5)′

$$E_p\left(\underset{j=1}{\overset{N}{*}} \chi_{x_jK_j}\right) = \left(\prod_{j=2}^{N} \mu_p(K_{1,\dots,j}^o)\right) \chi_{x_{1,\dots,N}K_{1,\dots,N:G_p}},$$

in $\mathcal{H}(G_p)$, for $N \in \mathbb{N} - \{1\}$.

**Notation 4.1.** In the rest of this paper, for convenience, we write the conditional relation (4.5) simply as (4.5)′. i.e., even though we simply write (4.5)′, it will actually mean (4.5) under conditions (4.4). □

By definition, it is clear that, for any compact-open subgroup $K$ of $G_p$,

$$E_p\left(\chi_K\right) = \chi_{K_{G_p}},$$

by (4.4). Notice that the linear morphism $E_p$ mapped $\mathcal{H}(G_p)$ onto $\mathcal{H}_{Y_p}$, again by (4.4). Note also that, even though the morphism $E_p$ is not a conditional expectation in the sense of [25], it is a well-defined surjective linear transformation, satisfying that:

$$E_p(\chi_{xK}) = \chi_{xK}, \text{ for all } \chi_{xK} \in \mathcal{H}_{Y_p},$$

and

$$E_p^2\left(f\right) = E_p(f), \text{ for all } f \in \mathcal{H}(G_p).$$

i.e., $E_p$is an *idempotent* surjective linear transformation, preserving (or fixing) the normal Hecke subalgebra $\mathcal{H}_{Y_p}$ in the Hecke algebra $\mathcal{H}(G_p)$.

Moreover, one can get that

(4.6)

$$E_p\left(e_{xK}\right) = E_p\left(\frac{1}{\mu_p(K)}\ \chi_{xK}\right)$$

$$= \frac{1}{\mu_p(K)}\chi_{xK_{G_p}} \neq e_{xK_{G_p}},$$

in general, for all $x \in G_p$ satisfying $xK = Kx$, for compact-open subgroups $K$ of $G_p$. However, if $K \triangleleft G_p$, then

$$E_p\left(e_{xK}\right) = e_{xK}.$$

Thus, under same hypothesis, we obtain that

$$E_p\left(e_{x_1 K_1} * ... * e_{x_N K_N}\right)$$

$$= E_p\left(\frac{\mu_p(K_{1,2}^o)...\mu_p(K_{1,...,N}^o)\mu_p(K_{1,...,N})}{\mu_p(K_1)\mu_p(K_2)...\mu_p(K_N)}\ e_{x_{1,...,N}K_{1,...,N}}\right)$$

$$= \left(\frac{\mu_p(K_{1,2}^o)...\mu_p(K_{1,...,N}^o)\mu_p(K_{1,...,N})}{\mu_p(K_1)\mu_p(K_2)...\mu_p(K_N)}\right)\left(\frac{1}{\mu_p(x_{1,...,N}K_{1,...,N})}\ \chi_{x_{1,...,N}K_{1,...,N:G_p}}\right)$$

(4.7)

$$= \frac{\mu_p(K_{1,2}^o)...\mu_p(K_{1,...,N}^o)}{\mu_p(K_1)...\mu_p(K_N)\mu_p(K_{1,...,N})}\chi_{x_{1,...,N}K_{1,...,N:G_p}},$$

by (4.5) and (4.6), under **Notation 4.1**.

PROPOSITION 4.1. *Let $f_j = \chi_{x_j K_j}$, $h_j = e_{x_j K_j}$ be generating elements of the Hecke algebra $\mathcal{H}(G_p)$, for $j = 1, ..., N$, for $N \in \mathbb{N}$, and let $E_p$ be the idempotent linear transformation (4.4) on $\mathcal{H}(G_p)$. Then*

*(4.8)*

$$E_p\left(\underset{j=1}{\overset{N}{*}}\ f_j\right) = \prod_{l=2}^{N}\mu_p(K_{1,...,j}^o)\ \chi_{x_{1,...,N}K_{1,...,N:G_p}},$$

*and*

*(4.9)*

$$E_p\left(\underset{j=1}{\overset{N}{*}}\ h_j\right) = \frac{\left(\prod_{l=2}^{N}\mu_p(K_{1,...,j}^o)\right)\mu_p(K_{1,...,N:G_p})}{\mu_p(K_1)...\mu_p(K_N)}\ \chi_{x_{1,...,N}K_{1,...,N:G_p}},$$

*under **Notation 4.1**, where $K_{1,...,N:G_p}$ is the normal core of $K_{1,...,N}$.*

PROOF. The proofs of (4.8) and (4.9) are done by (4.5), (4.6) and (4.7). See [12] for more details. □

As we discussed above, the idempotent linear transformation $E_p$ maps $\mathcal{H}(G_p)$ onto the normal Hecke subalgebra $\mathcal{H}_{Y_p}$, by (4.4) and (4.5).

DEFINITION 4.1. *We will call the morphism $E_p$ of (4.4), the normal-coring on $\mathcal{H}(G_p)$.*

Define now a linear functional $\psi_p$ on the Hecke algebra $\mathcal{H}(G_p)$ by
(4.10)

$$\psi_p \overset{def}{=} \varphi_p \circ E_p \text{ on } \mathcal{H}(G_p),$$

where $\varphi_p$ is the linear functional (3.3.1) on the normal Hecke subalgebra $\mathcal{H}_{Y_p}$ of $\mathcal{H}(G_p)$, and $E_p$ is the normal-coring (4.4).

We call this linear functional $\psi_p$ of (4.10), the *normal-core linear functional on* $\mathcal{H}(G_p)$. So, the pair $(\mathcal{H}(G_p), \psi_p)$ forms a noncommutative probability space.

DEFINITION 4.2. *The noncommutative probability space $(\mathcal{H}(G_p), \psi_p)$ of the Hecke algebra $\mathcal{H}(G_p)$, and the normal-core linear functional $\psi_p$ of (4.10) is said to be the normal-core Hecke probability space (for a prime p).*

Generally, we obtain the following joint free-moment computations.

THEOREM 4.2. *Let $(\mathcal{H}(G_p), \psi_p)$ be the normal-core Hecke probability space, and let $f_j = \chi_{x_jK_j}$ and $h_j = e_{x_jK_j}$ be generating elements of $(\mathcal{H}(G_p), \psi_p)$, for $j \in \mathbb{N}$. Then*
*(4.11)*

$$\psi_p\left(\overset{N}{\underset{j=1}{*}} f_j\right) = \frac{\left(\prod_{l=2}^{N}\mu_p(K^o_{1,\ldots,l})\right)\mu_p\left(x_{1,\ldots,N}K_{1,\ldots,N:G}\cap K_{1,\ldots,N:G_p}\right)}{\mu_p\left(K_{1,\ldots,N:G_p}\right)},$$

*and*
*(4.12)*

$$\psi_p\left(\overset{N}{\underset{j=1}{*}} h_j\right) = \frac{\left(\prod_{l=2}^{N}\mu_p(K^o_{1,\ldots,l})\right)\mu_p\left(x_{1,\ldots,N}K_{1,\ldots,N:G}\cap K_{1,\ldots,N:G_p}\right)}{\left(\prod_{l=1}^{N}\mu_p(K_l)\right)\left(\mu_p(K_{1,\ldots,N:G_p})\right)},$$

*for all $N \in \mathbb{N}-\{1\}$, under* ***Notation 4.1****, where $K_{1,\ldots,N:G_p}$ is in the sense of (4.2).*
*If $N = 1$, then*

$$\psi_p\left(f_j\right) = \chi_{x_j(K_j)_{G_p}}, \text{ and } \psi_p\left(h_j\right) = e_{x_j(K_j)_{G_p}},$$

*under* ***Notation 4.1****, for all $j \in \mathbb{N}$.*

PROOF. The proof of (4.11) and (4.12) are done by (4.8), respectively, by (4.9), under **Notation 4.1**. See [14] for more details. □

One can realize that: if all compact-open subgroups $K_1$, ..., $K_N$ are "normal" in $G_p$, then the formula (4.11) (resp., (4.12)) is identified with the formula (3.3.2) (resp., (3.3.3)), for all $N \in \mathbb{N} - \{1\}$.

So, we have that
(4.13)

$$\psi_p\left(\underset{j=1}{\overset{N}{*}}\chi_{K_j}\right) = \frac{\mu_p(K^o_{1,2})\mu_p(K^o_{1,2,3})...\mu_p(K^o_{1,...,N})\mu_p(K_{1,...,N:G_p})}{\mu_p(K_{1,...,N:G_p})}$$

$$= \mu_p(K^o_{1,2,})\mu_p(K^o_{1,2,3,})...\mu_p(K^o_{1,...,N}),$$

by (4.11), and
(4.14)

$$\psi_p\left(\underset{j=1}{\overset{N}{*}}e_{K_j}\right) = \frac{\mu_p(K^o_{1,2})\mu_p(K^o_{1,2,3})...\mu_p(K^o_{1,...,N})\mu_p(K_{1,...,N:G_p})}{\mu_p(K_1)\mu_p(K_2)...\mu_p(K_N)\mu_p(K_{1,...,N:G_p})}$$

$$= \frac{\mu_p(K^o_{1,2})\mu_p(K^o_{1,2,3})...\mu_p(K^o_{1,...,N})}{\mu_p(K_1)\mu_p(K_2)...\mu_p(K_N)},$$

by (4.12).

Now, let $K_1$ and $K_2$ be compact-open subgroups of $G_p$, and let $\chi_{K_j}$ be corresponding free random variables in the normal-core Hecke probability space $(\mathcal{H}(G_p), \psi_p)$. Suppose $k_N(...)$ is the free cumulant on $\mathcal{H}(G_p)$ for the normal-core linear functional $\psi_p$. Then, for any $(i_1, ..., i_N) \in \{1, 2\}^N$, for all $N \in \mathbb{N}$, we obtain the following free cumulants;
(4.15)

$$k_N\left(\chi_{K_{i_1}}, ..., \chi_{K_{i_N}}\right) = \sum_{\pi\in NC((i_1,...,i_N))}\left(\prod_{V\in\pi}\mu_p(V)\right)\mu(\pi, 1_N),$$

with

$$\mu_p(V) = \mu_p(K^o_{i_{j_1},i_{j_2}})\mu_p\left(K^o_{i_{j_1},i_{j_2},i_{j_3}}\right)...\mu_p\left(K^o_{i_{j_1},...,i_{j_k}}\right),$$

by (4.11), under **Notation 4.1**, whenever $V = (j_1, ..., j_k) \in \pi$, for all $\pi \in NC(N)$, for all $N \in \mathbb{N}$, where $\mu_p(V)$ are the $V$-block-depending free moments.

By (4.15), we obtain the following freeness condition on the normal-core Hecke probability space $(\mathcal{H}(G_p), \psi_p)$.

PROPOSITION 4.3. *Let $f_j = \chi_{K_j}$ and $h_j = e_{K_j}$ be free random variables in the normalized Hecke probability space $(\mathcal{H}(G_p), \psi_p)$, for $j = 1, 2$. Then;*
*(4.16) $f_1$ and $f_2$ are free in $(\mathcal{H}(G_p), \psi_p) \Leftrightarrow \mu_p(K^o_{1,2}) = \mu_p(K_1)\mu_p(K_2)$.*
*(4.17) $h_1$ and $h_2$ are free in $(\mathcal{H}(G_p), \mu_p) \Leftrightarrow \mu_p(K^o_{1,2}) = \mu_p(K_1)\mu_p(K_2)$.*

PROOF. The statements (4.16) and (4.17) are proven by (4.15). See [14] for more details. □

## 5. Representations on normal-core Hecke Probability Spaces

Remark that the representations of Hecke algebras, and Hecke-type algebras are introduced in various different ways (e.g., [1], [2], [7], [22] and [24]). In this paper, we introduce different representations of Hecke algebras under our free-probabilistic settings (e.g., also see [12], [14] and [15]).

Let $p$ be a fixed prime, and let $(\mathcal{H}(G_p), \psi_p)$ be the normal-core Hecke probability space. Define a form

$$[,]_p : \mathcal{H}(G_p) \times \mathcal{H}(G_p) \to \mathbb{C},$$

on the Hecke algebra $\mathcal{H}(G_p)$ by

(5.1)

$$[f_1, f_2]_p \overset{def}{=} \psi_p(f_1 * f_2^*), \text{ for all } f_1, f_2 \in \mathcal{H}(G_p),$$

where

$$f^*(x) \overset{def}{=} \overline{f(x)} \text{ in } \mathbb{C}, \text{ for all } x \in G_p,$$

where $\overline{z}$ mean the *conjugates of* $z$, for all $z \in \mathbb{C}$. We call the unary operation

(5.2)

$$f \in \mathcal{H}(G_p) \longmapsto f^* \in \mathcal{H}(G_p),$$

the *adjoint* on $\mathcal{H}(G_p)$. And the element $f^*$ is said to be the *adjoint of* $f$.

Since the adjoint (5.2) is well-defined on $\mathcal{H}(G_p)$, we from now regard our Hecke algebra $\mathcal{H}(G_p)$ as a $*$-*algebra* over $\mathbb{C}$.

PROPOSITION 5.1. *(See [14]) The sesqui-linear form* $[,]_p$ *of (5.1) on the Hecke algebra* $\mathcal{H}(G_p)$ *forms a pseudo-inner product on* $\mathcal{H}(G_p)$. *i.e.,*

*(5.3)* $(\mathcal{H}(G_p), [,]_p)$ *is a pseudo-inner-product space over* $\mathbb{C}$. □

Suppose $K$ is a proper normal compact-open subgroup of $G_p$, and assume that

$$xK \neq Kx, \text{ for some } x \in G_p.$$

Then

$$[\chi_{xK}, \chi_{xK}]_p = 0,$$

under **Notation 4.1**. i.e., there do exist nonzero elements $f$ of $\mathcal{H}(G_p)$, such that

$$[f, f]_p = 0.$$

So, the pseudo-inner product space $(\mathcal{H}(G_p), [,]_p)$ of (5.3) is not an inner product space.

**Notation.** When we understand our Hecke algebra $\mathcal{H}(G_p)$, as a pseudo-inner product space (5.3), we denote it by $\mathcal{H}_p$. □

On the pseudo-inner product space $\mathcal{H}_p$ of (5.6), define an *equivalence relation* $\mathcal{R}_p$ by

(5.4)

$$f_1 \ \mathcal{R}_p \ f_2 \overset{def}{\Longleftrightarrow} [f_1, f_1]_p = [f_2, f_2]_p.$$

DEFINITION 5.1. *Let* $\mathcal{H}_p$ *be the pseudo-inner product space (5.3), and let* $\mathcal{R}_p$ *be the equivalence relation (5.4) on* $\mathcal{H}_p$. *Define a new vector space* $\mathfrak{H}_p^a$ *by the quotient space,*

*(5.5)*

$$\mathfrak{H}_p^a = \mathcal{H}_p \ / \ \mathcal{R}_p,$$

*equipped with the corresponding pseudo-inner product, also denoted by* $[,]_p$, *on it. i.e.,*

$$\left[[f_1]_{\mathcal{R}_p},\ [f_2]_{\mathcal{R}_p}\right]_p = [f_1,\ f_2]_p\,,$$

*for the equivalence classes*

$$[f_j]_{\mathcal{R}_p} = \{g \in \mathcal{H}_p : g\ \mathcal{R}_p\ f_j\},$$

*for all* $j = 1,\ 2$. *Then the pair*

$$\mathfrak{H}_p^a \overset{denote}{=} (\mathfrak{H}_p^a,\ [,]_p) = (\mathcal{H}_p\ /\ \mathcal{R}_p,\ [,]_p)$$

*is called the (normal-core) Hecke inner product space.*

Indeed, our Hecke inner product space $\mathfrak{H}_p^a$ of (5.5) forms a well-determined *inner product space* by $\mathcal{R}_p$ of (5.4).

PROPOSITION 5.2. *The Hecke inner product space* $\mathfrak{H}_p^a$ *of (5.5) is an inner product space over* $\mathbb{C}$. □

**Notation.** For convenience, we denote equivalence classes

$$[f]_{\mathcal{R}_p} = \{h \in \mathcal{H}_p : h\ \mathcal{R}_p\ f\}$$

of our Hecke inner product space $\mathfrak{H}_p^a$, simply by $f$, if there is no confusion. □

For the given inner product space $\mathfrak{H}_p^a$, one can define the corresponding *norm* $\|.\|_p$ *on* $\mathfrak{H}_p^a$ by

(5.6)

$$\|f\|_p \overset{def}{=} \sqrt{[f,\ f]_p},\ \text{for all } f \in \mathfrak{H}_p^a,$$

and the corresponding *metric* $d_p$ *on* $\mathfrak{H}_p^a$ by

(5.7)

$$d_p\,(f_1,\ f_2) = \|f_1 - f_2\|_p\,,\ \text{for all } f_1,\ f_2 \in \mathfrak{H}_p^a.$$

DEFINITION 5.2. *Construct the* $d_p$*-metric topology closure* $\mathfrak{H}_p$ *of* $\mathfrak{H}_p^a$, *where* $d_p$ *is the metric (5.7) induced by the norm* $\|.\|_p$ *of (5.6). We call it the (normal-core) Hecke Hilbert space (for a prime p).*

Then, by the very construction of the Hecke Hilbert space $\mathfrak{H}_p$, the Hecke algebra $\mathcal{H}(G_p)$ acts on $\mathfrak{H}_p$ via an $*$-*algebra-action* $\alpha^p$,

(5.8)

$$\alpha^p(f)(h) = f * h,\ \text{for all } h \in \mathfrak{H}_p,$$

for all $f \in \mathcal{H}(G_p)$. More precisely,

(5.9)

$$\alpha^p(f)(h) = \alpha^p(f)\left([h]_{\mathcal{R}_p}\right) = [f * h]_{\mathcal{R}_p} = f * h,$$

in $\mathfrak{H}_p$. i.e., the action $\alpha^p$ of (5.8) assigns $f \in \mathcal{H}(G_p)$ to the *convolution operator* $\alpha^p(f)$ on the Hecke Hilbert space $\mathfrak{H}_p$.

**Notation.** For convenience, we denote $\alpha^p(f)$ by $\alpha_f^p$, for all $f \in \mathcal{H}(G_p)$. □

The above morphism $\alpha^p$ of (5.8) or (5.9) is indeed a well-defined $*$-algebra-action of $\mathcal{H}(G_p)$ acting on $\mathfrak{H}_p$, since

(5.10)

$$\alpha_{f_1 * f_2}^p = \alpha_{f_1}^p \alpha_{f_2}^p \text{ on } \mathfrak{H}_p,$$

for all $f_1,\ f_2 \in \mathcal{H}(G_p)$, and

(5.11)

$$\left(\alpha_f^p\right)^* = \alpha_{f^*}^p \text{ on } \mathfrak{H}_p, \text{ for all } f \in \mathcal{H}(G_p).$$

PROPOSITION 5.3. *The pair $(\mathfrak{H}_p,\ \alpha^p)$ of the Hecke Hilbert space $\mathfrak{H}_p$ and the $*$-algebra-action $\alpha^p$ of (5.8) forms a Hilbert-space representation of the Hecke algebra $\mathcal{H}(G_p)$.*

PROOF. By (5.10) and (5.11), the bounded linear morphism $\alpha^p$ of (5.8) is a $*$-algebra-action of $\mathcal{H}(G_p)$ acting on $\mathfrak{H}_p$. Equivalently, the pair $(\mathfrak{H}_p,\ \alpha^p)$ forms a Hilbert-space representation of $\mathcal{H}(G_p)$. □

We call the $*$-algebra-action $\alpha^p$ of (5.15), the (*normal-core*) *Hecke*(*-algebra*) *action of $\mathcal{H}(G_p)$ acting on $\mathfrak{H}_p$*.

DEFINITION 5.3. *The Hilbert-space representation $(\mathfrak{H}_p,\ \alpha^p)$ of the Hecke Hilbert space $\mathfrak{H}_p$, and the Hecke action $\alpha^p$ of the Hecke algebra $\mathcal{H}(G_p)$ acting on $\mathfrak{H}_p$ is called the (normal-core) Hecke representation (of the normal-core Hecke probability space $(\mathcal{H}(G_p),\ \psi_p)$).*

## 6. $C^*$-Probability Spaces Induced by normal-core Hecke Probability Spaces

Let $(\mathcal{H}(G_p),\ \psi_p)$ be the normal-core Hecke probability space for a prime $p$, and $(\mathfrak{H}_p,\ \alpha^p)$, the Hecke representation of $\mathcal{H}(G_p)$, equipped with the Hecke Hilbert space $\mathfrak{H}_p$, and the Hecke action $\alpha^p$ of (5.8).

We now understand each element $f$ of the Hecke algebra $\mathcal{H}(G_p)$ as a *Hilbert-space operator* $\alpha_f^p$ on $\mathfrak{H}_p$ (under quotient by $\mathcal{R}_p$). Moreover, it is acting as a *convolution operator on* $\mathfrak{H}_p$, satisfying (5.10) and (5.11).

Let $B(\mathfrak{H}_p)$ be the *operator algebra* consisting of all bounded linear operators on $\mathfrak{H}_p$, equipped with its *operator-norm* $\|.\|$,

(6.1)

$$\|T\| = \sup\left\{\|Th\|_p \middle| h \in \mathfrak{H}_p,\ \ \|h\|_p = 1\right\},$$

for all $T \in B(\mathfrak{H}_p)$, where $\|.\|_p$ is the Hilbert-space norm (5.6) on $\mathfrak{H}_p$. Then, under the $\|.\|$-norm topology, the operator algebra $B(\mathfrak{H}_p)$ forms a $C^*$-*algebra*.

DEFINITION 6.1. *Let $\mathcal{H}(G_p)$ be the Hecke algebra, and $(\mathfrak{H}_p,\ \alpha^p)$, the Hecke representation of $\mathcal{H}(G_p)$, and let $B(\mathfrak{H}_p)$ on $\mathfrak{H}_p$. Define the $C^*$-subalgebra $\mathbb{H}(G_p)$ of $B(\mathfrak{H}_p)$ by*

*(6.2)*

$$\mathbb{H}(G_p) = C^*(\mathcal{H}(G_p)) = \overline{\mathbb{C}\left[\alpha^p\left(\mathcal{H}(G_p)\right)\right]}^{\|\cdot\|},$$

*where $\overline{Y}^{\|\cdot\|}$ mean the $\|.\|$-closures of subsets $Y$ of $B(\mathfrak{H}_p)$, where $\|.\|$ is the operator norm (6.1). This $C^*$-subalgebra $\mathbb{H}(G_p)$ of $B(\mathfrak{H}_p)$ is called the p-Hecke $C^*$-algebra (generated by $(\mathcal{H}(G_p),\ \psi_p)$ in $B(\mathfrak{H}_p)$).*

Let $\mathbb{H}(G_p)$ be the $p$-Hecke $C^*$-algebra (6.2) in $B(\mathfrak{H}_p)$, and let $T \in \mathbb{H}(G_p)$. Then

(6.3)

$$T = \sum_{x \in G_p,\ K \leq G_p:\ \text{compact-open}} t_{xK}\ \alpha_{\chi_{xK}}^p, \text{ with } t_{xK} \in \mathbb{C},$$

where $\sum$ in (6.3) means a finite, or an infinite (limit of finite) sum(s) under $C^*$-topology for $\mathbb{H}(G_p)$.

i.e.,

$$T(h) = \sum_{x \in G_p,\ K \leq G_p:\ \text{compact-open}} t_{xK}\ (\chi_{xK} * h) \text{ in } \mathfrak{H}_p,$$

for all $h \in \mathfrak{H}_p$, whenever $T$ is in the sense of (6.3) in $\mathbb{H}(G_p)$.

**Notation.** For convenience, we will denote the generating operators $\alpha^p_{\chi_{xK}}$ of the $p$-Hecke $C^*$-algebra $\mathbb{H}(G_p)$, simply by $\alpha^p_{xK}$, for all $x \in G_p$, and $K$, compact-open subgroups of $G_p$. □

The operator-theoretic properties of generating operators $\alpha^p_{xK}$ of the $p$-Hecke $C^*$-algebra $\mathbb{H}(G_p)$ was studied in [12], [13] and [14].

Let $Z_p$ be a (non-normal) compact-open subgroup $G_p$,

(6.4)

$$Z_p = GL_2(\mathbb{Z}_p) \text{ in } GL_2(\mathbb{Q}_p) = G_p,$$

satisfying

$$\mu_p\,(Z_p) = 1.$$

For a fixed compact-open subgroup $Z_p$ of (6.4), define a linear functional $\tau_p$ on $\mathbb{H}(G_p)$ by

(6.5)

$$\tau_p\left(\alpha^p_f\right) \overset{def}{=} \left[\alpha^p_f\,(\chi_{Z_p}),\ \chi_{Z_p}\right]_p,$$

for all $f \in \mathcal{H}(G_p)$, where $[,]_p$ is the inner product (5.1) (under $\mathcal{R}_p$) on the Hecke Hilbert space $\mathfrak{H}_p$.

Note that, for any compact-open subgroup $K$ of $G_p$, one can define

$$e_K = \frac{1}{\mu_p(K)}\ \chi_K \in \mathcal{H}(G_p),$$

satisfying

$$e_K * e_K = e_K.$$

Remark that, for a compact-open subgroup $Z_p$ of (6.4) in $G_p$,

$$e_{Z_p} = \frac{1}{\mu_p(Z_p)}\ \chi_{Z_p} = \frac{1}{1}\ \chi_{Z_p} = \chi_{Z_p},$$

and hence,

$$e_{Z_p} * e_{Z_p} = \chi_{Z_p} * \chi_{Z_p} = \chi_{Z_p} = e_{Z_p},$$

i.e., one obtains

(6.6)

$$\chi^{(k)}_{Z_p} = e^{(k)}_{Z_p} = e_{Z_p} = \chi_{Z_p}, \text{ for all } k \in \mathbb{N}.$$

By the definition (6.5) of $\tau_p$, one has

$$\tau_p\,(\alpha^p_{xK}) = \left[\alpha^p_{xK}\,(\chi_{Z_p}),\ \chi_{Z_p}\right]_p$$

$$= \left[\chi_{xK} * \chi_{Z_p},\ \chi_{Z_p}\right]_p = \psi_p\left(\chi_{xK} * \chi_{Z_p} * \chi_{Z_p}\right)$$

$$= \psi_p\left(\chi_{xK} * \chi_{Z_p}\right)$$

(6.6)

$$= \psi_p\left(\mu_p(xK \cap Z_p)\ \chi_{xKZ_p}\right)$$

under **Notation 4.1**

(6.7)

$$= \mu_p (xK \cap Z_p) \left( \frac{\mu_p\left(xKZ_p \cap KZ_{p:G_p}\right)}{\mu_p\left(KZ_{p:G_p}\right)} \right),$$

under **Notation 4.1**, for all $x \in G_p$, and compact-open subgroups $K$ of $G_p$, where $KZ_{p:G_p}$ is the normal core of the product group $KZ_p$.

The above computation (6.7) is non-vanishing, only if

$$xK = Kx, \text{ and } xZ_p = Z_p x,$$

for $x \in G_p$, and $K$, a compact-open subgroup of $G_p$.

Thus, we will restrict our interests to the possible non-vanishing cases where $x$ are from the *centralizer* $C_{G_p}(Z_p)$ *of* $Z_p$ *in* $G_p$ (e.g., [3]), where

(6.8)

$$C_{G_p}(Z_p) = \{x \in G_p : xZ_p = Z_p x\}.$$

Indeed, if $x \notin C_{G_p}(Z_p)$ in $G_p$, then

$$\tau_p\left(\alpha_{xK}^{p}\right) = 0, \forall \text{compact-open subgroup } K \text{ of } G_p$$

under **Notation 4.1**.

Thus it suffices to restrict our interests to the cases where the generating operators $\alpha_{xK}^{p}$ of $\mathbb{H}(G_p)$ satisfy $x \in C_{G_p}(Z_p)$, to obtain possible non-vanishing free-distributional data. More precisely, one obtains the following proposition, refining (6.7).

THEOREM 6.1. *Let* $x \in C_{G_p}(K) \cap C_{G_p}(Z_p)$, *where* $C_{G_p}(K)$ *and* $C_{G_p}(Z_p)$ *are the centralizer of* $K$, *respectively, that of* $Z_p$, *in* $G_p$. *Then*

*(6.9)*

$$\tau_p\left(\left(\alpha_{xK}^{p}\right)^n\right) = \frac{\mu_p(xK)^{n-1}\mu_p(x^nK \cap Z_p)\mu_p\left(x^nKZ_p \cap KZ_{p:G_p}\right)}{\mu_p\left(x^nKZ_{p:G_p}\right)} \neq 0,$$

*for all* $n \in \mathbb{N}$, *where* $\tau_p$ *is the linear functional (6.5) on the p-Hecke* $C^*$*-algebra* $\mathbb{H}(G_p)$, *and* $KZ_{p:G_p}$ *is the normal core of* $KZ_p$ *in* $G_p$.

*Meanwhile, if* $x \notin C_{G_p}(Z_p) \cap C_{G_p}(K)$, *then*

*(6.10)*

$$\tau_p\left(\left(\alpha_{xK}^{p}\right)^n\right) = 0, \textit{ for all } n \in \mathbb{N}.$$

PROOF. First, assume that

(6.11)

$$x \in C_{G_p}(K) \cap C_{G_p}(Z_p), \text{ in } G_p.$$

Then, we obtain that

(6.12)

$$\tau_p\left(\alpha_{xK}^{p}\right) = \frac{\mu_p(xK \cap Z_p)\mu_p\left(xKZ_p \cap KZ_{p:G_p}\right)}{\mu_p\left(KZ_{p:G_p}\right)} \neq 0,$$

by (6.7).

Now, suppose $n > 1$ in $\mathbb{N}$. Then

$$\tau_p\left((\alpha^p_{xK})^n\right) = \tau_p\left(\alpha^p(\chi^{(n)}_{xK})\right)$$

where $f^{(n)} = f * f * ... * f$ ($n$-times) in $\mathcal{H}(G_p)$, for all $f \in \mathcal{H}(G_p)$

$$= \tau_p\left(\alpha^p\left(\mu_p(xK)^{n-1}\chi_{x^n K}\right)\right)$$

by the assumption (6.11)

$$= \mu_p\left(xK\right)^{n-1}\ \tau_p\left(\alpha^p_{x^n K}\right)$$

$$= \mu_p\left(xK\right)^{n-1}\left(\frac{\mu_p(x^n K \cap Z_p)\mu_p\left(x^n K Z_p \cap K Z_{p:G_p}\right)}{\mu_p\left(K Z_{p:G_p}\right)}\right)$$

by (6.7)

(6.13)

$$= \frac{\mu_p(xK)^{n-1}\mu_p(x^n K \cap Z_p)\mu_p\left(x^n K Z_p \cap K Z_{p:G_p}\right)}{\mu_p\left(K Z_{p:G_p}\right)} \neq 0,$$

by (6.11).

Therefore, the formula (6.9) holds true, by (6.12) and (6.13).

Assume now that the condition (6.11) does not hold. i.e., suppose

$$x \notin C_{G_p}(K) \cap C_{G_p}(Z_p) \text{ in } G_p,$$

equivalently, either

$$x \in G_p - C_{G_p}(K), \text{ or } x \in G_p - C_{G_p}(Z_p).$$

Then, by (4.5), (4.5)′, (4.10), (5.1), and (6.5), we obtain that

$$\tau_p\left((\alpha^p_{xK})^n\right) = 0, \text{ for all } n \in \mathbb{N}.$$

Therefore, the formula (6.10) holds. □

By the above free-distributional data (6.9) and (6.10), we obtain the following result.

THEOREM 6.2. *Let $x_1, ..., x_N \in G_p$, and $K_1, ..., K_N$, compact-open subgroups of $G_p$, for $N \in \mathbb{N} - \{1\}$, and let $\alpha^p_{x_j K_j}$ be generating operators of our $p$-Hecke $C^*$-algebra $\mathbb{H}(G_p)$, for $j = 1, ..., N$. Then*

$$\tau_p\left(\prod_{j=1}^{N}\alpha^p_{x_j K_j}\right)$$

*(6.14)*

$$= \frac{\left(\prod_{j=2}^{N}\mu_p\left(K^o_{1,\ldots,j}\right)\right)\mu_p(x_{1,\ldots,N}K_{1,\ldots,N}\cap Z_p)\mu_p\left(x_{1,\ldots,N}K_{1,\ldots,N}Z_p\cap K_{1,\ldots,N}Z_{p:G_p}\right)}{\mu_p\left(K_{1,\ldots,N}Z_{p:G_p}\right)}.$$

*The above joint free-moment formula (6.14) is non-vanishing, "only if"*

$$x_j \in \left(\bigcap_{j=1}^{N} C_{G_p}\left(K_j\right)\right) \cap C_{G_p}(Z_p) \text{ in } G_p,$$

*for all $j = 1, ..., N$.*

PROOF. Let $\alpha_{x_j K_j}^p$ be generating operators of $\mathbb{H}(G_p)$, for $j = 1, ..., N$, for $N \in \mathbb{N}$. Assume first that

$$x_j \in \left( \underset{j=1}{\overset{N}{\cap}} C_{G_p}(K_j) \right) \cap C_{G_p}(Z_p) \text{ in } G_p,$$

for all $j = 1, ..., N$.

Then, for any $j = 1, ..., N$, one has

(6.15)

$$x_j K_i = K_i x_j, \text{ and } x_j Z_p = Z_p x_j,$$

for all $i = 1, ..., N$. So,

$$\tau_p \left( \underset{j=1}{\overset{N}{\Pi}} \alpha_{x_j K_j}^p \right) = \tau_p \left( \alpha^p \left( \underset{j=1}{\overset{N}{*}} \chi_{x_j K_j} \right) \right)$$

$$= \tau_p \left( \alpha^p \left( \left( \underset{j=2}{\overset{N}{\Pi}} \mu_p \left( K_{1,...,j}^o \right) \right) \chi_{x_{1,...,N} K_{1,...,N}} \right) \right)$$

$$= \left( \underset{j=2}{\overset{N}{\Pi}} \mu_p \left( K_{1,...,j}^o \right) \right) \tau_p \left( \alpha_{x_{1,...,N} K_{1,...,N}}^p \right)$$

$$= \left( \underset{j=2}{\overset{N}{\Pi}} \mu_p \left( K_{1,...,j}^o \right) \right) \cdot \left( \frac{\mu_p(x_{1,...,N} K_{1,...,N} \cap Z_p) \mu_p \left( x_{1,...,N} K_{1,...,N} Z_p \cap K_{1,...,N} Z_{p:G_p} \right)}{\mu_p \left( K_{1,...,N} Z_{p:G_p} \right)} \right)$$

by (6.7)

$$\neq 0,$$

by the assumption (6.15).

Therefore, the joint free-moment formula (6.14) holds, and it provides non-vanishing value under (6.15).

Indeed, if

$$x \notin \left( \underset{j=1}{\overset{N}{\cap}} C_{G_p}(K_j) \right) \cap C_{G_p}(Z_p) \text{ in } G_p,$$

and hence, if there exists at least one $j_0 \in \{j_1, ..., j_N\}$, such that

$$x \in G_p - C_{G_p}(K_{j_0}), \text{ or } x \in C_{G_p}(Z_p),$$

in $G_p$. Then

$$\tau_p \left( \underset{j=1}{\overset{N}{\Pi}} \alpha_{x_j K_j}^p \right) = 0,$$

by (4.5), (4.5)′, (4.10), (5.1) and (6.5). □

By our main results (6.9), (6.10) and (6.14), we obtain the following result.

COROLLARY 6.3. *Let $\alpha_K^p$, $\alpha_{K_j}^p$ be generating operators of the p-Hecke $C^*$-algebra $\mathbb{H}(G_p)$, for $j = 1, ..., N$, for $N \in \mathbb{N} - \{1\}$. Then*

*(6.16)*

$$\tau_p \left( (\alpha_K^p)^n \right) = \mu_p (K)^{n-1} \mu_p (K \cap Z_p),$$

*for all $n \in \mathbb{N}$, and*

*(6.17)*

$$\tau_p \left( \underset{j=1}{\overset{N}{\Pi}} \alpha_{K_j}^p \right) = \left( \underset{j=2}{\overset{N}{\Pi}} \mu_p \left( K_{1,...,j}^o \right) \right) \left( \mu_p \left( K_{1,...,N} \cap Z_p \right) \right).$$

PROOF. Note that $K = u_p K$, and $K_j = u_p K_j$, for all $j = 1, ..., N$, where $u_p$ is the group-identity of $G_p$, satisfying

(6.18)

$$u_p \in C_{G_p}(K), \text{ and } u_p \in C_{G_p}(K_j), \text{ in } G_p,$$

for all $j = 1, ..., N$.

Observe first that

$$\tau_p\left((\alpha_K^p)^n\right) = \frac{\mu_p(K)^{n-1}\mu_p(K \cap Z_p)\mu_p\left(KZ_p \cap KZ_{p:G_p}\right)}{\mu_p\left(KZ_{p:G_p}\right)}$$

by (6.9) and (6.10), under (6.18)

$$= \frac{\mu_p(K)^{n-1}\mu_p(K \cap Z_p)\mu_p\left(KZ_{p:G_p}\right)}{\mu_p\left(KZ_{p:G_p}\right)}$$

since $KZ_p \cap KZ_{p:G_p} = KZ_{p:G_p}$ in $G_p$

$$= \mu_p(K)^{n-1}\mu_p(K \cap Z_p),$$

for all $n \in \mathbb{N}$. Therefore, the free-moment formula (6.16) holds.

Consider now that,

$$\tau_p\left(\prod_{j=1}^{N} \alpha_{K_j}^p\right) = \frac{\left(\prod_{j=2}^{N}\mu_p(K_{1,...,j}^o)\right)\mu_p(K_{1,...,N} \cap Z_p)\mu_p\left(K_{1,...,N}Z_p \cap K_{1,...,N}Z_{p:G_p}\right)}{\mu_p\left(K_{1,...,N}Z_{p:G_p}\right)}$$

by (6.14), under (6.18)

$$= \left(\prod_{j=2}^{N}\mu_p\left(K_{1,...,j}^o\right)\right)\mu_p\left(K_{1,...,N} \cap Z_p\right)\frac{\mu_p\left(K_{1,...,N}Z_{p:G_p}\right)}{\mu_p\left(K_{1,...,N}Z_{p:G_p}\right)}$$

since $K_{1,...,N}Z_p \cap K_{1,...,N}Z_{p:G_p} = K_{1,...,N}Z_{p:G_p}$ in $G_p$

$$= \left(\prod_{j=2}^{N}\mu_p\left(K_{1,...,j}^o\right)\right)\mu_p\left(K_{1,...,N} \cap Z_p\right).$$

Therefore, the joint free-moment formula (6.17) holds true. □

As we have seen above, for a $p$-Hecke $C^*$-algebra $\mathbb{H}(G_p)$, one can define a linear functional $\tau_p$ of (6.5), determined by the Hecke representation $(\mathfrak{H}_p,\ \alpha^p)$ of the Hecke probability space $(\mathcal{H}(G_p),\ \psi_p)$ for a fixed prime $p$, for all $p \in \mathcal{P}$, where $\mathcal{P}$ is the subset consisting of all primes in $\mathbb{N}$. It means that one can get $C^*$-probability spaces

(6.19)

$$\{(\mathbb{H}(G_p),\ \tau_p) : p \in \mathcal{P}\}.$$

DEFINITION 6.2. *Let $p \in \mathcal{P}$, and let $(\mathbb{H}(G_p),\ \tau_p)$ be the corresponding $C^*$-probability space in (6.19), where $\mathbb{H}(G_p)$ is the p-Hecke $C^*$-algebra of $\mathcal{H}(G_p)$, and $\tau_p$ is in the sense of (6.5). Then we call $(\mathbb{H}(G_p),\ \tau_p)$, the p-Hecke $C^*$-probability space.*

## 7. Free Product $C^*$-Probability Spaces over $\{(\mathbb{H}(G_p),\ \tau_p)\}_{p\in\mathcal{P}}$

Let $(\mathbb{H}(G_p),\ \tau_p)$ be the $p$-Hecke $C^*$-probability spaces, for $p \in \mathcal{P}$, and

$$\{(\mathbb{H}(G_p),\ \tau_p)\}_{p\in\mathcal{P}},$$

the system (6.19) of these $C^*$-probability spaces over $\mathcal{P}$.

Recall that, for any fixed prime $p \in \mathcal{P}$, if $\alpha_{x_jK_j}^p \in (\mathbb{H}(G_p),\ \tau_p)$ are generating operators, for $j = 1, ..., N$, for $N \in \mathbb{N} - \{1\}$, then

(7.0.1)

$$\tau_p\left(\left(\alpha_{x_jK_j}^p\right)^n\right) = \frac{\left(\mu_p(K_j)^{n-1}\right)\mu_p(x^nK_j \cap Z_p)\mu_p\left(x_j^nK_jZ_p \cap K_jZ_{p:G_p}\right)}{\mu_p\left(K_jZ_{p:G_p}\right)} \neq 0,$$

for all $n \in \mathbb{N}$, only if

$$x_j \in \left( \overset{N}{\underset{j=1}{\cap}} C_{G_p}(K_j) \right) \cap C_{G_p}(Z_p),$$

and

$$\tau_p \left( \prod_{j=1}^{N} \alpha^p_{x_j K_j} \right) \tag{7.0.2}$$

$$= \frac{\left( \prod_{j=2}^{N} \mu_p\left(K^o_{1,\ldots,j}\right) \right) \left(\mu_p(x_{1,..,N} K_{1,\ldots,N} \cap Z_p)\right) \left(\mu_p\left(x_{1,..,N} K_{1,..,N} Z_p \cap K_{1,..,N} Z_{p:G_p}\right)\right)}{\mu_p\left(K_{1,\ldots,N} Z_{p:G_p}\right)}$$

is non-vanishing, only if

$$x_j \in \left( \overset{N}{\underset{j=1}{\cap}} C_{G_p}(K_j) \right) \cap C_{G_p}(Z_p),$$

for all $j = 1, \ldots, N$, for all $N \in \mathbb{N} - \{1\}$, by (6.9), (6.10) and (6.14).

**7.1. Free Product $C^*$-Probability Spaces.** Let $(A_k, \varphi_k)$ be arbitrary $C^*$-probability spaces, consisting of $C^*$-algebras $A_k$, and linear functionals $\varphi_k$ on $A_k$, for $k \in \Delta$, where $\Delta$ is an arbitrary countable (finite or infinite) index set. The *free product $C^*$-algebra $A$*,

$$A = \underset{l \in \Delta}{\star} A_l$$

is the $C^*$-algebra generated by the *noncommutative reduced words* in $\overset{N}{\underset{l=1}{\cup}} A_l$, having a new linear functional

$$\varphi = \underset{l \in \Delta}{\star} \varphi_l.$$

The $C^*$-algebra $A$ is understood as a *Banach space*,

$$\mathbb{C} \oplus \left( \overset{\infty}{\underset{n=1}{\oplus}} \left( \underset{(i_1,\ldots,i_n) \in alt(\Delta^n)}{\oplus} \left( \overset{n}{\underset{k=1}{\otimes}} A^o_{i_k} \right) \right) \right) \tag{7.1.1}$$

with

$$A^o_{i_k} = A_{i_k} \ominus \mathbb{C}, \text{ for all } k = 1, \ldots, n,$$

where

$$alt(\Delta^n) = \left\{ (i_1, \ldots, i_n) \;\middle|\; \begin{array}{c} (i_1, \ldots, i_n) \in \Delta^n \\ i_1 \neq i_2, \; i_2 \neq i_3, \\ \ldots.., \; i_{n-1} \neq i_n \end{array} \right\},$$

for all $n \in \mathbb{N}$, and where the *direct product* $\oplus$, and the *tensor product* $\otimes$ are topological on Banach spaces.

In particular, if an element $a \in A$ is a *free "reduced" word*,

$$\prod_{l=1}^{n} a_{i_l} \text{ in } A,$$

then one can understand $a$ as an equivalent Banach-space vector,

$$\overset{n}{\underset{l=1}{\otimes}} a_{i_l} \text{ in the Banach space } A \text{ of (7.1.1)},$$

contained in a direct summand, $\overset{n}{\underset{k=1}{\otimes}} A^o_{i_k}$ of (7.1.1), for

$$(i_1, \ldots, i_n) \in alt(\Delta^n), \text{ for } n \in \mathbb{N}.$$

Here, one can understand $\overset{n}{\underset{l=1}{\otimes}} a_{i_l}$ as an operator in the tensor product $C^*$-algebra,

$$\mathbb{C} \oplus \left( \overset{n}{\underset{k=1}{\otimes}} A_{i_k}^{o} \right) = \overset{n}{\underset{k=1}{\otimes_{\mathbb{C}}}} \left(\mathbb{C} \oplus A_{i_k}^{o}\right) = \overset{n}{\underset{k=1}{\otimes_{\mathbb{C}}}} A_{i_k} \text{ in } A,$$

which is called *the minimal free summand of* $A$ *containing* $a$, where $\otimes_{\mathbb{C}}$ means the *tensor product of* $C^*$*-algebras.*

We denote this relation by

(7.1.2)

$$a \overset{\text{equi}}{=} \overset{n}{\underset{l=1}{\otimes}} a_{i_l}$$

in the minimal free summand of $A$ containing $a$, where "$\overset{\text{equi}}{=}$" means "being equivalent to."

Remark that, if $a$ is a free reduced word in $A$, and if it is regarded as an operator under the equivalence (7.1.2) in the minimal free summand of $A$ containing $a$, then

(7.1.3)

$$a^k \overset{\text{equi}}{=} \left( \overset{n}{\underset{l=1}{\otimes}} a_{i_l} \right)^k = \overset{n}{\underset{l=1}{\otimes}} a_{i_l}^k \overset{\text{equi}}{=} \prod_{l=1}^{n} a_{i_l}^k,$$

in its minimal free summand of $A$, for all $k \in \mathbb{N}$.

**Notation and Remark (NR 7.1.1).** Let $a = \prod_{l=1}^{n} a_{i_l}$ be a free reduced word in $A$, for

$$(i_1, \ldots, i_n) \in alt\left(\Delta^n\right), \text{ for } n \in \mathbb{N}.$$

The operator $a^k$ in (7.1.3) means the $k$-th power of $a$ "in the minimal free summand,"

$$\overset{n}{\underset{l=1}{\otimes_{\mathbb{C}}}} A_{i_l} \text{ of } A.$$

To avoid the confusion, we use the notation $a^{(k)}$, as a new free "*non-reduced*" word (which can be reduced in some cases),

$$a^{(k)} = \underbrace{a \cdot a \cdots a}_{k\text{-times}} \quad \text{in } A,$$

of $a$ in $A$, for all $k \in \mathbb{N}$. Remark that, $a^{(k)}$ can again be a free reduced word in $A$, but not in general.

For example, let $a = a_{i_1} a_{i_2} a_{i_1}$ be a free reduced word, where

$$(i_1, i_2, i_1) \in alt\left(\Delta^3\right),$$

equivalent to,

$$a_{i_1} \otimes a_{i_2} \otimes a_{i_1} \text{ in its minimal free summand of } A.$$

Then

$$a^3 \overset{\text{equi}}{=} (a_{i_1} \otimes a_{i_2} \otimes a_{i_3})^3 \overset{\text{equi}}{=} a_{i_1}^3 a_{i_2}^3 a_{i_1}^3,$$

in the minimal free summand of $A$ containing $a$, but

$$\begin{aligned} a^{(3)} &= (a_{i_1} a_{i_2} a_{i_1})^{(3)} \\ &= a_{i_1} a_{i_2} a_{i_1} a_{i_1} a_{i_2} a_{i_1} a_{i_1} a_{i_2} a_{i_1} \quad \text{(non-reduced)} \\ &= a_{i_1} a_{i_2} a_{i_1}^2 a_{i_2} a_{i_1}^2 a_{i_2} a_{i_1}, \quad \text{(reduced)} \end{aligned}$$

i.e.,

$$a^{(3)} = a_{i_1} a_{i_2} a_{i_1}^2 a_{i_2} a_{i_1}^2 a_{i_2} a_{i_1},$$

in $A$. Meanwhile, if $a_{i_1}$ is a free reduced word with its length 1 in the free block $A_{i_1}$ of $A$, then, definitely,

$$a_{i_1}^k = a_{i_1}^{(k)} \text{ for all } k \in \mathbb{N},$$

in both the minimal free summand $A_{i_1}$ of $A$ containing $a_{i_1}$, and $A$.

At this moment, we emphasize the difference between $a^k$ (in the minimal free summand of $A$), and $a^{(k)}$ (in $A$), in $A$.

In the similar manner, we can have

$$a^* \overset{\text{equi}}{=} \left( \underset{l=1}{\overset{n}{\otimes}} a_{i_l} \right)^* = \underset{l=1}{\overset{n}{\otimes}} a_{i_l}^* \overset{\text{equi}}{=} \prod_{l=1}^{n} a_{i_l}^*,$$

in its minimal free summand of $A$, and

$$a^{(*)} = (a_{i_1} a_{i_2} ... a_{i_n})^{(*)} = a_{i_n}^* a_{i_{n-1}}^* ... a_{i_2}^* a_{i_1}^*, \text{ in } A,$$

where "$*$" means the adjoint. □

The free product linear functional $\varphi = \underset{i \in \Delta}{\star} \varphi_i$ on $A$ satisfies that, whenever $a$ is a reduced free word in $A$ satisfying (7.1.2), then

(7.1.4)

$$\varphi(a^k) = \varphi\left( \prod_{l=1}^{n} a_{i_l}^k \right) = \prod_{l=1}^{n} \left( \varphi_{i_l}(a_{i_l}^k) \right),$$

by (7.1.3), for all $k \in \mathbb{N}$. By abusing (7.1.3), one can / may write

$$\varphi(a^k) \overset{\text{equi}}{=} \varphi\left( \underset{l=1}{\overset{n}{\otimes}} a_{i_l}^k \right) = \prod_{l=1}^{n} \varphi(a_{i_l}^k),$$

whenever $a = \prod_{l=1}^{n} a_{i_l}$ is a free reduced word in $A$, for all $k \in \mathbb{N}$.

Now, let

$$b = \sum_{l=1}^{n} b_{i_l} \in (A, \varphi),$$

where $b_{i_l}$ are free reduced words in $A$, for all $l = 1, ..., n$.

We say that $b$ is a *free sum in* $A$, if all summands $b_{i_1}, ..., b_{i_n}$ of $b$ are contained in "mutually-distinct" minimal free summands of $A$ containing them, as free reduced words. Then, similar to the above terminology, one can write

(7.1.5)

$$b \overset{\text{equi}}{=} \overset{n}{\underset{l=1}{\oplus}} b_{i_l} \text{ in the minimal free summand of } A \text{ containing } b,$$

satisfying

$$\varphi\left(b^k\right) \overset{\text{equi}}{=} \varphi\left(\left(\overset{n}{\underset{l=1}{\oplus}} b_{i_l}\right)^k\right) = \varphi\left(\overset{n}{\underset{l=1}{\oplus}} b_{i_l}^k\right)$$

$$\overset{\text{equi}}{=} \varphi\left(\sum_{l=1}^n b_{i_l}^k\right) = \sum_{l=1}^n \varphi\left(b_{i_l}^k\right),$$

for all $k \in \mathbb{N}$.

Here, remark that each summand $\varphi\left(b_{i_l}^k\right)$ of (7.1.5) satisfies (7.1.4).

**Notation and Remark (NR 7.1.2).** Similar to the free-reduced-word case, if $b$ is a free sum in $A$, then one can consider

$$\begin{aligned} b^{(k)} &= \left(\sum_{l=1}^n b_{i_l}\right)^{(k)} \\ &= \sum_{(l_1,...,l_k)\in\{1,...,n\}^k} \left(b_{i_{l_1}} b_{i_{l_2}} ... b_{i_{l_k}}\right), \end{aligned}$$

in $A$, where the summands are free non-reduced words in $A$. Also, note that

$$b^* = \sum_{l=1}^n b_{i_l}^* \text{ in the minimal free summand of } A \text{ containing } b,$$

and

$$b^{(*)} = \sum_{l=1}^n b_{i_l}^{(*)} \text{ in } A.$$

□

**7.2. The Adelic Hecke $C^*$-Probability Space.** Let $\{(\mathbb{H}(G_p),\ \tau_p)\}_{p\in\mathcal{P}}$ be the system (7.0.1) of our $p$-Hecke $C^*$-probability spaces over $\mathcal{P}$. Construct the free product $C^*$-probability space $(\mathbb{H}_{\mathcal{P}},\ \tau)$ by

(7.2.1)

$$(\mathbb{H}_{\mathcal{P}}, \tau) \overset{def}{=} \underset{p\in\mathcal{P}}{\star} (\mathbb{H}(G_p),\ \tau_p)$$

$$= \left(\underset{p\in\mathcal{P}}{\star} \mathbb{H}(G_p),\ \underset{p\in\mathcal{P}}{\star} \tau_p\right).$$

Definition 7.1. *Let $(\mathbb{H}_{\mathcal{P}},\ \tau)$ be the free product $C^*$-probability space (7.2.1) induced by the system (7.0.1). We call it the Adelic Hecke $C^*$-probability space. For convenience, we denote $(\mathbb{H}_{\mathcal{P}},\ \tau)$ simply by $\mathbb{H}_{\mathcal{P}}$. So, the term "$\mathbb{H}_{\mathcal{P}}$" will mean both the free product $C^*$-algebra, and the free product $C^*$-probability space, case-by-case.*

Let $N \in \mathbb{N}$, and let

(7.2.2)

$$T_{p_l} = \alpha_{x_l K_l}^{p_l} \in (\mathbb{H}(G_{p_l}),\ \tau_{p_l})$$

be generating operators of $\mathbb{H}(G_{p_l})$, with

$$x_l \in C_{G_{p_l}}(K_l) \cap C_{G_{p_l}}(Z_{p_l}) \text{ in } G_{p_l},$$

and $K_l$, compact-open subgroups of $G_{p_l}$, for all $l = 1, ..., N$.

Construct a free random variable $T$ in the Adelic Hecke $C^*$-probability space $\mathbb{H}_{\mathcal{P}}$,

(7.2.3)

$$T = \prod_{l=1}^{N} T_{p_l}^{n_l} \in \mathbb{H}_{\mathcal{P}},$$

where $T_{p_l}$ are in the sense of (7.2.2), for $n_l \in \mathbb{N}$, for all $l = 1, ..., N$.

Let's consider the free-distributional data of a free random variable $T$ of (7.2.3) in $\mathbb{H}_{\mathcal{P}}$.

THEOREM 7.1. *Let $T$ be in the sense of (7.2.3) in the Adelic Hecke $C^*$-probability space $\mathbb{H}_{\mathcal{P}}$, where "$N \in \mathbb{N} - \{1\}$." Assume that*

$$(p_1, \ ..., \ p_N) \in alt\left(\mathcal{P}^N\right)$$

*is an alternating $N$-tuple. Then*

*(7.2.4)*

$$\tau\left(T^n\right) = \prod_{l=1}^{N} \left(\frac{\mu_{p_l}(K_l)^{nn_l - 1}\mu_{p_l}(K_l \cap Z_p)\mu_{p_l}\left(x_l^{nn_l} K_l Z_p \cap K_l Z_{p:G_p}\right)}{\mu_{p_l}\left(K_l Z_{p:G_p}\right)}\right),$$

*in the minimal free summand of $\mathbb{H}_{\mathcal{P}}$ containing $T$, for all $n \in \mathbb{N}$.*

*Furthermore, if $p_1 \neq p_N$ in $\mathcal{P}$, then*

*(7.2.5)*

$$\tau\left(T^{(n)}\right) = \prod_{l=1}^{N} \left(\frac{\mu_p(K_l)^{n_l - 1}\mu_p(K_l \cap Z_p)\mu_p\left(x_l^{n_l} K_l Z_p \cap K_l Z_{p:G_p}\right)}{\mu_p\left(K_l Z_{p:G_p}\right)}\right)^n,$$

*in $\mathbb{H}_{\mathcal{P}}$, for all $n \in \mathbb{N}$.*

PROOF. Let $T$ be a free random variable (7.2.3) in the Adelic Hecke $C^*$-probability space $\mathbb{H}_{\mathcal{P}}$, and suppose the $N$-tuple $(p_1, \ ..., \ p_N)$ is alternating in $\mathcal{P}$. Then the operator $T$ forms a free reduced word in $\mathbb{H}_{\mathcal{P}}$. Thus, one can get that

(7.2.6)

$$T^n \overset{\text{equi}}{=} \left(\bigotimes_{l=1}^{N} T_{p_l}^{n_l}\right)^n = \bigotimes_{l=1}^{N} T_{p_l}^{n_l n} \overset{\text{equi}}{=} \prod_{l=1}^{N} T_{p_l}^{n_l n},$$

in the minimal free summand of $\mathbb{H}_{\mathcal{P}}$ containing $T$, for all $n \in \mathbb{N}$, by (7.1.3). So, for any $n \in \mathbb{N}$, we obtain that

$$\tau\left(T^n\right) = \tau\left(\prod_{l=1}^{N} T_{p_l}^{n_l n}\right) = \prod_{l=1}^{N} \tau_{p_l}\left(T_{p_l}^{n_l n}\right)$$

$$= \prod_{l=1}^{N} \left(\frac{\mu_{p_l}(K_l)^{n_l n - 1}\mu_{p_l}(K_l \cap Z_p)\mu_{p_l}\left(x_l^{n_l n} K_l Z_p \cap K_l Z_{p:G_p}\right)}{\mu_{p_l}\left(K_l Z_{p:G_p}\right)}\right),$$

by (7.2.6), (7.0.1) and (7.2.2). Therefore, the free-moment formula (7.2.4) holds in the minimal free summand of $\mathbb{H}_{\mathcal{P}}$ containing $T$.

Now, assume that $p_1 \neq p_N$ in $\mathcal{P}$. Then, for any $n \in \mathbb{N}$, the $nN$-tuple,

$$W_n = \left(\underbrace{(p_1, ..., p_N), \ (p_1, ..., p_N), \ ..., \ (p_1, ..., p_N)}_{n\text{-times}}\right)$$

is an alternating $nN$-tuple in $\mathcal{P}$, i.e.,

(7.2.7)

$$W_n \in alt\left(\mathcal{P}^{nN}\right), \text{ for all } n \in \mathbb{N}.$$

Thus, the free random variables $T^{(n)}$ are free reduced words

$$T^{(n)} = \underbrace{\left(T_{p_1}^{n_1} \cdots T_{p_N}^{n_N}\right)\left(T_{p_1}^{n_1} \cdots T_{p_N}^{n_N}\right) \cdots \left(T_{p_1}^{n_1} \cdots T_{p_N}^{n_N}\right)}_{n\text{-times}}$$

in $\mathbb{H}_{\mathcal{P}}$, for all $n \in \mathbb{N}$, by (7.2.7). So, one obtains that

$$\tau\left(T^{(n)}\right) = \tau\left((T_{p_1}^{n_1}...T_{p_N}^{n_N})...(T_{p_1}^{n_1}...T_{p_N}^{n_N})\right)$$

$$= \left(\prod_{l=1}^{N} \tau_{p_l}\left(T_{p_l}^{n_l}\right)\right) \cdots \left(\prod_{l=1}^{N} \tau_{p_l}\left(T_{p_l}^{n_l}\right)\right)$$

$$= \left(\prod_{l=1}^{N} \tau_{p_l}\left(T_{p_l}^{n_l}\right)\right)^n$$

$$= \left(\prod_{l=1}^{N} \left(\frac{\mu_p(K_l)^{n_l-1}\mu_p(K_l \cap Z_p)\mu_p\left(x_l^{n_l} K_l Z_p \cap K_l Z_{p:G_p}\right)}{\mu_p\left(K_l Z_{p:G_p}\right)}\right)\right)^n$$

by (7.0.1) and (7.2.2)

$$= \prod_{l=1}^{N} \left(\frac{\mu_p(K_l)^{n_l-1}\mu_p(K_l \cap Z_p)\mu_p\left(x_l^{n_l} K_l Z_p \cap K_l Z_{p:G_p}\right)}{\mu_p\left(K_l Z_{p:G_p}\right)}\right)^n,$$

for all $n \in \mathbb{N}$. Therefore, the free-moment formula (7.2.5) holds true in $\mathbb{H}_{\mathcal{P}}$. □

The above two free-distributional data (7.2.4) and (7.2.5) provides free-distributional data of free reduced words generated by the generating operators in our Adelic Hecke $C^*$-probability space $\mathbb{H}_{\mathcal{P}}$. Also, they illustrate the differences between the data,

$$\left(\tau(T^n)\right)_{n=1}^{\infty}, \text{ and } \left(\tau(T^{(n)})\right)_{n=1}^{\infty},$$

where $T^n$ and $T^{(n)}$ are from **NR 7.1.1**, for all $n \in \mathbb{N}$. Of course, if $N = 1$ in the above theorem, then

(7.2.5)′

$$\tau\left(T^n\right) = \tau\left(T^{(n)}\right), \text{ for all } n \in \mathbb{N}, \text{ in } \mathbb{H}_{\mathcal{P}},$$

because

$$T^{(n)} = T^n \text{ in } (\mathbb{H}(G_{p_1}), \tau_{p_1}), \text{ in } \mathbb{H}_{\mathcal{P}},$$

since $T$ is a length-1 free reduced word in $\mathbb{H}_{\mathcal{P}}$.

THEOREM 7.2. *Let* $S_k = \prod_{l=1}^{N_k} T_{p_{l_k}}^{n_{l_k}}$ *be free random variables (7.2.3) of the Adelic Hecke $C^*$-probability space $\mathbb{H}_{\mathcal{P}}$, where $T_{p_{l_k}} \in \mathbb{H}(G_{p_{l_k}})$ are in the sense of (7.2.2), for all $k = 1, ..., N$, for $N \in \mathbb{N} - \{1\}$. Also, let*

$$T = \sum_{k=1}^{N} S_k \in \mathbb{H}_{\mathcal{P}}.$$

*Suppose*
*(7.2.8)*

$$W_k = (p_{1_k},\ p_{2_k},\ ...,\ p_{N_k}) \in alt\left(\mathcal{P}^{N_k}\right),$$

*for all* $k = 1, ..., N$, *and*

$$W_{k_1} \neq W_{k_2} \text{ in } \mathcal{P}^\infty, \text{ whenever } k_1 \neq k_2 \text{ in } \{1, ..., N\}.$$

*Then*
*(7.2.9)*

$$\tau\left(T^n\right) = \sum_{k=1}^N \left( \prod_{l_k=1}^{N_k} \left( \frac{\mu_p\left(K_{l_k}\right)^{n-1} \mu_p\left(x_{l_k}^n K_{l_k} \cap Z_p\right) \mu_p\left(x_{l_k}^n K_{l_k} Z_p \subset K_{l_k} Z_{p:G_p}\right)}{\mu_p\left(K_{l_k} Z_{p:G_p}\right)} \right) \right),$$

*in the minimal free summand of* $\mathbb{H}_\mathcal{P}$ *containing* $T$, *for all* $n \in \mathbb{N}$.

PROOF. Let $S_k$ be in the sense of (7.2.3) in $\mathbb{H}_\mathcal{P}$, for all $k = 1, ..., N$. Then, by the first assumption of (7.2.8), every operator $S_k$ is a free reduced word in $\mathbb{H}_\mathcal{P}$, for all $k = 1, ..., N$. Moreover, by the second assumption of (7.2.8), the operator $T = \sum_{k=1}^N S_k$ forms a free sum in $\mathbb{H}_\mathcal{P}$, satisfying

$$T^n \overset{\text{equi}}{=} \left( \underset{k=1}{\overset{N}{\oplus}} S_k \right)^n = \underset{k=1}{\overset{N}{\oplus}} S_k^n \overset{\text{equi}}{=} \sum_{k=1}^N S_k^n,$$

in the minimal free summand of $\mathbb{H}_\mathcal{P}$ containing $T$, for all $n \in \mathbb{N}$ (See **NR 7.1.2**).

So, one obtains that

$$\tau\left(T^n\right) = \sum_{k=1}^N \tau\left(S_k^n\right), \text{ for all } n \in \mathbb{N},$$

where

$$\tau\left(S_k^n\right) = \prod_{l_k=1}^{N_k} \left( \frac{\mu_p\left(K_{l_k}\right)^{n-1} \mu_p\left(x_{l_k}^n K_{l_k} \cap Z_p\right) \mu_p\left(x_{l_k}^n K_{l_k} Z_p \cap K_{l_k} Z_{p:G_p}\right)}{\mu_p\left(K_{l_k} Z_{p:G_p}\right)} \right),$$

for all $n \in \mathbb{N}$, by (7.2.4).

Therefore, the free-moment formula (7.2.9) holds in the minimal free summand of $\mathbb{H}_\mathcal{P}$ containing $T$. □

## 8. On a Certain $C^*$-Subalgebra $\mathfrak{S}$ of $\mathbb{H}_\mathcal{P}$

In this section, we consider a certain $C^*$-subalgebra $\mathfrak{S}$ of the Adelic Hecke $C^*$-algebra $\mathbb{H}_\mathcal{P}$, having an interesting free-probabilistic properties. For any $p \in \mathcal{P}$, define a subset $\mathcal{K}_p$ of the $\sigma$-algebra $\sigma(G_p)$ of the general linear group $G_p = GL_2(\mathbb{Q}_p)$ by

(8.1)

$$\mathcal{K}_p = \{K \lhd G_p : K \text{ is compact-open in } G_p\}.$$

Now, define the $C^*$-subalgebras $\mathfrak{S}_p$ of the $p$-Hecke $C^*$-algebras $\mathbb{H}(G_p)$ by the $C^*$-algebras generated by $\mathcal{K}_p$ of (8.1), i.e.,

(8.2)

$$\mathfrak{S}_p \overset{def}{=} C^*\left(\{\alpha_K^p \in \mathbb{H}(G_p) : K \in \mathcal{K}_p\}\right) \text{ in } \mathbb{H}(G_p),$$

for all $p \in \mathcal{P}$.

Then the $C^*$-algebras $\mathfrak{S}_p$ of (8.2) are well-defined $C^*$-subalgebras of $\mathbb{H}(G_p)$, inducing the free-probabilistic sub-structures,

(8.3)

$$\mathfrak{S}_p \overset{denote}{=} (\mathfrak{S}_p,\ \tau_p),\ \text{with } \tau_p = \tau_p\mid_{\mathfrak{S}_p},$$

in the $p$-Hecke $C^*$-probability spaces $(\mathbb{H}(G_p),\ \tau_p)$, where $\tau_p$ are in the sense of (6.5), for all $p \in \mathcal{P}$.

Then, from the system

$$\{\mathfrak{S}_p = (\mathfrak{S}_p,\ \tau_p) : p \in \mathcal{P}\},$$

consisting of all $C^*$-probability spaces $\mathfrak{S}_p$ of (8.3), one can obtain the free-product $C^*$-probability space

(8.4)

$$\begin{aligned}\mathfrak{S} &= (\mathfrak{S},\ \tau) \overset{def}{=} \underset{p\in\mathcal{P}}{\star} \mathfrak{S}_p \\ &= \left(\underset{p\in\mathcal{P}}{\star} \mathfrak{S}_p,\ \underset{p\in\mathcal{P}}{\star} \tau_p\right).\end{aligned}$$

Then this new $C^*$-probability space $\mathfrak{S}$ of (8.4) is a well-determined $C^*$-probability space embedded in our Adelic Hecke $C^*$-probability space

$$\mathbb{H}_{\mathcal{P}} = (\mathbb{H}_{\mathcal{P}},\ \tau)$$

of (7.2.1).

DEFINITION 8.1. *Let $\mathfrak{S}$ be the free-product $C^*$-probability space (8.4) of the system $\{\mathfrak{S}_p\}_{p\in\mathcal{P}}$ of $C^*$-probability spaces $\mathfrak{S}_p$'s in the sense of (8.3). Then it is called the (Adelic-normal-compact-open) sub-Hecke $C^*$-probability space of the Adelic Hecke $C^*$-probability space $\mathbb{H}_{\mathcal{P}}$.*

Then, by the main results of Section 7, we obtain the following free-distributional data on the sub-Hecke $C^*$-probability space $\mathfrak{S}$ of $\mathbb{H}_{\mathcal{P}}$.

COROLLARY 8.1. *Let $\alpha_K^p$ be a generating operator of the sub-Hecke $C^*$-probability space $\mathfrak{S}$ of $\mathbb{H}_{\mathcal{P}}$, for $K \in \mathcal{K}_p$, for $p \in \mathcal{P}$. Then*

*(8.5)*

$$\begin{aligned}\tau\left((\alpha_K^p)^n\right) &= \mu_p(K)^{n-1}\,\mu_p(K \cap Z_p) \\ &= \tau\left((\alpha_K^p)^{(n)}\right),\end{aligned}$$

*for all $n \in \mathbb{N}$.*

PROOF. Observe that , by (7.2.4), (7.2.5) and (7.2.5)′

$$\tau\left(\left(\alpha_K^p\right)^n\right) = \tau_p\left(\left(\alpha_K^p\right)^n\right)$$

$$= \left(\frac{\mu_p(K)^{n-1}\mu_p(K \cap Z_p)\mu_p\left(KZ_p \cap KZ_{p:G_p}\right)}{\mu_p\left(KZ_{p:G_p}\right)}\right)$$

$$= \frac{\mu_p(K)^{n-1}\mu_p(K \cap Z_p)\mu_p\left(KZ_{p:G_p}\right)}{\mu_p\left(KZ_{p:G_p}\right)}$$

$$= \mu_p(K)^{n-1}\mu_p\left(K \cap Z_p\right)$$

$$= \tau_p\left(\left(\alpha_K^p\right)^{(n)}\right) = \tau\left(\left(\alpha_K^p\right)^{(n)}\right),$$

for all $K \in \mathcal{K}_p$, for all $n \in \mathbb{N}$. □

More general to (8.5), one obtains the following result.

COROLLARY 8.2. *Let* $(p_1, ..., p_N) \in alt\left(\mathcal{P}^N\right)$, *for* $N \in \mathbb{N} - \{1\}$, *and let*
*(8.6)*

$$T = \prod_{l=1}^{N} \left(\alpha_{K_l}^{p_l}\right)^{n_l} \in \mathfrak{S},$$

*where* $K_l \in \mathcal{K}_{p_l}$, *and* $n_l \in \mathbb{N}$, *for all* $l = 1, ..., N$. *Then*
*(8.7)*

$$\tau(T^n) = \prod_{l=1}^{N} \left(\mu_{p_l}\left(K_l\right)^{n_l-1}\mu_{p_l}\left(K_l \cap Z_{p_l}\right)\right),$$

*in the minimal free summand of* $\mathbb{H}_{\mathcal{P}}$ *containing* $T$, *for all* $n \in \mathbb{N}$.
*Assume further that* $p_N \neq p_1$ *in* $\mathcal{P}$. *Then*
*(8.8)*

$$\tau(T^{(n)}) = \prod_{l=1}^{N} \left(\mu_{p_l}\left(K_l\right)^{nn_l-1}\mu_{p_l}\left(K_l \cap Z_{p_l}\right)\right)^n,$$

*in* $\mathbb{H}_{\mathcal{P}}$, *for all* $n \in \mathbb{N}$.

PROOF. By the assumption that the $N$-tuple $(p_1, ..., p_N)$ is alternating in $\mathcal{P}^N$, for $N \in \mathbb{N} - \{1\}$, the operator $T$ of (8.6) is a free reduced word in the sub-Hecke $C^*$-probability space $\mathfrak{S}$ of the Adelic Hecke $C^*$-probability space $\mathbb{H}_{\mathcal{P}}$, by (8.4). So, it satisfies that

$$\tau\left(T^n\right) \overset{\text{equi}}{=} \tau\left(\prod_{l=1}^{N} \left(\alpha_{K_l}^{p_l}\right)^{nn_l}\right)$$

$$= \prod_{l=1}^{N} \tau_{p_l}\left(\left(\alpha_{K_l}^{p_l}\right)^{nn_l}\right)$$

by (7.2.4)

$$= \prod_{l=1}^{N} \left(\mu_{p_l}\left(K_l\right)^{nn_l-1}\mu_{p_l}\left(K_l \cap Z_{p_l}\right)\right)$$

by (8.5), for all $n \in \mathbb{N}$. Therefore, the formula (8.7) holds.

Assume now that $(p_1, ..., p_N)$ is an alternating $N$-tuple, and suppose $p_N \neq p_1$ in $\mathcal{P}$. Then the given operator $T$ is a free reduced word, and the free (non-reduced) word $T^{(n)}$ become free reduced words in the sub-Hecke $C^*$-probability space $\mathfrak{S}$ of $\mathbb{H}_{\mathcal{P}}$, for all $n \in \mathbb{N}$. Therefore, by (7.2.5), one obtains that

$$\tau(T^{(n)}) = (\tau(T))^n \text{, for all } n \in \mathbb{N}.$$

So, by the formula (8.7), we obtain

$$\tau\left(T^{(n)}\right) = \left(\prod_{l=1}^{N} \left(\mu_{p_l}\left(K_l\right)^{nn_l - 1} \mu_{p_l}\left(K_l \cap Z_{p_l}\right)\right)\right)^n$$

for all $n \in \mathbb{N}$. Thus, the formula (8.8) holds true. □

The free-moment formulas (8.5) and (8.7) characterize the free distributions of free reduced words in their minimal free summands of the sub-Hecke $C^*$-probability space $\mathfrak{S}$. And the formulas (8.5) and (8.8) provides free-distributional information of free reduced words in $\mathfrak{S}$.

For $K \in \mathcal{K}_p$, for any $p \in \mathcal{P}$, one can take the operator

(8.9)

$$\beta_K^p \overset{def}{=} \tfrac{1}{\mu_p(K)}\, \alpha_K^p \in \mathfrak{S}_p \text{ in } \mathfrak{S}.$$

Observe that if $\beta_K^p$ is in the sense of (8.9) in $\mathfrak{S}$, then

(8.10)

$$\begin{aligned}\beta_K^p &= \tfrac{1}{\mu_p(K)}\alpha_K^p = \tfrac{1}{\mu_p(K)}\alpha_{\chi_K}^p \\ &= \alpha^p\left(\tfrac{1}{\mu_p(K)}\chi_K\right) = \alpha_{e_K}^p,\end{aligned}$$

in the free block $\mathbb{H}(G_p)$ of $\mathbb{H}_{\mathcal{P}}$, where $e_K$ are in the sense of (2.2.5) in $(\mathcal{H}(G_p),\ \psi_p)$, for all $K \in \mathcal{K}_p$, for all $p \in \mathcal{P}$.

So, the operators $\beta_K^p$ of (8.9) are well-determined in $\mathfrak{S}$, for all $K \in \mathcal{K}_p$, for all $p \in \mathcal{P}$.

PROPOSITION 8.3. *Let $\beta_K^p$ be an operator (8.9) in the sub-Hecke $C^*$-probability space $\mathfrak{S}$ of $\mathbb{H}_{\mathcal{P}}$, for $K \in \mathcal{K}_p$, for $p \in \mathcal{P}$. Then $\beta_K^p$ is a projection in $\mathfrak{S}$.*

PROOF. Let $\beta_K^p$ be an operator (8.9) in $\mathfrak{S}$. Then

$$(\beta_K^p)^* = \left(\tfrac{1}{\mu_p(K)}\ \alpha_K^p\right)^* = \tfrac{1}{\mu_p(K)}\ \alpha_K^p = \beta_K^p, \text{ in } \mathfrak{S},$$

and

$$\begin{aligned}\left(\beta_K^p\right)^2 &= \tfrac{1}{\mu_p(K)^2}\alpha^p\left(\chi_K * \chi_K\right) \\ &= \tfrac{1}{\mu_p(K)^2}\alpha^p\left(\mu_p(K)\ \chi_K\right) \\ &= \tfrac{\mu_p(K)}{\mu_p(K)^2}\alpha_K^p = \tfrac{1}{\mu_p(K)}\alpha_K^p = \beta_K^p,\end{aligned}$$

in $\mathfrak{S}$.

Therefore, the operator $\beta_K^p$ is a projection in $\mathfrak{S}$. □

The above proposition shows that the sub-Hecke $C^*$-algebra $\mathfrak{S}$ is understood as a $C^*$-subalgebra of the Adelic Hecke $C^*$-algebra $\mathbb{H}_{\mathcal{P}}$, generated by projections

$$\{\beta_K^p \in \mathbb{H}_{\mathcal{P}} \,|K \in \mathcal{K}_p, \ \forall p \in \mathcal{P}\},$$

since

$$\alpha_K^p = \mu_p(K)\ \beta_K^p \text{ in } \mathbb{H}_{\mathcal{P}},$$

for all $K \in \mathcal{K}_p$, for all $p \in \mathcal{P}$.

Note that

$$\beta_{K_1}^{p_1}\beta_{K_2}^{p_2} = \begin{cases} \beta_{K_1}^{p_1}\beta_{K_2}^{p_2} & \text{(reduced word)} \quad \text{if } p_1 \neq p_2 \\ \frac{\mu_{p_1}(K_1 \cap K_2)}{\mu_{p_1}(K_1)\mu_p(K_2)}\ \alpha_{K_1K_2}^{p_1} & \text{(in } \mathfrak{S}_{p_1} \text{ of } \mathfrak{S}) \quad \text{if } p_1 = p_2, \end{cases}$$

for $K_l \in \mathcal{K}_{p_l}$, for $l = 1, 2$, where $p_1, p_2 \in \mathcal{P}$.

Note also that $\beta_{K_1}^p\beta_{K_2}^p$ is not a projection in general in $\mathfrak{S}$.

COROLLARY 8.4. *Let $\beta_K^p$ be a projection (8.9) in the free block $\mathfrak{S}_p$ of the sub-Hecke $C^*$-probability space $\mathfrak{S}$ of $\mathbb{H}_{\mathcal{P}}$, for $K \in \mathcal{K}_p$, for $p \in \mathcal{P}$. Then*
*(8.11)*

$$\tau\left(\left(\beta_K^p\right)^n\right) = \frac{\mu_p(K \cap Z_p)}{\mu_p(K)} = \tau\left(\left(\beta_K^p\right)^{(n)}\right),$$

*for all $n \in \mathbb{N}$.*

PROOF. Observe that

$$\tau\left(\left(\beta_K^p\right)^n\right) = \tau_p\left(\left(\beta_K^p\right)^n\right) = \left(\frac{1}{\mu_p(K)^n}\right)\tau_p\left(\left(\alpha_K^p\right)^n\right)$$

by (8.10)

$$= \left(\frac{1}{\mu_p(K)^n}\right)\left(\mu_p(K)^{n-1}\mu_p\left(K \cap Z_p\right)\right)$$

by (8.5)

$$= \frac{\mu_p(K \cap Z_p)}{\mu_p(K)},$$

in the free block $\mathfrak{S}_p$ (as the minimal free summand of $\mathfrak{S}$ containing $\beta_K^p$) of $\mathfrak{S}$, for all $n \in \mathbb{N}$.

Similarly, one obtains that

$$\tau\left(\left(\beta_K^p\right)^{(n)}\right) = \tau\left(\left(\beta_K^p\right)^n\right) = \frac{\mu_p(K \cap Z_p)}{\mu_p(K)},$$

by (8.5), for all $n \in \mathbb{N}$.

Therefore, we obtain the free-distributional data (8.11) of projections $\beta_K^p$, for all $K \in \mathcal{K}_p$, for all $p \in \mathcal{P}$, in $\mathfrak{S}$. □

Also, one obtains the following free-probabilistic information.

COROLLARY 8.5. *Let $\beta_{K_l}^{p_l}$ be free random variables (8.9) in the free blocks $\mathfrak{S}_{p_l}$ of the sub-Hecke $C^*$-probability space $\mathfrak{S}$ of $\mathbb{H}_{\mathcal{P}}$, for $l = 1, \ldots, N$, for $N \in \mathbb{N} - \{1\}$. Let*

(8.12)

$$T = \prod_{l=1}^{N} \left(\beta_{K_l}^{p_l}\right)^{n_l} \in \mathfrak{S},$$

*where $n_1, \ldots, n_N \in \mathbb{N}$. If $(p_1, \ldots, p_N) \in alt\left(\mathcal{P}^N\right)$, then*

(8.13)

$$\tau\left(T^n\right) = \prod_{l=1}^{N} \left(\frac{\mu_{p_l}(K_l \cap Z_p)}{\mu_{p_l}(K_l)}\right), \text{ for all } n \in \mathbb{N},$$

*in the minimal free summand of $\mathfrak{S}$ containing $T$ of (8.12).*

*Suppose $(p_1, \ldots, p_N) \in alt\left(\mathcal{P}^N\right)$, and assume $p_N \neq p_1$ in $\mathcal{P}$. If $T$ is in the sense of (8.12), then*

(8.14)

$$\tau(T^{(n)}) = \prod_{l=1}^{N} \left(\frac{\mu_{p_l}(K_l \cap Z_p)}{\mu_{p_l}(K_l)}\right)^n, \text{ for all } n \in \mathbb{N}, \text{ in } \mathfrak{S}.$$

PROOF. Let $T$ be in the sense of (8.12) in $\mathfrak{S}$. Then, by the assumption that the $N$-tuple $(p_1, \ldots, p_N)$ is alternating in $\mathcal{P}$, the operator $T$ is a free reduced word in $\mathfrak{S}$. So, one has

$$\tau(T^n) \overset{\text{equi}}{=} \tau\left(\prod_{l=1}^{N} \left(\left(\beta_{K_l}^{p_l}\right)^{nn_l}\right)\right)$$

$$= \prod_{l=1}^{N} \tau_{p_l}\left(\left(\beta_{K_l}^{p_l}\right)^{nn_l}\right) = \prod_{l=1}^{N} \left(\frac{\mu_{p_l}(K \cap Z_p)}{\mu_{p_l}(K)}\right),$$

in the minimal free summand of $\mathfrak{S}$ containing $T$, for all $n \in \mathbb{N}$, by (7.2.4) and (8.11). Therefore, the free-distributional data (8.13) holds.

Now, for the fixed operator $T$ of (8.12), suppose $(p_1, \ldots, p_N)$ is an alternating $N$-tuple, satisfying $p_N \neq p_1$ in $\mathcal{P}$. Then the operators $T^{(n)}$ are free reduced words in $\mathfrak{S}$, for all $n \in \mathbb{N}$. So, by (7.2.5),

(8.15)

$$\tau\left(T^{(n)}\right) = \left(\tau(T)\right)^n, \text{ for all } n \in \mathbb{N}.$$

Thus, one obtains the free-moment formula (8.14), by (8.13) and (8.15). □

Now, we concentrate on the free-distributional data of the projection $\beta_K^p \in \mathfrak{S}$, for $K \in \mathcal{K}_p$, for $p \in \mathcal{P}$. As we have seen in (8.11),

$$\tau\left(\left(\beta_K^p\right)^n\right) = \tau\left(\left(\beta_K^p\right)^{(n)}\right) = \frac{\mu_p(K \cap Z_p)}{\mu_p(K)},$$

for all $n \in \mathbb{N}$.

In the rest of this section, let's fix $K \in \mathcal{K}_p$ and $p \in \mathcal{P}$, and the corresponding free random variable $\beta_K^p$ in the sub-Hecke $C^*$-probability space $\mathfrak{S}$. One can easily verify that

(8.16)

$$\mu_p\left(K \cap Z_p\right) \leq \mu_p\left(Z_p\right) = 1.$$

Now, let $r_K^p \in \mathbb{R}$ in $\mathbb{C}$, such that
(8.17)

$$0 < r_K^p \overset{denote}{=} \mu_p (K \cap Z_p) \leq 1.$$

The quantity $r_K^p$ of (8.17) is well-determined in the interval (0, 1] of $\mathbb{R}$, by (8.16).

By (8.11),
(8.18)

$$\tau\left((\beta_K^p)^n\right) = \frac{r_K^p}{\mu_p(K)}, \text{ for all } n \in \mathbb{N},$$

where $r_K^p$ is in the sense of (8.17) in (0, 1].

Note that
(8.19)

$$r_K^p = \mu_p (K \cap Z_p) \leq \mu_p (K).$$

It shows that the quantities in (8.18) are contained in (0, 1] in $\mathbb{R}$, by (8.19).

So, by (8.17) and (8.19), we obtain the following estimations for the free distribution (8.11) of the operator $\beta_K^p$ in $\mathfrak{S}$, for all $K \in \mathcal{K}_p$, and $p \in \mathcal{P}$.

THEOREM 8.6. *Let $\beta_K^p = \frac{1}{\mu_p(K)} \alpha_K^p$ be free random variables (8.9) of sub-Hecke $C^*$-probability space $\mathfrak{S}$ of $\mathbb{H}_{\mathcal{P}}$, for all $K \in \mathcal{K}_p$, for all $p \in \mathcal{P}$. Then*
*(8.20)*

$$0 < \tau\left((\beta_K^p)^n\right) = \tau\left((\beta_K^p)^{(n)}\right) \leq 1 \text{ in } \mathbb{R},$$

*for all $n \in \mathbb{N}$.*

PROOF. Let $\beta_K^p$ be in the sense of (8.9) in $\mathfrak{S}$, for $K \in \mathcal{K}_p$, for $p \in \mathcal{P}$. Then,

$$0 < \tau\left((\beta_K^p)^n\right) = \frac{\mu_p(K \cap Z_p)}{\mu_p(K)} = \tau\left((\beta_K^p)^{(n)}\right) \leq 1,$$

in $\mathbb{R}$, for all $n \in \mathbb{N}$, by (8.11), (8.17), (8.18) and (8.19). □

More general to (8.20), the following estimation is obtained.

THEOREM 8.7. *Let $K_l \in \mathcal{K}_{p_l}$, for $p_l \in \mathcal{P}$, where $l = 1, \ldots, N$, for $N \in \mathbb{N} - \{1\}$. Suppose the $N$-tuple $(p_1, \ldots, p_N)$ is alternating in $\mathcal{P}^N$, and let*
*(8.21)*

$$T = \prod_{l=1}^{N} \left(\beta_{K_l}^{p_l}\right)^{n_l} \in \mathfrak{S}, \text{ with } \beta_{K_l}^{p_l} = \frac{1}{\mu_{p_l}(K_l)} \alpha_{K_l}^{p_l} \in \mathfrak{S}_{p_l},$$

*where $n_l \in \mathbb{N}$, for all $l = 1, \ldots, N$. If $T$ is in the sense of (8.21) in $\mathfrak{S}$, then we obtain*
*(8.22)*

$$0 < \tau\left(T^n\right) \leq 1, \text{ in its minimal free summand of } \mathfrak{S}$$

*and*

$$0 < \tau\left(T^{(n)}\right) \leq 1, \text{ in } \mathfrak{S},$$

*for all $n \in \mathbb{N}$.*

PROOF. Let $T$ be a free random variable (8.21) in $\mathfrak{S}$. Then, by the assumption that the $N$-tuple $(p_1, ..., p_N)$ is an alternating $N$-tuple, the operator $T$ is a free reduced word in $\mathfrak{S}$, satisfying

$$\tau(T^n) = \prod_{l=1}^{N} \left( \frac{\mu_{p_l}(K_l \cap Z_p)}{\mu_{p_l}(K_l)} \right),$$

and

$$\tau(T^{(n)}) = \prod_{l=1}^{N} \left( \frac{\mu_{p_l}(K_l \cap Z_p)}{\mu_{p_l}(K_l)} \right)^n,$$

for all $n \in \mathbb{N}$, by (8.13) and (8.14), respectively.

The factors of $\tau(T^n)$ satisfy

$$0 < \frac{\mu_{p_l}(K_l \cap Z_p)}{\mu_{p_l}(K_l)} = \tau\left(\beta_{K_l}^{p_l}\right) \leq 1, \text{ in } \mathbb{R},$$

by (8.20), for all $n \in \mathbb{N}$, and hence, the factors of $\tau\left(T^{(n)}\right)$ satisfy

$$0 < \left( \frac{\mu_{p_l}(K_l \cap Z_p)}{\mu_{p_l}(K_l)} \right)^n = \left(\tau\left(\beta_{K_l}^{p_l}\right)\right)^n \leq 1, \text{ in } \mathbb{R},$$

too, for all $n \in \mathbb{N}$.

Since all factors of $\tau(T^n)$ and $\tau\left(T^{(n)}\right)$ are contained in the interval $(0, 1]$ in $\mathbb{R}$,

$$\tau(T^n), \tau(T^{(n)}) \in (0, 1], \text{ in } \mathbb{R},$$

for all $n \in \mathbb{N}$.

Therefore, the estimations in (8.22) hold. □

Based on the above estimations (8.20) and (8.22), we obtain the following interesting result in terms of the freeness condition on the Hecke probability spaces $(\mathcal{H}(G_p), \psi_p)$ for primes $p \in \mathcal{P}$. First, recall that, for an arbitrarily fixed prime $p \in \mathcal{P}$, two free random variables $\chi_{K_1}$ and $\chi_{K_2}$ are free in the Hecke probability space $(\mathcal{H}(G_p), \psi_p)$, if and only if

(8.23)

$$\mu_p(K_1 \cap K_2) = \mu_p(K_1)\, \mu_p(K_2),$$

by (4.16). So, the two free random variables $e_{K_1}$ and $e_{K_2}$ are free in $(\mathcal{H}(G_p), \psi_p)$, if and only if the condition (8.23) holds, by (4.17).

THEOREM 8.8. *Let $K_l \in \mathcal{K}_p$, for $p_l \in \mathcal{P}$, and let $\beta_{K_l}^{p_l} = \frac{1}{\mu_{p_l}(K_l)} \alpha_{K_l}^{p_l} \in \mathfrak{S}_{p_l}$ be generating operators (8.9) of the sub-Hecke probability space $\mathfrak{S}$ of the Adelic Hecke $C^*$-probability space $\mathbb{H}_{\mathcal{P}}$, for $l = 1, ..., N$, for $N \in \mathbb{N} - \{1\}$. Assume that the $\mu_{p_l}$-measurable functions $\chi_{K_l}$ and $\chi_{Z_{p_l}}$ are free in the Hecke probability space $(\mathcal{H}(G_{p_l}), \psi_{p_l})$, for all $l = 1, ..., N$, where $Z_{p_l} = GL_2(\mathbb{Z}_{p_l})$ in $G_{p_l}$. Then*

*(8.24)*

$$\tau\left(\left(\beta_{K_l}^{p_l}\right)^n\right) = 1 = \tau\left(\left(\beta_{K_l}^{p_l}\right)^{(n)}\right), \textit{ for all } n \in \mathbb{N},$$

*for all $l = 1, ..., N$. More precisely, if $(p_1, ..., p_N)$ is alternating in $\mathcal{P}$, and if*

$$T = \prod_{l=1}^{N} \left(\beta_{K_l}^{p_l}\right)^{n_l} \in \mathfrak{S}, \textit{ for } n_1, ..., n_N \in \mathbb{N},$$

*then*

*(8.25)*

$$\tau(T^n) = 1 = \tau\left(T^{(n)}\right), \textit{ for all } n \in \mathbb{N}.$$

PROOF. Recall first that $\beta_K^p = \alpha_{e_K}^p$ in $\mathfrak{S}_p \subset \mathfrak{S}$, for all $K \in \mathcal{K}_p$, for all $p \in \mathcal{P}$.

Let $K_1$, ..., $K_N$ be given as above, and suppose $\chi_{K_l}$ and $\chi_{Z_{p_l}}$ are free in $(\mathcal{H}(G_{p_l}),\ \psi_{p_l})$, for all $l = 1$, ..., $N$. Then, by (4.16) and (8.23),

$$\mu_{p_l}(K_l \cap Z_{p_l}) = \mu_{p_l}(K_l)\ \mu_{p_l}(Z_{p_l}) = \mu_{p_l}(K_l),$$

for all $l = 1$, ..., $N$, because $\mu_{p_l}(Z_{p_l}) = 1$, for all $p_l \in \mathcal{P}$. i.e., $\chi_{K_l}$ and $\chi_{Z_{p_l}}$ are free in $(\mathcal{H}(G_{p_l}),\ \psi_{p_l})$, if and only if

(8.26)

$$\mu_{p_l}(K_l \cap Z_{p_l}) = \mu_{p_l}(K_l),$$

if and only if $e_{K_l}$ and $e_{Z_{p_l}}$ are free in $(\mathcal{H}(G_{p_l}),\ \psi_{p_l})$, for all $l = 1$, ..., $N$.

Under hypothesis, one has

(8.27)

$$\begin{aligned}\tau\left(\left(\beta_{K_l}^{p_l}\right)^n\right) &= \tau\left(\left(\beta_{K_l}^{p_l}\right)^{(n)}\right)\\ &= \frac{\mu_{p_l}(K_l \cap Z_p)}{\mu_{p_l}(K_l)} = \frac{\mu_{p_l}(K_l)}{\mu_{p_l}(K_l)} = 1,\end{aligned}$$

by (8.11) and (8.26), for all $n \in \mathbb{N}$. Therefore, the formula (8.24) holds true.

If the $N$-tuple $(p_1, \ ..., \ p_N)$ is an alternating sequence, then the operator $T = \prod_{l=1}^{N} \left(\beta_{K_l}^{p_l}\right)^{n_l}$ is a free reduced word in $\mathfrak{S}$, satisfying

$$\tau(T^n) = \prod_{l=1}^{N} \left(\frac{\mu_{p_l}(K_l \cap Z_{p_l})}{\mu_{p_l}(K_l)}\right),$$

in its minimal free summand of $\mathfrak{S}$, and

$$\tau(T^{(n)}) = \prod_{l=1}^{N} \left(\frac{\mu_{p_l}(K_l \cap Z_{p_l})}{\mu_{p_l}(K_l)}\right)^n,$$

in $\mathfrak{S}$, for all $n \in \mathbb{N}$, by (8.13) and (8.14), respectively. So, if $\chi_{K_l}$ and $\chi_{Z_{p_l}}$ are free in $(\mathcal{H}(G_{p_l}),\ \psi_{p_l})$, for all $l = 1$, ..., $N$, then

(8.28)

$$\tau(T^n) = 1^N = 1, \text{ and } \tau(T^{(n)}) = 1^{nN} = 1,$$

for all $n \in \mathbb{N}$, by (8.27).

Therefore, the formula (8.25) holds true by (8.28). □

The following corollary is a direct consequence of (8.20), (8.22), (8.24) and (8.25).

COROLLARY 8.9. *Let $\beta_{K_l}^{p_l}$ be the operators (8.9) in $\mathfrak{S}$, for $l = 1$, ..., $N$, for $N \in \mathbb{N} - \{1\}$, and let*

$$T = \prod_{l=1}^{N} \left(\beta_{K_l}^{p_l}\right)^{n_l} \in \mathfrak{S}, \text{ for } n_1, \ ..., \ n_N \in \mathbb{N}.$$

*If two $\mu_{p_l}$-measurable functions $\chi_{K_l}$ and $\chi_{Z_{p_l}}$ are free in the Hecke probability space $(\mathcal{H}(G_{p_l}),\ \psi_{p_l})$, where $Z_{p_l} = GL_2(\mathbb{Z}_{p_l})$ in $G_{p_l}$, for all $l = 1, ..., N$, then, for any arbitrarily fixed $n \in \mathbb{N}$,*

*(8.29)* $\tau\left(\left(\beta_{K_l}^{p_l}\right)^n\right) = \tau\left(\left(\beta_{K_l}^{p_l}\right)^{(n)}\right)$ *is maximal.*

*(8.30)* $\tau(T^n)$, *and* $\tau\left(T^{(n)}\right)$ *are maximal.*

PROOF. Under hypothesis, we have

$$0 < \tau(\left(\beta_{K_l}^{p_l}\right)^n) = \tau\left(\left(\beta_{K_l}^{p_l}\right)^{(n)}\right) \leq 1,$$

and

$$0 < \tau(T^n) \leq 1, \text{ and } 0 < \tau(T^{(n)}) \leq 1,$$

for any fixed $n \in \mathbb{N}$, by (8.20) and (8.22), respectively.

It is shown in (8.24) and (8.25) that

$$\tau\left(\left(\beta_{K_l}^{p_l}\right)^n\right) = 1 = \tau\left(\left(\beta_{K_l}^{p_l}\right)^{(n)}\right),$$

and

$$\tau(T^n) = 1 = \tau(T^{(n)}),$$

if $\chi_{K_l}$ and $\chi_{Z_{p_l}}$ are free in $(\mathcal{H}(G_{p_l}),\ \psi_{p_l})$, for all $l = 1, ..., N$.

Therefore, the freeness of $\chi_{K_l}$ and $\chi_{Z_{p_l}}$ in $(\mathcal{H}(G_{p_l}),\ \psi_{p_l})$, for all $l = 1, ..., N$, implies the maximality of linear-functional values of such operators. □

## 9. Acting $\mathfrak{S}$ on $C^*$-Probability Spaces

Throughout this section, let's fix an arbitrary $C^*$-probability space $(A, \varphi)$ of a $C^*$-algebra $A$, which is a $C^*$-subalgebra of an *operator algebra* $B(H)$, consisting of all bounded linear operators on a Hilbert space $H$, and a bounded linear functional $\varphi$ on $A$. Also, let $\mathfrak{S}$ be the sub-Hecke $C^*$-probability space $(\mathfrak{S}, \tau)$ of the Adelic Hecke $C^*$-probability space $\mathbb{H}_{\mathcal{P}}$.

As we discussed in Section 8, the $C^*$-algebra $\mathfrak{S}$ is generated by the projections

(9.1)

$$\{\beta_K^p \in \mathbb{H}_{\mathcal{P}} : K \in \mathcal{K}_p, \forall p \in \mathcal{P}\},$$

satisfying that

(9.2)

$$\beta_{K_1}^{p_1}\beta_{K_2}^{p_2} = \begin{cases} \beta_{K_1}^{p_1}\beta_{K_2}^{p_2} & \text{(reduced word)} \quad \text{if } p_1 \neq p_2 \\ \\ \frac{\mu_{p_1}(K_1 \cap K_2)}{\mu_{p_1}(K_1)\mu_{p_2}(K_2)}\alpha_{K_1K_2}^{p_1} & (\text{in } \mathfrak{S}_{p_1} \subset \mathfrak{S}) \quad \text{if } p_1 = p_2, \end{cases}$$

in $\mathfrak{S}$, for all $p_1, p_2 \in \mathcal{P}$. Remark again that if $p_1 = p = p_2$ in $\mathcal{P}$, then $\beta_{K_1}^p\beta_{K_2}^p$ is not a projection in $\mathfrak{S}$, in general, by (9.2).

LEMMA 9.1. *Let $K_l \in \mathcal{K}_{p_l}$, for $p_l \in \mathcal{P}$, for $l = 1, ..., N$, where $N \in \mathbb{N} - \{1\}$, and let $\beta_{K_l}^{p_l} \in \mathfrak{S}$ in $\mathbb{H}_{\mathcal{P}}$, for all $l = 1, ..., N$. If $(p_1, ..., p_N)$ is alternating in $\mathcal{P}$, then the product $T = \prod_{l=1}^{N} \beta_{K_l}^{p_l}$ forms a free reduced word in $\mathfrak{S}$. Meanwhile, if*

$$p_1 = p_2 = ... = p_N = p \text{ in } \mathcal{P},$$

*for some $p \in \mathcal{P}$, Then the product $T$ is identical to*

$$T = \frac{\prod_{l=2}^{N} \mu_p(K_{1,...,l}^o)}{\prod_{l=1}^{N} \mu_p(K_l)} \alpha_{K_{1,...,N}}^p,$$

*in the free block $\mathfrak{S}_p$ in $\mathfrak{S}$.*

PROOF. This lemma is proven by the induction on (9.2). □

Now, let $\mathfrak{H}_p$ be the $p$-Hecke Hilbert spaces of Section 5, where the free blocks $\mathfrak{S}_p$ of $\mathfrak{S}$ act, for all $p \in \mathcal{P}$. Let $\mathfrak{H}$ be a Hilbert space, isomorphic to
(9.3)

$$\mathfrak{H} = \mathbb{C} \oplus \left( \bigoplus_{n \in \mathbb{N}} \left( \bigotimes_{(p_1, ..., p_n) \in alt(\mathcal{P}^n)} \left( \bigotimes_{l=1}^{n} \mathfrak{H}_{p_l}^o \right) \right) \right),$$

with

$$\mathfrak{H}_{p_l}^o = \mathfrak{H}_{p_l} \ominus \mathbb{C}, \text{ for all } l = 1, ..., n,$$

where $\oplus$ (and hence, $\ominus$) and $\otimes$ are the direct product, respectively, the tensor product of topological vector spaces.

Then our sub-Hecke $C^*$-algebra $\mathfrak{S}$ is acting on the Hilbert space $\mathfrak{H}$ of (9.3) (e.g., see [25] and [27]). Now, define a new Hilbert space,
(9.4)

$$\mathcal{H} = H \otimes \mathfrak{H},$$

where $H$ is the Hilbert space where a fixed $C^*$-algebra $A$ is acting, and $\mathfrak{H}$ is the Hilbert space (9.3) where $\mathfrak{S}$ acts.

Then the *tensor product $C^*$-algebra*
(9.5)

$$\mathfrak{S}_A = A \otimes_{\mathbb{C}} \mathfrak{S}$$

is acting on the Hilbert space $\mathcal{H}$ of (9.4).

DEFINITION 9.1. *Let $\mathfrak{S}_A$ be the tensor product $C^*$-algebra (9.5) of a $C^*$-algebra $A$, and our sub-Hecke $C^*$-algebra $\mathfrak{S}$, acting on the Hilbert space $\mathcal{H}$ of (9.4). We call it $A$-tensor (normal-compact-open-)sub-Hecke $C^*$-algebra.*

Let $\mathfrak{S}_A$ be the $A$-tensor sub-Hecke $C^*$-algebra of (9.5). Define a bounded linear transformation

$$E_A : \mathfrak{S}_A \to A$$

by a linear morphism satisfying
(9.6)

$$E_A(a \otimes T) = \tau(T)\, a,$$

for all $a \in A$, and $T \in \mathfrak{S}$, where $\tau$ is the linear functional (8.4) on $\mathfrak{S}$.

The linear morphism (9.6) is indeed a well-defined bounded linear transformation from $\mathfrak{S}_A$ onto $A$. Now, define a linear functional $\tau_A$ on $\mathfrak{S}_A$ by

(9.7)

$$\tau_A = \varphi \circ E_A \text{ on } \mathfrak{S}_A,$$

where $\varphi$ is the given linear functional on $A$, and $E_A$ is the linear transformation (9.6) on the $A$-tensor sub-Hecke $C^*$-algebra $\mathfrak{S}_A$. Then the $C^*$-probability space $(\mathfrak{S}_A,\ \tau_A)$ is well-determined.

DEFINITION 9.2. *Let $\mathfrak{S}_A$ be the $A$-tensor sub-Hecke $C^*$-algebra (9.5), and let $\tau_A = \varphi \circ E_A$ be the linear functional (9.7) on $\mathfrak{S}_A$. The corresponding $C^*$-probability space*

*(9.8)*

$$\mathfrak{S}_A \overset{denote}{=} (\mathfrak{S}_A,\ \tau_A)$$

*is called the $A$-(tensor-)sub-Hecke $C^*$-probability space for a $C^*$-probability space $(A,\ \varphi)$.*

Let $\mathfrak{S}_A = (\mathfrak{S}_A,\ \tau_A)$ be the $A$-sub-Hecke $C^*$-probability space (9.8) for a fixed $C^*$-probability space $(A,\ \varphi)$. By the very definition (9.7), one has that

(9.9)

$$\begin{aligned}\tau_A\,(a \otimes T) &= \varphi\,(E_A(a \otimes T)) \\ &= \varphi\,(\tau(T)a) = \tau(T)\varphi(a) \\ &= \varphi(a)\tau(T),\end{aligned}$$

for all $a \in A$, and $T \in \mathfrak{S}$.

By abusing notation, one may write the linear functional $\tau_A$ of (9.7) by

(9.10)

$$\tau_A = \varphi \otimes \tau \text{ on } A \otimes_{\mathbb{C}} \mathfrak{S} = \mathfrak{S}_A,$$

by (9.9).

In the following, we will use the following notations;

(9.11)

$$(a \otimes T)^{(n)} = a^n \otimes T^{(n)}, \text{ in } \mathfrak{S}_A,$$

and

$$(a \otimes T)^n = a^n \otimes T^n,$$

in the minimal free summand of $\mathfrak{S}_A$ containing $a \otimes T$, for all $n \in \mathbb{N}$ (See **NR 7.1.1**).

Since $A$ and $(A,\ \varphi)$ are arbitrarily fixed, let's focus on the notations (9.11). However, if $A$ is fixed to be a free product $C^*$-algebra, then the notations (9.11) will / can be refined by

$$(a \otimes T)^{(n)} = a^{(n)} \otimes T^{(n)}, \text{ and } (a \otimes T)^n = a^n \otimes T^n,$$

for all $n \in \mathbb{N}$, under **NR 7.1.1** and **NR 7.1.2**.

Under (9.11), we obtain the following result.

THEOREM 9.2. *Let $\mathfrak{S}_A = (\mathfrak{S}_A,\ \tau_A)$ be the $A$-sub-Hecke $C^*$-probability space (9.8) for a $C^*$-probability space $(A,\ \varphi)$. Suppose $K_l \in \mathcal{K}_{p_l}$, for $p_l \in \mathcal{P}$, for $l = 1$, ..., $N$, for $N \in \mathbb{N} - \{1\}$, and let*

$$T = \prod_{l=1}^{N} \left(\beta_{K_l}^{p_l}\right)^{n_l} \text{ in } \mathfrak{S}, \text{ for } n_1, \ldots, n_N \in \mathbb{N}.$$

*If $(p_1, ..., p_N)$ is alternating in $\mathcal{P}$, then*
*(9.12)*

$$\tau_A\left((a \otimes T)^n\right) = \varphi(a^n)\left(\prod_{l=1}^{N} \frac{\mu_{p_l}(K_l \cap Z_{p_l})}{\mu_{p_l}(K_l)}\right),$$

*and*

$$\tau_A\left((a \otimes T)^{(n)}\right) = \varphi(a^n)\left(\prod_{l=1}^{N}\left(\frac{\mu_{p_l}(K_l \cap Z_{p_l})}{\mu_{p_l}(K_l)}\right)^n\right),$$

*for all $n \in \mathbb{N}$.*

*Meanwhile, if $p_1 = ... = p_N = p$ in $\mathcal{P}$, for some $p \in \mathcal{P}$, then*
*(9.13)*

$$\tau_A\left((a \otimes T)^n\right) = \varphi(a^n)\left(\frac{\left(\prod_{l=2}^{N}\mu_p(K^o_{1,\ldots,l})\right)(\mu_p(K_{1,\ldots,N} \cap Z_p))}{\left(\prod_{l=1}^{N}\mu_p(K_l)\right)(\mu_p(K_{1,\ldots,N}))}\right)$$

$$= \tau_A\left((a \otimes T)^{(n)}\right),$$

*for all $n \in \mathbb{N}$.*

PROOF. Suppose $(p_1, ..., p_N)$ is contained in $alt\left(\mathcal{P}^N\right)$, for the fixed $N \in \mathbb{N} - \{1\}$. Then the operator $T = \prod_{l=1}^{N} \left(\beta_{K_l}^{p_l}\right)^{n_l}$ forms a free reduced word in the sub-Hecke $C^*$-probability space $\mathfrak{S}$. So, if

$$a \otimes T \in \mathfrak{S}_A, \text{ for } a \in (A, \varphi),$$

then one obtains

$$\tau_A\left((a \otimes T)^n\right) = \tau_A\left(a^n \otimes T^n\right)$$

by (9.11)

$$= \tau\left(T^n\right)\varphi(a^n)$$

by (9.9) and (9.10)

$$= \varphi(a^n)\left(\prod_{l=1}^{N} \frac{\mu_{p_l}(K_l \cap Z_p)}{\mu_{p_l}(K_l)}\right),$$

by (8.13), for all $n \in \mathbb{N}$. Therefore, the first free-moment formula of (9.12) holds in the minimal free summand of $\mathfrak{S}_A$ containing $a \otimes T$ in $\mathfrak{S}_A$.

Observe now that

$$\tau_A\left((a \otimes T)^{(n)}\right) = \tau_A\left(a^n \otimes T^{(n)}\right)$$

by (9.11)

$$= \tau\left(T^{(n)}\right)\varphi(a^n)$$

by (9.9) and (9.10)

$$= \varphi(a^n)\left(\prod_{l=1}^{N}\left(\frac{\mu_{p_l}(K_l \cap Z_{p_l})}{\mu_{p_l}(K_l)}\right)^n\right),$$

by (8.14), for all $n \in \mathbb{N}$. Therefore, the second free-moment formula of (9.12) holds in $\mathfrak{S}_A$.

Assume now that $p_1 = p_2 = ... = p_N = p$ in $\mathcal{P}$, for some $p \in \mathcal{P}$. Then, $\beta_{K_l}^p$ are contained in the same free block $\mathfrak{S}_p$ (regarded as the minimal free summand of $\mathfrak{S}$ containing it) of $\mathfrak{S}$, and hence,

$$T \in \mathfrak{S}_p \text{ in } \mathfrak{S},$$

satisfying

$$T^{(n)} = T^n \in \mathfrak{S}_p \text{ in } \mathfrak{S}, \text{ for all } n \in \mathbb{N}.$$

Thus, by (9.11), one has

$$\tau_A\left((a \otimes T)^{(n)}\right) = \tau_A\left((a \otimes T)^n\right),$$

for all $n \in \mathbb{N}$.

Observe that

$$\tau_A\left((a \otimes T)^n\right) = \tau_A\left(a^n \otimes T^n\right)$$
$$= \tau(T^n)\ \varphi(a^n) = \tau_p\left(T^n\right)\ \varphi(a^n)$$

by (9.9), (9.10) and (8.4)

$$= \varphi(a^n)\ \tau_p\left(\frac{\prod_{l=2}^{N}\mu_p(K_{1,\ldots,l}^o)}{\prod_{l=1}^{N}\mu_p(K_l)}\ \alpha_{K_{1,\ldots,N}}^p\right)$$

by Lemma 9.1

$$= \varphi(a^n)\left(\frac{\prod_{l=2}^{N}\mu_p(K_{1,\ldots,l}^o)}{\prod_{l=1}^{N}\mu_p(K_l)}\ \frac{\mu_p(K_{1,\ldots,N} \cap Z_p)}{\mu_p(K_{1,\ldots,N})}\right)$$

by (8.9)

$$= \varphi(a^n)\left(\frac{\left(\prod_{l=2}^{N}\mu_p(K_{1,\ldots,l}^o)\right)(\mu_p(K_{1,\ldots,N} \cap Z_p))}{\left(\prod_{l=1}^{N}\mu_p(K_l)\right)(\mu_p(K_{1,\ldots,N}))}\right),$$

for all $n \in \mathbb{N}$. Therefore, the free-moment formula (9.13) holds true in $\mathfrak{S}_A$. □

Independently, one also obtains the following free-probabilistic information.

THEOREM 9.3. *Let $K \in \mathcal{K}_p$, for $p \in \mathcal{P}$, and let $\beta_K^p$ be the corresponding projection in the free block $\mathfrak{S}_p$ of $\mathfrak{S}$. For $a \in (A, \varphi)$, let*

$$T = a \otimes \beta_K^p$$

*be a free random variable in the A-sub-Hecke $C^*$-probability space $\mathfrak{S}_A$. Then*

(9.14)

$$\tau_A\left(T^n\right) = \varphi(a^n)\left(\frac{\mu_p(K \cap Z_p)}{\mu_p(K)}\right) = \tau_A(T^{(n)}),$$

*for all $n \in \mathbb{N}$.*

PROOF. Let $T$ be given as above in $\mathfrak{S}_A$. Then

$$T^n = a^n \otimes T^n = a^n \otimes T^{(n)} = T^{(n)},$$

for all $n \in \mathbb{N}$, by (9.11).

Thus, we have

$$\tau_A(T^n) = \tau\left(\left(\beta_K^p\right)^n\right)\varphi(a^n) = \tau_p\left(\left(\beta_K^p\right)^n\right)\varphi(a^n)$$
$$= \varphi(a^n)\ \tau_p\left(\beta_K^p\right) = \varphi(a^n)\left(\frac{\mu_p(K \cap Z_p)}{\mu_p(K)}\right),$$

by (8.11), for all $n \in \mathbb{N}$. Therefore, the free-distributional data (9.14) holds in $\mathfrak{S}_A$. □

By (9.12), (9.14), (8.20) and (8.22), the following estimations are verified.

THEOREM 9.4. *Let $K \in \mathcal{K}_p$, for $p \in \mathcal{P}$, and $K_l \in \mathcal{K}_{p_l}$, for $p_l \in \mathcal{P}$, for $l = 1, ..., N$, where $N \in \mathbb{N} - \{1\}$, and let $\beta_K^p$, $\beta_{K_l}^{p_l}$ be corresponding projections in the sub-Hecke $C^*$-probability space $\mathfrak{S}$ of $\mathbb{H}_{\mathcal{P}}$, for $l = 1, ..., N$. For a fixed $C^*$-probability space $(A, \varphi)$, let*

*(9.15)*

$$T_K^p = a \otimes \beta_K^p, \text{ and } T = a \otimes \left( \prod_{l=1}^{N} \beta_{K_l}^{p_l} \right)$$

*be free random variables in the A-sub-Hecke $C^*$-probability space $\mathfrak{S}_A$. If $T_K^p$ is as in (9.15), then*

*(9.16)*

$$0 \leq \left| \tau_A \left( (T_K^p)^n \right) \right| = \left| \tau_A \left( (T_K^p)^{(n)} \right) \right| \leq |\varphi(a^n)|,$$

*in $\mathfrak{S}_A$, for all $n \in \mathbb{N}$.*

*If $(p_1, ..., p_N) \in alt\left(\mathcal{P}^N\right)$ and if $T$ is in the sense of (9.15), then*

*(9.17)*

$$0 \leq |\tau_A (T^n)| \leq |\varphi(a^n)|,$$

*and*

$$0 \leq \left|\tau_A(T^{(n)})\right| \leq |\varphi(a^n)|,$$

*for all $n \in \mathbb{N}$.*

PROOF. Recall that,

$$0 < \tau \left( (\beta_K^p)^n \right) = \tau \left( (\beta_K^p)^{(n)} \right) \leq 1,$$

by (8.20), and if $S = \prod_{l=1}^{N} \beta_{K_l}^{p_l}$ is a free reduced word in $\mathfrak{S}$, then

$$0 < \tau (S^n) \leq 1, \text{ and } 0 < \tau(S^{(n)}) \leq 1,$$

by (8.22), for all $n \in \mathbb{N}$. Thus, if $T_K^p$ is in the sense of (9.15) in $\mathfrak{S}_A$, then

$$0 \leq \left| \tau_A \left( (T_K^p)^n \right) \right| = \left| \varphi(a^n) \tau \left( (\beta_K^p)^n \right) \right| = |\varphi(a^n)| \left| \tau \left( (\beta_K^p)^{(n)} \right) \right| \leq |\varphi(a^n)|,$$

by (9.14), for all $n \in \mathbb{N}$.

Similarly, if $T$ is in the sense of (9.15) in $\mathfrak{S}_A$, then, for the free reduced word $S = \prod_{l=1}^{N} \beta_{K_l}^{p_l}$ of $\mathfrak{S}$,

$$0 \leq |\tau_A(T^n)| = |\varphi(a^n)\tau(S^n)| \leq |\varphi(a^n)|,$$

and

$$0 \leq \left|\tau_A(T^{(n)})\right| = \left|\varphi(a^n)\tau(S^{(n)})\right| \leq |\varphi(a^n)|,$$

by (9.12) and (9.13), for all $n \in \mathbb{N}$. So, the estimations (9.16) and (9.17) holds. □

The above theorem shows that the free distributions of free random variables $T_K^p$, or $T$, in the sense of (9.15), are bounded by those of $a$ in $(A, \varphi)$, by (9.16) and (9.17). Therefore, one can obtain the following maximality of the estimations (9.16) and (9.17).

THEOREM 9.5. *Let $K \in \mathcal{K}_p$, for $p \in \mathcal{P}$, and $K_l \in \mathcal{K}_{p_l}$, for $p_l \in \mathcal{P}$, for $l = 1, \ldots,$ $N$, where $N \in \mathbb{N} - \{1\}$, and let $\beta_K^p$, $\beta_{K_l}^{p_l}$ be free random variables in the sub-Hecke $C^*$-probability space $\mathfrak{S}$ of $\mathbb{H}_{\mathcal{P}}$, for $l = 1, \ldots, N$. For a fixed $C^*$-probability space $(A,$ $\varphi)$, let*

$$T_K^p = a \otimes \beta_K^p, \text{ and } T = a \otimes \left( \prod_{l=1}^{N} \beta_{K_l}^{p_l} \right)$$

*be free random variables (9.15) of the $A$-sub-Hecke $C^*$-probability space $\mathfrak{S}_A$.*

*(9.18) If $\chi_K$ and $\chi_{Z_p}$ are free in the $p$-prime normal-core Hecke probability space $(\mathcal{H}(G_p), \psi_p)$, then*

$$\left|\tau_A\left((T_K^p)^n\right)\right| = \left|\tau_A\left((T_K^p)^{(n)}\right)\right| = |\varphi(a^n)|,$$

*for all $n \in \mathbb{N}$.*

*(9.19) Suppose $(p_1, \ldots, p_N)$ is an alternating $N$-tuple in $\mathcal{P}$. If $\chi_{K_l}$ and $\chi_{Z_{p_l}}$ are free in the $p_l$-prime normal-core Hecke probability spaces $(\mathcal{H}(G_{p_l}), \psi_{p_l})$, for all $l = 1, \ldots, N$, then*

$$\left|\tau_A\left(T^n\right)\right| = |\varphi(a^n)| = \left|\tau_A(T^{(n)})\right|,$$

*for all $n \in \mathbb{N}$.*

PROOF. The proofs of (9.18) and (9.19) are done by (9.16), (9.17), (8.24), (8.25), (8.29) and (8.30), with help of freeness characterizations (4.16) and (4.17). □

The very above two theorems show that the free distributions of the operators $T_K^p$ and $T$ of (9.15) in the $A$-sub-Hecke $C^*$-probability space $\mathfrak{S}_A$ are bounded by those of $a$ in a fixed $C^*$-probability space $(A, \varphi)$, and they can be maximal under the freeness on $p$-prime normal-core Hecke probability spaces $(\mathcal{H}(G_p), \psi_p)$, for $p \in \mathcal{P}$.

## 10. Applications

In this last section, we apply the main results of Sections 7, 8, and 9. Throughout this section, we let $\mathfrak{S}$ be the sub-Hecke $C^*$-probability space of the Adelic Hecke $C^*$-probability space $\mathbb{H}_{\mathcal{P}}$. We will use same notations and terminologies used in the above sections.

**10.1. Cases where $(A, \varphi) = (C^*(G), \tau_G)$.** Throughout this section, let's fix a *countable discrete group* $G$. For a fixed group $G$, one can construct the *group Hilbert space* $H_G$ of $G$ by

(10.1.1)

$$H_G = l^2(G), \text{ the } l^2\text{-space of } G,$$

equipped with its *orthonormal basis* $\mathcal{B}_G$,

$$\mathcal{B}_G = \{\eta_g : g \in G - \{e_G\}\},$$

where $e_G$ is the *group-identity of* $G$. The $l^2$-*inner product* $<,>_G$ on $H_G$ is canonically defined by the orthogonality, i.e.,

(10.1.2)

$$< \eta_{g_1}, \eta_{g_2} >_G = \delta_{g_1, g_2},$$

for all $g_1, g_2 \in G - \{e_G\}$.

Different from the usual cases, on the orthonormal basis $\mathcal{B}_G$ of (10.1.1), one has the natural multiplication rule in $H_G$;

$$\eta_{g_1}\eta_{g_2} = \eta_{g_1 g_2}, \forall g_1, g_2 \in G - \{e_G\},$$

moreover, for any $g_1, g_2 \in G$, the above multiplication rule holds, under axiomatization:

(10.1.3)

$$\eta_{e_G} = 1_G, \text{ the identity element in } H_G,$$

satisfying

$$1_G \, \eta = \eta = \eta \, 1_G, \text{ for all } \eta \in H_G.$$

i.e., we obtain that the multiplication rule on $H_G$,

(10.1.4)

$$\eta_{g_1} \, \eta_{g_2} = \eta_{g_1 g_2}, \text{ for all } g_1, g_2 \in G,$$

by (10.1.3), because $\eta_g \eta_{g^{-1}} = \eta_{g^{-1}} \eta_g = 1_G$ in $H_G$.

By the definition (10.1.1) of the group Hilbert space $H_G$ of $G$, the group $G$ is acting on $H_G$ as *multiplication operators*;

(10.1.5)

$$u : g \in G \longmapsto u_g \in B(H_G),$$

such that

$$u_g \left(\eta_{g'}\right) = \eta_g \eta_{g'} = \eta_{gg'}, \text{ for all } g' \in G,$$

by (10.1.4).

Note that such operators $\{u_g\}_{g \in G}$ under the action (10.1.5) satisfy

(10.1.6)

$$u_{g_1} u_{g_2} = u_{g_1 g_2}, \text{ and } u_g^* = u_{g^{-1}},$$

for all $g_1, g_2, g \in G$, where $g^{-1}$ are the *group-inverses of* $g$ in $G$.

By (10.1.6), one can get that

(10.1.7)

$$u_g^* u_g = u_{g^{-1}g} = u_{e_G} = 1_{H_G} = u_{gg^{-1}} = u_g u_g^*,$$

on $H_G$, where $1_{H_G}$ means the *identity operator on* $H_G$ in the operator algebra $B(H_G)$. Equivalently, each multiplication operator $u_g$ is a *unitary on* $H_G$, satisfying

$$u_g^* = u_g^{-1} = u_{g^{-1}}, \text{ by (10.1.7)},$$

where $u_g^{-1}$ means the *operator-inverse of* $u_g$ in $B(H_G)$.

So, the pair $(H_G, u)$ of the group Hilbert space $H_G$ and the group-action $u$ of (10.1.5) forms a well-determined Hilbert-space representation of the group $G$. This representation is said to be the *left-regular unitary representation of* $G$.

On the representation $(H_G, u)$, define a $C^*$-subalgebra $A_G$ of $B(H_G)$ by
(10.1.8)
$$A_G = C^*(G) = \overline{\mathbb{C}[u(G)]},$$
where $\overline{X}$ mean the operator-norm-topology closures of subsets $X$ of $B(H_G)$. This $C^*$-subalgebra $A_G$ of $B(H_G)$ is called the *group $C^*$-algebra of $G$.*

Remark that every element $T$ of the group $C^*$-algebra $A_G$ of (10.1.8) is expressed by
$$T = \sum_{g \in G} t_g\, u_g, \text{ with } t_g \in \mathbb{C},$$
where $\sum$ is a finite or an infinite (limits of finite) sum(s) under the $C^*$-topology of $A_G$.

On the group $C^*$-algebra $A_G$ of $G$, define a *trace* $\tau_G$ by
(10.1.9)
$$\tau_G\left(\sum_{g \in G} t_g\, u_g\right) = t_{e_G},$$
which is indeed a well-defined bounded linear functional on $A_G$, satisfying
$$\tau_G(T_1T_2) = \tau_G(T_2T_1), \text{ for all } T_1,\, T_2 \in A_G.$$
Such a linear functional $\tau_G$ of (10.1.9) is said to be the *canonical trace on $A_G$.*

DEFINITION 10.1. *Let $A_G$ be the group $C^*$-algebra (10.1.8), and $\tau_G$, the canonical trace (10.1.9) on $A_G$. The $C^*$-probability space $(A_G, \tau_G)$ is called the group $C^*$-probability space of a countable discrete group $G$.*

Let $(A_G, \tau_G)$ be the group $C^*$-probability space of a group $G$, and let $\mathfrak{S}$ be the sub-Hecke $C^*$-probability space of the Adelic Hecke $C^*$-probability space $\mathbb{H}_{\mathcal{P}}$. Let
(10.1.10)
$$\mathfrak{S}_G \overset{denote}{=} (\mathfrak{S}_{A_G},\, \tau_{A_G})$$
be the $A_G$-sub-Hecke $C^*$-probability space in the sense of (9.8).

PROPOSITION 10.1. *Let $T = u_g \otimes \beta_K^p$ be a free random variable of the $A_G$-sub-Hecke $C^*$-probability space $\mathfrak{S}_G$ of (10.1.10), where $g \in G$, and $\beta_K^p$ is in the sense of (8.9) in $\mathfrak{S}$. Then*
*(10.1.11)*
$$\tau_{A_G}(T^n) = \tau_{A_G}\left(T^{(n)}\right) = \delta_{e_G, g^n}\left(\tfrac{\mu_p(K \cap Z_p)}{\mu_p(K)}\right),$$
*for all $n \in \mathbb{N}$.*

PROOF. Observe that
$$\tau_{A_G}(T^n) = \tau_{A_G}(T^{(n)}) = \tau_{A_G}\left(u_{g^n} \otimes (\beta_K^p)^n\right)$$
$$= \tau_G(u_{g^n})\, \tau\left((\beta_K^p)^n\right) = \delta_{e_G, g^n}\, \tau_p(\beta_K^p)$$
$$= \delta_{e_G, g^n}\left(\tfrac{\mu_p(K \cap Z_p)}{\mu_p(K)}\right),$$
by (10.1.4), for all $n \in \mathbb{N}$. So, the free-distributional data (10.1.11) holds in $\mathfrak{S}_G$. □

Similarly, one obtains the following free-distributional data on $\mathfrak{S}_G$ of (10.1.10).

PROPOSITION 10.2. *Let $g \in G$, and $u_g \in (A_G, \tau_G)$, and let $\beta_{K_l}^{p_l}$ be in the sense of (8.9) in $\mathfrak{S}$, for $l = 1, ..., N$, for $N \in \mathbb{N} - \{1\}$. Suppose*

$$S = \prod_{l=1}^{N} \beta_{K_l}^{p_l} \in \mathfrak{S}.$$

*If $(p_1, ..., p_N) \in alt\left(\mathcal{P}^N\right)$, and if $T = u_g \otimes S \in \mathfrak{S}_G$, then*
*(10.1.12)*

$$\tau_{A_G}\left(T^n\right) = \delta_{e_G, g^n}\left(\prod_{l=1}^{N} \frac{\mu_{p_l}(K_l \cap Z_{p_l})}{\mu_{p_l}(K_l)}\right),$$

*and*

$$\tau_{A_G}\left(T^{(n)}\right) = \delta_{e_G, g^n}\left(\prod_{l=1}^{N}\left(\frac{\mu_{p_l}(K_l \cap Z_{p_l})}{\mu_{p_l}(K_l)}\right)^n\right),$$

*for all $n \in \mathbb{N}$, under (9.11), (9.11)′ and (9.11)″.*

PROOF. Let $u_g \in (A_G, \tau_G)$ and $T \in \mathfrak{S}$ be given as above inducing a free random variable $S$ in the $A_G$-sub-Hecke $C^*$-probability space $\mathfrak{S}_G$ of (10.1.10). Then

$$\tau_{A_G}\left(S^n\right) = \tau_G\left(u_{g^n}\right)\tau\left(T^n\right),$$

and

$$\tau_{A_G}\left(S^{(n)}\right) = \tau_G\left(u_{g^n}\right)\tau(T^{(n)}),$$

for all $n \in \mathbb{N}$.

So, the free-distributional data (10.1.12) is obtained. □

The above free distributions (10.1.11) and (10.1.12) illustrate that if $u_g \otimes T$ is a free random variable in the $A_G$-sub-Hecke $C^*$-probability space $\mathfrak{S}_G$ of (10.1.10), for $g \in G$, and $T \in \mathfrak{S}$, then the free distribution

$$\left\{\tau_{A_G}\left((u_g \otimes T)^n\right)\right\}_{n=1}^{\infty}, \text{ or } \left\{\tau_{A_G}\left((u_g \otimes T)^{(n)}\right)\right\}_{n=1}^{\infty}$$

of $u_g \otimes T$ is *non-vanishing* in the sense that:
(10.1.13)

$$\tau_{A_G}\left((u_g \otimes T)^n\right) \neq 0, \text{ for some } n \in \mathbb{N},$$

respectively,

$$\tau_{A_G}\left((u_g \otimes T)^{(n)}\right) \neq 0, \text{ for some } n \in \mathbb{N},$$

if and only if there exists the minimal quantity $n_0 \in \mathbb{N}$, such that

$$g^{n_0} = e_G \text{ in } G,$$

i.e., either $g = e_G$ in $G$, or there is a *relator on* $G$, saying $g^{n_0} = e_G$.

THEOREM 10.3. *Let $(A_G, \tau_G)$ be the group $C^*$-probability space of a countable discrete group $G$, and $\mathfrak{S}$, the sub-Hecke $C^*$-probability space of the Adelic Hecke $C^*$-probability space $\mathbb{H}_{\mathcal{P}}$, and let $\mathfrak{S}_G$ be the $A_G$-sub-Hecke $C^*$-probability space (10.1.10). For any $g \in G$, the operator $u_g \otimes T$ of $\mathfrak{S}_G$ has its non-vanishing free distribution, if and only if either*

$$g = e_G \text{ in } G,$$

*or there is an relator on $G$, guaranteeing the existence of the minimal quantity $n_0 \in \mathbb{N}$, satisfying*

$$g^{n_0} = e_G \text{ in } G.$$

PROOF. ($\Leftarrow$) If either $g = e_G$ in $G$, or there exists the minimal quantity $n_0 \in \mathbb{N}$, such that

$$g^{n_0} = e_G \text{ (and hence } g^{kn_0} = e_G, \forall k \in \mathbb{Z}),$$

then, by (10.1.11) and (10.1.12), the operators $u_g \otimes T$ have non-vanishing free distributions in the sense of (10.1.13), for $g \in G$, for all $T \in \mathfrak{S}$.

($\Rightarrow$) Suppose now $T_g = u_g \otimes T$ has its non-vanishing free distribution in $\mathfrak{S}_G$ in the sense of (10.1.13), and assume that neither

$$\text{“}g = e_G,\text{” nor “}g^n = e_G, \text{ for some } n \in \mathbb{N}.\text{”}$$

Then, by (10.1.11) and (10.1.12),

$$\tau_{A_G}\left(T_g^n\right) = 0 = \tau_{A_G}\left(T_g^{(n)}\right), \text{ for “all” } n \in \mathbb{N}.$$

It contradicts our assumption that $T_g$ has its non-vanishing free distribution. □

Now, fix $g \in G$, and corresponding self-adjoint operator $x_g$,
(10.1.14)

$$x_g = u_g + u_g^* = u_g + u_{g^{-1}}$$

in $(A_G, \tau_G)$.
Since $\tau_G$ is a trace on $A_G$, one can get that

$$\tau_G\left(x_g^n\right) = \tau_G\left((u_g + u_{g^{-1}})^n\right)$$

$$= \sum_{k=0}^{n} \binom{n}{k} \tau_G\left(u_g^k u_{g^{-1}}^{n-k}\right)$$

$$= \sum_{k=0}^{n} \binom{n}{k} \tau_G\left(u_{g^k} u_{g^{-n+k}}\right)$$

(10.1.15)

$$= \sum_{k=0}^{n} \binom{n}{k} \tau_G\left(u_{g^{2k-n}}\right),$$

for all $n \in \mathbb{N}$, where

$$\binom{n}{k} = \frac{n!}{k!(n-k)!}, \text{ for all } k \leq n \in \mathbb{N}.$$

PROPOSITION 10.4. *Let $x_g = u_g + u_{g^{-1}}$ be in the sense of (10.1.14) in $(A_G, \tau_G)$, and let $\beta_K^p$ be in the sense of (8.9) in $\mathfrak{S}$. Now, let $T = x_g \otimes \beta_K^p$ in the $A_G$-sub-Hecke $C^*$-probability space $\mathfrak{S}_G$ of (10.1.10). Suppose $g \in G$ satisfies that*

$$g^n \neq e_G \text{ in } G, \text{ for all } n \in \mathbb{N}.$$

*Then*
*(10.1.16)*

$$\tau_{A_G}\left(T^n\right) = \tau_{A_G}\left(T^{(n)}\right) = \omega_n \begin{pmatrix} n \\ \frac{n}{2} \end{pmatrix} \left(\frac{\mu_p(K \cap Z_p)}{\mu_p(K)}\right)$$

*where*

$$\omega_n = \begin{cases} 1 & \textit{if } n \textit{ is even} \\ 0 & \textit{if } n \textit{ is odd,} \end{cases}$$

*for all* $n \in \mathbb{N}$.

PROOF. Under hypothesis,

$$\tau_{A_G}\left(T^n\right) = \tau_{A_G}\left(T^{(n)}\right) = \tau_G\left(x_g^n\right)\tau\left(\left(\beta_K^p\right)^n\right)$$

$$= \left(\sum_{k=0}^{n} \begin{pmatrix} n \\ k \end{pmatrix} \tau_G\left(u_{g^{2k-n}}\right)\right)\left(\frac{\mu_p(K \cap Z_p)}{\mu_p(K)}\right)$$

$$= \left(\sum_{k=0}^{n} \begin{pmatrix} n \\ k \end{pmatrix} \delta_{e_G, g^{2k-n}}\right)\left(\frac{\mu_p(K \cap Z_p)}{\mu_p(K)}\right)$$

$$= \left(\sum_{k=0}^{n} \begin{pmatrix} n \\ k \end{pmatrix} \delta_{0, 2k-n}\right)\left(\frac{\mu_p(K \cap Z_p)}{\mu_p(K)}\right)$$

$$= \omega_n \begin{pmatrix} n \\ \frac{n}{2} \end{pmatrix} \left(\frac{\mu_p(K \cap Z_p)}{\mu_p(K)}\right),$$

where

$$\omega_n = \begin{cases} 1 & \text{if } n \text{ is even} \\ 0 & \text{if } n \text{ is odd,} \end{cases}$$

by (10.1.15), for all $n \in \mathbb{N}$.

Remark here that the above condition on $\omega_n$ is satisfied because of the assumption that $g^n \neq e_G$ in $G$, for all $n \in \mathbb{N}$. □

So, one can get the following result, too.

PROPOSITION 10.5. *Let* $x_g$ *be in the sense of (10.1.14) in* $(A_G, \tau_G)$, *and let* $\beta_{K_l}^{p_l}$ *be in the sense of (8.9) in* $\mathfrak{S}$, *and let*

$$T = x_g \otimes \left(\prod_{l=1}^{N} \beta_{K_l}^{p_l}\right)$$

*be an element of* $\mathfrak{S}_G$. *Suppose* $(p_1, \ldots, p_N)$ *is alternating in* $\mathcal{P}$, *and assume that*

$$g^n \neq e_G \textit{ in } G, \textit{ for all } n \in \mathbb{N}.$$

*Then we have that*
*(10.1.17)*

$$\tau_{A_G}\left(T^n\right) = \omega_n \begin{pmatrix} n \\ \\ \frac{n}{2} \end{pmatrix} \left( \prod_{l=1}^{N} \frac{\mu_{p_l}(K_l \cap Z_{p_l})}{\mu_{p_l}(K_l)} \right),$$

*and*

$$\tau_{A_G}\left(T^{(n)}\right) = \omega_n \begin{pmatrix} n \\ \\ \frac{n}{2} \end{pmatrix} \left( \prod_{l=1}^{N} \left( \frac{\mu_{p_l}(K_l \cap Z_{p_l})}{\mu_{p_l}(K_l)} \right)^n \right),$$

*for all $n \in \mathbb{N}$, where $\omega_n$ are in the sense of (10.1.16).*

PROOF. By the assumption that $(p_1, ..., p_N) \in alt\left(\mathcal{P}^N\right)$, the operator $\prod_{l=1}^{N} \beta_{K_l}^{p_l}$ forms a free reduced word in $\mathfrak{S}$. And, since $g^n \neq e_G$, for all $n \in \mathbb{N}$,

$$\tau_G\left(x_g^n\right) = \omega_n \begin{pmatrix} n \\ \\ \frac{n}{2} \end{pmatrix}, \text{ for all } n \in \mathbb{N},$$

by (10.1.15).

So, we obtain the free-distributional data (10.1.17) by (9.12). ☐

Therefore, the following estimations are obtained.

PROPOSITION 10.6. *Let $g \in G$, and $u_g$, $x_g \in (A_G, \tau_G)$, the operators as above, and let $\beta_K^p$, $\beta_{K_l}^{p_l}$ be in the sense of (8.9) in the sub-Hecke $C^*$-probability space $\mathfrak{S}$ of the Adelic Hecke $C^*$-probability space $\mathbb{H}_{\mathcal{P}}$, for $l = 1, ..., N$, for $N \in \mathbb{N} - \{1\}$. Suppose*

$$S_1 = u_g \otimes \beta_K^p, \text{ and } S_2 = u_g \otimes \left( \prod_{l=1}^{N} \beta_{K_l}^{p_l} \right),$$

*and*

$$T_1 = x_g \otimes \beta_K^p, \text{ and } T_2 = x_g \otimes \left( \prod_{l=1}^{N} \beta_{K_l}^{p_l} \right),$$

*in the $A_G$-sub-Hecke $C^*$-probability space $\mathfrak{S}_G$ of (10.1.10). Then*
*(10.1.18)*

$$0 \leq \left|\tau_{A_G}\left(S_1^n\right)\right| = \left|\tau_{A_G}\left(S_1^{(n)}\right)\right| \leq \delta_{e_G, g^n},$$

*and*

$$0 \leq \left|\tau_{A_G}\left(S_2^n\right)\right| \leq \delta_{e_G, g^n},$$

*and*

$$0 \leq \left|\tau_{A_G}\left(S_2^{(n)}\right)\right| \leq \delta_{e_G, g^n},$$

*and*

$$0 \leq \left|\tau_{A_G}\left(T_1^n\right)\right| = \left|\tau_{A_G}\left(T_1^{(n)}\right)\right| \leq \omega_n \begin{pmatrix} n \\ \\ \frac{n}{2} \end{pmatrix},$$

*and*

$$0 \leq \left|\tau_{A_G}\left(T_2^n\right)\right| \leq \omega_n \begin{pmatrix} n \\ \\ \frac{n}{2} \end{pmatrix},$$

*and*

$$0 \leq \left|\tau_{A_G}\left(T_2^{(n)}\right)\right| \leq \omega_n \begin{pmatrix} n \\ \frac{n}{2} \end{pmatrix},$$

*for all* $n \in \mathbb{N}$.

PROOF. The proofs of the estimations in (10.1.18) are done by (9.16) and (9.17). ☐

**10.2. Cases where** $(A, \varphi) = (A_X, \varphi_X)$. In this section, we let $X = (X, \sigma(X), \mu_X)$ be a *bounded measure space*, inducing a natural Hilbert space $H_X$,

(10.2.1)

$$H_X = L^2(X, \mu_X) = L^2(X),$$

the $L^2$-*space* consisting of all *square-integrable functions on* $X$, equipped with its inner product $<,>_X$,

$$< h_1, h_2 >_X = \int_X h_1 \, h_2^* \, d\mu_X,$$

for all $h_1, h_2 \in H_X$.

Recall that all Hilbert-space vectors $h \in H_X$ have their expressions,

$$h = \sum_{S \in \sigma(X)} t_S \, \chi_S, \text{ with } t_S \in \mathbb{C},$$

where $\sum$ is a finite or infinite (limits of finite) sum(s) under $L^2$-topology for $H_X$ of (10.2.1).

Let $M_X$ be the (pure-algebraic) $*$-algebra consisting of all measurable functions generated by

$$\{\chi_S : S \in \sigma(X)\}.$$

Then all elements $f \in M_X$ act on $H_X$ as *multiplication operators*, i.e., there exists an *action* $\pi$ *of* $M_X$,

(10.2.2)

$$\pi : f \in M_X \longmapsto \pi_f \in B(H_X),$$

such that

$$\pi_f(h) = fh \in H_X, \text{ for all } h \in H_X.$$

The linear morphism (10.2.2) is indeed a well-determined $*$-algebra action of $M_X$ acting on the Hilbert space $H_X$ of (10.2.1), since

(10.2.3)

$$\pi_{f_1 f_2} = \pi_{f_2}\pi_{f_2} \text{ on } H_X,$$

and

$$\pi_f^* = \pi_{f^*} \text{ on } H_X,$$

for all $f_1, f_2, f \in M_X$.

Thus, the pair $(H_X, \pi)$ of the Hilbert space $H_X$ of (10.2.1), and the action $\pi$ of (10.2.2) forms a well-defined Hilbert-space representation of the $*$-algebra $M_X$, by (10.2.3). So, one can define the operator-norm closure $A_X$ of $M_X$ in the operator algebra $B(H_X)$,

(10.2.4)

$$A_X = \overline{\mathbb{C}\left[\pi(M_X)\right]} \text{ in } B(H_X).$$

Remark that all elements $\pi_f \in A_X$ has their expressions,

$$\pi_f = \sum_{S \in \sigma(X)} t_S \, \pi_S,$$

with $t_S \in \mathbb{C}$, and $\pi_S = \pi_{\chi_S}$, for all $S \in \sigma(X)$, where $\sum$ is a finite or an infinite (limits of finite) sum(s) under the $C^*$-topology on $A_X$ of (10.2.4).

Let $A_X$ be the $C^*$-algebra (10.2.4) induced by a fixed measure space $X$. Define a linear functional $\varphi_X$ on $A_X$ by the bounded linear transformation satisfying

(10.2.5)

$$\varphi_X\,(\pi_S) = \int_X \chi_S \, d\mu_X = \mu_X(S),$$

for all $S \in \sigma(X)$, where $\pi_S = \pi_{\chi_S}$, as above.

Then the pair $(A_X, \, \varphi_X)$ is a well-defined $C^*$-probability space.

DEFINITION 10.2. *The $C^*$-probability space $(A_X, \varphi_X)$ of the $C^*$-algebra $A_X$ induced by a bounded measure space $X$, and the linear functional $\varphi_X$ of (10.2.5) on $A_X$ is called the (measure-space-)$C^*$-probability space of $X$.*

Let $(A_X, \varphi_X)$ be the $C^*$-probability space of a measure space $X$, and let

(10.2.6)

$$\mathfrak{S}_X = (\mathfrak{S}_{A_X}, \, \tau_{A_X})$$

be the $A_X$-sub-Hecke $C^*$-probability space in the sense of (9.5).

On the $C^*$-probability space $\mathfrak{S}_X$ of (10.2.6), one can get the following free-distributional data.

PROPOSITION 10.7. *Let $\pi_S \in (A_X, \varphi_X)$, where $S \in \sigma(X)$, and let $\beta_K^p$, $\beta_{K_l}^{p_l}$ be in the sense of (8.5) in the sub-Hecke $C^*$-probability space $\mathfrak{S}$ of the Adelic Hecke $C^*$-probability space $\mathbb{H}_{\mathcal{P}}$, for $l = 1, \ldots, N$, for $N \in \mathbb{N} - \{1\}$. Also, let*

$$T_1 = \pi_S \otimes \beta_K^p, \text{ and } T_2 = \pi_S \otimes \left(\prod_{l=1}^{N} \beta_{K_l}^{p_l}\right)$$

*be operators in the $A_X$-sub-Hecke $C^*$-probability space $\mathfrak{S}_X$ of (10.2.6). Then*

*(10.2.7)*

$$\tau_{A_X}\,(T_1^n) = \tau_{A_X}\left(T_1^{(n)}\right) = \mu_X(S)\,\left(\tfrac{\mu_p(K \cap \mathbb{Z}_p)}{\mu_p(K)}\right),$$

*for all $n \in \mathbb{N}$.*

*Suppose now that $(p_1, \ldots, p_N)$ is alternating in $\mathcal{P}$, and assume further that $p_N \neq p_1$ in $\mathcal{P}$. Then*

*(10.2.8)*

$$\tau_{A_X}\,(T_2^n) = \mu_X(S)\left(\prod_{l=1}^{N} \tfrac{\mu_{p_l}(K_l \cap \mathbb{Z}_{p_l})}{\mu_{p_l}(K_l)}\right),$$

*and*

$$\tau_{A_X}\left(T_2^{(n)}\right) = \mu_X(S)\left(\prod_{l=1}^{N}\left(\tfrac{\mu_{p_l}(K_l \cap \mathbb{Z}_{p_l})}{\mu_{p_l}(K_l)}\right)^n\right),$$

*for all $n \in \mathbb{N}$.*

PROOF. Observe that

$$\varphi_X\left(\pi_S^n\right) = \varphi_X\left(\pi_{\chi_S^n}\right) = \varphi_X\left(\pi_{\chi_{S\cap S\cap \ldots \cap S}}\right)$$

$$= \varphi_X\left(\pi_S\right) = \int_X \chi_S d\mu_X = \mu_X(S),$$

by (10.2.5), for all $n \in \mathbb{N}$, for all $S \in \sigma(X)$.

Thus, by (9.14), we obtain the free-distributional data (10.2.7); and, by (9.12), one can get the free-moment formulas in (10.2.8). □

Therefore, one automatically obtains the following estimations.

PROPOSITION 10.8. *Let $\pi_S \in (A_X, \varphi_X)$, where $S \in \sigma(X)$, and let $\beta_K^p$, $\beta_{K_l}^{p_l}$ be in the sense of (8.5) in the sub-Hecke $C^*$-probability space $\mathfrak{S}$ of the Adelic Hecke $C^*$-probability space $\mathbb{H}_{\mathcal{P}}$, for $l = 1, \ldots, N$, for $N \in \mathbb{N} - \{1\}$. Also, let*

$$T_1 = \pi_S \otimes \beta_K^p, \text{ and } T_2 = \pi_S \otimes \left(\prod_{l=1}^{N} \beta_{K_l}^{p_l}\right)$$

*be operators in the $A_X$-sub-Hecke $C^*$-probability space $\mathfrak{S}_X$ of (10.2.6). Then*
*(10.2.9)*

$$0 \leq \tau_{A_X}\left(T_1^n\right) = \tau_{A_X}\left(T_1^{(n)}\right) \leq \mu_X(S),$$

*for all $n \in \mathbb{N}$.*

*Suppose now $(p_1, \ldots, p_N)$ is an alternating $N$-tuple in $\mathcal{P}$, and assume $p_N \neq p_1$ in $\mathcal{P}$. Then*
*(10.2.10)*

$$0 \leq \tau_{A_X}\left(T_2^n\right) \leq \mu_X(S),$$

*and*

$$0 \leq \tau_{A_X}\left(T_2^{(n)}\right) \leq \mu_X(S),$$

*for all $n \in \mathbb{N}$.*

PROOF. The estimations (10.2.9) and (10.2.10) are obtained by (10.2.7), (10.2.8), (9.16) and (9.17). □

## References


[1] Dean Alvis, *A class of representations of Hecke algebras*, Bull. Inst. Math. Acad. Sin. (N.S.) **11** (2016), no. 2, 301–342. MR3558782

[2] Susumu Ariki and Euiyong Park, *Representation type of finite quiver Hecke algebras of type $C_l^{(1)}$*, Osaka J. Math. **53** (2016), no. 2, 463–488. MR3492809

[3] M. Aschbacher, *Finite group theory*, 2nd ed., Cambridge Studies in Advanced Mathematics, vol. 10, Cambridge University Press, Cambridge, 2000. MR1777008

[4] Ole E. Barndorff-Nielsen and Steen Thorbjørnsen, *Classical and free infinite divisibility and Lévy processes*, Quantum independent increment processes. II, Lecture Notes in Math., vol. 1866, Springer, Berlin, 2006, pp. 33–159. MR2213448

[5] Ole E. Barndorff-Nielsen and Steen Thorbjørnsen, *Regularizing mappings of Lévy measures*, Stochastic Process. Appl. **116** (2006), no. 3, 423–446. MR2199557

[6] Michael Brannan and Kay Kirkpatrick, *Quantum groups and generalized circular elements*, Pacific J. Math. **282** (2016), no. 1, 35–61. MR3463424

[7] Alexander Braverman, David Kazhdan, and Manish M. Patnaik, *Iwahori-Hecke algebras for p-adic loop groups*, Invent. Math. **204** (2016), no. 2, 347–442. MR3489701

[8] J.-B. Bost and A. Connes, *Hecke algebras, type III factors and phase transitions with spontaneous symmetry breaking in number theory*, Selecta Math. (N.S.) **1** (1995), no. 3, 411–457. MR1366621

[9] Daniel Bump, *Automorphic forms and representations*, Cambridge Studies in Advanced Mathematics, vol. 55, Cambridge University Press, Cambridge, 1997. MR1431508

[10] Ilwoo Cho, *Operators induced by prime numbers*, Methods Appl. Anal. **19** (2012), no. 4, 313–339. MR3056525

[11] Ilwoo Cho, *p-adic Banach space operators and adelic Banach space operators*, Opuscula Math. **34** (2014), no. 1, 29–65. MR3162799

[12] Ilwoo Cho, *Free probability on Hecke algebras and certain group $C^*$-algebras induced by Hecke algebras*, Opuscula Math. **36** (2016), no. 2, 153–187. MR3437213

[13] Ilwoo Cho, *Free probability on $W^*$-dynamical systems determined by $GL_2(\mathbb{Q}_p)$: generalized Hecke algebras*, Boll. Unione Mat. Ital. **10** (2017), no. 4, 725–764. MR3736717

[14] Ilwoo Cho, *Representations and corresponding operators induced by Hecke algebras*, Complex Anal. Oper. Theory **10** (2016), no. 3, 437–477. MR3463809

[15] Ilwoo Cho and Timothy L. Gillespie, *Free probability on Hecke algebras*, Complex Anal. Oper. Theory **9** (2015), no. 7, 1491–1531. MR3395230

[16] Ilwoo Cho and Palle E. T. Jorgensen, *Krein-space operators induced by Dirichlet characters*, Commutative and noncommutative harmonic analysis and applications, Contemp. Math., vol. 603, Amer. Math. Soc., Providence, RI, 2013, pp. 3–33. MR3204024

[17] Kurusch Ebrahimi-Fard and Frédéric Patras, *The splitting process in free probability theory*, Int. Math. Res. Not. IMRN **9** (2016), 2647–2676. MR3519126

[18] T. Gillespie, Superposition of Zeroes of Automorphic $L$-Functions and Functoriality, Univ. of Iowa, (2010) PhD Thesis.

[19] Tim Gillespie and GuangHua Ji, *Prime number theorems for Rankin-Selberg L-functions over number fields*, Sci. China Math. **54** (2011), no. 1, 35–46. MR2764782

[20] David Hilbert, *Mathematical problems*, Bull. Amer. Math. Soc. **8** (1902), no. 10, 437–479. MR1557926

[21] Tobias Mai, Roland Speicher, and Moritz Weber, *Absence of algebraic relations and of zero divisors under the assumption of full non-microstates free entropy dimension*, Adv. Math. **304** (2017), 1080–1107. MR3558227

[22] Eric Opdam, *Spectral correspondences for affine Hecke algebras*, Adv. Math. **286** (2016), 912–957. MR3415700

[23] Florin Rădulescu, *Random matrices, amalgamated free products and subfactors of the von Neumann algebra of a free group, of noninteger index*, Invent. Math. **115** (1994), no. 2, 347–389. MR1258909

[24] Tobias Schmidt, *Hecke algebras and affine flag varieties in characteristic p*, J. Pure Appl. Algebra **220** (2016), no. 9, 3233–3247. MR3486299

[25] Roland Speicher, *Combinatorial theory of the free product with amalgamation and operator-valued free probability theory*, Mem. Amer. Math. Soc. **132** (1998), no. 627, x+88. MR1407898

[26] V. S. Vladimirov, I. V. Volovich, and E. I. Zelenov, *p-adic analysis and mathematical physics*, Series on Soviet and East European Mathematics, vol. 1, World Scientific Publishing Co., Inc., River Edge, NJ, 1994. MR1288093

[27] D. V. Voiculescu, K. J. Dykema, and A. Nica, *Free random variables*, CRM Monograph Series, vol. 1, American Mathematical Society, Providence, RI, 1992. A noncommutative probability approach to free products with applications to random matrices, operator algebras and harmonic analysis on free groups. MR1217253

[28] Dan-Virgil Voiculescu, *Free probability for pairs of faces III: 2-variables bi-free partial $S$- and $T$-transforms*, J. Funct. Anal. **270** (2016), no. 10, 3623–3638. MR3478868



Department of Mathematics & Statistics, Saint Ambrose University, 421 Ambrose Hall, 518 W. Locust Street, Davenport, Iowa 52803

*Email address*: `choilwoo@sau.edu`

Department of Mathematics, University of Iowa, 14 MacLean Hall, Iowa City, Iowa 52242

*Email address*: `palle-jorgensen@uiowa.edu`

Contemporary Mathematics
Volume **714**, 2018
https://doi.org/10.1090/conm/714/14377

# Atomic decompositions of mixed norm Bergman spaces on tube type domains

Jens Gerlach Christensen

ABSTRACT. We use the author's previous work on atomic decompositions of Besov spaces with spectrum on symmetric cones, to derive new atomic decompositions for Bergman spaces on tube type domains. It is related to work by Ricci and Taibleson who derived decompositions for classical Besov spaces from atomic decompositions of Bergman spaces on the upper half plane. Moreover, for this class of domains our method is an alternative to classical results by Coifman and Rochberg, and it works for a larger range of Bergman spaces.

## 1. Introduction

In this paper we suggest a new approach to atomic decompositions for the Bergman spaces on tube type domains. Atomic decompositions have previously been obtained for the upper half plane (unit disc) in [**8**, **19**, **20**], and for the unit ball [**7**, **21**] as well as other bounded symmetric domains [**10**, **17**]. These decompositions have typically been attained by investigating oscillations of the Bergman kernel and in most cases the atoms are samples of the Bergman kernel. In the case of the unit ball such oscillations can be estimated for all parameters, but for higher rank spaces it is more complicated. The issue is connected to the question of boundedness of the Bergman projection on bounded symmetric domains, which is still an open problem. Advances to answer this problem have recently been made in the case of tube type domains in [**1**, **11**] and for general domains in [**18**]. In the case of tube type domains over forward light cones the problem is now solved [**5**].

In this paper we will concentrate on tube type domains, and we use Fourier-Laplace extensions to transfer known atomic decompositions for Besov spaces [**6**] to Bergman spaces. This allows us to narrow the gap in the atomic decompositions from [**10**] in the case of tube type domains. Moreover, these decompositions are for mixed norm Bergman spaces which have not previously been dealt with. We would like to mention that this work seems to be in the reverse direction of the paper [**20**] which uses atomic decompositions for the mixed norm Bergman spaces on the upper half plane to get atomic decompositions for the Besov spaces. See also [**15**] for related work for Bergman spaces on the unit ball in relation to Besov spaces on the Heisenberg group.

---

2010 *Mathematics Subject Classification.* Primary 32A36, 41A65.

## 2. Symmetric cones

For an introduction to symmetric cones we refer to the book [**13**]. Let $V$ be a Euclidean vector space over the real numbers of finite dimension $n$. A subset $\Omega$ of $V$ is a cone if $\lambda\Omega \subseteq \Omega$ for all $\lambda > 0$. Assume $\Omega$ is open and convex, and define the open dual cone $\Omega^*$ by

$$\Omega^* = \{y \in V \mid (x,y) > 0 \text{ for all non-zero } x \in \overline{\Omega}\}.$$

The cone $\Omega$ is called symmetric if $\Omega = \Omega^*$, and the automorphism group

$$\mathcal{G}(\Omega) = \{g \in \mathrm{GL}(V) \mid g\Omega = \Omega\}$$

acts transitively on $\Omega$. Notice that the group $\mathcal{G}(\Omega)$ is semisimple. Define the characteristic function of $\Omega$ by

$$\varphi(x) = \int_{\Omega^*} e^{-(x,y)}\, dy,$$

then

$$\varphi(gx) = |\det(g)|^{-1}\varphi(x).$$

Also,

$$f \mapsto \int_\Omega f(x)\varphi(x)\, dx \tag{1}$$

defines a $\mathcal{G}(\Omega)$-invariant measure on $\Omega$. The connected component $\mathcal{G}_0(\Omega)$ of $\mathcal{G}(\Omega)$ has Iwasawa decomposition

$$\mathcal{G}_0(\Omega) = \mathcal{KAN}$$

where $\mathcal{K} = \mathcal{G}_0(\Omega) \cap \mathrm{O}(V)$ is compact, $\mathcal{A}$ is abelian and $\mathcal{N}$ is nilpotent. The unique fixed point in $\Omega$ for the mapping $x \mapsto \nabla \log \varphi(x)$ will be denoted $e$, and we note that $\mathcal{K}$ fixes $e$. The connected solvable subgroup $\mathcal{H} = \mathcal{AN}$ of $\mathcal{G}_0(\Omega)$ acts simply transitively on $\Omega$ and the integral (1) thus also defines the left-Haar measure on $\mathcal{H}$. Denote by $\mathcal{S}(V)$ the space of rapidly decreasing smooth functions with topology induced by the semi-norms

$$\|f\|_k = \sup_{|\alpha| \le k} \sup_{x \in V} |\partial^\alpha f(x)|(1+|x|)^k.$$

Here $\alpha$ is a multi-index, $\partial^\alpha$ denotes usual partial derivatives of functions, and $k \ge 0$ is an integer. For $f \in \mathcal{S}(V)$ the Fourier transform is defined by

$$\widehat{f}(w) = \frac{1}{(2\pi)^{n/2}} \int_V f(x) e^{-i(x,w)}\, dx \text{ for } w \in V.$$

The convolution

$$f * g(x) = \int_V f(y) g(x-y)\, dy$$

of functions $f, g \in \mathcal{S}(V)$ satisfies

$$\widehat{f * g}(w) = \widehat{f}(w)\widehat{g}(w).$$

The space $\mathcal{S}'(V)$ of tempered distributions is the linear dual of $\mathcal{S}(V)$. For functions on $V$ define $\tau_x f(y) = f(y-x)$, $f^\vee(y) = f(-y)$ and $f^*(y) = \overline{f(-y)}$. Convolution of $f \in \mathcal{S}'(V)$ and $\phi \in \mathcal{S}(V)$ is defined by

$$f * \phi(x) = f(\tau_x \phi^\vee).$$

As usual, the Fourier transform extends to tempered distributions by duality. The space of rapidly decreasing smooth functions with Fourier transform vanishing on

$\Omega$ is denoted $\mathcal{S}_\Omega$. It is a closed subspace of $\mathcal{S}(V)$ and will be equipped with the subspace topology.

The space $V$ can be equipped with a Jordan algebra structure such that $\overline{\Omega}$ is identified with the set of all squares. This gives rise to the notion of a determinant $\Delta(x)$. We only need the fact that the determinant is related to the characteristic function $\varphi$ by

$$\varphi(x) = \varphi(e)\Delta(x)^{-n/R},$$

where $R$ denotes the rank of the cone. If $x = ge$ we have

$$\Delta(x) = \Delta(ge) = |\mathrm{Det}(g)|^{R/n}. \tag{2}$$

## 3. Besov spaces related to symmetric cones

The cone $\Omega$ can be identified as a Riemannian manifold $\Omega = \mathcal{G}_0(\Omega)/\mathcal{K}$ where $\mathcal{K}$ is the compact group fixing $e$. The Riemannian metric in this case is defined by

$$\langle u, v\rangle_y = (g^{-1}u, g^{-1}v)$$

for $u, v$ tangent vectors to $\Omega$ at $y = ge$. Denote the balls of radius $\delta$ centered at $x$ by $B_\delta(x)$. For $\delta > 0$ and $\lambda \geq 2$ the points $\{x_j\}$ are called a $(\delta,\lambda)$-lattice if

(1) $\{B_\delta(x_j)\}$ are disjoint, and
(2) $\{B_{\lambda\delta}(x_j)\}$ cover $\Omega$.

We now fix a $(\delta,\lambda)$-lattice $\{x_j\}$ with $\delta = 1/2$ and $\lambda = 2$. Then there are functions $\psi_j \in \mathcal{S}_\Omega$, such that $0 \leq \widehat{\psi_j} \leq 1$, $\mathrm{supp}(\widehat{\psi_j}) \subseteq B_2(x_j)$, $\widehat{\psi_j}$ is one on $B_{1/2}(x_j)$ and $\sum_j \widehat{\psi_j} = 1$ on $\Omega$. Using this decomposition of the cone, the Besov space norm for $1 \leq p, q < \infty$ and $\nu \in \mathbb{R}$ is defined in [**1**] by

$$\|f\|_{B_\nu^{p,q}} = \left(\sum_j \Delta(x_j)^{-\nu}\|f * \psi_j\|_p^q\right)^{1/q}.$$

The Besov space $B_\nu^{p,q}$ consists of the equivalence classes of tempered distributions $f$ in $(\mathcal{S}_\Omega)' \simeq \{f \in \mathcal{S}'(V) \mid \mathrm{supp}(\widehat{f}) \subseteq \overline{\Omega}\}/\mathcal{S}'_{\partial\Omega}$ for which $\|f\|_{B_s^{p,q}} < \infty$.

Define the index

$$\widetilde{q}_{\nu,p} = \frac{\nu + n/R - 1}{(n/Rp') - 1}$$

if $n/R > p'$ and set $\widetilde{q}_{\nu,p} = \infty$ if $n/R \leq p'$. The following results from [**1**] states when the Besov spaces are included in the space of tempered distributions $\mathcal{S}'(V)$.

LEMMA 3.1. *Let $\nu > 0$, $1 \leq p < \infty$ and $1 \leq q < \widetilde{q}_{\nu,p}$. Then for every $f \in B_\nu^{p,q}$ the series $\sum_j f * \psi_j$ converges in the space $\mathcal{S}'(V)$, and the correspondence*

$$f + \mathcal{S}'_{\partial\Omega} \mapsto f^\sharp = \sum_j f * \psi_j$$

*is continuous, injective and does not depend on the particular choice of $\{\psi_j\}$.*

The main result from [**6**] is that the quasiregular representation of the group $\mathcal{H} \rtimes V$ can be used to obtain atomic decompositions for these Besov spaces. We summarize the result in

THEOREM 3.2. *Let $\psi$ in $\mathcal{S}_\Omega$ be such that $\widehat{\psi}$ is compactly supported. Then there is an index set $I$, a set $\{(h_i, x_i)\}_{i\in I} \subseteq \mathcal{H} \times V$, a Banach sequence space $b^{p,q}_\nu(I)$, a set of continuous functionals $\{c_i : B^{p,q}_\nu \to \mathbb{C}\}_{i\in I}$, and a constant $C > 0$ such that*

(1) $f(x) = \sum_{i\in I} c_i(f)\frac{1}{\sqrt{\det(h_i)}}\psi(h_i^{-1}(x - x_i))$ *with convergence in norm in* $B^{p,q}_\nu$

(2) $\|c_i(f)\|_{b^{p,q}_\nu} \le C\|f\|_{B^{p,q}_\nu}$

(3) *if* $\{\lambda_i\} \in b^{p,q}_\nu$ *then*

$$f(x) = \sum_i \lambda_i \frac{1}{\sqrt{\det(h_i)}}\psi(h_i^{-1}(x - x_i))$$

*is in* $B^{p,q}_\nu$ *and* $\|f\|_{B^{p,q}_\nu} \le C\|\{\lambda_i\}_{b^{p,q}_\nu}\|$.

REMARK 3.3. At this stage it is appropriate to describe the sequence of points $\{(h_i, x_i)\}_{i\in I}$ and the sequence space $b^{p,q}_\nu$ in some detail. For this one chooses a covering $\{U_i\}_{i\in I}$ of the space $\Omega \times V$. This covering is chosen such that each $U_i$ is a translate of a fixed relatively compact neighbourhood $U$ of $\{e\} \times \{0\}$ by some element $(h_i, x_i)$ of the semidirect product $\mathcal{H} \rtimes V$. Moreover, the sets $U_i$ have the finite overlapping property, that is, there is an $N$ such that each set $U_i$ overlap at most $N$ others. A sequence $\{\lambda_i\}$ is in $b^{p,q}_\nu$ if

$$\|\{\lambda_i\}\|_{b^{p,q}_\nu} := \left(\int_\Omega \left(\int_V \sum_{i\in I} |\lambda_i| 1_{U_i}(x,t)^p\, dt\right)^{q/p} \Delta(x)^{\nu - qn/(2R) - n/R}\, dx\right)^{1/q}$$

is finite. If $p = q$ this is a $\Delta^{\nu - pn/(2R) - n/R}$-weighted $\ell^p$-space.

## 4. Bergman spaces on tube type domains.

In this section we introduce the Bergman spaces on the tube type domains, and describe the isomorphism between a range of Besov spaces and Bergman spaces.

Let $T = \{z = x + iy \mid x \in V, y \in \Omega\}$ be the tube type domain related to the symmetric cone $\Omega$. For $1 \le p, q < \infty$ and $\nu > 0$ define the weighted Lebesgue space $L^{p,q}_\nu$ on the tube type domain to consist of the equivalence classes of measurable functions on $\Omega$ for which the norm

$$\|F\|_{L^{p,q}_\nu} := \left(\int_\Omega \left(\int_V |F(x+iy)|^p\, dx\right)^{q/p} \Delta^{\nu - n/r}(y)\, dy\right)^{1/q}$$

is finite. Here $dx$ and $dy$ denote the usual Lebesgue measures on $V$ and $\Omega$. The mixed norm Bergman space $\mathcal{A}^{p,q}_\nu$ on $\Omega$ consists of the holomorphic functions in $L^{p,q}_\nu$. It is well-known that this is a reproducing kernel Banach space, that is, for every $z \in T$ the mapping $F \mapsto F(z)$ is continuous from $\mathcal{A}^{p,q}_\nu$ to $\mathbb{C}$.

The special case $p = q = 2$ and $\nu = n/R$ is the usual Bergman space and the reproducing kernel in this case is

$$B(z, w) = B_{n/r}(z, w) = c(\nu)\Delta\left(\frac{z - \overline{w}}{i}\right)^{-2n/r},$$

which will be called the Bergman kernel.

Following [**1**] we now define the Fourier-Laplace extensions of elements in the Besov spaces. This extension only works for Besov spaces which can be naturally imbedded in the usual space of tempered distributions. This introduces a restriction in the range of indices that can be used. We summarize the results from [**1**] that we need.

Define the Fourier-Laplace extension of a tempered distribution $f$ whose Fourier transform $\widehat{f}$ is supported on $\overline{\Omega}$ by

$$\mathcal{E}f = \int \widehat{f}(w)e^{iz\cdot w}\, dw$$

for $z \in T$. For $1 \leq q < \widetilde{q}_{\nu,p}$ the Besov space can be indentified with a space of such distributions, and therefore we can define

$$\widetilde{\mathcal{E}}f = \mathcal{E}\sum_j f * \psi_j$$

for $f \in B_\nu^{p,q}$. Define the index

$$q_{\nu,p} = \min(p,p')\frac{\nu + n/R - 1}{n/R - 1}$$

when $n > R$ and set $q_{\nu,p} = \infty$ when $n = 1$. Notice that $2 < q_{\nu,p} \leq \widetilde{q}_{\nu,p}$ so when $1 \leq q < q_{\nu,p}$ all elements of the Besov space $B_\nu^{p,q}$ can be identified with tempered distributions whose Fourier-Laplace extensions are in the Bergman space $\mathcal{A}_\nu^{p,q}$. We have

THEOREM 4.1. *If $\nu > n/R - 1$, $1 \leq p < \infty$ and $1 \leq q < q_{\nu,p}$, then the mapping $\widetilde{\mathcal{E}} : B_\nu^{p,q} \to \mathcal{A}_\nu^{p,q}$ is an isomorphism. Moreover,*

$$\lim_{y\to 0} F(x+iy) = f(x)$$

*in both $\mathcal{S}'(V)$ and $B_\nu^{p,q}$.*

## 5. Atomic decomposition of Bergman spaces

In this section we merge the results from [**6**] and [**1**] to obtain atomic decompositions for Bergman spaces on the tube type domains. This will give an alternate approach to the atomic decompositions found in [**10**]. This new approach allows for a large class of atoms, and moreover, through the Paley-Wiener theorem the decay properties of these atoms are quite well known. Note, that the decomposition from [**10**] uses samples of the Bergman kernel, but the Bergman kernel is not among the possible atoms with the new approach. The paper [**20**] goes in the opposite direction and uses [**10**] to obtain atomic decompositions the Besov spaces. It would of course be interesting to investigate how to completely align the two methods and to determine exactly which atoms can be moved from the Besov setting and to the Bergman setting. In this paper we are clearly only dealing with a subset of possible atoms.

Let $F \in \mathcal{A}_\nu^{p,q}$ then $f = \widetilde{\mathcal{E}}^{-1}F$ is in $B_\nu^{p,q}$ with equivalent norms and can be decomposed as

$$f(x) = \sum_{i\in I} c_i(f)\frac{1}{\sqrt{\det(h_i)}}\psi(h_i^{-1}(x - x_i)).$$

Since $\psi \in \mathcal{S}_0$ with compactly supported Fourier transform is in every Besov space, we get

$$\psi_i(z) := \widetilde{\mathcal{E}}(\psi(h_i^{-1}(\cdot - x_i)))(z) = \widetilde{\mathcal{E}}\psi(h^{-1}(z - x_i)).$$

This results in the following atomic decompositions.

THEOREM 5.1. *Let $\nu > n/R - 1$, $1 \leq p < \infty$ and $1 \leq q < q_{\nu,p}$. There is a sequence $\{d_i\}$ of functionals and atoms $\{\psi_i\}$ parameterized by appropriate $\{(h_i, x_i)\} \subseteq \mathcal{H} \rtimes V$ such that*

(1) *if $F \in \mathcal{A}_\nu^{p,q}$ then*

$$F(z) = \sum_{i \in I} d_i(F) \frac{1}{\sqrt{\det(h_i)}} \psi_i(z).$$

*where $d_i(F) = c_i(\widetilde{\mathcal{E}}^{-1}F)$ satisfy that $\|\{d_i(F)\}\|_{b_\nu^{p,q}} \leq C\|F\|_{\mathcal{A}_\nu^{p,q}}$,*

(2) *and if $\{d_i\} \in b_\nu^{p,q}$, then*

$$F(z) = \sum_{i \in I} d_i \frac{1}{\sqrt{\det(h_i)}} \psi_i(z).$$

*is in $\mathcal{A}_\nu^{p,q}$ and $\|F\|_{\mathcal{A}_\nu^{p,q}} \leq C\|\{d_i\}\|_{b_\nu^{p,q}}$.*

REMARK 5.2. Notice that for the case of cones of rank 2 (for example the forward light cones) this theorem can be extended to the larger range $1 \leq q < \widetilde{q}_{\nu,p}$ which is the entire range of $q$ for which Laplace extensions can be defined. See [**5**].

## 6. Comparison with previous results and some open problems

To demonstrate how this work extendeds the range of Bergman spaces for which atomic decompositions can be found, we now compare our atomic decompositions to classical results due to Coifman and Rochberg [**10**] in the special case of tube type domains. Their results only work for rank one spaces, but with minor modifications this issue can be addressed via the Forelli-Rudin estimates from Theorem 4.1 in [**12**] or Corollary II.4 in [**4**]. The latter result was used in [**3**] to correct and generalize the atomic decompositions of Coifman and Rochberg to also include two non-symmetric domains.

We first summarize the atomic decompsitions from [**3, 10**]. Let $V$ be an open convex cone in $\mathbb{R}^m$ and let $F$ be a $V$-valued Hermitian form on $\mathbb{C}^n$. The open subset of of elements $(z, w)$ in $\mathbb{C}^m \times \mathbb{C}^n$ for which $\mathrm{Im}(z) - F(w, w) \in V$ is called a Siegel domain of type II. The domain is called symmetric if it is also a symmetric space. Let $D$ be a symmetric Siegel domain of type II and let $B$ denote the associated Bergman kernel. Coifman and Rochberg use a parametrization of Bergman spaces that differs from the one used earlier in this paper. In their notation the Bergman space $\widetilde{\mathcal{A}}_r^p$ consists of holomorphic functions for which the following norm is bounded

$$\|F\|_{\widetilde{\mathcal{A}}_r^p} = \int_D |F(z)|^p B(z, z)^{-r} dz.$$

THEOREM 6.1. *Let $p \geq 1$ and assume that $-\epsilon_D + \gamma_D(p-1) < r < \infty$. Given $\theta > p(1 - \epsilon_D) + \epsilon_D + \gamma_D - 2 - r$ there is a lattice $\{\xi_i\}$ in $D$ and a constant $C > 0$ such that for $F$ in $\widetilde{\mathcal{A}}_r^p$ we have*

$$F(z) = \sum_i \lambda_i(F) \left(\frac{B(z, \xi_i)^2}{B(\xi_i, \xi_i)}\right)^{\frac{1+r}{p}} \left(\frac{B(z, \xi_i)}{B(\xi_i, \xi_i)}\right)^{\frac{\theta}{p}},$$

*and* $\sum_i |\lambda_i(F)|^p \leq C\|F\|^p_{\widetilde{\mathcal{A}}^p_r}$. *Moreover, if* $\{\lambda_i\} \in \ell^p$ *then the series*

$$F(z) = \sum_i \lambda_i \left(\frac{B(z,\xi_i)^2}{B(\xi_i,\xi_i)}\right)^{\frac{1+r}{p}} \left(\frac{B(z,\xi_i)}{B(\xi_i,\xi_i)}\right)^{\frac{\theta}{p}}.$$

*defines a function in* $\widetilde{\mathcal{A}}^p_r$ *and* $\|F\|_{\mathcal{A}^p_r} \leq C\|\{\lambda_i\}\|_{\ell^p}$.

REMARK 6.2. The constants are given by $\epsilon_D = 1/G$ and $\gamma_D = (R-1)a/(2G)$, where $G$ is the genus and $a$ is another structural constant. See, for example, [**12**] for a full explanation of these constants.

We will now establish the range of $r$ which work for tube type domains. The connection between the number $\nu$ from Theorem 5.1 in the special case of $p = q$ and the number $r$ from Theorem 6.1 is

$$\nu = \frac{2nr}{R} + \frac{n}{R}.$$

Moreover, the structural constants $\epsilon_D$ and $\gamma_D$ for tube type domains are

$$\epsilon_D = \frac{R}{2n} \qquad \text{and} \qquad \gamma_D = \frac{1}{2} - \epsilon_D.$$

With this in mind the conditions $\nu > \frac{n}{R} - 1$ and $p < \frac{\nu + n/R - 1}{n/R - 1} + 1$ from Theorem 5.1 rewrite into

$$r > \max\{-\epsilon_D, -\frac{3}{2} + p(1-\epsilon_D) + \epsilon_D - \frac{p}{2}\}.$$

The result from Theorem 6.1 works for

$$r > \max\{-\epsilon_D + \gamma_D, -\frac{3}{2} + p(1-\epsilon_D)\},$$

when restricted to the case $\theta = 0$. Since for tube type domains over cones $\epsilon_D = R/(2n) \leq 1/2$ and $p \geq 1$ we see that the atomic decompositions in Theorem 5.1 work for a larger range than those of Theorem 6.1.

REMARK 6.3. The reason we restrict to the $\theta = 0$ when comparing the two methods is, that the the atomic decompositions provided by Laplace extensions is connected to the discrete series representation of the automorphism group on the tube type domain, and we might as well transfer this result to the bounded realization of the domain. Therefore Theorem 5.1 can be transfered to Bergman spaces on the bounded symmetric domain. It thus makes sense to compare with the version of Theorem 6.1 which also can be transfered to the bounded realization, that is, $\theta = 0$.

We finish this section with a list of open problems connected to the results of this paper.

PROBLEM 1. The atoms from Theorem 5.1 do not include samples of the Bergman kernel as in [**3**, **10**]. The reason is that the atoms we use are extensions of compactly supported smooth functions, and by the Paley-Wiener Theorem these cannot include the Bergman kernel. This means that Theorem 5.1 does not include Theorem 6.1 as a special case. We believe that it is possible to overcome this issue. The atomic decompositions in [**6**] build on irreducible, unitary, and integrable group representations, and therefore a much larger class of atoms for the Besov spaces can be used via [**14**]. It would be interesting to see if the Laplace extensions of this larger class of atoms for the Besov spaces would include the Bergman kernel in order to obtain Theorem 6.1 as a special case of Theorem 5.1. This question would be of interest even on the upper half plane.

REMARK 6.4. Upon completion of this work the author was made aware of a related paper by D. Békollé, J. Gonessa and C. Nana [**2**]. They obtain atomic decompositions for the exact same range of Bergman spaces. In their work the atoms are indeed samples of the Bergman kernel. Problem 1 thus formulates one possible approach to uncovering the connection between their result and the present paper.

PROBLEM 2. Another way to get a larger set of atoms including samples of the Bergman kernel is to apply the coorbit theory [**9**] as has been done for the unit ball in [**7**, **16**]. The integral operator with positive kernel from Theorem II.7 in [**4**] which was used to derive Theorem 6.1 would play a crucial role in this approach (as it did on the unit ball [**7**, **16**]), so we predict this approach would work for the same range of parameters as Theorem 6.1. It would be interesting to see if the use of Theorem II.7 in [**4**] could be avoided or refined in the context of coorbits in order to get to the same range of Bergman spaces as in Theorem 5.1.

PROBLEM 3. The approach highlighted in this paper could be used in the setting of the unit ball by using Laplace extensions of Besov spaces on the Heisenberg group mentioned in [**15**]. It also seems possible to extend our approach to all bounded symmetric domains via work in [**18**].

## References


[1] D. Békollé, A. Bonami, G. Garrigós, and F. Ricci, *Littlewood-Paley decompositions related to symmetric cones and Bergman projections in tube domains*, Proc. London Math. Soc. (3) **89** (2004), no. 2, 317–360. MR2078706

[2] D. Békollé, J. Gonessa, and C. Nana. Atomic decomposition and interpolation via the complex method for mixed norm Bergman space on tube domains over symmetric cones. *ArXiv e-prints*, Mar. 2017.

[3] David Bekolle and Anatole Temgoua Kagou, *Molecular decompositions and interpolation*, Integral Equations Operator Theory **31** (1998), no. 2, 150–177. MR1623528

[4] David Békollé and Anatole Temgoua Kagou, *Reproducing properties and $L^p$-estimates for Bergman projections in Siegel domains of type II*, Studia Math. **115** (1995), no. 3, 219–239. MR1351238

[5] Aline Bonami and Cyrille Nana, *Some questions related to the Bergman projection in symmetric domains*, Adv. Pure Appl. Math. **6** (2015), no. 4, 191–197. MR3403436

[6] Jens Gerlach Christensen, *Atomic decompositions of Besov spaces related to symmetric cones*, Geometric analysis and integral geometry, Contemp. Math., vol. 598, Amer. Math. Soc., Providence, RI, 2013, pp. 97–110. MR3156440

[7] Jens Gerlach Christensen, Karlheinz Gröchenig, and Gestur Ólafsson, *New atomic decompositions for Bergman spaces on the unit ball*, Indiana Univ. Math. J. **66** (2017), no. 1, 205–235. MR3623408

[8] Jens Gerlach Christensen and Gestur Ólafsson, *Examples of coorbit spaces for dual pairs*, Acta Appl. Math. **107** (2009), no. 1-3, 25–48. MR2520008

[9] Jens Gerlach Christensen and Gestur Ólafsson, *Coorbit spaces for dual pairs*, Appl. Comput. Harmon. Anal. **31** (2011), no. 2, 303–324. MR2806486

[10] R. R. Coifman and R. Rochberg, *Representation theorems for holomorphic and harmonic functions in $L^p$*, Representation theorems for Hardy spaces, Astérisque, vol. 77, Soc. Math. France, Paris, 1980, pp. 11–66. MR604369

[11] Daniele Debertol, *Besov spaces and the boundedness of weighted Bergman projections over symmetric tube domains*, Publ. Mat. **49** (2005), no. 1, 21–72. MR2140199

[12] J. Faraut and A. Korányi, *Function spaces and reproducing kernels on bounded symmetric domains*, J. Funct. Anal. **88** (1990), no. 1, 64–89. MR1033914

[13] Jacques Faraut and Adam Korányi, *Analysis on symmetric cones*, Oxford Mathematical Monographs, The Clarendon Press, Oxford University Press, New York, 1994. Oxford Science Publications. MR1446489

[14] Hans G. Feichtinger and K. H. Gröchenig, *Banach spaces related to integrable group representations and their atomic decompositions. I*, J. Funct. Anal. **86** (1989), no. 2, 307–340. MR1021139

[15] Mark Feldman, *Mean oscillation, weighted Bergman spaces, and Besov spaces on the Heisenberg group and atomic decompositions*, J. Math. Anal. Appl. **158** (1991), no. 2, 376–395. MR1117569

[16] J. Gerlach Christensen, A. H. Darweesh, and G. Olafsson. Coorbits for projective representations with an application to Bergman spaces. *ArXiv e-prints*, Apr. 2017.

[17] Daniel H. Luecking, *Representation and duality in weighted spaces of analytic functions*, Indiana Univ. Math. J. **34** (1985), no. 2, 319–336. MR783918

[18] Cyrille Nana, *$L^{p,q}$-boundedness of Bergman projections in homogeneous Siegel domains of type II*, J. Fourier Anal. Appl. **19** (2013), no. 5, 997–1019. MR3110589

[19] Margit Pap, *Properties of the voice transform of the Blaschke group and connections with atomic decomposition results in the weighted Bergman spaces*, J. Math. Anal. Appl. **389** (2012), no. 1, 340–350. MR2876501

[20] Fulvio Ricci and Mitchell Taibleson, *Boundary values of harmonic functions in mixed norm spaces and their atomic structure*, Ann. Scuola Norm. Sup. Pisa Cl. Sci. (4) **10** (1983), no. 1, 1–54. MR713108

[21] Kehe Zhu, *Spaces of holomorphic functions in the unit ball*, Graduate Texts in Mathematics, vol. 226, Springer-Verlag, New York, 2005. MR2115155

Department of Mathematics, Colgate University, 13 Oak Drive, Hamilton, New York 13346

*Email address*: jchristensen@colgate.edu

*URL*: http://www.math.colgate.edu/~jchristensen

Contemporary Mathematics
Volume **714**, 2018
https://doi.org/10.1090/conm/714/14375

# C*-algebraic normalization and Godement-Jacquet factors

Pierre Clare

*To Gestur Ólafsson on the occasion of his 65th birthday, in admiration.*

Abstract. We observe how certain distribution on GL($n$) that generates Godement-Jacquet $\gamma$-factors appears naturally in the C*-algebraic normalization process for standard intertwining integrals on SL($n+1$).

## 1. Introduction

The purpose of this note is to point out a relationship between well-known factors introduced by R. Godement and H. Jacquet in the 1970's and distributions that occur in the study of the tempered dual of reductive groups in the framework of Noncommutative Geometry. The functions $\gamma$ considered here appear in the functional equations satisfied by the zeta functions associated with irreducible representations of GL($n$) in [**GJ72**]. They are instances of more general factors involved in the local Langlands conjectures: general $\gamma$-factors appear in connection with the lifting map in Langlands functoriality; see for instance the introduction of [**BK00**].

The other facet of Representation Theory to which we shall connect these objects is the C*-algebraic picture of the tempered dual of reductive groups. In the past few years, various aspects of the tempered theory have been recast in the language of Noncommutative Geometry, thus making the machinery of operator algebras, known to be efficient in the general study of locally compact groups, available to study unitary representations of Lie groups. In particular, the study of parabolic induction by means of Hilbert modules and operator spaces allows to handle generalized principal series representations in families (see Section 2.2.1 below), which has proven useful in the study of adjunction properties for real groups (see [**CCH16b, CCH16a**]).

In this picture, the analogues of the meromorphic factors used in Knapp-Stein theory to normalize intertwining integrals are distributions, conjecturally defined on the Levi components of the inducing parabolic subgroups. We observe in Section 2.3 that, when these distributions are known, they can be used to generate $\gamma$-factors in the sense of Braverman and Kazhdan, who conjectured in [**BK00**], for any quasi-split reductive algebraic group $G$, the existence of a distribution acting in every irreducible representation of $G$ as the multiplication by the associated $\gamma$-factor.

---

2010 *Mathematics Subject Classification.* Primary 22E50, 11S40, 22D25, 46L08.

In the case of GL($n$), where the $\gamma$-factors coincide with those of Godement and Jacquet, this distribution is well-known, and our contribution is the observation that this distribution is essentially the inverse for the convolution product of the C*-algebraic normalizing distribution associated with a maximal parabolic subgroup of SL($n$+1). The explicit formula relating the two distributions is given in Proposition 2.9 and we indicate further connections with the Langlands-Shahidi method in Section 3.

**Notation.** In what follows $F$ is a local field, that is, $\mathbb{R}$, $\mathbb{C}$ or a finite extension of $\mathbb{Q}_p$ or of $\mathbb{F}_p((t))$. We fix a non-trivial continuous character $\psi : F \to \mathrm{U}(1)$. Then the map $a \mapsto \psi(a\,\cdot)$ is a group isomorphism between $F$ and its group of continuous unitary characters. We also fix an additive Haar measure $dx$ on $F$ so that the Fourier transform $\mathcal{F}_\psi$ defined by

$$\mathcal{F}_\psi f(x) = \int_F \psi(xy) f(y)\, dy$$

is an isometry of $L^2(F)$ with inverse

$$f \longmapsto [x \mapsto \int_F \psi(-xy) f(y)\, dy].$$

Throughout the paper, additive Haar measures are denoted by $dx$ and multiplicative Haar measures are denoted by $d^\times x$. In particular, we write $dg$ for the additive Haar measure on the space $\mathrm{M}_n(F)$ of $n \times n$ matrices with entries in $F$ and $d^\times g$ for the Haar measure on $\mathrm{GL}(n, F)$. The two measures are related by

$$dg = |\det(g)|^n\, d^\times g$$

where $|\cdot|$ denotes the norm on $F$ that coincides with the modular function associated with the chosen Haar measure. In the non-Archimedean case, it is specified by the relation $|\varpi| = q^{-1}$, where $\varpi$ is a uniformizer and $q$ is the cardinality of the residue field.

## 2. Generating Godement-Jacquet factors and the inverse normalizing distribution

**2.1. $\gamma$-factors and the generating problem.** In 1972, Godement and Jacquet introduced zeta functions associated with representations of $\mathrm{GL}(n, F)$ see [**GJ72,Jac79,Jac97**], generalizing the work of J. Tate on the case of $n = 1$ [**Tat67**]. These functions can be defined as follows:

$$Z(\Phi, s, f_\pi) = \int_{\mathrm{GL}(n,F)} \Phi(g) f_\pi(g) |\det(g)|^s\, d^\times g, \tag{2.1}$$

where $s$ is a complex parameter[1], $\Phi$ is a Schwartz function on $\mathrm{M}_n(F)$ and $f_\pi$ is a matrix coefficient of an admissible representation $\pi$ of $G$, that is, a function of the form

$$g \longmapsto \langle \pi(g)\xi, \eta\rangle$$

with $\xi$ in the carrying space $V_\pi$ of $\pi$ and $\eta$ in its dual $V_\pi^*$. The convergence of the integral in (2.1) is guaranteed for $\Re(s)$ large enough by [**GJ72**, Theorem 3.3] and the expression $Z(\Phi, s, f_\pi)$ depends meromorphically on $s$.

[1]Here, we adopt the convention of [**BK00**]. The notation originally used in [**GJ72**] can be recovered by a simple shift, namely replacing $s$ with $s - \frac{n-1}{2}$.

An important property of these functions is the functional equation they satisfy, namely:

$$Z(\hat{\Phi}, n-s, \check{f}_\pi) = \gamma(s,\pi) \cdot Z(\Phi, s, f_\pi). \tag{2.2}$$

Here, $\check{f}_\pi(g) = f_\pi(g^{-1})$, so that $\check{f}_\pi$ is a matrix coefficient of the contragredient representation $\pi^*$, and $\hat{\Phi}$ denotes the Fourier transform of $\Phi$. We recall that the Fourier transform on $\mathrm{M}_n(F)$ is the operator

$$\begin{array}{ccc} \mathcal{S}(\mathrm{M}_n(F)) & \longrightarrow & \mathcal{S}(\mathrm{M}_n(F)) \\ \Phi & \longmapsto & \hat{\Phi} \end{array}$$

with

$$\hat{\Phi}(x) = \int_{\mathrm{M}_n(F)} \Phi(y)\psi(\mathrm{tr}(xy))\, dy. \tag{2.3}$$

The meromorphic functions $\gamma(s,\pi)$ that appear in the functional equation (2.2) are examples of $\gamma$-factors. These objects can be defined in a more general setting, which we briefly describe now.

2.1.1. *$\gamma$-factors.* Let $G$ be a split reductive group over $F$ and let $\rho$ be a representation of dimension $n$ of its Langlands dual $G^\vee$. To every irreducible representation $\pi$ of $G$, local factors $\gamma_{\psi,\rho}(s,\pi)$, $L_\rho(s,\pi)$ and $\varepsilon_{\psi,\rho}(s,\pi)$ are conjecturally attached, satisfying the relation

$$\gamma_{\psi,\rho}(s,\pi) = \varepsilon_{\psi,\rho}(s,\pi)\frac{L_\rho(1-s,\pi^*)}{L_\rho(s,\pi)}.$$

It is worth noting that, unlike the factors $\gamma$ and $\varepsilon$, the functions $L_\rho$ do not depend on the choice of $\psi$ and that the $\gamma$-factors are meromorphic functions of $s$. They are in fact quotients of products of $\Gamma$ functions if $F$ is Archimedean, and rational functions of $q^s$ in the non-Archimedean case, where $q$ is the cardinality of the residue field.

The Godement-Jacquet factors $\gamma(s,\pi)$ involved in the functional equation (2.2) are obtained by considering $G = \mathrm{GL}(n,F)$ and letting $\rho = \mathrm{St}_n$ be the standard representation of $G^\vee = \mathrm{GL}(n,F)$ on $F^n$.

REMARK 2.1. An explicit description of the Godement-Jacquet factors $\gamma$, $L$ and $\varepsilon$ in terms of the classification of irreducible representations of $\mathrm{GL}(n,F)$ can be found in [**Kna94**] for the Archimedean case and in [**Kud94**] in the non-Archimedean case.

REMARK 2.2. The Rankin-Selberg local factors (see [**JPSS83**]) correspond to the case of $G = \mathrm{GL}(m) \times \mathrm{GL}(n)$ with $\rho = \mathrm{St}_m \otimes \mathrm{St}_n$.

In [**BK00**], A. Braverman and D. Kazhdan studied the problem of constructing the factors $\gamma_{\psi,\rho}$ explicitly. They suppose given a character $\sigma$ of $G$ such that the cocharacter

$$\sigma^* : \mathbb{G}_m \to \mathrm{Z}(G^\vee)$$

of the center of $G^\vee$ satisfies the relation $\rho \circ \sigma^* = \mathrm{Id}$. Under this assumption, one has

$$\gamma_{\psi,\rho}(s,\pi) = \gamma_{\psi,\rho}(0, \pi|\sigma|^s)$$

so that the factor $\gamma_{\psi,\rho}$ is determined by $\gamma_{\psi,\rho}(0,\pi)$, thought of as a meromorphic function on the space of irreducible representations of $G$ (see Section 3.4 in [**BK00**] for a precise definition of this notion).

2.1.2. *Generating distribution.* According to Braverman and Kazhdan, $\gamma$-factors should be generated by a single distribution on $G$. Indeed, they prove in some cases the existence of a central stable distribution $\Delta_{\psi,\rho}$ on $G$ whose action in every irreducible representation $\pi$ of $G$ is given by multiplication by $\gamma_{\psi,\rho}$. More precisely, convolution with the distribution[2] $\Delta_{\psi,\rho}$ acts in $\pi$ by $\gamma_{\psi,\rho}(0,\pi)$ and in $\pi|\sigma|^s$ by $\gamma_{\psi,\rho}(s,\pi)$, that is,

$$(\dagger) \qquad \Delta_{\psi,\rho} * (f_\pi|\sigma|^s) = \gamma_{\psi,\rho}(s,\pi) \cdot f_\pi|\sigma|^s$$

for every matrix coefficient $f_\pi$ of $\pi$.

REMARK 2.3. The shifted convention for the complex parameter used in [**BK00**] and here has the merit of leading to the relation

$$\Delta_{\psi,\rho} = \Delta_{\psi,\rho_1} * \Delta_{\psi,\rho_2}$$

for $\rho = \rho_1 \oplus \rho_2$.

In the case of $G = \mathrm{GL}(n)$ and $\rho = \mathrm{St}_n$, we let

$$\sigma = \det : \mathrm{GL}(n) \longrightarrow \mathbb{G}_m$$

so that the defining property ($\dagger$) becomes

$$(\ddagger) \qquad \Delta_{\psi,\mathrm{St}_n} * (f_\pi|\det|^s) = \gamma(s,\pi) \cdot f_\pi|\det|^s$$

with the notation of (2.2).

The determination of a distribution $\Delta_{\psi,\mathrm{St}_n}$ satisfying ($\ddagger$) is very straightforward (see Section 2.3 below). The purpose of the present note is to point out a natural relation with the approach to the representation theory of algebraic groups *via* C*-algebras and Hilbert modules.

**2.2. C*-normalization for degenerate principal series of** $\mathrm{SL}(n)$**.** The basic principle underlying the Noncommutative Geometry approach to the study of tempered representations of real reductive groups is that the tempered dual of a topological group can be understood as a *noncommutative space* in the sense of A. Connes [**Con94**]. If $G$ is a locally compact group, its reduced C*-algebra $\mathrm{C}^*_r(G)$ is the norm-closure of the convolution algebra $C_c(G)$ of continuous compactly supported functions on $G$ in the algebra of bounded operator on $L^2(G)$. One can show that the spectrum of $\mathrm{C}^*_r(G)$ coincides with the tempered dual of $G$, making this algebra a useful proxy to study tempered representations of $G$ (see [**Dix77**]).

Following Harish-Chandra's philosophy of cusp forms, particular attention has been devoted in recent work to the description of principal series representations in the C*-algebraic language. We shall recall here only what is needed to establish a connection with the problem of generating $\gamma$-factors. The reader is referred to [**Cla13**] and [**CCH16b**] for proofs and details.

2.2.1. *Parabolic induction in the C*-algebraic framework.* Let $G$ be a reductive group and $P$ a parabolic subgroup with Levi component $L$ and unipotent radical $N$ so that $P = LN$. The generalized principal series representations induced from $P$ can be described in terms of certain $\mathrm{C}^*_r(G)$-$\mathrm{C}^*_r(L)$-bimodules (or correspondences) $\mathrm{C}^*_r(G/N)$ obtained as completions[3] of $C_c(G/N)$.

[2] This distribution is denoted by $\Phi_{\psi,\rho}$ in [**BK00**].

[3] These modules were denoted by $\mathcal{E}(G/N)$ in [**Cla13**] where they were introduced, and by $\mathrm{C}^*_r(G/N)$ in [**CCH16b**] where they were analyzed in further detail.

The most important property of the module $C^*_r(G/N)$ is that its induces the principal series attached to $P$ in the sense of M. A. Rieffel [**Rie74**]:

PROPOSITION 2.4. *The Hilbert* $C^*_r(L)$*-module* $C^*_r(G/N)$ *is equipped with an action of* $C^*_r(G)$ *through compact operators and yields a unitary equivalence of G-representations*

$$C^*_r(G/N) \otimes_{C^*_r(L)} \mathcal{H}_\lambda \simeq \mathrm{Ind}_P^G \lambda \otimes 1_N$$

*for any tempered unitary representation* $\lambda$ *of* $L$ *and* $1_N$ *the trivial representation of* $N$.

See [**Cla13**] and [**CCH16b**] for a proof and further discussion of this result. In view of this property, the bimodules $C^*_r(G/N)$ are sometimes referred to as *C*-algebraic universal principal series*, adopting the terminology used in the context of $p$-adic groups, for instance in the the geometric approach to second adjointness (see [**BK15**]), which was a part of the original motivation to study these modules. See [**CCH16a**] for general adjunction properties of the C*-algebraic functors.

2.2.2. *Intertwining integrals and C*-normalization.* Another aspect of the C*-algebraic picture concerns the description of intertwining relations between principal series at the level of the Hilbert bimodules $C^*_r(G/N)$. The approach to this problem adopted in [**Cla15**] and [**Cla14**] is directly inspired by the work of A. Knapp and E. M. Stein on intertwining integrals and their normalization [**KS71, KS80**].

With $G$ and $P$ as before, we consider the parabolic subgroup $\bar{P}$ opposite to $P$. It decomposes as $\bar{P} = L\bar{N}$, giving rise to another $C^*_r(G)$-$C^*_r(L)$-correspondence $C^*_r(G/\bar{N})$. The C*-algebraic analogues of intertwining operators between principal series representations are defined as operators between correspondences:

DEFINITION 2.5. A *C*-algebraic intertwiner* is a bounded adjointable $C^*_r(L)$-linear operator

$$C^*_r(G/N) \longrightarrow C^*_r(G/\bar{N})$$

that commutes with the $C^*_r(G)$-actions.

The basic observation made in [**Cla13**] is that the standard intertwining integral

$$\mathcal{I}(f)(g\bar{N}) = \int_{\bar{N}} f(g\bar{n})\, d\bar{n} \tag{2.4}$$

converges for functions in $C_c(G/N) \subset C^*_r(G/N)$. However, although this expression formally satisfies good equivariance properties with respect to convolution actions of $C_c(G) \subset C^*_r(G)$ and $C_c(L) \subset C^*_r(L)$, it does not extend to a C*-algebraic intertwiner in the sense of Definition 2.5.

To circumvent this issue, we proposed in [**Cla15, Cla14**] a framework to adapt the normalization procedure used in Knapp-Stein theory.

DEFINITION 2.6 (C*-normalization). A map $\mathcal{U} : C^*_r(G/N) \longrightarrow C^*_r(G/\bar{N})$ is said to *normalize* the standard intertwining integral $\mathcal{I}$ in (2.4) if

(i) $\mathcal{U}$ is a C*-algebraic intertwiner that preserves $C^*_r(L)$-norms;
(ii) there exists a distribution $\Gamma$ on $L$ such that

$$\mathcal{U} = \mathcal{I} \circ C_\Gamma$$

on a dense subspace of functions in $C^*_r(G/N)$.

Here, $C_\Gamma$ denotes the convolution operator $f \mapsto f * \Gamma$.

A distribution $\Gamma$ satisfying (ii) in the definition above is said to *normalize* the standard integral $\mathcal{I}$.

The problem of normalizing standard intertwining integrals in the above sense has been solved so far in the particular case of maximal parabolic subgroup of the the special linear group. Both the unitary $\mathcal{U}$ and the normalizing distribution have been calculated explicitly, but we shall only need the latter for our current purposes:

PROPOSITION 2.7. [**Cla14**] *Let $G = \mathrm{SL}(n+1, F)$ and $P$ be a maximal parabolic subgroup $P$ with Levi component $L \simeq \mathrm{GL}(n, F)$. The distribution*

$$\Gamma = |\det(g)|^{\frac{n+1}{2}} \psi(\mathrm{tr}(g^{-1}))\, d^{\times} g$$

*normalizes $\mathcal{I}$.*

REMARK 2.8. The existence of C*-algebraic intertwiners in the sense of Definition 2.5 is crucial in the study of the structure of the reduced C*-algebra $C^*_r(G)$. It can be obtained in general by reformulating the results of [**KS71, KS80**]; see Theorem 6.1 in [**CCH16b**].

**2.3. Relationship in the case of** $\mathrm{GL}(n)$**.** We are now in a position to establish a connection between the problem of explicitly generating $\gamma$-factors in the sense of Braverman-Kazhdan and that of calculating normalizing distributions for C*-algebraic intertwining integrals. We consider the case of $\mathrm{GL}(n)$, where both problems have known answers.

We shall say that two distributions $S$ and $T$ on a group are *inverse* to each other if their convolution is the Dirac measure at the neutral element, that is,

$$S * T = \delta_{\mathrm{Id}}.$$

Moreover, if $T$ is a distribution on $\mathrm{GL}(n, F)$ of the form $u(g)\, d^{\times} g$, we write

$$\tilde{T} = u(-g)\, d^{\times} g.$$

PROPOSITION 2.9. *The distribution $\Delta_{\psi,\mathrm{St}_n}$ generating the Godement-Jacquet $\gamma$-factors is related to the C*-algebraic normalizing distribution $\Gamma$ associated to a parabolic subgroup of* $\mathrm{SL}(n+1, F)$ *with Levi component* $\mathrm{GL}(n, F)$ *by*

$$\Delta_{\psi,\mathrm{St}_n} = |\det|^{\frac{n+1}{2}}\, \tilde{\Gamma}^{-1}.$$

PROOF. The expression of $\Gamma$ given in Proposition 2.7 shows that

$$\tilde{\Gamma} = |\det(g)|^{\frac{n+1}{2}} \psi(\mathrm{tr}(-g^{-1}))\, d^{\times} g$$

A straightforward calculation shows that the inverse for convolution of a distribution of the form

$$|\det(g)|^{\alpha}\, \psi(\mathrm{tr}(-g^{-1}))\, d^{\times} g$$

is

$$|\det(g)|^{n-\alpha}\, \psi(\mathrm{tr}(g))\, d^{\times} g.$$

It follows that

$$|\det|^{\frac{n+1}{2}}\, \tilde{\Gamma}^{-1} = |\det(g)|^{n}\, \psi(\mathrm{tr}(g))\, d^{\times} g \tag{2.5}$$

and it suffices to check that this distribution satisfies the Braverman-Kazhdan relation (‡).

Indeed, given an irreducible representation $\pi$ of $\mathrm{GL}(n,F)$ and a matrix coefficient $f_\pi$, the functional equation (2.2) can be written as

$$\int_{\mathrm{GL}(n,F)} \hat{\Phi}(g) f_\pi(g^{-1}) |\det(g)|^{n-s}\, d^\times g \\ = \gamma(s,\pi) \int_{\mathrm{GL}(n,F)} \Phi(g) f_\pi(g) |\det(g)|^{s}\, d^\times g. \tag{2.6}$$

By definition of the Fourier transform (2.3), the left-hand side equals

$$\int_{\mathrm{GL}(n,F)} \int_{\mathrm{M}_n} \Phi(x)\, \psi(\mathrm{tr}(xy))\, f_\pi(y^{-1})\, |\det(y)|^{n-s}\, dx\, d^\times y$$

and since $dx = |\det(x)|^n\, d^\times x$, the fact that (2.6) holds for any $\Phi$ is equivalent to having

$$\int_{\mathrm{GL}(n,F)} |\det(x)|^n\, \psi(\mathrm{tr}(xy))\, f_\pi(y^{-1})\, |\det(y)|^{n-s}\, d^\times y \\ = \gamma(s,\pi) \cdot f_\pi(x) |\det(x)|^{s} \tag{2.7}$$

for any matrix coefficient $f_\pi$. Letting $g = xy$, the left-hand side becomes

$$\int_{\mathrm{GL}(n,F)} |\det(g)|^n\, \psi(\mathrm{tr}(g))\, (f_\pi \cdot |\det|^s)\, (g^{-1}x)\, d^\times z,$$

that is, according to (2.5)

$$\left(|\det|^{\frac{n+1}{2}}\, \tilde{\Gamma}^{-1} * (f_\pi \cdot |\det|^s)\right)(x).$$

This shows that the functional equation (2.6) is equivalent to

$$|\det|^{\frac{n+1}{2}}\, \tilde{\Gamma}^{-1} * (f_\pi \cdot |\det|^s) = \gamma(s,\pi) \cdot f_\pi |\det|^s$$

which is exactly the relation (‡), thus concluding the proof. □

REMARK 2.10. A similar method to normalize intertwiners (at the level of $L^2$ spaces) by convolution was used in [**BK02**]. See in particular Example 4.4.

## 3. Concluding remarks

To conclude this note, we shall point out that the existence of a relationship between $\gamma$-functions and normalization factors for intertwining integrals has been well-known for a long time. In particular the observation of Proposition 2.9 can be interpreted as a consequence of the Langlands-Shahidi method (see [**Sha10**] for a general exposition).

Assume that $F$ is non-Archimedean and consider the parabolic subgroup $P$ of type $(p,q)$ in $G = \mathrm{SL}(p+q,F)$, that is, the group of block upper-triangular matrices of the form

$$\begin{bmatrix} \alpha & * \\ 0 & \beta \end{bmatrix}$$

with $\alpha$ in $\mathrm{GL}(p,F)$ and $\beta$ in $\mathrm{GL}(q,F)$ satisfying $\det(\alpha)\cdot\det(\beta) = 1$. Denote by $L$ the Levi subgroup of block diagonal matrices and by $N$ the unipotent radical of $P$. The opposite parabolic subgroup containing $L$ is denoted by $\bar{P} = L\bar{N}$ and $\bar{N}$ consists of matrices of the form

$$\begin{bmatrix} \mathrm{Id}_p & 0 \\ * & \mathrm{Id}_q \end{bmatrix}.$$

In this setting, the standard integral (2.4) defines an unbounded operator

$$A : L^2(G/N) \longrightarrow L^2(G/\bar{N})$$

and F. Shahidi has proved in [**Sha84**] that the Rankin-Selberg $\gamma$-factors (called *local coefficients for* $P$ there) defined in [**JPSS83**] and mentioned in Remark 2.2 above normalize this operator in the sense that the resulting operator $\mathcal{A}$ is equivariant with respect to the left and right actions of $G$ and $L$ on the spaces $L^2(G/N)$ and $L^2(G/\bar{N})$ and is compatible with the Whittaker functionals defined on the Schwartz Spaces of $G/N$ and $G/\bar{N}$. The operator $\mathcal{A}$ is obtained as the convolution of $A$ and the distribution on $L$ that generates the Rankin-Selberg factors as in (†).

REMARK 3.1. A comparison between the Braverman-Kazhdan approach and the Langlands-Shahidi method can be found in the recent paper [**Sha17b**]. See also [**Sha17a**] and similar questions in connection with the doubling method in [**GL17**].

REMARK 3.2. Concerning real groups, results were obtained in [**Sha85**] and [**Art89**], where unitarity of $\mathcal{A}$ is established.

The case treated in the C*-algebraic framework in Proposition 2.7 above corresponds to the situation when $p = 1$. The result in Proposition 2.9 suggests that studying intertwining relations at the level of C*-algebraic universal principal series and performing C*-algebraic normalization explicitly may help shed some light on the problem of generating $\gamma$-factors in the sens of Braverman and Kazhdan. Of particular interest is the question of determining an explicit formula for the kernel of the normalized operator $\mathcal{A}$ or, equivalently, for the central distribution satisfying (†) beyond the Godement-Jacquet case.

**Acknowledgments.** The remark presented here was first made in the context of a seminar held in the Mathematics Department of the University of Orléans, involving A. Alvarez, P. Julg and V. Lafforgue. The author thanks Vincent Lafforgue for many enlightening discussions about $\gamma$-factors and Wen-Wei Li for bringing recent literature on the subject to his attention.

## References

[Art89] James Arthur, *Intertwining operators and residues. I. Weighted characters*, J. Funct. Anal. **84** (1989), no. 1, 19–84. MR999488

[BK00] A. Braverman and D. Kazhdan, *$\gamma$-functions of representations and lifting*, Geom. Funct. Anal. **Special Volume** (2000), 237–278. With an appendix by V. Vologodsky; GAFA 2000 (Tel Aviv, 1999). MR1826255

[BK02] *Dedicated to Yuri I. Manin on the occasion of his 65th birthday*, Independent University of Moscow, Moscow, 2002. Mosc. Math. J. **2** (2002), no. 2. MR1944504

[BK15] Roman Bezrukavnikov and David Kazhdan, *Geometry of second adjointness for p-adic groups*, Represent. Theory **19** (2015), 299–332. With an appendix by Yakov Varshavsky, Bezrukavnikov and Kazhdan. MR3430373

[CCH16a] Pierre Clare, Tyrone Crisp, and Nigel Higson, *Adjoint functors between categories of C*-modules*, J. Inst. Math. Jussieu **17** (2018), no. 2, 453–488, DOI 10.1017/S1474748016000074. MR3773277

[CCH16b] Pierre Clare, Tyrone Crisp, and Nigel Higson, *Parabolic induction and restriction via C*-algebras and Hilbert C*-modules*, Compos. Math. **152** (2016), no. 6, 1286–1318. MR3518312

[Cla13] Pierre Clare, *Hilbert modules associated to parabolically induced representations*, J. Operator Theory **69** (2013), no. 2, 483–509. MR3053351

[Cla14] Pierre Clare, C*-*algebraic intertwiners for degenerate principal series of special linear groups*, Chin. Ann. Math. Ser. B **35** (2014), no. 5, 691–702. MR3246931

[Cla15] Pierre Clare, *C*-algebraic intertwiners for principal series: case of* SL(2), J. Noncommut. Geom. **9** (2015), no. 1, 1–19. MR3337952

[Con94] Alain Connes, *Noncommutative geometry*, Academic Press, Inc., San Diego, CA, 1994. MR1303779

[Dix77] Jacques Dixmier, *$C^*$-algebras*, North-Holland Publishing Co., Amsterdam-New York-Oxford, 1977. Translated from the French by Francis Jellett; North-Holland Mathematical Library, Vol. 15. MR0458185

[GJ72] Roger Godement and Hervé Jacquet, *Zeta functions of simple algebras*, Lecture Notes in Mathematics, Vol. 260, Springer-Verlag, Berlin-New York, 1972. MR0342495

[GL17] J. R. Getz and B. Liu, *A refined Poisson summation formula for certain Braverman-Kazhdan spaces*, 2017, preprint arXiv:1707.06091.

[Jac79] Hervé Jacquet, *Principal L-functions of the linear group*, Automorphic forms, representations and $L$-functions (Proc. Sympos. Pure Math., Oregon State Univ., Corvallis, Ore., 1977), Proc. Sympos. Pure Math., XXXIII, Amer. Math. Soc., Providence, R.I., 1979, pp. 63–86. MR546609

[Jac97] Hervé Jacquet, *Principal L-functions for* GL$(n)$, Representation theory and automorphic forms (Edinburgh, 1996), Proc. Sympos. Pure Math., vol. 61, Amer. Math. Soc., Providence, RI, 1997, pp. 321–329. MR1476503

[JPSS83] H. Jacquet, I. I. Piatetskii-Shapiro, and J. A. Shalika, *Rankin-Selberg convolutions*, Amer. J. Math. **105** (1983), no. 2, 367–464. MR701565

[Kna94] A. W. Knapp, *Local Langlands correspondence: the Archimedean case*, Motives (Seattle, WA, 1991), Proc. Sympos. Pure Math., vol. 55, Amer. Math. Soc., Providence, RI, 1994, pp. 393–410. MR1265560

[KS71] A. W. Knapp and E. M. Stein, *Intertwining operators for semisimple groups*, Ann. of Math. (2) **93** (1971), 489–578. MR0460543

[KS80] A. W. Knapp and E. M. Stein, *Intertwining operators for semisimple groups. II*, Invent. Math. **60** (1980), no. 1, 9–84. MR582703

[Kud94] Stephen S. Kudla, *The local Langlands correspondence: the non-Archimedean case*, Motives (Seattle, WA, 1991), Proc. Sympos. Pure Math., vol. 55, Amer. Math. Soc., Providence, RI, 1994, pp. 365–391. MR1265559

[Rie74] Marc A. Rieffel, *Induced representations of $C^*$-algebras*, Advances in Math. **13** (1974), 176–257. MR0353003

[Sha84] Freydoon Shahidi, *Fourier transforms of intertwining operators and Plancherel measures for* GL$(n)$, Amer. J. Math. **106** (1984), no. 1, 67–111. MR729755

[Sha85] Freydoon Shahidi, *Local coefficients as Artin factors for real groups*, Duke Math. J. **52** (1985), no. 4, 973–1007. MR816396

[Sha10] Freydoon Shahidi, *Eisenstein series and automorphic L-functions*, American Mathematical Society Colloquium Publications, vol. 58, American Mathematical Society, Providence, RI, 2010. MR2683009

[Sha17a] ———, *Local factors, reciprocity and Vinberg monoids*, Prime Numbers and Representation Theory, Lecture Series of Modern Number Theory, vol. 2, Science Press, Beijing, 2017, pp. 200–256.

[Sha17b] ———, *On generalized Fourier transforms for standard L-functions (with an appendix by Wen-Wei Li)*, 2017, preprint arXiv:1710.06841.

[Tat67] J. T. Tate, *Fourier analysis in number fields, and Hecke's zeta-functions*, Algebraic Number Theory (Proc. Instructional Conf., Brighton, 1965), Thompson, Washington, D.C., 1967, pp. 305–347. MR0217026

Department of Mathematics, Dartmouth College, HB 6188, Hanover, New Hampshire 03755-3551

*Email address*: clare@math.dartmouth.edu

Contemporary Mathematics
Volume **714**, 2018
https://doi.org/10.1090/conm/714/14383

# Radial Toeplitz operators on the weighted Bergman spaces of Cartan domains

Matthew Dawson and Raúl Quiroga-Barranco

ABSTRACT. Let $D$ be an irreducible bounded symmetric domain with biholomorphism group $G$ with maximal compact subgroup $K$. For the Toeplitz operators with $K$-invariant symbols we provide explicit simultaneous diagonalization formulas on every weighted Bergman space. The expressions are given in the general case, but are also worked out explicitly for every irreducible bounded symmetric domain including the exceptional ones.

## 1. Introduction

Irreducible bounded symmetric domains and the weighted Bergman spaces that they support constitute a fundamental object in functional analysis. These allow us to consider Toeplitz operators and irreducible unitary representations of Lie groups. In the last two decades, it has been observed that there is a natural and far reaching relationship between these operators and representations. This work continues the study of Toeplitz operators derived from such interaction.

For $D$ a bounded symmetric domain with biholomorphism group $G$, it was proved in [**1**] that there are plenty of closed subgroups $H$ in $G$ such that the Toeplitz operators with $H$-invariant symbols pairwise commute. This yields commutative $C^*$-algebras generated by Toeplitz operators. Ultimately, this result is a consequence of multiplicity-free restriction results. Among the latter, the very first one is found in [**15**] for the maximal compact subgroup $K$ in $G$ that fixes a given point.

We consider the multiplicity-free decomposition for the restriction to $K$ of the unitary representations of $G$ on the Bergman spaces, as found in [**15**] or [**18**]. This is used to achieve our main goal: an explicit simultaneous diagonalization of Toeplitz operators with $K$-invariant symbols. A general but also very explicit formula is obtained in Theorem 4.12, where every $K$-invariant symbol is really seen as a complex-valued function defined on $[0,1)^r$, where $r$ is the rank of $D$. Correspondingly, our formulas are given by integrals computed in this set. Our computations make use of the spherical and conical polynomials for the representation of $K$ on the space of polynomials on the domain $D$. Hence, Theorem 4.11 provides the simultaneous diagonalization in terms of these polynomials. Since there are explicit expressions for the conical polynomials (see [**8**]) this should prove to be useful in further developments.

---

2010 *Mathematics Subject Classification.* Primary 47B35, 22D10; Secondary 32M15, 22E46.
Research supported by SNI and Conacyt.

On the other hand, we rely strongly on the Jordan structures associated to the irreducible bounded symmetric domains. For example, the preliminary computation given by Theorem 4.9 reduces the diagonalization of a Toeplitz operator with $K$-invariant symbol to integrals over the cone of the largest tube-type domain contained in $D$. Most importantly, the Jordan structure allows us to consider primitives (a sort of minimal tripotent) thus making possible to obtain the formula in Theorem 4.12 that involves integrals over $[0,1)^r$. This integration is in fact given by taking coordinates with respect to a maximal collection of primitives (in the Jordan setup) or a basis of a maximal flat (in the Lie setup). In the last section we exhibit examples of these sets for every irreducible bounded symmetric domain, including the exceptional ones, which leads us to completely explicit formulas that provide the diagonalization of Toeplitz operators with $K$-invariant symbols.

The properties of Jordan pairs and analysis on irreducible bounded symmetric domains that we require are reviewed in Sections 2 and 3, respectively. We obtain our main results for irreducible bounded symmetric domains in the last two sections: Section 4 with general expressions for an arbitrary domain and Section 5 with explicit formulas for every irreducible case.

## 2. Jordan structures associated to bounded symmetric domains

In this section we collect some standard facts on bounded symmetric domains. The standard references are [**2, 3, 5, 6, 12, 18, 19**] where the reader will find further details and proofs.

Let $D$ be an irreducible Hermitian symmetric space of non-compact type, and denote by $G$ the connected component of the automorphism group of $D$ and by $K$ the subgroup of $G$ fixing some point. Consider the corresponding Cartan decomposition given by

$$\mathfrak{g} = \mathfrak{k} \oplus \mathfrak{p}$$

where $\mathfrak{g}$ and $\mathfrak{k}$ are the Lie algebras of $G$ and $K$, respectively. Choose $\mathfrak{h}$ a Cartan subalgebra of $\mathfrak{k}$, which is thus a Cartan subalgebra of $\mathfrak{g}$. For the complexified Lie algebras, we recall that a root $\alpha$ of $\mathfrak{g}^{\mathbb{C}}$ with respect to $\mathfrak{h}^{\mathbb{C}}$ is called compact if the corresponding root space $\mathfrak{g}_\alpha$ is contained in $\mathfrak{k}^{\mathbb{C}}$. Let us fix an order and let us denote by $\Phi^+$ the space of all positive noncompact roots. We consider the complex space

$$\mathfrak{p}_+ = \sum_{\alpha \in \Phi^+} \mathfrak{g}_\alpha.$$

Then, it is well known that $D$ admits a canonical embedding as a circled bounded domain in $\mathfrak{p}_+$ such that $K$ is precisely the isotropy subgroup of the origin: this is the Harish-Chandra realization of $D$. The complex space $\mathfrak{p}_+$ carries a natural Hermitian inner product obtained from the Killing form of $\mathfrak{g}^{\mathbb{C}}$ and the conjugation $\tau$ with respect to the real form $\mathfrak{k} \oplus i\mathfrak{p}$.

Every element $X \in \mathfrak{p}_+$ can be considered as a holomorphic vector field over $D$ whose one-parameter group of transformations is $t \mapsto \exp(tX)$ acting on $D$, and so $X$ can be seen as a holomorphic map $D \to \mathfrak{p}_+$. Then, the vector field $X$ can be written as

$$X(z) = y - Q_y(z,z),$$

for every $z \in D$, where $y = X(0)$ and $Q_y(z,w)$ is a complex bilinear form of $(z,w)$. Furthermore, for every $z, w$ the assignment $y \mapsto Q_y(z,w)$ is complex anti-linear.

This yields a triple product

$$\begin{aligned} \mathfrak{p}_+ \times \mathfrak{p}_+ \times \mathfrak{p}_+ &\rightarrow \mathfrak{p}_+ \\ \{zyw\} &= Q_y(z,w), \end{aligned}$$

that defines a Jordan pair which also carries a natural involutive anti-linear map $z \mapsto z^*$. From now on, we will consider $\mathfrak{p}_+$ endowed with this algebraic structure.

For this setup, an element $z \in \mathfrak{p}_+$ is called tripotent if and only if $\{zzz\} = z$, and two tripotent elements $z, w$ are called orthogonal when $\{zwy\} = 0$ for all $y \in \mathfrak{p}_+$. A tripotent element is called primitive if and only if it is not the sum of two orthogonal tripotents. Let $e_1, \dots, e_r$ be a maximal collection of mutually orthogonal primitive tripotents. It follows from [**12**] that $r$ is precisely the rank of $D$. It will be useful to consider the element

$$e = e_1 + \cdots + e_r,$$

that is a tripotent element in $\mathfrak{p}_+$ as well.

For every tripotent $z$ there is a so-called Peirce decomposition

$$\mathfrak{p}_+ = V_1(z) \oplus V_{\frac{1}{2}}(z) \oplus V_0(z),$$

where

$$V_\alpha(z) = \{w \in \mathfrak{p}_+ \mid \{zzw\} = \alpha w\}.$$

Then, the space $V_1(z)$ is a Jordan algebra with the product $x \circ y = \{xzy\}$ and for this structure $z$ is the identity.

From now on, we will denote

$$V_j = V_1(e_1 + \cdots + e_j)$$

for $j = 1, \dots, r$, which is thus a Jordan algebra with identity $e_1 + \cdots + e_j$. We observe that

$$V_1 \subset \cdots \subset V_r.$$

It follows from [**2**] that $D_j = V_j \cap D$ is a bounded symmetric domain of tube-type whose associated symmetric cone is given by

$$\Omega_j = \{x^2 \mid x \in V_j,\ x^* = x\}.$$

Let us denote by $L_j$ the identity component of the isotropy subgroup of automorphisms of $\Omega_j$ that fix $e_1 + \cdots + e_j$. By the results from [**2, 11**] there is a unique polynomial $\Delta_j(z)$ in $V_j$ that is $L_j$-invariant and that satisfies

$$\Delta_j(t_1 e_1 + \cdots + t_j e_j) = t_1 \dots t_j. \tag{2.1}$$

We will consider the polynomials $\Delta_1(z), \dots, \Delta_r(z)$ as defined on $\mathfrak{p}_+$ by orthogonal projection.

For the case $j = r$ we drop the index, and so we have $V = V_r$, $\Omega = \Omega_r$, $\Delta(z) = \Delta_r(z)$. We will also denote

$$D_T = V \cap D,$$

which is a bounded symmetric domain of tube-type. We will call $D_T$ the tube-type domain associated to $D$. In fact, $D_T$ is the largest tube-type domain obtained as the intersection of $D$ with a complex subspace of $\mathfrak{p}_+$.

On the other hand, there is a joint Peirce decomposition associated to $e_1, \dots, e_r$ given by

$$\mathfrak{p}_+ = \bigoplus_{0 \le j,k \le r} V_{jk},$$

where we define

$$\begin{aligned} V_{jj} &= V_1(e_j), \quad \text{for } j = 1, \ldots, r, \\ V_{jk} &= V_{\frac{1}{2}}(e_j) \cap V_{\frac{1}{2}}(e_k), \quad \text{for } j, k = 1, \ldots, r, \\ V_{j0} = V_{0j} &= V_{\frac{1}{2}}(e_j) \cap \bigcap_{k \neq j} V_0(e_k), \quad \text{for } j = 1, \ldots, r, \\ V_{00} &= V_0(e_1) \cap \cdots \cap V_0(e_r). \end{aligned}$$

We note that since $e_1, \ldots, e_r$ is maximal, it turns out that $V_{00} = 0$. It is also well known that there exists a pair of integers $a > 0$ and $b \geq 0$ such that

$$\begin{aligned} \dim V_{jj} &= 1, \quad \text{for } j = 1, \ldots, r, \\ \dim V_{jk} &= a, \quad \text{for } j, k = 1, \ldots, r, \\ \dim V_{j0} &= b, \quad \text{for } j = 1, \ldots, r. \end{aligned}$$

The numbers $a, b$ are the characteristic multiplicities of $D$. We observe that $D$ is of tube-type if and only if $b = 0$. If we denote by $n$ and $n_T$ the dimensions of $D$ and $D_T$, respectively, then we have

$$n = r + \frac{r(r-1)}{2}a + rb, \quad n_T = r + \frac{r(r-1)}{2}a.$$

The genus of the domain $D$ is defined by

$$p = \frac{n + n_T}{r} = 2 + (r-1)a + b.$$

Finally, it is also well known that there is a set of strongly orthogonal roots $\gamma_1, \ldots, \gamma_r$ in the sense of Harish-Chandra such that, for every $j = 1, \ldots, r$, the tripotent $e_j$ belongs to the root space $\mathfrak{g}_{\gamma_j}$. Without loss of generality, we can assume that $\gamma_1 > \cdots > \gamma_r > 0$.

## 3. Analysis on bounded symmetric domains

With the setup provided by Section 2 we have the following integral formula (see [**2**, **6**, **18**])

$$(3.1) \quad \int_{\mathfrak{p}_+} f(z)\, \mathrm{d}z$$
$$= c \int_{[0,\infty)^r} \int_K f\Big(g \cdot \sum_{j=1}^r t_j e_j\Big)\, \mathrm{d}g \prod_{1 \leq j < k \leq r} |t_j^2 - t_k^2|^a \prod_{j=1}^r t_j^{2b+1}\, \mathrm{d}t_1 \ldots \mathrm{d}t_r$$

where $\mathrm{d}z$ is the Lebesgue measure on $\mathfrak{p}_+$ corresponding to its Hermitian structure, $\mathrm{d}g$ is the probability Haar measure of $K$ and $c$ is a positive constant whose value will not be used.

We also have the following integral formula.

$$(3.2) \quad \int_\Omega f(x)\, \mathrm{d}x$$
$$= c' \int_{[0,\infty)^r} \int_L f\Big(s \cdot \sum_{j=1}^r x_j e_j\Big)\, \mathrm{d}s \prod_{1 \leq j < k \leq r} |x_j^2 - x_k^2|^a\, \mathrm{d}x_1 \ldots \mathrm{d}x_r,$$

where $\mathrm{d}x$ is the Lebesgue measure on $\Omega$, $\mathrm{d}g$ is the Haar measure of $L$ and $c'$ is a constant whose value we will not need.

On the other hand, for the Lebesgue measure on $\mathfrak{p}_+$, we have (see [**10**])

$$\mathrm{vol}(D) = \pi^n \prod_{j=1}^{r} \frac{\Gamma(p - \frac{n}{r} - (j-1)\frac{a}{2})}{\Gamma(p - (j-1)\frac{a}{2})}.$$

Hence, we consider the normalized Lebesgue measure

$$\mathrm{d}v(z) = \frac{1}{\mathrm{vol}(D)}\,\mathrm{d}z = \frac{1}{\pi^n} \prod_{j=1}^{r} \frac{\Gamma(p - (j-1)\frac{a}{2})}{\Gamma(p - \frac{n}{r} - (j-1)\frac{a}{2})}\,\mathrm{d}z.$$

The (weigthless) Bergman space is the closed subspace $\mathcal{A}^2(D)$ of $L^2(D, v)$ that consists of holomorphic functions. The reproducing kernel of this space can be described as follows.

Let us define the Bergman endomorphisms

$$\begin{aligned} B(x,y) &: \mathfrak{p}_+ \to \mathfrak{p}_+ \\ B(x,y)z &= z - 2\{xybz\} + \{x\{yzy\}x\}, \end{aligned}$$

for any $x, y \in \mathfrak{p}_+$. Then, the reproducing kernel $K_D$ of $\mathcal{A}^2(D)$ is given by (see [**12**])

$$\begin{aligned} K_D &: D \times D \to \mathbb{C} \\ K_D(z,w) &= \det(B(z,w))^{-1}. \end{aligned}$$

Furthermore, as noted in [**2**] there is a real polynomial $h(z,w)$ on $\mathfrak{p}_+ \times \mathfrak{p}_+$, holomorphic in $z$ and anti-holomorphic in $w$, such that

$$h(z,w)^p = \det(B(z,w)).$$

It can also be described as the unique $K$-invariant polynomial such that

$$h(x,x) = \Delta(e - x^2) = \prod_{j=1}^{r} (1 - x_j^2) \tag{3.3}$$

for every $x = x_1 e_1 + \cdots + x_r e_r$ ($x_j \in \mathbb{R}$) and with the square $x^2$ computed in the Jordan algebra $V$. Note that we have used equation (2.1) with $j = r$.

Hence, the Bergman kernel can also be computed as

$$K_D(z,w) = h(z,w)^{-p},$$

for every $z, w \in D$.

The following weighted volume computation can also be obtained (see [**18**])

$$\int_D h(z,z)^{\lambda - p}\,\mathrm{d}z = \pi^n \prod_{j=1}^{r} \frac{\Gamma(\lambda - \frac{n}{r} - (j-1)\frac{a}{2})}{\Gamma(\lambda - (j-1)\frac{a}{2})},$$

which holds for every $\lambda > p - 1$. Hence, for every such $\lambda$ we consider the weighted measure

$$\mathrm{d}v_\lambda(z) = \frac{1}{\pi^n} \prod_{j=1}^{r} \frac{\Gamma(\lambda - (j-1)\frac{a}{2})}{\Gamma(\lambda - \frac{n}{r} - (j-1)\frac{a}{2})} h(z,z)^{\lambda - p}\,\mathrm{d}z,$$

which thus satisfies $v_\lambda(D) = 1$.

For every $\lambda > p - 1$, the weighted Bergman space $\mathcal{A}^2_\lambda(D)$ corresponding to $\lambda$ is given by

$$\mathcal{A}^2_\lambda(D) = \{f \in L^2(D, v_\lambda) \mid f \text{ is holomorphic}\}.$$

This is a closed subspace of $L^2(D, v_\lambda)$ and a reproducing kernel space with Bergman kernel given by

$$K_{D,\lambda} : D \times D \to \mathbb{C}$$
$$K_{D,\lambda}(z,w) = h(z,w)^{-\lambda}.$$

In particular, the corresponding Bergman projection $B_{D,\lambda} : L^2(D, v_\lambda) \to \mathcal{A}^2_\lambda(D)$ satisfies

$$\begin{aligned} B_{D,\lambda}(f)(z) &= \int_D f(w)h(z,w)^{-\lambda}\, \mathrm{d}v_\lambda(w) \\ &= \frac{1}{\pi^n}\prod_{j=1}^{r}\frac{\Gamma(\lambda-(j-1)\frac{a}{2})}{\Gamma(\lambda-\frac{n}{r}-(j-1)\frac{a}{2})}\int_D \frac{f(w)h(w,w)^{\lambda-p}\,\mathrm{d}w}{h(z,w)^\lambda}. \end{aligned}$$

We note that $\mathcal{A}^2_p(D) = \mathcal{A}^2(D)$ is the Bergman space corresponding to the (weightless) Lebesgue measure.

For any function $\varphi \in L^\infty(D)$, the Toeplitz operator on $\mathcal{A}^2_\lambda(D)$ with bounded symbol $\varphi$ is defined by

$$T_\varphi : \mathcal{A}^2_\lambda(D) \to \mathcal{A}^2_\lambda(D)$$
$$T_\varphi(f) = B_{D,\lambda}(\varphi f).$$

On the other hand, for every $\lambda > p-1$ there is a unitary representation of the universal covering group $\widetilde{G}$ of $G$

$$\pi_\lambda : \widetilde{G} \times \mathcal{A}^2_\lambda(D) \to \mathcal{A}^2_\lambda(D)$$
$$(\pi_\lambda(g)f)(z) = J(g^{-1}, z)^{\frac{\lambda}{p}} f(g^{-1}z),$$

where $J(g,z)$ denotes the complex Jacobian of the transformation $g$ at the point $z$. These yield the analytic continuation of the scalar-type holomorphic discrete series representations of $G$.

Since $K$ is the subgroup of linear unitary transformations on $\mathfrak{p}_+$ that belong to $G$ we have $J(g,z) \equiv 1$ for all $g \in K$. In particular, for every $\lambda > p-1$ the above representation $\pi_\lambda$ yields a unitary representation of $K$ given by

$$\pi_\lambda|_K : K \times \mathcal{A}^2_\lambda(D) \to \mathcal{A}^2_\lambda(D)$$
$$(\pi_\lambda(g)f)(z) = f(g^{-1}z).$$

## 4. Radial operators and the isotypic decomposition of $\pi_\lambda|_K$

Let $\pi : H \to \mathrm{U}(\mathcal{H})$ be a unitary representation of a Lie group $H$. We recall that a bounded operator $T : \mathcal{H} \to \mathcal{H}$ is called $\pi$-intertwining if we have

$$T \circ \pi(h) = \pi(h) \circ T$$

for every $h \in H$. The algebra of $\pi$-intertwining bounded operators is denoted by $\mathrm{End}_H(\mathcal{H})$. This algebra is a fundamental invariant to understand the decomposition of $\mathcal{H}$ into irreducible subspaces. In particular, we have the following well known result where we have considered the case of compact groups.

PROPOSITION 4.1. *The unitary representation $\pi$ of the group $H$ on the Hilbert space $\mathcal{H}$ is multiplicity-free if and only if $\mathrm{End}_H(\mathcal{H})$ is commutative. For $H$ compact, this occurs exactly when $\mathcal{H}$ can be decomposed as a direct sum of inequivalent irreducible $H$-invariant closed subspaces.*

By focusing our attention to the subgroup $K$ of $G$ and the holomorphic discrete series we obtain a special type of operators.

DEFINITION 4.2. A bounded operator $T : \mathcal{A}^2_\lambda(D) \to \mathcal{A}^2_\lambda(D)$ is called radial if it is $\pi_\lambda|_K$-intertwining, i.e. if $T$ satisfies

$$T \circ \pi_\lambda(g) = \pi_\lambda(g) \circ T$$

for every $g \in K$. The algebra of bounded radial operators on $\mathcal{A}^2_\lambda(D)$ is denoted by $\mathrm{End}_K(\mathcal{A}^2_\lambda(D))$.

It is well known that the unitary representations $\pi_\lambda$ are irreducible highest weight representations of the group $\widetilde{G}$. The fact that $K$ is a maximal compact subgroup of $G$ thus implies that the restrictions $\pi_\lambda|_K$ are multiplicity-free: this follows, for example, from the results in [**9**] and [**13**].

As a consequence of the previous remarks we obtain the following.

PROPOSITION 4.3. *For every $\lambda > p - 1$, the algebra $\mathrm{End}_K(\mathcal{A}^2_\lambda(D))$ is commutative.*

We now consider the corresponding invariance property for symbols and their associated Toeplitz operators.

DEFINITION 4.4. A bounded symbol $\psi \in L^\infty(D)$ is called *radial* if it is $K$-invariant. In other words, if for every $g \in K$ we have

$$\psi(gz) = \psi(z)$$

for almost every $z \in D$.

We recall the following well known result. It follows from [**5**] by using the fact that the real span of $e_1, \dots, e_r$ is a maximal Abelian subspace of $\mathfrak{p}_+$.

PROPOSITION 4.5. *For every $z \in D$ there exist $g \in K$ and $t \in [0,1)^r$ such that*

$$z = g \sum_{j=1}^{r} t_j e_j.$$

*In particular, a bounded symbol $\psi \in L^\infty(D)$ is radial if and only if we have*

$$\psi(z) = \psi\Big(\sum_{j=1}^{r} t_j e_j\Big),$$

*when the above relation holds between $z \in D$ and $t \in [0,1)^r$.*

It was observed in [**1**] the equivalence between the invariance of a symbol and of its associated Toeplitz operators. In particular, we have the following result which is a particular case of Corollary 3.3 and Theorem 5.2 from [**1**].

PROPOSITION 4.6 (DOQ [**1**]). *For a given bounded symbol $\psi \in L^\infty(D)$ and a fixed $\lambda > p - 1$ the following conditions are equivalent*

(1) *The symbol $\psi$ is radial.*
(2) *The Toeplitz operator $T_\psi$ on $\mathcal{A}^2_\lambda(D)$ is radial.*

*Hence, if we denote by $\mathcal{A}^R$ the space of bounded radial symbols and by $\mathcal{T}^{(\lambda)}(\mathcal{A}^R)$ the $C^*$-algebra generated by the Toeplitz operators on $\mathcal{A}^2_\lambda(D)$ with bounded radial symbols, then we have*

$$\mathcal{T}^{(\lambda)}(\mathcal{A}^R) \subset \mathrm{End}_K(\mathcal{A}^2_\lambda(D)).$$

*In particular, for every $\lambda > p - 1$ the $C^*$-algebra $\mathcal{T}^{(\lambda)}(\mathcal{A}^R)$ is commutative.*

We now describe the multiplicity-free decomposition of $\mathcal{A}^2_\lambda(D)$ with respect to the representation $\pi_\lambda|_K$.

Recall that the space of polynomials $\mathcal{P}(\mathfrak{p}_+)$ is a dense $\pi_\lambda|_K$-invariant subspace of $\mathcal{A}^2_\lambda(D)$ for every $\lambda > p-1$. Hence, it is enough to describe the isotypic decomposition of the representation of $K$ on $\mathcal{P}(\mathfrak{p}_+)$ (see [**15**] and [**18**]).

Let us denote by $\overrightarrow{\mathbb{N}}^r$ the space of $r$-tuples of integers $(\alpha_1, \dots, \alpha_r)$ satisfying the condition $\alpha_1 \geq \cdots \geq \alpha_r \geq 0$. Then, the highest weights of the irreducible $K$-submodules contained in $\mathcal{P}(\mathfrak{p}_+)$ are precisely those of the form

$$-\alpha_1\gamma_1 - \cdots - \alpha_r\gamma_r \tag{4.1}$$

for some $\alpha \in \overrightarrow{\mathbb{N}}^r$. Furthermore, for each $\alpha \in \overrightarrow{\mathbb{N}}^r$ there is exactly one such $K$-submodule that we will denote by $\mathcal{P}^\alpha(\mathfrak{p}_+)$. In particular, we have a $K$-invariant algebraic direct sum

$$\mathcal{P}(\mathfrak{p}_+) = \bigoplus_{\alpha \in \overrightarrow{\mathbb{N}}^r} \mathcal{P}^\alpha(\mathfrak{p}_+),$$

that induces the multiplicity-free isotypic decomposition for the representation $\pi_\lambda|_K$ given by

$$\mathcal{A}^2_\lambda(D) = \bigoplus_{\alpha \in \overrightarrow{\mathbb{N}}^r} \mathcal{P}^\alpha(\mathfrak{p}_+).$$

In particular, for every weight of the form (4.1) the corresponding space of highest vectors in $\mathcal{P}(\mathfrak{p}_+)$ is 1-dimensional.

The following is an immediate consequence of the previous discussion.

PROPOSITION 4.7. *For every radial operator $T : \mathcal{A}^2_\lambda(D) \to \mathcal{A}^2_\lambda(D)$ and for every $\alpha \in \overrightarrow{\mathbb{N}}^r$ let us denote by $c_\alpha(T)$ the complex number such that*

$$T|_{\mathcal{P}^\alpha(\mathfrak{p}_+)} = c_\alpha(T) I_{\mathcal{P}^\alpha(\mathfrak{p}_+)}.$$

*Then, the map defined by*

$$\begin{aligned} \mathrm{End}_{\pi_\lambda|_K}(\mathcal{A}^2_\lambda(D)) &\to \ell_\infty(\overrightarrow{\mathbb{N}}^r) \\ T &\mapsto (c_\alpha(T))_{\alpha \in \overrightarrow{\mathbb{N}}^r}, \end{aligned}$$

*is an isomorphism of $C^*$-algebras. Furthermore, for every radial operator $T$ we have*

$$c_\alpha(T) = \frac{\langle T\phi, \phi\rangle_\lambda}{\langle \phi, \phi\rangle_\lambda},$$

*for every non-zero $\phi \in \mathcal{P}^\alpha(\mathfrak{p}_+)$ and for every $\alpha \in \overrightarrow{\mathbb{N}}^r$.*

On the other hand (see [**2, 8, 18**]), it is well known that the polynomial $\Delta_j(z)$ on $\mathfrak{p}_+$ is a highest weight vector for $\pi_\lambda|_K$ with corresponding weight $-\gamma_1 - \cdots - \gamma_j$, for every $j = 1, \dots, r$. In particular, the space of highest weight vectors is the free polynomial algebra $\mathbb{C}[\Delta_1(z), \dots, \Delta_r(z)]$. As a consequence of this, a highest weight vector corresponding to (4.1) is given by

$$\Delta_\alpha(z) = \Delta_1(z)^{\alpha_1 - \alpha_2}\Delta_2(z)^{\alpha_2-\alpha_3} \dots \Delta_r(z)^{\alpha_r},$$

for every $\alpha \in \overrightarrow{\mathbb{N}}^r$. These are called the conical polynomials associated to $\pi_\lambda|_K$.

We have the following naturally defined polynomial

$$\phi_\alpha(z) = \int_L \Delta_\alpha(sz)\, \mathrm{d}s,$$

for every $\alpha \in \overrightarrow{\mathbb{N}}^r$, which belongs to $\mathcal{P}^\alpha(\mathfrak{p}_+)$ and that is ($L$-)spherical for the representation $\pi_\lambda|_K$ restricted to $\mathcal{P}^\alpha(\mathfrak{p}_+)$.

The next result is proved, for example, in **[18]** (see Theorem 2.8.10 therein).

LEMMA 4.8. *With the above notation we have*

$$\int_K |\phi_\alpha(kz)|^2 \, \mathrm{d}k = \frac{1}{d_\alpha} \phi_\alpha(z^2),$$

*for every $\alpha \in \overrightarrow{\mathbb{N}}^r$ and $z \in V$, where $d_\alpha = \dim \mathcal{P}^\alpha(\mathfrak{p}_+)$ and for $z^2$ computed using the Jordan algebra structure of $V$ described in Section 2.*

The previous results allow us to compute the coefficients described in Proposition 4.7 for Toeplitz operators. The first step is given by the following result. We recall that for every $x$ in the cone $\Omega$ there exists a unique $y \in \Omega$ such that $y^2 = x$, and we denote $y = \sqrt{x}$.

THEOREM 4.9. *Let $D$ be a bounded symmetric domain and $\psi \in L^\infty(D)$ a radial symbol. Then, for every $\lambda > p - 1$ and for every $\alpha \in \overrightarrow{\mathbb{N}}^r$ we have*

$$\begin{aligned}\langle T_\psi \phi_\alpha, \phi_\alpha \rangle_\lambda &= \langle \psi \phi_\alpha, \phi_\alpha \rangle_\lambda \\ &= C \int_{\Omega \cap (e - \Omega)} \psi(\sqrt{x}) \Delta_\alpha(x) \Delta(e - x)^{\lambda - p} \Delta(x)^b \, \mathrm{d}x,\end{aligned}$$

*for some constant $C$ independent of $\psi$ and $\alpha$.*

PROOF. We fix $\psi$, $\lambda$ and $\alpha$ to compute as follows.

First, we apply the integral formula (3.1) to obtain

$$\begin{aligned}&\langle \psi \phi_\alpha, \phi_\alpha \rangle_\lambda = c_\lambda \int_D \psi(z) |\phi_\alpha(z)|^2 h(z, z)^{\lambda - p} \, \mathrm{d}z \\ &= c c_\lambda \int_{[0,1)^r} \int_K \psi\Big(g \cdot \sum_{j=1}^r t_j e_j\Big) \Big|\phi_\alpha\Big(g \cdot \sum_{j=1}^r t_j e_j\Big)\Big|^2 h\Big(g \cdot \sum_{j=1}^r t_j e_j, g \cdot \sum_{j=1}^r t_j e_j\Big)^{\lambda - p} \mathrm{d}g \\ &\qquad \times \prod_{1 \le j < k \le r} |t_j^2 - t_k^2|^a \prod_{j=1}^r t_j^{2b+1} \, \mathrm{d}t_1 \ldots \mathrm{d}t_r,\end{aligned}$$

where we take

$$c_\lambda = \frac{1}{\pi^n} \prod_{j=1}^r \frac{\Gamma(\lambda - (j-1)\frac{a}{2})}{\Gamma(\lambda - \frac{n}{r} - (j-1)\frac{a}{2})}$$

Note that we have restricted the integral to $[0,1)^r$ in order to integrate over $D$. Since $\psi, h$ are $K$-invariant and by Lemma 4.8 we now obtain

$$\begin{aligned}&\langle \psi \phi_\alpha, \phi_\alpha \rangle_\lambda = \\ &= \frac{c c_\lambda}{d_\alpha} \int_{[0,1)^r} \psi\Big(\sum_{j=1}^r t_j e_j\Big) \phi_\alpha\Big(\sum_{j=1}^r t_j^2 e_j\Big) h\Big(\sum_{j=1}^r t_j e_j, \sum_{j=1}^r t_j e_j\Big)^{\lambda - p} \\ &\times \prod_{1 \le j < k \le r} |t_j^2 - t_k^2|^a \prod_{j=1}^r t_j^{2b+1} \, \mathrm{d}t_1 \ldots \mathrm{d}t_r\end{aligned}$$

and using equation (3.3) we get

$$
\begin{aligned}
&= \frac{cc_\lambda}{d_\alpha} \int_{[0,1)^r} \psi\Big(\sum_{j=1}^r t_j e_j\Big) \phi_\alpha\Big(\sum_{j=1}^r t_j^2 e_j\Big) \Delta\Big(e - \sum_{j=1}^r t_j^2 e_j\Big)^{\lambda - p} \\
&\quad \times \prod_{1 \le j < k \le r} |t_j^2 - t_k^2|^a \prod_{j=1}^r t_j^{2b+1} \, \mathrm{d}t_1 \dots \mathrm{d}t_r \\
&= \frac{cc_\lambda}{d_\alpha} \int_{[0,1)^r} \psi\Big(\sum_{j=1}^r \sqrt{x_j} e_j\Big) \phi_\alpha\Big(\sum_{j=1}^r x_j e_j\Big) \Delta\Big(e - \sum_{j=1}^r x_j e_j\Big)^{\lambda - p} \\
&\quad \times \prod_{1 \le j < k \le r} |x_j - x_k|^a \prod_{j=1}^r x_j^b \, \mathrm{d}x_1 \dots \mathrm{d}x_r
\end{aligned}
$$

using equation (2.1) we get

$$
\begin{aligned}
&= \frac{cc_\lambda}{d_\alpha} \int_{[0,1)^r} \psi\Big(\sum_{j=1}^r \sqrt{x_j} e_j\Big) \phi_\alpha\Big(\sum_{j=1}^r x_j e_j\Big) \Delta\Big(e - \sum_{j=1}^r x_j e_j\Big)^{\lambda - p} \\
&\quad \times \Delta\Big(\sum_{j=1}^r x_j e_j\Big)^b \prod_{1 \le j < k \le r} |x_j - x_k|^a \, \mathrm{d}x_1 \dots \mathrm{d}x_r
\end{aligned}
$$

and then applying the definition of $\varphi_\alpha$ we obtain

$$
\begin{aligned}
&= \frac{cc_\lambda}{d_\alpha} \int_{[0,1)^r} \psi\Big(\sum_{j=1}^r \sqrt{x_j} e_j\Big) \Delta\Big(e - \sum_{j=1}^r x_j e_j\Big)^{\lambda - p} \Delta\Big(\sum_{j=1}^r x_j e_j\Big)^b \\
&\quad \times \int_L \Delta_\alpha\Big(s \cdot \sum_{j=1}^r x_j e_j\Big) \, \mathrm{d}s \prod_{1 \le j < k \le r} |x_j - x_k|^a \, \mathrm{d}x_1 \dots \mathrm{d}x_r.
\end{aligned}
$$

Since $\psi$, $\Delta$ and $e$ are $L$-invariant, we are now in position to apply equation (3.2). To do so, we note that integration over $[0,1)^r$ for

$$
x = \sum_{J=1}^r x_j e_j,
$$

corresponds to the set $\Omega \cap (e - \Omega)$. Hence, we obtain

$$
\langle \psi\phi_\alpha, \phi_\alpha \rangle_\lambda = C \int_{\Omega \cap (e - \Omega)} \psi(\sqrt{x}) \Delta_\alpha(x) \Delta(e - x)^{\lambda - p} \Delta(x)^b \, \mathrm{d}x,
$$

where $C$ is a constant that is independent of $\psi$ and $\alpha$. Note that we have used

$$
\sqrt{\sum_{j=1}^r x_j e_j} = \sum_{j=1}^r \sqrt{x_j} e_j,
$$

which follows from the mutual orthogonality of $e_1, \dots, e_r$. □

As a consequence of the proof of Theorem 4.9 we have the following.

COROLLARY 4.10. *With the assumptions of Theorem 4.9 we have for every* $\lambda > p-1$ *and* $\alpha \in \overrightarrow{\mathbb{N}}^r$

$$\begin{aligned}\langle T_\psi \phi_\alpha, \phi_\alpha\rangle_\lambda &= \langle \psi\phi_\alpha, \phi_\alpha\rangle_\lambda \\ &= C\int_{[0,1)^r} \psi\Big(\sum_{j=1}^r \sqrt{x_j} e_j\Big)\Delta_\alpha\Big(\sum_{j=1}^r x_j e_j\Big)\Delta\Big(\sum_{j=1}^r (1-x_j)e_j\Big)^{\lambda-p}\Delta\Big(\sum_{j=1}^r x_j e_j\Big)^b \\ &\qquad\times \prod_{1\le j<k\le r} |x_j - x_k|^a \,\mathrm{d}x_1 \ldots \mathrm{d}x_r,\end{aligned}$$

*for some constant* $C$ *independent of* $\psi$ *and* $\alpha$.

A direct application of Theorem 4.9 yields the following expression of the coefficients in Proposition 4.7 for radial Toeplitz operators in terms of the polynomials $\Delta_\alpha(z)$ and $\Delta_j(z)$. We recall that for our previous notation we have $\Delta(z) = \Delta_r(z)$ for a domain of rank $r$.

THEOREM 4.11. *Let* $D$ *be a bounded symmetric domain and* $\psi \in L^\infty(D)$ *a radial symbol. Then, for every* $\lambda > p-1$ *and for every* $\alpha \in \overrightarrow{\mathbb{N}}^r$ *the coefficients* $c_\alpha(T_\psi)$ *from Proposition 4.7 satisfy*

$$c_\alpha(T_\psi) = \frac{\langle T_\psi\phi_\alpha, \phi_\alpha\rangle_\lambda}{\langle \phi_\alpha, \phi_\alpha\rangle_\lambda} = \frac{\displaystyle\int_{\Omega\cap(e-\Omega)} \psi(\sqrt{x})\Delta_\alpha(x)\Delta(e-x)^{\lambda-p}\Delta(x)^b\,\mathrm{d}x}{\displaystyle\int_{\Omega\cap(e-\Omega)} \Delta_\alpha(x)\Delta(e-x)^{\lambda-p}\Delta(x)^b\,\mathrm{d}x}$$

$$= \frac{\displaystyle\int_{\Omega\cap(e-\Omega)} \psi(\sqrt{x})\Delta_1(x)^{\alpha_1-\alpha_2}\Delta_2(x)^{\alpha_2-\alpha_3}\ldots\Delta_r(x)^{\alpha_r+b}\Delta_r(e-x)^{\lambda-p}\,\mathrm{d}x}{\displaystyle\int_{\Omega\cap(e-\Omega)} \Delta_1(x)^{\alpha_1-\alpha_2}\Delta_2(x)^{\alpha_2-\alpha_3}\ldots\Delta_r(x)^{\alpha_r+b}\Delta_r(e-x)^{\lambda-p}\,\mathrm{d}x}.$$

We can now apply Corollary 4.10 and use (2.1) to write down the formulas from Theorem 4.11 in terms of Lebesgue integrals over $[0,1)^r$ to obtain the following result.

THEOREM 4.12. *Let* $D$ *be a bounded symmetric domain and* $\psi \in L^\infty(D)$ *a radial symbol. Then, for every* $\lambda > p-1$ *and for every* $\alpha \in \overrightarrow{\mathbb{N}}^r$ *the coefficients* $c_\alpha(T_\psi)$ *from Proposition 4.7 satisfy*

$$c_\alpha(T_\psi) = \frac{\langle T_\psi\phi_\alpha, \phi_\alpha\rangle_\lambda}{\langle \phi_\alpha, \phi_\alpha\rangle_\lambda} =$$

$$\frac{\displaystyle\int_{[0,1)^r} \psi\Big(\sum_{j=1}^r \sqrt{x_j}e_j\Big)\prod_{j=1}^r x_j^{\alpha_j+b}\prod_{j=1}^r (1-x_j)^{\lambda-p}\prod_{1\le j<k\le r}|x_j-x_k|^a\,\mathrm{d}x_1\ldots\mathrm{d}x_r}{\displaystyle\int_{[0,1)^r}\prod_{j=1}^r x_j^{\alpha_j+b}\prod_{j=1}^r (1-x_j)^{\lambda-p}\prod_{1\le j<k\le r}|x_j-x_k|^a\,\mathrm{d}x_1\ldots\mathrm{d}x_r}.$$

## 5. Radial Toeplitz operators on the Bergman spaces

All the data associated to the domain $D$ involved in Theorem 4.12 is readily available in the references (see [**5, 18**]). We collect this information and provide specific formulas for all irreducible bounded symmetric domains. Note that all the information for the classical cases are explicitly contained in the references, but not so in the exceptional cases. For the latter we provide the required arguments.

**5.1. Type I.** Up to finite index covering maps the domain $D^I_{m,n}$ is given by the groups

$$G = \mathrm{SU}(m,n), \quad K = \mathrm{S}(\mathrm{U}(m)\times \mathrm{U}(n)).$$

For simplicity and without loss of generality we will assume that $m \leq n$. The elements of $K$ are matrices of the form

$$\begin{pmatrix} A & 0 \\ 0 & B \end{pmatrix},$$

where $A \in \mathrm{U}(m), B \in \mathrm{U}(n)$ and $\det(A)\det(B) = 1$.

The domain in this case is given by

$$D^I_{m,n} = \{Z \in M_{m\times n}(\mathbb{C}) \mid ZZ^* < I_m\},$$

where the $K$-action has the expression

$$\begin{pmatrix} A & 0 \\ 0 & B \end{pmatrix} Z = AZB^{-1},$$

where $A, B$ are as above and $Z \in D^I_{m,n}$. And we have

$$\dim D^I_{m,n} = mn, \quad r = m, \quad a = 2, \quad b = n-m, \quad p = n+m.$$

A maximal collection of mutually orthogonal primitives is given by the $m\times n$ matrices $E_1, \dots, E_m$, where $E_j$ has 1 at the $(j,j)$-th entry and 0 elsewhere. Hence, we have

$$E = (I_m, 0),$$

where 0 denotes here a $m \times (n-m)$ matrix. Proposition 4.5 states that for every $Z \in D^I_{m,n}$ there exist $A \in \mathrm{U}(m), B \in \mathrm{U}(n)$ and a diagonal matrix $D$ of size $m\times m$ with entries in $[0,1)$ such that

$$Z = A(D,0)B^{-1}.$$

It follows that for every $\lambda > n+m-1$, $\psi$ bounded radial symbol and $\alpha \in \overrightarrow{\mathbb{N}}^m$ we have

$$c_\alpha(T_\psi) = \frac{\langle T_\psi \phi_\alpha, \phi_\alpha\rangle_\lambda}{\langle \phi_\alpha, \phi_\alpha\rangle_\lambda} =$$

$$\frac{\displaystyle\int_{[0,1)^m} \psi(D(\sqrt{x}),0) \prod_{j=1}^m x_j^{\alpha_j+n-m} \prod_{j=1}^m (1-x_j)^{\lambda-m-n} \prod_{1\leq j<k\leq m} |x_j - x_k|^2 \, dx_1 \dots dx_m}{\displaystyle\int_{[0,1)^m} \prod_{j=1}^m x_j^{\alpha_j+n-m} \prod_{j=1}^m (1-x_j)^{\lambda-m-n} \prod_{1\leq j<k\leq m} |x_j - x_k|^2 \, dx_1 \dots dx_m},$$

where $D(\sqrt{x})$ denotes the $m\times m$ diagonal matrix whose entries along the diagonal are $\sqrt{x_1}, \dots, \sqrt{x_m}$.

If we assume that $m = 1$, then the domain is the unit ball $\mathbb{B}^n$ in $\mathbb{C}^n$ and the previous formula reduces to Theorem 6.2 from [**14**].

**5.2. Type II.** Up to finite index covering maps the domain $D^{II}_m$ is given by the groups

$$G = \mathrm{SO}^*(2m) = \left\{ \begin{pmatrix} A & B \\ -\overline{B} & \overline{A} \end{pmatrix} \Big| A, B \in M_m(\mathbb{C}) \right\} \cap \mathrm{SU}(m,m)$$

$$K = \mathrm{U}(n) \cong \left\{ \begin{pmatrix} A & 0 \\ 0 & \overline{A} \end{pmatrix} \Big| A \in \mathrm{U}(m) \right\}.$$

The domain is give by

$$D^{II}_m = \{ Z \in M_m(\mathbb{C}) \mid ZZ^* < I_m, Z^t = -Z \},$$

with $K$-action

$$A \cdot Z = AZA^t,$$

where $A \in \mathrm{U}(n)$ and $Z \in D^{II}_m$. The properties of this domain depend on the parity of $m$. Hence, we will write $m = 2n + \epsilon$ where $\epsilon \in \{0, 1\}$. With this notation we now have

$$\dim D^{II}_m = n(2n + 2\epsilon - 1), \quad r = n, \quad a = 4, \quad b = 2\epsilon, \quad p = 4n + 2\epsilon - 2.$$

For every $j = 1, \dots, n$ let us denote by $E_j \in M_m(\mathbb{C})$ the block diagonal matrix of the form

$$\begin{pmatrix} 0 & \cdots & 0 & & \cdots \\ \vdots & \ddots & \vdots & & \\ 0 & \cdots & \begin{matrix} 0 & 1 \\ -1 & 0 \end{matrix} & & \cdots \\ \vdots & & \vdots & & \ddots \end{pmatrix},$$

where the only non-zero block is the $(j, j)$-th one. In this description the blocks are $2 \times 2$ from the upper left corner and the $j$-th diagonal block has the pictured entry. Note that for $\epsilon = 1$, we have zero blocks of size $2 \times 1$ at the far right, of size $1 \times 2$ at the lower bottom and a single 0 at the lower right corner. Hence, $E_1, \dots, E_n$ is a maximal collection of mutually orthogonal primitives. From this it is easy to write down $E$.

For this case, Proposition 4.5 states that for every $Z \in D^{II}_m$ there exist $A \in \mathrm{U}(n)$, $t_1, \dots, t_n \in [0, 1)$ such that

$$Z = AD(t)A^t,$$

where $D(t) = D(t_1, \dots, t_n) = \mathrm{Diag}(d(t_1), \dots, d(t_j), \dots)$ is a $m \times m$ matrix that is block diagonal with $2 \times 2$ blocks along the diagonal starting from the upper left corner and given by

$$d(t_j) = \begin{pmatrix} 0 & t_j \\ -t_j & 0 \end{pmatrix}.$$

As before, we have to complete the matrix $D(t)$ with some smaller zero blocks for the case $\epsilon = 1$.

It follows that for $\lambda > 4n + 2\epsilon - 3$, $\psi$ bounded radial symbol and $\alpha \in \overrightarrow{\mathbb{N}}^n$ we have

$$c_\alpha(T_\psi) = \frac{\langle T_\psi \phi_\alpha, \phi_\alpha \rangle_\lambda}{\langle \phi_\alpha, \phi_\alpha \rangle_\lambda} =$$

$$\frac{\displaystyle\int_{[0,1)^n} \psi(D(\sqrt{x})) \prod_{j=1}^n x_j^{\alpha_j + 2\epsilon} \prod_{j=1}^n (1 - x_j)^{\lambda - 4n - 2\epsilon + 2} \prod_{1 \le j < k \le n} |x_j - x_k|^4 \, \mathrm{d}x_1 \dots \mathrm{d}x_n}{\displaystyle\int_{[0,1)^n} \prod_{j=1}^n x_j^{\alpha_j + 2\epsilon} \prod_{j=1}^n (1 - x_j)^{\lambda - 4n - 2\epsilon + 2} \prod_{1 \le j < k \le n} |x_j - x_k|^4 \, \mathrm{d}x_1 \dots \mathrm{d}x_n},$$

where the block diagonal matrix $D(\sqrt{x}) = D(\sqrt{x_1}, \dots, \sqrt{x_n})$ is given as before.

**5.3. Type III.** Up to finite index covering maps the domain $D_n^{III}$ is given by the groups

$$G = \mathrm{Sp}(n, \mathbb{R}),$$
$$K = \mathrm{U}(n) = \left\{ \begin{pmatrix} A & 0 \\ 0 & \overline{A} \end{pmatrix} \Big| A \in \mathrm{U}(n) \right\}.$$

The domain is naturally identified is given by

$$D_n^{III} = \{ Z \in M_n(\mathbb{C}) \mid ZZ^* < I_n, Z^t = Z \},$$

with $K$-action

$$A \cdot Z = AZA^t,$$

where $A \in \mathrm{U}(n)$ and $Z \in D_n^{III}$. For this domain we have

$$\dim D_n^{III} = \frac{n(n+1)}{2}, \quad r = n, \quad a = 1, \quad b = 0, \quad p = n + 1.$$

A maximal collection of mutually orthogonal primitives is given by $E_1, \dots, E_n$, the canonical basis for the space of diagonal matrices in $M_n(\mathbb{C})$. In particular, we have $E = I_n$. Proposition 4.5 states that for every $Z \in D_n^{III}$ there exist $A \in \mathrm{U}(n)$ and a diagonal matrix $D$ with entries in $[0, 1)$ such that

$$Z = ADA^t.$$

It follows that for every $\lambda > n$, $\psi$ a bounded radial symbol and $\alpha \in \overrightarrow{\mathbb{N}}^n$ we have

$$c_\alpha(T_\psi) = \frac{\langle T_\psi \phi_\alpha, \phi_\alpha \rangle_\lambda}{\langle \phi_\alpha, \phi_\alpha \rangle_\lambda} =$$

$$\frac{\displaystyle\int_{[0,1)^n} \psi(D(\sqrt{x})) \prod_{j=1}^n x_j^{\alpha_j} \prod_{j=1}^n (1 - x_j)^{\lambda - n - 1} \prod_{1 \le j < k \le r} |x_j - x_k| \, \mathrm{d}x_1 \dots \mathrm{d}x_n}{\displaystyle\int_{[0,1)^n} \prod_{j=1}^n x_j^{\alpha_j} \prod_{j=1}^n (1 - x_j)^{\lambda - n - 1} \prod_{1 \le j < k \le m} |x_j - x_k| \, \mathrm{d}x_1 \dots \mathrm{d}x_m},$$

where $D(\sqrt{x})$ denotes the $n \times n$ diagonal matrix whose entries along the diagonal are $\sqrt{x_1}, \dots, \sqrt{x_n}$.

**5.4. Type IV.** The groups associated to the domain $D_n^{IV}$ can be taken to be

$$G = \mathrm{SO}(n,2)$$
$$K = \mathrm{SO}(n) \times \mathrm{SO}(2).$$

For this domain we have the realization

$$D_n^{IV} = \left\{ z \in \mathbb{C}^n \middle| |z|^2 < 2,\ |z|^2 < 1 + \left| \frac{1}{2} \sum_{j=1}^n z_j^2 \right| \right\}$$

with the $K$-action

$$\begin{pmatrix} A & 0 \\ 0 & B \end{pmatrix} \cdot z = e^{-i\theta} Az,$$

where $z \in D_n^{IV}$, $A \in \mathrm{SO}(n)$ and $B \in SO(2)$ has the form

$$B = \begin{pmatrix} \cos(\theta) & -\sin(\theta) \\ \sin(\theta) & \cos(\theta) \end{pmatrix}.$$

This domain has the following data

$$\dim D_n^{IV} = n, \quad r = 2, \quad a = n-1, \quad b = 0, \quad p = n.$$

A maximal collection of mutually orthogonal primitives is given by

$$e_1 = \left(\frac{1}{2}, \frac{i}{2}, 0, \dots, 0\right)$$
$$e_2 = \left(\frac{1}{2}, -\frac{i}{2}, 0, \dots, 0\right),$$

and so we have $e = (1, 0, \dots, 0)$. Proposition 4.5 says that for every $z \in D_n^{IV}$ there is $A \in \mathrm{SO}(n)$, $\theta \in \mathbb{R}$ and $t_1, t_2 \in [0,1)$ such that

$$e^{i\theta} Az = t_1 e_1 + t_2 e_2.$$

We now have that for every $\lambda > n-1$, $\psi$ a bounded radial symbol and $\alpha \in \vec{\mathbb{N}}^2$ we have

$$c_\alpha(T_\psi) = \frac{\langle T_\psi \phi_\alpha, \phi_\alpha \rangle_\lambda}{\langle \phi_\alpha, \phi_\alpha \rangle_\lambda} =$$
$$\frac{\displaystyle\int_{[0,1)^2} \psi(\sqrt{x_1} e_1 + \sqrt{x_2} e_2) x_1^{\alpha_1} x_2^{\alpha_2} (1-x_1)^{\lambda-n} (1-x_2)^{\lambda-n} |x_1 - x_2|^{n-1} \, dx_1 \, dx_2}{\displaystyle\int_{[0,1)^2} x_1^{\alpha_1} x_2^{\alpha_2} (1-x_1)^{\lambda-n} (1-x_2)^{\lambda-n} |x_1 - x_2|^{n-1} \, dx_1 \, dx_2}.$$

**5.5. Exceptional type V.** In the notation of [**5**], the symmetric pair of Lie algebras that realize this domain is $(\mathfrak{e}_{6(-14)}, \mathfrak{so}(10) + \mathbb{R})$, also known as $EIII$.

The properties of this domain are better described in terms of the octonians $\mathbb{O}$. We recall that $\mathbb{O}$ is an 8-dimensional real composition algebra whose basic properties can be found in the literature (see [**16**]). We consider the complexified octonions $\mathbb{O}_\mathbb{C} = \mathbb{O} \otimes_\mathbb{R} \mathbb{C}$ and the 16-dimensional complex vector space $\mathbb{O}_\mathbb{C}^2$. According to [**12**], the latter carries a Jordan pair structure that such that

$$\{xyx\} = x(\widetilde{y}^t x),$$

for every $x, y \in \mathbb{O}_\mathbb{C}^2$. Here, $\widetilde{y}$ is computed component-wise and comes from the the involution on $\mathbb{O}_\mathbb{C}$ obtained by complexifying the canonical conjugation on $\mathbb{O}$.

The exceptional domain $D^V$ has $\mathbb{O}_{\mathbb{C}}^2$ as associated Jordan pair (see [**12**]) and so it is a bounded domain in this complex vector space. The data for this domain is the following

$$\dim D^V = 16, \quad r = 2, \quad a = 6, \quad b = 4, \quad p = 12.$$

Furthermore, using the Jordan pair structure described it is easy to check that a maximal collection of mutually orthogonal primitives is given by

$$e_1 = (1,0), \quad e_2 = (0,1),$$

and so $e = (1,1)$. In this case, Proposition 4.5 states that for every $(a,b) \in D^V$ there exists $g \in K$ (where $K \simeq \mathrm{SO}(10) \times \mathbb{T}$) and $(t_1, t_2) \in [0,1)^2$ such that

$$(a,b) = g(t_1, t_2).$$

With respect to Toeplitz operators we now conclude that for every $\lambda > 11$, $\psi$ a bounded radial symbol and $\alpha \in \overrightarrow{\mathbb{N}}^2$ we have

$$c_\alpha(T_\psi) = \frac{\langle T_\psi \phi_\alpha, \phi_\alpha \rangle_\lambda}{\langle \phi_\alpha, \phi_\alpha \rangle_\lambda} =$$

$$\frac{\displaystyle\int_{[0,1)^2} \psi(\sqrt{x_1}, \sqrt{x_2}) x_1^{\alpha_1+4} 2_1^{\alpha_2+4} (1-x_1)^{\lambda-12}(1-x_2)^{\lambda-12}|x_1 - x_2|^6 \, dx_1 \, dx_2}{\displaystyle\int_{[0,1)^2} x_1^{\alpha_1+4} 2_1^{\alpha_2+4} (1-x_1)^{\lambda-12}(1-x_2)^{\lambda-12}|x_1 - x_2|^6 \, dx_1 \, dx_2}.$$

**5.6. Exceptional type VI.** In the notation of [**5**], the symmetric pair of Lie algebras that realize this domain is $(\mathfrak{e}_{7(-25)}, \mathfrak{e}_6 + \mathbb{R})$ which is also known as $EVII$.

Let us consider the 27-dimensional complex vector space $H_3(\mathbb{O}_{\mathbb{C}})$ of Hermitian $3 \times 3$ matrices with entries in $\mathbb{O}_{\mathbb{C}}$. By Hermitian we mean to satisfy

$$\widetilde{x}^t = x.$$

In particular, the elements of $H_3(\mathbb{O}_{\mathbb{C}})$ are of the form

$$\begin{pmatrix} z_1 & a & b \\ \widetilde{a} & z_2 & c \\ \widetilde{b} & \widetilde{c} & z_3 \end{pmatrix},$$

where $z_1, z_2, z_3 \in \mathbb{C}$ and $a, b, c \in \mathbb{O}_{\mathbb{C}}$. For $x, y \in H_3(\mathbb{O}_{\mathbb{C}})$ denote the commutative product

$$x \circ y = xy + yx.$$

Then, $H_3(\mathbb{O}_{\mathbb{C}})$ is a Jordan pair for the operation

$$\{xyx\} = \frac{1}{2}(x \circ (x \circ y) - x^2 \circ y)$$

for every $x, y \in H_3(\mathbb{O}_{\mathbb{C}})$, where $x^2$ is just the square with the respect to product of matrices. As shown in [**12**], this is the Jordan pair associated to the domain $D^{VI}$. In particular, $D^{VI}$ has a realization as a circled bounded domain in $H_3(\mathbb{O}_{\mathbb{C}})$. The data for this domain is given by

$$\dim D^{VI} = 27, \quad r = 3, \quad a = 8, \quad b = 0, \quad p = 26.$$

From the previous Jordan pair structure it is easy to see that a maximal collection of mutually orthogonal primitives is

$$e_1 = E_{1,2} + E_{2,1}, \quad e_2 = E_{1,3} + E_{3,1}, \quad e_3 = E_{2,3} + E_{3,2},$$

where as usual $E_{j,k}$ denotes the $3\times 3$ matrix with all entries 0 except at the position $(j,k)$ where it is 1. For this case, Proposition 4.5 implies that for every $x \in D^{VI}$ there are $g \in K$ (where $K \simeq E_6 \times \mathbb{T}$) and $(t_1,t_2,t_3) \in [0,1)^3$ such that

$$x = g(t_1e_1 + t_2e_2 + t_3e_3) = g\begin{pmatrix} 0 & t_1 & t_2 \\ t_1 & 0 & t_3 \\ t_2 & t_3 & 0 \end{pmatrix}.$$

And for the Toeplitz operators we have that for every $\lambda > 25$, $\psi$ a bounded radial symbol and $\alpha \in \overrightarrow{\mathbb{N}}^3$ we have

$$c_\alpha(T_\psi) = \frac{\langle T_\psi \phi_\alpha, \phi_\alpha\rangle_\lambda}{\langle \phi_\alpha, \phi_\alpha\rangle_\lambda} =$$

$$\frac{\displaystyle\int_{[0,1)^3} \psi\Big(\sum_{j=1}^3 \sqrt{x_j}e_j\Big) \prod_{j=1}^3 x_j^{\alpha_j} \prod_{j=1}^3 (1-x_j)^{\lambda-26} \prod_{1\le j<k\le 3} |x_j - x_k|^8 \, dx_1\, dx_2\, dx_3}{\displaystyle\int_{[0,1)^3} \prod_{j=1}^3 x_j^{\alpha_j} \prod_{j=1}^3 (1-x_j)^{\lambda-26} \prod_{1\le j<k\le 3} |x_j - x_k|^8 \, dx_1\, dx_2\, dx_r}.$$

## References


[1] Matthew Dawson, Gestur Ólafsson, and Raúl Quiroga-Barranco, *Commuting Toeplitz operators on bounded symmetric domains and multiplicity-free restrictions of holomorphic discrete series*, J. Funct. Anal. **268** (2015), no. 7, 1711–1732. MR3315576

[2] J. Faraut and A. Korányi, *Function spaces and reproducing kernels on bounded symmetric domains*, J. Funct. Anal. **88** (1990), no. 1, 64–89. MR1033914

[3] ———, Analysis on Symmetric Cones, Oxford Science Publications, Clarendon Press, Oxford, 1994.

[4] Roe Goodman and Nolan R. Wallach, *Symmetry, representations, and invariants*, Graduate Texts in Mathematics, vol. 255, Springer, Dordrecht, 2009. MR2522486

[5] Sigurdur Helgason, *Differential geometry, Lie groups, and symmetric spaces*, Graduate Studies in Mathematics, vol. 34, American Mathematical Society, Providence, RI, 2001. Corrected reprint of the 1978 original. MR1834454

[6] Sigurdur Helgason, *Groups and geometric analysis*, Mathematical Surveys and Monographs, vol. 83, American Mathematical Society, Providence, RI, 2000. Integral geometry, invariant differential operators, and spherical functions; Corrected reprint of the 1984 original. MR1790156

[7] L. K. Hua, *Harmonic analysis of functions of several complex variables in the classical domains*, Translations of Mathematical Monographs, vol. 6, American Mathematical Society, Providence, R.I., 1979. Translated from the Russian, which was a translation of the Chinese original, by Leo Ebner and Adam Korányi; With a foreword by M. I. Graev; Reprint of the 1963 edition. MR598469

[8] Kenneth D. Johnson, *On a ring of invariant polynomials on a Hermitian symmetric space*, J. Algebra **67** (1980), no. 1, 72–81. MR595020

[9] Toshiyuki Kobayashi, *Multiplicity-free theorems of the restrictions of unitary highest weight modules with respect to reductive symmetric pairs*, Representation theory and automorphic forms, Progr. Math., vol. 255, Birkhäuser Boston, Boston, MA, 2008, pp. 45–109. MR2369496

[10] Adam Korányi, *Analytic invariants of bounded symmetric domains*, Proc. Amer. Math. Soc. **19** (1968), 279–284. MR0224864

[11] Adam Korányi and Stephen Vági, *Rational inner functions on bounded symmetric domains*, Trans. Amer. Math. Soc. **254** (1979), 179–193. MR539914

[12] Ottmar Loos, *Jordan pairs*, Lecture Notes in Mathematics, Vol. 460, Springer-Verlag, Berlin-New York, 1975. MR0444721

[13] Gestur Ólafsson and Bent Ørsted, *Generalizations of the Bargmann transform*, Lie theory and its applications in physics (Clausthal, 1995), World Sci. Publ., River Edge, NJ, 1996, pp. 3–14. MR1634480

[14] Raul Quiroga-Barranco, *Separately radial and radial Toeplitz operators on the unit ball and representation theory*, Bol. Soc. Mat. Mex. (3) **22** (2016), no. 2, 605–623. MR3544156

[15] Wilfried Schmid, *Die Randwerte holomorpher Funktionen auf hermitesch symmetrischen Räumen* (German), Invent. Math. **9** (1969/1970), 61–80. MR0259164

[16] Tonny A. Springer and Ferdinand D. Veldkamp, *Octonions, Jordan algebras and exceptional groups*, Springer Monographs in Mathematics, Springer-Verlag, Berlin, 2000. MR1763974

[17] Harald Upmeier, *Toeplitz $C^*$-algebras on bounded symmetric domains*, Ann. of Math. (2) **119** (1984), no. 3, 549–576. MR744863

[18] Harald Upmeier, *Toeplitz operators and index theory in several complex variables*, Operator Theory: Advances and Applications, vol. 81, Birkhäuser Verlag, Basel, 1996. MR1384981

[19] Nolan R. Wallach, *The analytic continuation of the discrete series. I, II*, Trans. Amer. Math. Soc. **251** (1979), 1–17, 19–37. MR531967

Centro de Investigación en Matemáticas, Mérida, Mexico
*Email address*: `matthew.dawson@cimat.mx`

Centro de Investigación en Matemáticas, Guanajuato, Mexico
*Email address*: `quiroga@cimat.mx`

Contemporary Mathematics
Volume **714**, 2018
https://doi.org/10.1090/conm/714/14382

# A rational theory of Clebsch-Gordan coefficients

Robert W. Donley, Jr. and Won Geun Kim

*Dedicated to Professor Ólafsson on the occasion of his sixty-fifth birthday.*

ABSTRACT. A theory of Clebsch-Gordan coefficients for $SL(2,\mathbb{C})$ is given using only rational numbers. Features include orthogonality relations, recurrence relations, and Regge's symmetry group. Results follow from elementary representation theory and properties of binomial coefficients. A computational algorithm is given based on Pascal's recurrence.

## 1. Introduction

In Wigner's formula for Clebsch-Gordan coefficients for $SU(2)$ (also called Wigner coefficients; see also 3-j symbols), the main sum may be expressed with terms consisting of triple products of binomial coefficients. Traditional approaches to the theory rely on special function methods: hypergeometric series, Jacobi polynomials [**16**], and Hahn polynomials [**8**]. On the other hand, an approach to Racah's alternative formula is given in [**4**], 10.14–10.16 using mostly combinatorial methods, but still rooted in use of the inner product. See also [**3**], Ch.5, or [**16**], Ch. III, Sec. 8.9, (10).

Several notions drive the present work: that the tensor product problem exists independent of orthonormal bases, that finite-dimensional representations of $SL(2,\mathbb{C})$ are structurally rational, and that finite-dimensional representations are somewhat combinatorial. Of course, the finite-dimensional representation theory of $SU(2)$ matches to that of $SL(2,\mathbb{C})$ by Weyl's Unitary Trick. Here a theory is presented with the following features:

(1) an explicit formula for weight vectors, whose integral coordinates may be expressed through a generating function in two variables (Definitions 4.5, 4.6),
(2) a summation formula for Clebsch-Gordan coefficients with rational values (Theorem 4.10, Corollary 8.5),
(3) orthogonality relations (Proposition 4.4, Corollary 4.11),
(4) an algorithm based on Pascal's recurrence for computer implementation (Theorem 6.1),
(5) recurrence relations (Propositions 7.1, 7.2, 7.3),

2010 *Mathematics Subject Classification.* Primary 22E70; Secondary 81R05.
*Key words and phrases.* Clebsch-Gordan coefficient, Wigner coefficient, 3-j symbol, coupling coefficient, angular momentum, Pascal's recurrence.
The first author was supported by PSC-CUNY Research Awards Program 46 (Trad. A).

(6) a version of Regge's group of order 72 (Section 8), and
(7) derivations of Wigner's and Racah's formulas (Section 9).

With a similar philosophy in mind, a projection operator method is given in Section 10. The connection to Pascal's Triangle follows empirically from consideration of these projection operators. Additionally formula (10.4) corrects the scaling factor in the Casimir extremal projectors in [**2**], Definitions 5.5 and 6.1, for the case of $SL(2,\mathbb{R})$; further adjustments are needed for general groups.

Since the techniques used are algebraic, all the results in this work hold without changes for $SL(2,F)$, where $F$ is any algebraically closed field of characteristic zero. For instance, see [**6**], VI.22, Problem 7. In fact, results follow over a general field of characteristic zero; the Clebsch-Gordan decomposition follows from the highest weight assumption.

The authors thank the referee for useful comments and suggesting the comprehensive handbook [**15**]. The authors also thank Bent Ørsted and Bart Van Steirteghem for helpful conversations.

## 2. Definitions and notation

Let $SL(2,\mathbb{C})$ be the linear semisimple Lie group of unimodular invertible matrices of size 2 over the complex numbers, let $sl(2)$ be the Lie algebra of traceless matrices of size 2 over $\mathbb{C}$, and let $U(sl(2))$ be the associated universal enveloping algebra. Define the standard basis for $sl(2)$ by

$$e = \begin{bmatrix} 0 & 1 \\ 0 & 0 \end{bmatrix}, \ f = \begin{bmatrix} 0 & 0 \\ 1 & 0 \end{bmatrix}, \ h = \begin{bmatrix} 1 & 0 \\ 0 & -1 \end{bmatrix} \tag{2.1}$$

Fix $n \geq 0$, and let $V(n)$ denote the irreducible finite-dimensional representation for $SL(2,\mathbb{C})$ with highest weight $n$ with respect to the $h$ and $e$. In this case, $\dim_{\mathbb{C}} V(n) = n+1$, and if $\phi_n$ is a nonzero highest weight vector for $V(n)$, then the set

$$B = \{\phi_n, \ f\phi_n, \ f^2\phi_n, \ \ldots, \ f^n\phi_n\} \tag{2.2}$$

forms a basis for $V(n)$. With respect to the Lie algebra action, one has

(LA.1) $h \cdot f^i\phi_n = (n-2i)f^i\phi_n$,
(LA.2) $e \cdot f^i\phi_n = i(n-i+1)f^{i-1}\phi_n$, and
(LA.3) $f \cdot f^i\phi_n = f^{i+1}\phi_n$.

In particular,

(LA.4) $f^i\phi_n$ is a weight vector of weight $n-2i$,
(LA.5) $e \cdot \phi_n = 0$ (highest weight vector with weight $n$), and
(LA.6) $f \cdot f^n\phi_n = 0$ (lowest weight vector with weight $-n$).

Note that, as a representation of $SL(2,\mathbb{C})$,

$$V(n) \cong Sym^n(V(1)), \tag{2.3}$$

where $V(1)$ arises from the standard action of $SL(2,\mathbb{C})$ on $\mathbb{C}^2$. The space $V(1)$ admits a non-degenerate, invariant bilinear form by

$$\langle (x_1,y_1),(x_2,y_2)\rangle = det \begin{bmatrix} x_1 & x_2 \\ y_1 & y_2 \end{bmatrix}. \tag{2.4}$$

If $V$ and $W$ are representations of $SL(2,\mathbb{C})$ with non-degenerate, invariant bilinear forms $\langle \cdot, \cdot \rangle_1$ and $\langle \cdot . \cdot \rangle_2$, we obtain a non-degenerate, invariant bilinear form on $V \otimes W$ by extending

$$\langle v_1 \otimes w_1, v_2 \otimes w_2 \rangle = \langle v_1, v_2 \rangle_1 \, \langle w_1, w_2 \rangle_2. \tag{2.5}$$

One may extend this definition to any finite number of factors in the tensor product. In this way, one obtains a non-degenerate, invariant bilinear form on $Sym^n(V(1))$, unique up to a complex scalar.

For a given $V(n)$-type, any invariant form satisfies

$$\langle X \cdot v, w \rangle_n = \langle v, -X \cdot w \rangle_n \tag{2.6}$$

for $X$ in $sl(2)$. Since

$$\langle f \cdot v, w \rangle_n = \langle v, -f \cdot w \rangle_n, \tag{2.7}$$

rescaling the form such that

$$\langle \phi_n, f^n \phi_n \rangle_n = 1 \tag{2.8}$$

determines a normalization for all basis vectors. That is,

$$\langle f^i \phi_n, f^{n-i} \phi_n \rangle_n = \langle \phi_n, (-1)^i f^n \phi_n \rangle_n = (-1)^i. \tag{2.9}$$

and, among basis vectors, $f^i \phi_n$ pairs nontrivially only with $f^{n-i}\phi_n$. Finally it follows that this form is symmetric when $n$ is even and alternating when $n$ is odd.

## 3. Tensor products

Notation for Clebsch-Gordan coefficients varies greatly in the literature, here presenting yet another variation. In addition to dropping the traditional half-integer notation associated with $SU(2)$, this variation facilitates an emphasis on exponents over weights as indices, an organizing principle based on matrices, and a convenient indexing of a five-dimensional polyhedron in subsequent work. Indices always satisfy

$$0 \le i \le m, \quad 0 \le j \le n, \quad 0 \le k \le \min(m,n), \quad k \le i+j \le m+n-k. \tag{3.1}$$

Unless stated otherwise, assume $m \le n$, so that $0 \le k \le m \le n$.

One obtains an action of $SL(2,\mathbb{C})$ on $V(m) \otimes V(n)$ by

$$g \cdot (v \otimes w) = gv \otimes gw \tag{3.2}$$

and an action of $sl(2)$ by

$$X \cdot (v \otimes w) = Xv \otimes w + v \otimes Xw. \tag{3.3}$$

With these actions, one obtains

Proposition 3.1 (Clebsch-Gordan Decomposition). *With respect to representations of $SL(2,\mathbb{C})$, one has*

$$V(m) \otimes V(n) \cong V(n-m) \oplus V(n-m+2) \oplus \cdots \oplus V(m+n) \tag{3.4}$$

Proof using Section 4. This proof works over a field of characteristic zero. Irreducible, finite-dimensional highest weight representations are determined by their weight sets as in LA.1-6. Explicit highest and lowest weight vectors for each summand are given in Propositions 4.1 and 4.9. That is, each summand occurs as a subrepresentation of the tensor product, and counting dimensions gives

$$\sum_{s=0}^{m}(n-m+2s+1)=(m+1)(n+1). \tag{3.5}$$

Since each irreducible subrepresentation has an invariant complement, the sum is direct. □

The weight set of $V(m)\otimes V(n)$ is all pairwise sums of weights for $V(m)$ with weights for $V(n)$. In particular,

$$h\cdot(f^i\phi_m\otimes f^j\phi_n)=(m-2i+n-2j)\ f^i\phi_m\otimes f^j\phi_n \tag{3.6}$$

One immediately notes that

$$\mathbf{B}=\{f^i\phi_m\otimes f^j\phi_n\},\qquad (0\le i\le m,\ 0\le j\le n) \tag{3.7}$$

forms a basis of weight vectors for $V(m)\otimes V(n)$, but in general this basis does not respect the decomposition given above. Furthermore, the subspace of weight vectors of weight $m+n-2p$ has basis

$$\mathbf{B}_p=\{f^i\phi_m\otimes f^j\phi_n\},\qquad (i+j=p). \tag{3.8}$$

In the classical case for $SU(2)$, one describes an orthonormal basis of weight vectors for each $V(m')$ in the decomposition of $V(m)\otimes V(n)$. Choose orthonormal bases $\{u_i\}$ (where $u_i$ has weight $m-2i$) and $\{v_j\}$ (where $v_j$ has weight $n-2j$) for $V(m)$ and $V(n)$, respectively. After setting conventions, one may define an orthonormal basis $\{w_{m,n,k}(p)\}$ (weight $m+n-2p$) for each $V(m+n-2k)$ in the tensor product such that

$$w_{m,n,k}(p)=\sum_{i+j=p}c'_{m,n,k}(i,j)\ u_i\otimes v_j \tag{3.9}$$

with each $c'_{m,n,k}(i,j)$ real.

In turn, the Clebsch-Gordan coefficients are defined by

$$u_i\otimes v_j=\sum_{k=0}^{m}C'_{m,n,k}(i,j)w_{m,n,k}(i+j). \tag{3.10}$$

After converting to the present notation for indices and changing $s$ to $i-s$ in the summation, one has Wigner's formula in binomial form:

PROPOSITION 3.2 ([**16**], Ch. III.8.3, Equation (14′)).

$$C'_{m,n,k}(i,j)\ =\ \sqrt{\frac{(m+n-2k+1)\binom{m+n-k}{m-i,\ n-j,\ *}}{(m+n-k+1)\binom{m+n-k}{m-k,\ n-k,\ k}\binom{m+n-k}{i,\ j,\ *}}}\cdot S$$

*where*

$$S=\sum_{s=0}^{k}(-1)^s\binom{i+j-k}{i-s}\binom{m-s}{k-s}\binom{n-k+s}{s}$$

For non-negative integers $a, b, c$, multinomial coefficients are defined by

$$(3.11) \qquad \binom{a+b}{a} = \frac{(a+b)!}{a!\,b!} \quad \text{and} \quad \binom{a+b+c}{a,\ b,\ c} = \frac{(a+b+c)!}{a!\,b!\,c!}.$$

When notation is cumbersome, the lower index $c$ is replaced with $*$. In this work, these coefficients are set to zero if negative entries occur.

## 4. Generating functions for weight vectors

Among conventions, of key importance is a fixed choice of weight vectors. Here conventions are set by a uniform description of highest weight vectors. This choice determines the weight vector bases for all $V(m')$-types that occur, and, by duality, fixes sign conventions for Clebsch-Gordan coefficients.

PROPOSITION 4.1 (Highest Weight Vector for $V(m+n-2k)$). *Define the highest weight vector $\phi_{m,n,k}$ of $V(m+n-2k)$ in $V(m) \otimes V(n)$ such that the coordinate vector $[\phi_{m,n,k}]_{\mathbf{B_k}}$ in $\mathbb{C}^{k+1}$ is given by*

$$\left(\binom{m}{k}\binom{n-k}{0}, \ldots, (-1)^l \binom{m-l}{k-l}\binom{n-k+l}{l}, \ldots, (-1)^k \binom{m-k}{0}\binom{n}{k}\right).$$

PROOF. See [5], Sec. II, Prop. 2.1.1. Assume the coordinate vector has the form

$$(c_0, c_1, c_2, \ldots, c_k).$$

with $c_0 = \binom{m}{k}$. For each basis vector, applying (LA.2) gives

$$e \cdot (f^i\phi_m \otimes f^{k-i}\phi_n) = i(m-i+1)f^{i-1}\phi_m \otimes f^{k-i}\phi_n + (k-i)(n-k+i+1)f^i \otimes f^{k-i-1}\phi_n.$$

Since $e \cdot \phi_{m,n,k} = 0$, each $c_i$ may be computed from $c_{i-1}$. For instance,

$$k(n-k+1)\binom{m}{k} + m\,c_1 = 0,$$

$$(k-1)(n-k+2)c_1 + 2(m-1)\,c_2 = 0,$$

and so on. Continuing in this manner, the proposition holds. □

Weight vectors for $V(m+n-2k)$ are obtained by repeated application of $f$ to $\phi_{m,n,k}$. Thus a weight vector basis for $V(m+n-2k)$ in $V(m) \otimes V(n)$ is given by

$$(4.1) \qquad \mathbf{B^k} = \{\phi_{m,n,k},\ f\phi_{m,n,k}, f^2\phi_{m,n,k}, \ldots, f^{m+n-2k}\phi_{m,n,k}\}$$

DEFINITION 4.2. Define the coordinate functions $c_{m,n,k}(i,j)$ by the formula

$$(4.2) \qquad f^{p-k}\phi_{m,n,k} = \sum_{i+j=p} c_{m,n,k}(i,j)\ f^i\phi_m \otimes f^j\phi_n.$$

DEFINITION 4.3. Define the Clebsch-Gordan coefficients $C_{m,n,k}(i,j)$ by the formula

$$(4.3) \qquad f^i\phi_m \otimes f^j\phi_n = \sum_{k=0}^{m} C_{m,n,k}(i,j)\ f^{i+j-k}\phi_{m,n,k}.$$

It follows immediately from these definitions that

PROPOSITION 4.4 (Orthogonality Relations 1).

$$\sum_{i+j=p} c_{m,n,k}(i,j)\; C_{m,n,k'}(i,j) = \delta_{k,k'} \tag{4.4}$$

$$\sum_{k=0}^{m} c_{m,n,k}(i,j)\; C_{m,n,k}(i',j') = \delta_{i,i'}\delta_{j,j'} \quad (i+j=i'+j') \tag{4.5}$$

Next recall the binomial series for positive integer $r$

$$\frac{1}{(1-x)^r} = \sum_{i=0}^{\infty} \binom{r-1+i}{i} x^i. \tag{4.6}$$

One observes that the coordinates of the highest weight vector are products of terms from ascending and descending binomial series, which yields

DEFINITION 4.5 (Generating Function for Highest Weight Vectors). The coordinate $c_{m,n,k}(i,k-i)$ for $f^i\phi_m \otimes f^{k-i}\phi_n$ in $[\phi_{m,n,k}]_{\mathbf{B_k}}$ is the coefficient of $x^{k-i}y^i$ in

$$\frac{1}{(1-x)^{m-k+1}(1+y)^{n-k+1}} \tag{4.7}$$

Noting that

$$f\cdot(f^i\phi_m \otimes f^j\phi_n) = f^{i+1}\phi_m \otimes f^j\phi_n + f^i\phi_m \otimes f^{j+1}\phi_n, \tag{4.8}$$

coordinate vectors for these weight vectors are obtained by binomial recursion of coordinates (Pascal's Recurrence). This effect translates immediately to generating functions.

DEFINITION 4.6 (Generating Function for Weight Vectors). Consider the weight vector $f^{p-k}\phi_{m,n,k}$ of weight $m+n-2p$ in the $V(m+n-2k)$-type. For $i+j=p$, the coordinate $c_{m,n,k}(i,j)$ of $f^i\phi_m \otimes f^j\phi_n$ in $f^{p-k}\phi_{m,n,k}$ is the coefficient of $x^jy^i$ in

$$\frac{(x+y)^{i+j-k}}{(1-x)^{m-k+1}(1+y)^{n-k+1}} \tag{4.9}$$

To obtain an explicit formula similar to Wigner's formula, one expands the binomial series as an alternating sum of triple products of binomial coefficients. In the following formulas, one uses the convention that binomial coefficients with negative entries are set to zero.

PROPOSITION 4.7 (Coordinate vector of weight $m+n-2i-2j$ in $V(m+n-2k)$).

$$c_{m,n,k}(i,j) = \sum_{l=0}^{k}(-1)^l \binom{i+j-k}{i-l}\binom{m-l}{k-l}\binom{n-k+l}{l} \tag{4.10}$$

The corresponding formula for Clebsch-Gordan coefficients follows from a variant of the classical Chu-Vandermonde identity ([**1**], p. 67) and an explicit form for the lowest weight vectors.

PROPOSITION 4.8. *Suppose $r,s \geq 0$ and $0 \leq k \leq \min(r,s)$. Then*

$$\sum_{l=0}^{k}(-1)^l \binom{k}{l}\binom{r+s-k+l}{s-k+l} = (-1)^k\binom{r+s-k}{s}. \tag{4.11}$$

PROOF. With $0 \leq k \leq r$, expand

$$\frac{(1-x)^k}{(1-x)^{r+1}} = \frac{1}{(1-x)^{r-k+1}}$$

into series

$$\sum_{l'=0}^{k} (-1)^{l'} \binom{k}{l'} x^{l'} \cdot \sum_{l''=0}^{\infty} \binom{r+l''}{l''} x^{l''} = \sum_{l=0}^{\infty} \binom{r-k+l}{l} x^l.$$

Comparing coefficients for $x^s$ yields

$$\sum_{l=0}^{k} (-1)^l \binom{k}{l} \binom{r+s-l}{s-l} = \binom{r-k+s}{s}.$$

Substituting $l' = k - l$ gives (4.11). □

PROPOSITION 4.9 (Lowest Weight Vector for $V(m+n-2k)$). *The coordinate vector of the lowest weight vector* $f^{m+n-2k}\phi_{m,n,k}$ *of* $V(m+n-2k)$ *in* $V(m) \otimes V(n)$ *is given by*

$$[f^{m+n-2k}\phi_{m,n,k}]_{\mathbf{B_{m+n-k}}} = \binom{m+n-2k}{m-k} (1, -1, 1, \ldots, (-1)^k) \quad \in \mathbb{C}^{k+1}.$$

PROOF. That $(1, -1, 1, \ldots, (-1)^k)$ is a lowest weight vector follows in a manner similar to the highest weight vector formula, only now using the effect of $f$ on $\mathbf{B_{m+n-k}}$. To find the scalar, the last entry is found using the coordinate vector formula with $i = m$ and $j = n - k$. This gives

$$\begin{aligned} c_{m,n,k}(m, n-k) &= \sum_{l=0}^{k} (-1)^l \binom{m+n-2k}{m-l} \binom{m-l}{k-l} \binom{n-k+l}{l} \\ &= \binom{m+n-2k}{m-k} \sum_{l=0}^{k} (-1)^l \binom{k}{l} \binom{n-k+l}{k} \end{aligned}$$

The proposition then follows from (4.11) with $r = k$ and $s = n - k$. □

THEOREM 4.10 (Clebsch-Gordan Coefficients).

$$C_{m,n,k}(i,j) = \frac{(-1)^k}{D(m,\, n,\, k)} \; c_{m,n,k}(m-i, n-j) \tag{4.12}$$

*where*

$$\begin{aligned} D(m,\, n,\, k) &= \binom{m+n-k+1}{k} \binom{m+n-2k}{m-k} \\ &= \frac{m+n-k+1}{m+n-2k+1} \binom{m+n-k}{m-k,\; n-k,\; k} \end{aligned}$$

PROOF. The invariant bilinear form on $V(m) \otimes V(n)$ gives

$$\langle f^i\phi_m \otimes f^j\phi_n, f^{m-i}\phi_m \otimes f^{n-j}\phi_n \rangle = (-1)^{i+j},$$

and

$$\langle f^i\phi_{m,n,k}, f^{m+n-2k-i}\phi_{m,n,k} \rangle = (-1)^i D(m,n,k),$$

where $D(m,n,k)$ depends only on $m, n$, and $k$. To compute $D(m,n,k)$, first note

$$\begin{aligned}(-1)^{i+j} c_{m,n,k}(i,j) &= \langle f^i\phi_m \otimes f^j\phi_n, f^{m+n-i-j-k}\phi_{m,n,k}\rangle \\ &= (-1)^{i+j-k} D(m,n,k)\; C_{m,n,k}(m-i,n-j)\end{aligned}$$

Substituting into (4.5) gives

$$\sum_{i+j=p} c_{m,n,k}(i,j)\; c_{m,n,k}(m-i,n-j) = (-1)^k\; D(m,n,k) \tag{4.13}$$

Since this sum is independent of $p$, choose $p = k$. That is, one dots the coordinate vector of the highest weight vector with the reverse of the lowest weight vector. Using Proposition 4.9, the dot product has the form

$$(-1)^k \binom{m+n-2k}{m-k} \sum_{l=0}^{k} \binom{m-l}{k-l} \binom{n-k+l}{l}.$$

The terms in the sum are the coordinates of the highest weight vector without signs. In turn, the sum occurs as a term in the convolution of binomial series; one obtains

$$\sum_{r+s=k} \binom{m-r}{m-k} \binom{n-s}{n-k} = \binom{m+n-k+1}{k}.$$

□

Finally, substituting into Proposition 4.4 gives

COROLLARY 4.11 (Orthogonality Relations 2).

$$\sum_{i+j=p} c_{m,n,k}(i,j)\; c_{m,n,k'}(m-i,n-j) = (-1)^k\; \delta_{k,k'}\; D(m,n,k) \tag{4.14}$$

$$\sum_{k=0}^{m} (-1)^k\; C_{m,n,k}(m-i,n-j)\; C_{m,n,k}(i',j')\; D(m,n,k) = \delta_{i,i'}\delta_{j,j'} \tag{4.15}$$

*where* $i + j = i' + j'$.

## 5. Example

The matrices in Example 5.1 follow from the methods of Section 4. The matrices on the left-hand side represent coordinates for weight vectors. Weights are in descending order, decreasing by 2 at each column. The highest weight vector is represented by the leftmost column (Proposition 4.1), and one moves to the right using Pascal's recurrence (uppercase L pattern). When $k = 0$, one obtains part of Pascal's Triangle.

The matrices on the right-hand side represent Clebsch-Gordan coefficients. Each column gives coordinates for $f^i\phi_m \otimes f^j\phi_n$ in terms of the corresponding weight vectors for each $V(m')$-type.

Consider the third columns for the $V(5)$-component in Example 5.1. One uses the basis

$$\{\phi_3 \otimes f^3\phi_4,\quad f\phi_3 \otimes f^2\phi_4,\quad f^2\phi_3 \otimes f^1\phi_4,\quad f^3\phi_3 \otimes \phi_4\}, \tag{5.1}$$

since the weight is $5-2(2) = 1$ The corresponding coordinate vector is $(3, 2, -5, -4)$, and

$$v = f^2\phi_{3,4,1} = 3\;\phi_3 \otimes f^3\phi_4 \;+\; 2\; f\phi_3 \otimes f^2\phi_4 \;-5\; f^2\phi_3 \otimes f\phi_4 \;-4\; f^3\phi_3 \otimes \phi_4.$$

The Clebsch-Gordan coefficients for weight 1 are $\frac{-1}{70}(-9,-3,5,3)$, and

$$P_{V(5)}(\phi_3 \otimes f^3\phi_4) = \frac{9}{70}\, v, \quad P_{V(5)}(f\phi_3 \otimes f^2\phi_4) = \frac{3}{70}\, v, \quad \text{etc.},$$

where $P_{V(5)}$ records the component in the $V(5)$-type.

EXAMPLE 5.1.

$$k=0:\ V(7) \subset V(3) \otimes V(4)$$

$$\begin{bmatrix} 1 & 1 & 1 & 1 & 1 & 0 & 0 & 0 \\ 0 & 1 & 2 & 3 & 4 & 5 & 0 & 0 \\ 0 & 0 & 1 & 3 & 6 & 10 & 15 & 0 \\ 0 & 0 & 0 & 1 & 4 & 10 & 20 & 35 \end{bmatrix}; \qquad \frac{1}{35}\begin{bmatrix} 35 & 20 & 10 & 4 & 1 & 0 & 0 & 0 \\ 0 & 15 & 10 & 6 & 3 & 1 & 0 & 0 \\ 0 & 0 & 5 & 4 & 3 & 2 & 1 & 0 \\ 0 & 0 & 0 & 1 & 1 & 1 & 1 & 1 \end{bmatrix}$$

$$k=1:\ V(5) \subset V(3) \otimes V(4)$$

$$\begin{bmatrix} 3 & 3 & 3 & 3 & 0 & 0 \\ 4 & -1 & 2 & 5 & 8 & 0 \\ 0 & -4 & -5 & -3 & 2 & 10 \\ 0 & 0 & -4 & -9 & -12 & -10 \end{bmatrix}; \qquad \frac{-1}{70}\begin{bmatrix} -10 & -12 & -9 & -4 & 0 & 0 \\ 10 & 2 & -3 & -5 & -4 & 0 \\ 0 & 8 & 5 & 2 & -1 & -4 \\ 0 & 0 & 3 & 3 & 3 & 3 \end{bmatrix}$$

$$k=2:\ V(3) \subset V(3) \otimes V(4)$$

$$\begin{bmatrix} 3 & 3 & 3 & 0 \\ -6 & -3 & 0 & 3 \\ 6 & 0 & -3 & -3 \\ 0 & 6 & 6 & 3 \end{bmatrix}; \qquad \frac{1}{45}\begin{bmatrix} 3 & 6 & 6 & 0 \\ -3 & -3 & 0 & 6 \\ 3 & 0 & -3 & -6 \\ 0 & 3 & 3 & 3 \end{bmatrix}$$

$$k=3:\ V(1) \subset V(3) \otimes V(4)$$

$$\begin{bmatrix} 1 & 1 \\ -2 & -1 \\ 3 & 1 \\ -4 & -1 \end{bmatrix}; \qquad \frac{-1}{10}\begin{bmatrix} -1 & -4 \\ 1 & 3 \\ -1 & -2 \\ 1 & 1 \end{bmatrix}$$

## 6. Algorithm with Pascal's recurrence

Section 5 provides a template for computer output as a polytope of values, with each horizontal slice representing an irreducible constituent. In terms of modeling, the methods of Section 4 enable a holistic alternative to the traditional approach using summation formulas and give a recursive method using only integer types. For sophisticated models that compute single summations with large integers, see, for instance, [**7**].

THEOREM 6.1 (Algorithm with Pascal's recurrence). *To calculate the coordinate vector matrix for $V(m+n-2k)$ in $V(m)\otimes V(n)$:*

(1) *initialize an $(m+1)$-by-$(m+n-2k+1)$ matrix,*
(2) *set up coordinates for the highest weight vector in the leftmost column using Proposition 4.1, and extend the top row value,*
(3) *apply Pascal's recurrence rightwards in an uppercase L pattern, extending by zero where necessary,*
(4) *for the zero entries in lower-left corner, corresponding entries in the upper-right corner are set to zero, and*
(5) *the $(i+1, i+j-k+1)$-th entry is $c_{m,n,k}(i,j)$.*

*To calculate the Clebsch-Gordan coefficients associated to the corresponding basis vectors:*

(1) *rotate the coordinate vector matrix by* 180 *degrees,*
(2) *divide by the normalizing factor* $(-1)^k D(m,n,k)$ *in (4.12), and*
(3) *the* $(i+1, i+j-k+1)$*-th entry is* $C_{m,n,k}(i,j)$.

As a check on the algorithm, the dot product of vectors in corresponding columns from these matrices always equals one, as is easily verified in Example 5.1.

## 7. Recurrence Relations

Several families of recurrences and symmetries apply to $C_{m,n,k}(i,j)$. These are often expressed as consequences of special function theory, in particular, hypergeometric series of type ${}_3F_2$. In the present setting, these may be recaptured through representation theory and binomial coefficient properties, in particular though generating functions or the summation formula. For the remainder, results are stated for $c_{m,n,k}(i,j)$, noting that those for $C_{m,n,k}(i,j)$ are simply recaptured using Corollary 8.5.

The first two recurrences follow by applying $f$ or $e$ to basis vectors. Compare with [**16**], Ch. III., Sec. 8.7, (4) and (5). Alternatively, Proposition 7.2 follows from applying (8.5) to (7.1).

PROPOSITION 7.1 (Pascal's Recurrence).

$$c_{m,n,k}(i,j) = c_{m,n,k}(i,j-1) + c_{m,n,k}(i-1,j) \tag{7.1}$$

PROPOSITION 7.2 (Reverse Recurrence).

$$\begin{aligned}(i'+1)(m+n-2k-i')\; c_{m,n,k}(i,j) &= \\ &(i+1)(m-i)\; c_{m,n,k}(i+1,j) + (j+1)(n-j)\; c_{m,n,k}(i,j+1)\end{aligned}$$

*where* $i' = i+j-k$.

The next set of recurrences follow from the generating function for $c_{m,n,k}(i,j)$ and Proposition 7.1.

PROPOSITION 7.3 (Outer Recurrences).

$$c_{m,n,k}(i,j) = c_{m-1,n,k}(i,j) + c_{m-1,n-1,k-1}(i,j-1) \tag{7.2}$$

$$c_{m,n,k}(i,j) = c_{m,n-1,k}(i,j) - c_{m-1,n-1,k-1}(i-1,j) \tag{7.3}$$

$$c_{m,n,k}(i,j) = c_{m-1,n,k}(i-1,j) + c_{m,n-1,k}(i,j-1) \tag{7.4}$$

$$c_{m,n,k}(i,j) = c_{m+1,n,k+1}(i,j) - c_{m,n+1,k+1}(i,j). \tag{7.5}$$

PROOF. (7.2) follows by noting that

$$(1-x)\cdot\frac{(x+y)^{i+j-k}}{(1-x)^{m-k+1}(1+y)^{n-k+1}} = \frac{(x+y)^{i+j-k}}{(1-x)^{m-k}(1+y)^{n-k+1}}.$$

(7.3) follows in a similar manner, instead using the factor $(1+y)$.

For (7.4), subtract 1 from $i$ and $j$ in (7.1) and (7.2), respectively. Adding both equations gives

$$c_{m,n,k}(i-1,j)+c_{m,n,k}(i,j-1)=c_{m-1,n,k}(i-1,j)+c_{m,n-1,k}(i,j-1),$$

and the result follows from Proposition 7.1.

The proof of (7.5) follows in a similar matter with subtraction after increasing $m$, $n$ and $k$ by 1. □

Alternatively, Proposition 7.1 and (7.2) follow directly from Proposition 4.7 and Pascal's identity

$$\binom{r}{s}=\binom{r-1}{s}+\binom{r-1}{s-1}. \tag{7.6}$$

## 8. Regge Symmetries

In the 1950s, Regge [**12**] added to known symmetries to obtain a group of order 72 associated to $C'_{m,n,k}(i,j)$ in the classical context. See [**8**], [**9**], or [**16**], Ch. III, Sec. 8.4 for detailed background. These symmetries are efficiently encoded by transfer of indices in accordance with the following matrix, called the Regge symbol:

$$\left\|\begin{matrix} n-k & m-k & k \\ i & j & m+n-i-j-k \\ m-i & n-j & i+j-k \end{matrix}\right\|. \tag{8.1}$$

Symmetries correspond to the usual symmetries for determinant; one may permute any rows, permute any columns, or perform matrix transpose. Thus, to describe all symmetries in this form, it is enough to describe a generating set of row switches and transpose. The column switch corresponding to tensor transpose is also included. Table 1 summarizes index transformations.

With the conventions above, these generating symmetries are given below. For abbreviations, for instance, R13 indicates the interchange of rows 1 and 3 in the Regge symbol.

Proposition 8.1 (C12 Symmetry - Tensor Transpose).

$$c_{n,m,k}(j,i)=(-1)^k c_{m,n,k}(i,j) \tag{8.2}$$

Proposition 8.2 (Modified Transpose Symmetry: *Transpose* ∘ *R*23).

$$\binom{m-k+j}{m-k}\, c_{m,\,n-k+i,\,i}(k,\ i+j-k)=\binom{m-k+j}{m-i}\, c_{m,n,k}(i,j) \tag{8.3}$$

Proposition 8.3 (R13 Symmetry).

$$\binom{m+n-k}{m-k,\ n-k,\ k}\, c_{m',n',k'}(i,j)=(-1)^i\binom{m+n-k}{m-i,\ n-j,\ k'}\, c_{m,n,k}(i,j) \tag{8.4}$$

*where* $m'=n-k+i,\ n'=m-k+j,$ *and* $k'=i+j-k.$

Proposition 8.4 (R23 Symmetry - Weyl Group).

$$\binom{m+n-k}{i,\ j,\ *}\, c_{m,n,k}(m-i,n-j)=(-1)^k\binom{m+n-k}{m-i,\ n-j,\ *}\, c_{m,n,k}(i,j) \tag{8.5}$$

TABLE 1. Regge Symmetries

| | m | n | i | j | k |
|---|---|---|---|---|---|
| C12 | $n$ | $m$ | $j$ | $i$ | $k$ |
| Transpose | $m$ | $m+n-i-k$ | $m-k$ | $j$ | $m-i$ |
| R13 | $n-k+i$ | $m-k+j$ | $i$ | $j$ | $i+j-k$ |
| R23 | $m$ | $n$ | $m-i$ | $n-j$ | $k$ |

PROOF. The C12 symmetry follows directly from the definition of the coordinate vector for highest weights. The R23 symmetry follows from the Weyl group action

$$w(f^p\phi_m) = (-1)^p \frac{p!}{(m-p)!} f^{m-p}\phi_m \tag{8.6}$$

where $w$ is the group element

$$w = \begin{bmatrix} 0 & -1 \\ 1 & 0 \end{bmatrix}. \tag{8.7}$$

The R13 and modified Transpose symmetries follow from direct substitution into Proposition 4.7. The modified Transpose symmetry follows immediately. For the R13 symmetry, one also changes the summation variable to $l' = i - l$. □

The R23 Symmetry and Theorem 4.10 immediately yield

COROLLARY 8.5.

$$C_{m,n,k}(i,j) = \frac{\binom{m+n-k}{m-i,\ n-j,\ *} c_{m,n,k}(i,j)}{\binom{m+n-k}{i,\ j,\ *} D(m,n,k)} \tag{8.8}$$

## 9. Normalization

Another feature of the theory concerns alternative expressions for Clebsch-Gordan coefficients, either in terms of binomial summations [**13**] or hypergeometric series [**11**]. For the most part, these expressions are unified through the Regge group, which in turn is essentially a mechanism from hypergeometric theory. Due to their ubiquity in the literature, brief derivations of Wigner's and Racah's formulas are given here.

PROOF. (Proposition 3.2) Extending coefficients from the $SL(2,\mathbb{R})$ theory with real vector spaces, one may choose $\phi_m$, and in turn all $f^i\phi_m$, to be real in $V(m)$. With this choice of basis for $V(m)$, one restricts the group action to

$$SU(2) = \{g \in SL(2,\mathbb{C}) \mid g \cdot g^* = I_2\},$$

where $g^*$ is the conjugate transpose of $g$ and $I_2$ is the identity matrix of size 2. Note that

$$w = \begin{bmatrix} 0 & -1 \\ 1 & 0 \end{bmatrix} \tag{9.1}$$

is in $SU(2)$, and, for $g$ in $SU(2)$,

$$wgw^{-1} = (g^{-1})^T = \bar{g}. \tag{9.2}$$

For $v_1, v_2$ in $V(m)$, define the invariant Hermitian inner product

$$\langle\langle v_1, v_2\rangle\rangle_m = \langle v_1, \overline{w \cdot v_2}\rangle_m. \tag{9.3}$$

Then

$$\begin{aligned}\langle\langle f^i\phi_m, f^i\phi_m\rangle\rangle_m &= \langle f^i\phi_m, \overline{w \cdot f^i\phi_m}\rangle_m \\ &= (-1)^i \frac{i!}{(m-i)!}\langle f^i\phi_m, \overline{f^{m-i}\phi_m}\rangle_m \\ &= (-1)^i \frac{i!}{(m-i)!}\langle f^i\phi_m, f^{m-i}\phi_m\rangle_m = \frac{i!}{(m-i)!}\end{aligned} \tag{9.4}$$

Likewise, in $V(m) \otimes V(n)$,

$$\langle\langle f^{i+j-k}\phi_{m,n,k}, f^{i+j-k}\phi_{m,n,k}\rangle\rangle = \frac{(i+j-k)!\ D(m,n,k)}{(m+n-i-j-k)!}. \tag{9.5}$$

Normalizing basis vectors, one recovers Wigner's Formula as

$$C'_{m,n,k}(i,j) = \frac{||f^{i+j-k}\phi_{m,n,k}||}{||f^i\phi_m||_m\ ||f^j\phi_n||_n} C_{m,n,k}(i,j). \tag{9.6}$$

□

Next one has Racah's formula

PROPOSITION 9.1 ([**16**], Ch. III.8.9, Equation (10)).

$$C'_{m,n,k}(i,j) = \sqrt{\frac{(m+n-2k+1)\binom{m+n-k}{m-k,\ n-k,\ k}}{(m+n-k+1)\binom{m+n-k}{m-i,\ n-j,\ *}\binom{m+n-k}{i,\ j,\ *}}} \cdot S'$$

*where*

$$S' = \sum_{s=0}^{k} (-1)^s \binom{k}{s}\binom{m-k}{i-s}\binom{n-k}{j-k+s}.$$

Comparing with Wigner's Formula, Proposition 9.1 follows immediately from

PROPOSITION 9.2.

$$c_{m,n,k}(i,j) = \sum_{l=0}^{k} (-1)^l \binom{i+j-k}{i-l}\binom{m-i}{k-l}\binom{n-j}{l}.$$

PROOF BY INDUCTION. First note that, for any $m$ and $n$ with $k = 0$, only one term contributes in either sum; that is,

$$c_{m,n,0}(i,j) = \binom{i+j}{i}.$$

The induction is indexed by $i' = i + j - k$ (all $m, n$). For the base case, when $i' = 0$, the first binomial coefficient only contributes when $l = i$, and the formula agrees with Proposition 4.1.

For the induction step, suppose the proposition holds for $i' = i + j - k - 1$. Using (7.6), one has

$$\sum_{l=0}^{k} (-1)^l \binom{i+j-k}{i-l}\binom{m-i}{k-l}\binom{n-j}{l}$$
$$= \sum_{l=0}^{k} (-1)^l \left[\binom{i+j-k-1}{i-l-1} + \binom{i+j-k-1}{i-l}\right]\binom{m-i}{k-l}\binom{n-j}{l}$$
$$= c_{m-1,n,k}(i-1,j) + c_{m,n-1,k}(i,j-1) = c_{m,n,k}(i,j).$$

The last equality holds by the recurrence relation (7.4). □

## 10. Appendix: Projection Operator Methods

The formulas in Section 4 were discovered after explicitly computing projection operators for the types $V(m')$ in $V(m) \otimes V(n)$. These methods depend on the multiplicity-free properties of both the weight vectors in $V(m')$ and the Clebsch-Gordan decomposition. Extremal projector methods for $SU(2)$ first appear in [**10**]; see [**14**] for background on extremal projectors. An algorithm for constructing these projection operators follows.

Define the Casimir element of $U(sl(2))$ by

$$8\Omega = 2ef + 2fe + h^2 = 4ef + h^2 - 2h. \tag{10.1}$$

This element lies in the center of $U(sl(2))$. In fact, $Z(U(sl(2)) = \mathbb{C}[\Omega]$. One notes two useful facts about $\Omega$:

(1) On $V(n)$, $8\Omega$ acts by the scalar $\lambda_n = n(n+2)$
(2) $\Omega$ preserves weight spaces: if $hv = cv$ then $h(\Omega v) = c(\Omega v)$

For computational efficiency, one may use $ef$ instead of $\Omega$, as $ef$ also preserves weight spaces.

Recall spectral theory for finite-dimensional vector spaces. Let $M$ be an $r \times r$ matrix, diagonalizable with distinct eigenvalues $\lambda_i$, $1 \leq i \leq r$. If one computes the projection operator for the eigenspace $W(\lambda_j)$ of $M$ then $W(\lambda_j)$ is spanned by any nonzero column of the projection operator. Fix $1 \leq j \leq r$. For $i \neq j$, consider the quotients

$$M_i(\lambda_j) = \frac{M - \lambda_i I}{\lambda_j - \lambda_i} \tag{10.2}$$

Then

$$M_i(\lambda_j) = \begin{cases} 1 & \text{on } W(\lambda_j) \\ 0 & \text{on } W(\lambda_i) \\ * & \text{on other } W(\lambda_k) \end{cases} \tag{10.3}$$

Now define

$$P(\lambda_j) = \prod_{i \neq j} M_i(\lambda_j) \ : \ \mathbb{C}^r \to W(\lambda_j) \subset \mathbb{C}^r. \tag{10.4}$$

Fix $V(m)$ and $V(n)$, and define

$$M = [ef]_{\mathbf{B_p}}, \tag{10.5}$$

where $\mathbf{B_p}$ is the basis for the weight space for $m+n-2p$ in $V(m)\otimes V(n)$. As noted, $M$ is diagonalizable with eigenvalues of multiplicity one. In fact, $ef$ acts on the weight vector in $V(m+n-2k)$ with eigenvalue

$$\lambda_{m,n,k,p} = (m+n-p-k)(p-k+1). \tag{10.6}$$

Next, using formulas from Section 2, the effect of $ef$ on the elements of $\mathbf{B}_p$ is given by

PROPOSITION 10.1 (Tridiagonal Formula).

$$\begin{aligned}(ef)(f^i\phi_m\otimes f^j\phi_n) &= (ef^{i+1}\phi_m)\otimes f^j\phi_n \;+\; f^i\phi_m\otimes(ef^{j+1}\phi_n)\\ &\quad+(ef^i\phi_m)\otimes f^{j+1}\phi_n \;+\; f^{i+1}\phi_m\otimes(ef^j\phi_n)\\ &=[(i+1)(m-i)\;+\;(j+1)(n-j)]\,f^i\phi_m\otimes f^j\phi_n\\ &\quad+\;i(m-i+1)\;f^{i-1}\phi_m\otimes f^{j+1}\phi_n\\ &\quad+\;j(n-j+1)\,f^{i+1}\phi_m\otimes f^{j-1}\phi_n\end{aligned} \tag{10.7}$$

EXAMPLE 10.2. Returning to Example 5.1, consider the subspace for weight 1 in $V(3)\otimes V(4)$, with basis

$$B_3 = \{\phi_3\otimes f^3\phi_4,\ f\phi_3\otimes f^2\phi_4,\ f^2\phi_3\otimes f\phi_4,\ f^3\phi_3\otimes\phi_4\}, \tag{10.8}$$

Using the Tridiagonal Formula, $ef$ acts on the this subspace in coordinates by

$$M=[ef]_{B_3}=\begin{bmatrix}7&3&0&0\\6&10&4&0\\0&6&9&3\\0&0&4&4\end{bmatrix} \tag{10.9}$$

On each $V(7-2k)$ with $k\in\{3,2,1,0\}$, $M$ has respective eigenvalues $\{1,4,9,16\}$. Using the projection operator formula, the corresponding projections are given by

$$P_{V(1)}=\frac{1}{10}\begin{bmatrix}1&-1&1&-1\\-2&2&-2&2\\3&-3&3&-3\\-4&4&-4&4\end{bmatrix},\quad P_{V(3)}=\frac{1}{5}\begin{bmatrix}2&-1&0&1\\-2&1&0&-1\\0&0&0&0\\4&-2&0&2\end{bmatrix},$$

$$P_{V(5)}=\frac{1}{70}\begin{bmatrix}27&9&-15&-9\\18&6&-10&-6\\-45&-15&25&15\\-36&-12&20&12\end{bmatrix},\quad P_{V(7)}=\frac{1}{35}\begin{bmatrix}4&6&4&1\\12&18&12&3\\12&18&12&3\\4&6&4&1\end{bmatrix}.$$

Consider the basis vector $f^2\phi_3\otimes f\phi_4$ from $\mathbf{B}_3$ in $V(3)\otimes V(4)$. In coordinates, $[f^2\phi_3\otimes f\phi_4]_{\mathbf{B_3}}=(0,0,1,0)$. The decomposition according to type follows from the third columns in the projection operators:

$$\begin{aligned}&(\mathrm{V}(1)) && (0,0,1,0)=(1/10,-1/5,3/10,-2/5)\\ &(\mathrm{V}(3)) && \qquad+(0,\quad 0,\quad 0,\quad 0)\\ &(\mathrm{V}(5)) && \qquad+(-3/14,-1/7,5/14,2/7)\\ &(\mathrm{V}(7)) && \qquad+(4/35,12/35,12/35,4/35)\end{aligned}$$

Finally the coordinate vectors and Clebsch-Gordan coefficients in Section 5 factor the projection operators. For instance,

$$P_{V(5)} = \begin{bmatrix} 3 & 0 & 0 & 0 \\ 0 & 2 & 0 & 0 \\ 0 & 0 & -5 & 0 \\ 0 & 0 & 0 & -4 \end{bmatrix} \begin{bmatrix} 1 & 1 & 1 & 1 \\ 1 & 1 & 1 & 1 \\ 1 & 1 & 1 & 1 \\ 1 & 1 & 1 & 1 \end{bmatrix} \begin{bmatrix} 9/70 & 0 & 0 & 0 \\ 0 & 3/70 & 0 & 0 \\ 0 & 0 & -5/70 & 0 \\ 0 & 0 & 0 & -3/70 \end{bmatrix}$$

## References


[1] George E. Andrews, Richard Askey, and Ranjan Roy, *Special functions*, Encyclopedia of Mathematics and its Applications, vol. 71, Cambridge University Press, Cambridge, 1999. MR1688958

[2] Robert W. Donley Jr., *A Plancherel formula for representative functions on semisimple Lie groups*, J. Lie Theory **23** (2013), no. 2, 493–505. MR3113520

[3] U. Fano and A. R. P. Rau, *Symmetries in quantum physics*, Academic Press, Inc., San Diego, CA, 1996. MR1422916

[4] Howard Georgi, *Lie algebras in particle physics*, Frontiers in Physics, vol. 54, Benjamin/Cummings Publishing Co., Inc., Advanced Book Program, Reading, Mass., 1982. From isospin to unified theories; With an introduction by Sheldon L. Glashow. MR644800

[5] Roger Howe and Eng-Chye Tan, *Nonabelian harmonic analysis*, Universitext, Springer-Verlag, New York, 1992. Applications of $\mathrm{SL}(2,\mathbf{R})$. MR1151617

[6] James E. Humphreys, *Introduction to Lie algebras and representation theory*, Graduate Texts in Mathematics, vol. 9, Springer-Verlag, New York-Berlin, 1978. Second printing, revised. MR499562

[7] H. T. Johansson and C. Forssén, *Fast and accurate evaluation of Wigner $3j$, $6j$, and $9j$ symbols using prime factorization and multiword integer arithmetic*, SIAM J. Sci. Comput. **38** (2016), no. 1, A376–A384. MR3455141

[8] T. H. Koornwinder, *Clebsch-Gordan coefficients for* SU(2) *and Hahn polynomials*, Nieuw Arch. Wisk. (3) **29** (1981), no. 2, 140–155. MR641673

[9] C. Krattenthaler and K. Srinivasa Rao, *On group theoretical aspects, hypergeometric transformations and symmetries of angular momentum coefficients*, Symmetries in science XI, Kluwer Acad. Publ., Dordrecht, 2004, pp. 355–376. MR2170819

[10] Per-Olov Löwdin, *Angular momentum wavefunctions constructed by projector operators*, Rev. Modern Phys. **36** (1964), 966–976. MR0191466

[11] Jacques Raynal, *On the definition and properties of generalized $3-j$ symbols*, J. Mathematical Phys. **19** (1978), no. 2, 467–476. MR0473252

[12] Regge, T., *Symmetry properties of Clebsch-Gordan's coefficients*, Nuovo Cimento **10** (1958), 296–297.

[13] T. Shimpuku, *On the expressions of Clebsch-Gordan coefficients*, J. Math. Anal. Appl. **7** (1963), 397–419. MR0157598

[14] Valeriy N. Tolstoy, *Fortieth anniversary of extremal projector method for Lie symmetries*, Noncommutative geometry and representation theory in mathematical physics, Contemp. Math., vol. 391, Amer. Math. Soc., Providence, RI, 2005, pp. 371–384. MR2184036

[15] D. A. Varshalovich, A. N. Moskalev, and V. K. Khersonskiĭ, *Quantum theory of angular momentum*, World Scientific Publishing Co., Inc., Teaneck, NJ, 1988. Irreducible tensors, spherical harmonics, vector coupling coefficients, $3nj$ symbols; Translated from the Russian. MR1022665

[16] N. Ja. Vilenkin, *Special functions and the theory of group representations*, Translated from the Russian by V. N. Singh. Translations of Mathematical Monographs, Vol. 22, American Mathematical Society, Providence, R. I., 1968. MR0229863



Department of Mathematics and Computer Science, Queensborough Community College (CUNY), Bayside, New York 11364

*Email address*: `rdonley@qcc.cuny.edu`

Department of Mathematics, The City College of New York (CUNY), New York, New York 10031

*Email address*: `wkim@ccny.cuny.edu`

Contemporary Mathematics
Volume **714**, 2018
https://doi.org/10.1090/conm/714/14329

# Radon-John transforms and spherical harmonics

Ricardo Estrada and Boris Rubin

*Dedicated to Professor Gestur Ólafsson on the occasion of his 65th birthday*

ABSTRACT. The $d$-plane Radon-John transform takes functions on $\mathbb{R}^n$ to functions on the set of all $d$-dimensional planes in $\mathbb{R}^n$ by integration over these planes. We study the action of this transform on degenerate functions of the form $f(x) = f_0(r)\,Y_k(\theta)$, where $r = |x| > 0$, $\theta = x/|x|$, and $Y_k$ is a spherical harmonic of degree $k$. It is shown that the results for $d < n-1$ are surprisingly different from those in the known case $d = n-1$.

## 1. Introduction

In the present article, we are dealing with degenerate functions $f$ on $\mathbb{R}^n \setminus \{0\}$ having the form

$$f(x) = f_0(r)\,Y_k(\theta), \qquad r = |x|, \quad \theta = x/|x|, \tag{1.1}$$

where $Y_k$ is a spherical harmonic of fixed degree $k \in \{0, 1, 2, \ldots\}$. An idea of separation of variables dates back to early history of Analysis. In particular, functions (1.1) play an important role in harmonic analysis of operators commuting with rotations. For example, by the classical Bochner identity, the Fourier transform on $\mathbb{R}^n$ takes functions (1.1) to functions of the same form $F(x) = F_0(r)\,Y_k(\theta)$, where $F_0$ is a Hankel-type transform of $f_0$; see Theorem 3.10 in [**17**, Chapter IV, Section 3]. Another important example is the hyperplane Radon transform

$$(Rf)(\theta, t) = \int\limits_{x\cdot\theta=t} f(x)\,dx, \quad \theta \in S^{n-1}, \quad t \in \mathbb{R}. \tag{1.2}$$

It is known (see, e.g., [**6**, p. 19], [**3**], [**15**, p. 226]) that $R$ acts in a similar way as

$$R: \; f_0(r)\,Y_k(\theta) \to F_0(t)\,Y_k(\theta), \tag{1.3}$$

for some function $F_0(t)$ represented by the Gegenbauer-Chebyshev integral of $f_0$. We shall recall the definition of the Gegenbauer-Chebyshev integrals in Subsection 2.2 below. Detailed historical notes related to (1.3) can be found in [**15**, p. 277].

Our consideration is motivated by the following questions.

QUESTION 1. What is the analogue of (1.3) for the more general $d$-plane Radon-John transform $R_d$ when $1 \le d \le n-1$ is an arbitrary fixed integer?

---

2010 *Mathematics Subject Classification.* Primary 44A12; Secondary 26A33, 44A15.
*Key words and phrases.* Gegenbauer-Chebyshev integrals, Radon-John transforms, Grassmann manifolds.

QUESTION 2. Is $R_d$ injective on the functions (1.1)?

Extensive information about $d$-plane transforms can be found, e.g., in [**5**, **7**, **11**, **12**]. Question 2 is motivated by the study of non-injectivity of the hyperplane Radon transform in the pioneering works by Zalcman [**18**], Armitage [**1**], Armitage and Goldstein [**2**]. The present article is an attempt to bring some light to the similar problem for the $d$-plane transforms with $d < n-1$ in the spirit of our previous publications [**3**], [**15**, Subsection 4.12.2].

We show that the results for $d = n-1$ and $d < n-1$ are surprisingly different. Specifically, in the second case, the product $F_0(t)\,Y_k(\theta)$ in (1.3) must be replaced by a linear combination of Gegenbauer-Chebyshev integrals with the coefficients represented by certain Grassmannian modifications of spherical harmonics. Moreover, unlike the case $d = n-1$ (see Theorem 2.4), the operator $R_d$ with $d < n-1$ is injective on functions $f_0(r)\,Y_k(\theta)$ under very general assumptions. The precise statements are given by Theorems 3.1 and 3.4.

The paper is organized as follows. Sections 2 contains necessary preliminaries. Main results are presented in Section 3. Section 4 contains concluding remarks and related open problems.

## 2. Preliminaries

**2.1. Notation.** In the following, $K = SO(n)$ is the group of rotations of the real Euclidean space $\mathbb{R}^n$, $[a]$ stands for the integer part of the real number $a$; $S^{n-1} = \{x \in \mathbb{R}^n : |x| = 1\}$ is the unit sphere in $\mathbb{R}^n$; $e_1, \ldots, e_n$ are the coordinate unit vectors. For $\theta \in S^{n-1}$, $d\theta$ denotes the surface element on $S^{n-1}$; $\sigma_{n-1} = 2\pi^{n/2}/\Gamma(n/2)$ is the surface area of $S^{n-1}$. The letter $c$ denotes a constant that may vary at each occurrence.

Given a positive integer $d < n$, let $\mathfrak{A}_{n,d}$ be the affine Grassmann manifold of all non-oriented $d$-dimensional planes in $\mathbb{R}^n$. There exist several ways to parameterize $d$-dimensional planes in $\mathbb{R}^n$. For example, if $\mathfrak{G}_{n,d}$ is the Grassmann manifold of $d$-dimensional linear subspaces of $\mathbb{R}^n$, then every plane $\tau \in \mathfrak{A}_{n,d}$ can be identified with the pair $(\zeta, u)$, so that

$$\tau = \zeta + u, \qquad \zeta \in \mathfrak{G}_{n,d}, \quad u \in \zeta^\perp, \tag{2.1}$$

where $\zeta^\perp$ is the orthogonal complement to $\zeta$ in $\mathbb{R}^n$. We can also consider $\tau \in \mathfrak{A}_{n,d}$ as a hyperplane in some $(d+1)$-dimensional subspace $\xi$. Then $\tau$ can be parameterized by the pair $(\xi, u)$, so that

$$\tau = (\xi \cap u^\perp) + u, \qquad \xi \in \mathfrak{G}_{n,d+1}, \quad u \in \xi. \tag{2.2}$$

One can readily see that both parameterizations are equivalent so that $\zeta = \xi \cap u^\perp$ and the manifolds of all pairs $(\zeta, u)$ and $(\xi, u)$ have the same dimension $(n-d)(d+1)$.[1]

Using parametrization (2.1), we endow $\mathfrak{A}_{n,d}$ with the product measure $d\tau = d_*\zeta du$, where $d_*\zeta$ is the normalized $K$-invariant measure on $\mathfrak{G}_{n,d}$ of total mass 1 and $du$ denotes the Euclidean measure on $\zeta^\perp$.

[1] We recall that $\dim \mathfrak{G}_{n,d} = d(n-d)$.

**2.2. Gegenbauer-Chebyshev Integrals.** We use the standard notation for the Gegenbauer polynomials $C_\ell^\lambda(t)$, $\lambda > -1/2$:

$$C_\ell^\lambda(t)=\sum_{j=0}^{[\ell/2]} c_{\ell,j}\, t^{\ell-2j}, \quad c_{\ell,j}=(-1)^j \frac{2^{\ell-2j}\,\Gamma(\ell-j+\lambda)}{\Gamma(\lambda)\, j!\, (\ell-2j)!}, \tag{2.3}$$

where $[\ell/2]$ is the integer part of $\ell/2$; see [**4**, Subsection 11.1.2]. In the case $\lambda = 0$, they are usually substituted by the Chebyshev polynomials $T_\ell(t)$.

The *right-sided Gegenbauer-Chebyshev integrals* of a function $v$ on $\mathbb{R}_+$ are defined by

$$(\mathcal{G}_-^{\lambda,\ell} v)(t) = \frac{1}{c_{\lambda,\ell}} \int_t^\infty (r^2 - t^2)^{\lambda-1/2}\, C_\ell^\lambda\left(\frac{t}{r}\right) v(r)\, r\, dr, \tag{2.4}$$

$$c_{\lambda,\ell} = \frac{\Gamma(2\lambda+\ell)\,\Gamma(\lambda+1/2)}{2\ell!\,\Gamma(2\lambda)}, \qquad \lambda > -1/2, \quad \lambda \neq 0. \tag{2.5}$$

In the case $\lambda = 0$, we set

$$(\mathcal{T}_-^{\ell} v)(t) = \frac{2}{\sqrt{\pi}} \int_t^\infty (r^2 - t^2)^{-1/2}\, T_\ell\left(\frac{t}{r}\right) v(r)\, r\, dr. \tag{2.6}$$

We refer to [**15**, Subsection 2.6.3] for the properties of these integrals and related references. In particular, the following statements hold.

PROPOSITION 2.1. [**15**, p. 69] *Let $a > 0$, $\lambda > -1/2$. The integral $(\mathcal{G}_-^{\lambda,\ell} v)(t)$ is finite for almost all $t > a$ provided that*

$$\int_a^\infty |v(t)|\, t^{2\lambda-\eta}\, dt < \infty, \qquad \eta = \begin{cases} 0 & \text{if } \ell \text{ is even}, \\ 1 & \text{if } \ell \text{ is odd}. \end{cases} \tag{2.7}$$

*The case $\lambda = 0$ gives the similar statements for $\mathcal{T}_-^{\ell} v$.*

The next proposition provides information about injectivity of the operator $\mathcal{G}_-^{\lambda,\ell}$.

PROPOSITION 2.2. [**15**, p. 71] *Let $\lambda > -1/2$. If $\ell = 0, 1$, then $\mathcal{G}_-^{\lambda,\ell}$ is injective on $\mathbb{R}_+$ in the class of functions satisfying (2.7) for all $a > 0$. If $\ell \geq 2$, then $\mathcal{G}_-^{\lambda,\ell}$ is non-injective in this class of functions. Specifically, let $v_j(t) = t^{-2\lambda-j-2}$, where $j$ is a nonnegative integer such that $\ell - j = 2, 4, \ldots$. Then $(\mathcal{G}_-^{\lambda,\ell} v_j)(t) = 0$ for all $t > 0$. The case $\lambda = 0$ gives the similar statement for $\mathcal{T}_-^{\ell} v$.*

The null space of the operator $\mathcal{G}_-^{\lambda,\ell}$ is characterized as follows; see [**3**, p. 1170] and [**15**, p. 80]. Let

$$\varkappa_{\lambda,\ell}(t) = \begin{cases} t^{2\lambda} & \text{if } \ell \text{ is even}, \\ t^{-1} & \text{if } \ell \text{ is odd}, -1/2 < \lambda < 0, \\ t^{-1}(1 + |\log t|) & \text{if } \ell \text{ is odd}, \lambda = 0, \\ t^{2\lambda-1} & \text{if } \ell \text{ is odd}, \lambda > 0. \end{cases} \tag{2.8}$$

Given $a \geq 0$, we denote by $L^1_\varkappa(a,\infty)$ the set of all function $v$ on $(a,\infty)$ such that

$$\int\limits_{a_1}^{\infty} |v(t)|\, \varkappa_{\lambda,\ell}(t)\, dt < \infty \quad \forall\, a_1 > a. \tag{2.9}$$

PROPOSITION 2.3. *Suppose that $\ell \geq 2$, $L = [\ell/2]$, $a \geq 0$. If $v \in L^1_\varkappa(a,\infty)$ and $(\mathcal{G}^{\lambda,\ell}_- v)(r) = 0$ for almost all $r > a$, then*

$$f(t) = \sum_{j=0}^{L-1} c_j t^{2j-\ell-2\lambda} \quad a.e. \text{ on } (a,\infty) \tag{2.10}$$

*with some coefficients $c_j$.*

The above lemma shows that the functions $v_j$ in Proposition 2.2 exhaust the kernel of the operator $\mathcal{G}^{\lambda,\ell}_-$ in the space $L^1_\varkappa(a,\infty)$, $a \geq 0$.

The following result for the Radon transform (1.2) follows from Lemmas 4.107, 4.108, and Theorem 4.113 in [**15**]; see also Theorem 3.7 in [**3**].

THEOREM 2.4. *Let $\lambda = (n-2)/2$, $f(x) = f_0(r)\, Y_k(\theta)$, $k \geq 0$. Given $0 \leq a < \infty$, suppose that*

$$\int\limits_{a_1}^{\infty} |f_0(r)|\, r^{2\lambda} dr < \infty \quad \text{for all } a_1 > a. \tag{2.11}$$

*Then*

$$(Rf)(\theta,t) = F_0(t)\, Y_k(\theta), \tag{2.12}$$

$$F_0(t) = \begin{cases} \pi^{\lambda+1/2}(\mathcal{G}^{\lambda,k}_- f_0)(t) & if \quad n \geq 3, \\ \pi^{1/2}(\mathcal{T}^k_- f_0)(t) & if \quad n = 2. \end{cases} \tag{2.13}$$

*If $(Rf)(\theta,t)=0$ for almost all $(\theta,t)$ with $t > a$, then, for almost all $r > a$,*

$$f_0(r) = \begin{cases} 0 & if \quad k=0,1, \\ \sum\limits_{j=1}^{[k/2]} c_j\, r^{2j-k-n} & if \quad k \geq 2, \end{cases} \tag{2.14}$$

*with some constants $c_j$. Conversely, if $c_j$ and $f_0$ satisfy (2.11) and (2.14), then $(Rf)(\theta,t) = 0$ for almost all $(\theta,t)$ with $t > a$.*

**2.3. The Radon-John $d$-plane Transforms.** Let $\mathfrak{A}_{n,d}$ be the affine Grassmann manifold of all non-oriented $d$-dimensional planes in $\mathbb{R}^n$. The *Radon-John $d$-plane transform* of a function $f$ on $\mathbb{R}^n$ is a map $f \to R_d f$, where

$$(R_d f)(\tau) = \int\limits_{\tau} f(x)\, d_\tau x, \qquad \tau \in \mathfrak{A}_{n,d}, \tag{2.15}$$

$d_\tau x$ being the Euclidean measure on $\tau$. Using the parametrization (2.1), we write (2.15) in the form

$$(R_d f)(\tau) \equiv (R_d f)(\zeta,u) = \int\limits_{\zeta} f(u+v)\, d_\zeta v, \tag{2.16}$$

where $\zeta \in \mathfrak{G}_{n,d}$, $u \in \zeta^\perp$ and $d_\zeta v$ is the Euclidean measure on $\zeta$. This integral is well-defined for every $\tau \in \mathfrak{A}_{n,d}$ if $f$ is continuous and $O(|x|^{-\mu})$ with $\mu > d$. Moreover, the following statement holds.

LEMMA 2.5. (cf. [**13**, Theorem 3.2]) *If*

$$\int_{|x|>a} \frac{|f(x)|}{|x|^{n-d}}\, dx < \infty \quad \forall a > 0, \tag{2.17}$$

*then $(R_d f)(\tau)$ is finite for almost all $\tau \in \mathfrak{A}_{n,d}$. If $f$ is nonnegative, radial, and (2.17) fails, then $(R_d f)(\tau) \equiv \infty$.*

We will need the following hemispherical representation of the $d$-plane transform which was obtained in [**14**, Lemma 3.5]. Let

$$\begin{gathered} \mathbb{R}^{d+1} = \mathbb{R}e_1 \oplus \cdots \oplus \mathbb{R}e_{d+1} = \mathbb{R}^d \oplus \mathbb{R}e_{d+1}, \\ S^d = S^{n-1} \cap \mathbb{R}^{d+1}, \\ S^d_+ = \{\theta = (\theta_1, \cdots, \theta_{d+1}) \in S^d : \theta_{d+1} > 0\}. \end{gathered} \tag{2.18}$$

LEMMA 2.6. *Suppose $u \neq 0$ and let $f_g(x) = f(gx)$, where $g \in K$ is a rotation satisfying $g\,\mathbb{R}^d = \zeta$, $g\,e_{d+1} = u/|u|$. Then*

$$(R_d f)(\zeta, u) = (R_d f_g)(\mathbb{R}^k, te_{d+1}) = t^d \int_{S^d_+} f_g\Big(\frac{t\theta}{\theta_{d+1}}\Big) \frac{d\theta}{\theta^{d+1}_{d+1}}, \qquad t = |u|. \tag{2.19}$$

**2.4. Spherical Harmonics.** The theory of spherical harmonics is presented in many sources. We recall some facts following detailed exposition in [**15**, Appendix A] and adapting notation to our case.

Let $S^d = S^{n-1} \cap \mathbb{R}^{d+1}$ be the lower-dimensional unit sphere, as in (2.18). We choose an orthonormal basis $\{Y^{(d)}_{\ell,\nu}(\theta)\}$ of spherical harmonics in $L^2(S^d)$. Here $\ell \in \{0, 1, 2, \dots\}$ and $\nu = 1, 2, \dots, \delta_{d+1}(\ell)$, where

$$\delta_{d+1}(\ell) = \frac{(d + 2\ell - 1)(d + \ell - 2)!}{\ell!\,(d-1)!} \tag{2.20}$$

is the dimension of the subspace of spherical harmonics of degree $\ell$. The following Funk-Hecke Theorem is the heart of analysis on the sphere; see [**15**, p. 485].

THEOREM 2.7. *Let $h(s)(1-s^2)^{(d-2)/2} \in L^1(-1,1)$. Then for every spherical harmonic $Y^{(d)}_\ell$ on $S^d$ of degree $\ell$ and every $\theta \in S^d$,*

$$\int_{S^d} h(\theta \cdot \sigma)\, Y^{(d)}_\ell(\sigma)\, d\sigma = \lambda_\ell\, Y^{(d)}_\ell(\theta), \tag{2.21}$$

$$\lambda_\ell = \sigma_{d-1} \int_{-1}^{1} h(s)\, P^{(d)}_\ell(s)\, (1-s^2)^{(d-2)/2}\, ds, \tag{2.22}$$

$$P^{(d)}_\ell(s) = \begin{cases} T_\ell(s) & \text{if } d = 1, \\[2ex] \dfrac{\ell!\,(d-2)!}{(\ell + d - 2)!}\, C^{(d-1)/2}_\ell(s) & \text{if } d \geq 2. \end{cases} \tag{2.23}$$

The spherical polynomials $P_\ell^{(d)}(s)$ are orthogonal with the weight $(1–s^2)^{(d-2)/2}$. Specifically (see formula (A.6.7) in [**15**]),

$$\int_{-1}^{1} P_j^{(d)}(s)\, P_\ell^{(d)}(s)\,(1-s^2)^{(d-2)/2} ds = \begin{cases} 0 & \text{if } j \neq \ell, \\ \dfrac{\sigma_d}{\delta_{d+1}(\ell)\,\sigma_{d-1}} & \text{if } j = \ell. \end{cases} \tag{2.24}$$

By the addition theorem for spherical harmonics (see formula (A.6.14) in [**15**]), for any $\theta, \sigma \in S^d$,

$$\sum_{\nu=1}^{\delta_{d+1}(\ell)} Y_{\ell,\nu}^{(d)}(\theta) Y_{\ell,\nu}^{(d)}(e_{d+1}) = \frac{\delta_{d+1}(\ell)}{\sigma_d}\, P_\ell^{(d)}(\theta\cdot e_{d+1}). \tag{2.25}$$

By the Gauss decomposition (see formula (A.4.9) in [**15**]), the restriction onto $S^d$ of every homogeneous polynomial $Q_k(x)$ on $\mathbb{R}^n$ of degree $k \geq 2$ can be decomposed uniquely as a finite sum of spherical harmonics on $S^d$. Specifically,

$$Q_k(\theta) = \sum_{i=0}^{[k/2]} Y_{k-2i}^{(d)}(\theta), \qquad \theta \in S^d. \tag{2.26}$$

## 3. Main Results

Below we study the action of the $d$-plane transform $R_d f$ on functions of the form $f(x) = f_0(|x|)\, Y_k(x/|x|)$, where $Y_k$ is the spherical harmonic of degree $k \geq 0$ on $S^{n-1}$. Given an integer $\ell \geq 0$, we denote

$$w_\ell(t) = \begin{cases} \pi^{d/2}\,(\ell+d-3)!\,(\mathcal{G}_-^{(d-1)/2,\ell} f_0)(t) & \text{if } \ d \geq 2, \\ \sqrt{\pi}\,(\mathcal{T}_-^{\ell} f_0)(t) & \text{if } \ d = 1, \end{cases} \tag{3.1}$$

where $(\mathcal{G}_-^{(d-1)/2,\ell} f_0)(t)$ and $(\mathcal{T}_-^{\ell} f_0)(t)$ are Gegenbauer-Chebyshev integrals (2.4) and (2.6) corresponding to $\lambda = (d-1)/2$. Given $\tau = \tau(\zeta, u) \in \mathfrak{A}_{n,d}$ with $\zeta \in \mathfrak{G}_{n,d}$, $u \in \zeta^\perp$, and $u' = u/|u| \in S^{n-1} \cap \zeta^\perp$, we also set

$$\mathcal{H}_{\ell,k}(\zeta, u') = \frac{\delta_{d+1}(\ell)}{\sigma_d} \int\limits_{S^{n-1}\cap(\zeta\oplus\mathbb{R}u)} Y_k(\omega)\, P_\ell^{(d)}(\omega\cdot u')\, d\omega, \tag{3.2}$$

where $P_\ell^{(d)}$ and $\delta_{d+1}(\ell)$ are defined by (2.23) and (2.20), respectively.

The following theorem is a counterpart of (1.3) for the $d$-plane transform with $d < n-1$.

THEOREM 3.1. *Let $f(x) = f_0(|x|)\, Y_k(x')$, $x' = x/|x| \in S^{n-1}$, where*

$$\int_a^\infty |f_0(r)|\, r^{d-1} dr < \infty \quad \forall a > 0. \tag{3.3}$$

*Then $(R_d f)(\tau)$ is finite for almost all $\tau \in \mathfrak{A}_{n,d}$ and*

$$(R_d f)(\zeta, u) = \sum_{\substack{\ell=0 \\ k-\ell \text{ even}}}^{k} w_\ell(|u|)\, \mathcal{H}_{\ell,k}(\zeta, u'), \qquad u' = u/|u|. \tag{3.4}$$

PROOF. The first statement is an immediate consequence of Lemma 2.5. The proof of the second statement relies on the hemispherical representation (2.19). Specifically, for $\tau = \tau(\zeta, u) \in \mathfrak{A}_{n,d}$, $\zeta \in \mathfrak{G}_{n,d}$, $u \in \zeta^{\perp}$, suppose that $u \neq 0$ and let $f_g(x) = f(gx)$, where $g \in K$ satisfies $g\,\mathbb{R}^d = \zeta$ and $g\,e_{d+1} = u/|u|$. Then

$$(3.5) \qquad (R_d f)(\zeta, u) = (R_d f_g)(\mathbb{R}^d, te_{d+1}) = t^d \int\limits_{S^d_+} f_g\Big(\frac{t\theta}{\theta_{d+1}}\Big) \frac{d\theta}{\theta_{d+1}^{d+1}},$$

$$f_g\left(\frac{t\theta}{\theta_{d+1}}\right) = f_0\left(\frac{t}{\theta_{d+1}}\right) Y_k(g\theta), \qquad t = |u|, \quad \theta \in S^d_+.$$

Let $\{Y^{(d)}_{\ell,\nu}(\theta)\}$ be an orthonormal basis of spherical harmonics in $L^2(S^d)$. Then for $\theta \in S^d_+$,

$$(3.6) \qquad Y_k(g\theta) = \sum_{\ell=0}^{\infty} \sum_{\nu=1}^{\delta_{d+1}(\ell)} h_{\ell,\nu,k}(g)\, Y^{(d)}_{\ell,\nu}(\theta),$$

$$(3.7) \qquad h_{\ell,\nu,k}(g) = \int\limits_{S^d} Y_k(g\theta)\, Y^{(d)}_{\ell,\nu}(\theta)\, d\theta.$$

Applying Gauss's decomposition (2.26) to $Y_k(g\theta)$, we obtain

$$h_{\ell,\nu,k}(g) = \sum_{i=0}^{[k/2]} \int\limits_{S^d} Y^{(d)}_{\ell,\nu}(\theta)\, Y^{(d)}_{k-2i,g}(\theta)\, d\theta,$$

where $Y^{(d)}_{k-2i,g}$ are spherical harmonics on $S^d$ having degree $k-2i$ and depending on $g$. By the orthogonality of spherical harmonics, the integral in the last sum is zero whenever $k - 2i \neq \ell$. In other words, it is zero for all $\ell > k$ and for all $\ell \leq k$, when $k - \ell$ is odd. Hence (3.6) becomes

$$(3.8) \qquad Y_k(g\theta) = \sum_{\substack{\ell=0 \\ k-\ell \ even}}^{k} \sum_{\nu=1}^{\delta_{d+1}(\ell)} h_{\ell,\nu,k}(g)\, Y^{(d)}_{\ell,\nu}(\theta).$$

Now, by (3.5) and (3.8),

$$(3.9) \qquad (R_d f)(\zeta, u) = t^d \sum_{\substack{\ell=0 \\ k-\ell \ even}}^{k} \sum_{\nu=1}^{\delta_{d+1}(\ell)} h_{\ell,\nu,k}(g) \int\limits_{S^d_+} f_0\Big(\frac{t}{\theta_{d+1}}\Big)\, Y^{(d)}_{\ell,\nu}(\theta)\, \frac{d\theta}{\theta_{d+1}^{d+1}}.$$

By the Funk-Hecke formula (2.21),

$$(3.10) \qquad t^d \int\limits_{S^d_+} f_0\Big(\frac{t}{\theta_{d+1}}\Big)\, Y^{(d)}_{\ell,\nu}(\theta)\, \frac{d\theta}{\theta_{d+1}^{d+1}} = w_\ell(t)\, Y^{(d)}_{\ell,\nu}(e_{d+1}),$$

where

$$\begin{aligned} w_\ell(t) &= t^d \sigma_{d-1} \int_0^1 f_0\Big(\frac{t}{s}\Big)\, P_\ell^{(d)}(s)\,(1-s^2)^{d/2-1}\,\frac{ds}{s^{d+1}} \\ &= \sigma_{d-1} \int_t^\infty f_0(r)\, P_\ell^{(d)}\Big(\frac{t}{r}\Big)(r^2-t^2)^{d/2-1}\, r\, dr. \end{aligned}$$

The latter coincides with (3.1). Combining (3.9) with (3.10) and (3.7), and using the addition theorem for spherical harmonics (see (2.25)), we obtain

$$\begin{aligned} (R_d f)(\zeta, u) &= \sum_{\substack{\ell=0\\ k-\ell\ even}}^{k} w_\ell(t) \sum_{\nu=1}^{\delta_{d+1}(\ell)} h_{\ell,\nu,k}(g)\, Y_{k,\nu}^{(d)}(e_{d+1}) \\ &= \sum_{\substack{\ell=0\\ k-\ell\ even}}^{k} w_\ell(t) \int_{S^d} Y_k(g\theta) \left[\sum_{\nu=1}^{\delta_{d+1}(\ell)} Y_{\ell,\nu}^{(d)}(\theta) Y_{\ell,\nu}^{(d)}(e_{d+1})\right] d\theta \\ &= \sum_{\substack{\ell=0\\ k-\ell\ even}}^{k} \frac{\delta_{d+1}(\ell)\, w_\ell(t)}{\sigma_d} \int_{S^d} Y_k(g\theta)\, P_\ell^{(d)}(\theta\cdot e_{d+1})\, d\theta. \end{aligned}$$

Changing variable $\omega = g\theta \in gS^d$, where

$$gS^d \subset g\mathbb{R}^{d+1} = g(\mathbb{R}^d \oplus \mathbb{R}e_{d+1}) = \zeta \oplus \mathbb{R}u,$$

we obtain (3.4). □

REMARK 3.2. If the $d$-plane $\tau$ is parameterized by the pair $(\xi, u)$ according to (2.2), that is, $\tau = (\xi \cap u^\perp) + u$, $\xi \in \mathfrak{G}_{n,d+1}$, $u \in \xi$, then, setting $S^d_\xi = S^{n-1} \cap \xi$ and abusing notation, we can write (3.2) as a spherical convolution

$$\mathcal{H}_{\ell,k}(\xi, u') = \frac{\delta_{d+1}(\ell)}{\sigma_d} \int_{S^d_\xi} Y_k(\omega)\, P_\ell^{(d)}(\omega\cdot u')\, d\omega, \quad u' \in S^d_\xi. \tag{3.11}$$

In this notation, (3.4) becomes

$$(R_d f)(\xi, u) = \sum_{\substack{\ell=0\\ k-\ell\ even}}^{k} w_\ell(|u|)\, \mathcal{H}_{\ell,k}(\xi, u'). \tag{3.12}$$

The system of functions $\mathcal{H}_{\ell,k}(\xi, u')$ is of independent interest and enjoys a number of remarkable properties.

LEMMA 3.3. *Given a nonnegative integer $k$, suppose that*

$$0 \le \ell \le k, \quad 0 \le \tilde\ell \le k, \quad k-\ell \text{ is even}, \quad k-\tilde\ell \text{ is even}. \tag{3.13}$$

*Then for any subspace $\xi \in \mathfrak{G}_{n,d+1}$,*

$$\int_{S^d_\xi} \mathcal{H}_{\ell,k}(\xi, u')\, \mathcal{H}_{\tilde\ell,k}(\xi, u')\, du' = 0 \quad \textit{if} \quad \ell \ne \tilde\ell. \tag{3.14}$$

PROOF. Due to the addition theorem for spherical harmonics, the functions $u' \to \mathcal{H}_{\ell,k}(\xi, u')$ are spherical harmonics of degree $\ell$ on $S^d_\xi$ and therefore, they are mutually orthogonal on $S^d_\xi$. This gives (3.14). □

THEOREM 3.4. *Let $f(x) = f_0(|x|)\, Y_k(x')$ be the same as in Theorem 3.1. If $(R_d f)(\tau) = 0$ for almost all d-planes $\tau$ away from the origin, then $f_0(r) = 0$ for almost all $r > 0$.*

PROOF. Suppose that $k$ is even. We multiply both sides of (3.12) by $\mathcal{H}_{0,k}(\xi, u')$ and integrate in $u' \in S^d_\xi$. Then (3.14) yields

$$w_0(|u|) \int\limits_{S^d_\xi} \mathcal{H}^2_{0,k}(\xi, u')\, du' = 0 \tag{3.15}$$

in the corresponding almost everywhere sense. The integral in this formula differs from zero. Indeed, assuming the contrary, from (3.11) we obtain

$$\mathcal{H}_{0,k}(\xi, u') \equiv c \int\limits_{S^d_\xi} Y_k(\omega)\, d\omega = 0, \qquad c \neq 0, \tag{3.16}$$

for all $u' \in S^{n-1}$ and all $\xi \ni u'$. Noting that $S^d_\xi = S^{n-1} \cap \xi$, we arrive at the totally geodesic Funk-type transform of the form

$$f(\omega) \to \int\limits_{S^{n-1}\cap\xi} f(\omega)\, d\omega, \qquad \xi \in \mathfrak{G}_{n,d+1}, \tag{3.17}$$

which is injective on functions $f \in C^\infty(S^{n-1})$ and represents a smooth function on $\mathfrak{G}_{n,d+1}$ (see, e.g., [**7**, Chapter III, Section 1(C)]). For any fixed $u' \in S^{n-1}$, the set

$$\{\xi \in \mathfrak{G}_{n,d+1} : \xi \ni u'\}$$

has measure zero in $\mathfrak{G}_{n,d+1}$ and therefore, (3.16) extends by continuity to all $\xi \in \mathfrak{G}_{n,d+1}$. By injectivity of (3.17), it follows that $Y_k(\omega) \equiv 0$ on $S^{n-1}$, that contradicts the assumption. Thus, from (3.15), we conclude that $w_0(t) = 0$ for almost all $t > 0$ and therefore, $f_0(r) = 0$ for almost all $r > 0$; see Proposition 2.2.

If $k$ is odd, multiplication by $\mathcal{H}_{1,k}(\xi, u')$ in (3.12) with subsequent integration in $u' \in S^d_\xi$ yields

$$w_1(|u|) \int\limits_{S^d_\xi} \mathcal{H}^2_{1,k}(\xi, u')\, du' = 0, \tag{3.18}$$

$$\mathcal{H}_{1,k}(\xi, u') = \frac{k+1}{\sigma_k} \int\limits_{S^d_\xi} Y_k(\omega)\, (\omega \cdot u')\, d\omega, \qquad u' \in S^d_\xi.$$

As above, it suffices to show that $\mathcal{H}_{1,k}(\xi, u') \neq 0$ for some pair $(\xi, u')$. Suppose the contrary. Then for all unit vectors $u'$ and all $\xi \ni u'$ we have

$$0 = \int\limits_{S^d_\xi} Y_k(\omega)\, (\omega \cdot u')\, d\omega = u' \cdot V_\xi, \tag{3.19}$$

where $V_\xi$ is the $n$-vector with components

$$\int_{S^d_\xi} Y_k(\omega)\,\omega_j\,d\omega, \qquad j=1,2,\dots,n. \tag{3.20}$$

Because $u' \in S^{n-1}$ is arbitrary, (3.19) implies that $V_\xi$ is the zero vector, that is, all integrals in (3.20) equal zero for all $u' \in S^{n-1}$ and all $\xi \ni u'$. As above, using extension by continuity and injectivity of the Funk-type transform (3.17), we obtain that $Y_k(\omega)\,\omega_j \equiv 0$ on $S^{n-1}$ for all $j$ and therefore,

$$\sum_{j=1}^n |Y_k(\omega)\,\omega_j|^2 = |Y_k(\omega)|^2 \sum_{j=1}^n |\omega_j|^2 = |Y_k(\omega)|^2 \equiv 0.$$

The latter implies $Y_k(\omega) \equiv 0$, which is not true.

□

## 4. Conclusion

**1.** We have shown that the action of the $d$-plane transform on degenerate functions (1.1) is essentially different when $d = n-1$ and $d < n-1$. Suppose that $\{Y_{k,j}(\theta)\}$ is an orthonormal basis of spherical harmonics in $L^2(S^{n-1})$. Here $k = 0,1,2,\dots$ and $j = 1,2,\dots \delta_n(k)$, where $\delta_n(k)$ is the dimension of the subspace of spherical harmonics of degree $k$. Then, formally,

$$f(x) \equiv f(r\theta) \sim \sum_{k,j} f_{k,j}(r) Y_{k,j}(\theta)$$

where $f_{k,j}(r)$ are the corresponding Fourier-Laplace coefficients of the function $\theta \to f(r\theta)$. By Theorem 3.1,

$$(R_d f)(\zeta, u) \sim \sum_{k,j} \sum_{\substack{\ell=0 \\ k-\ell\ even}}^{k} (W_\ell f_{k,j})(|u|)\,(\mathcal{H}_{\ell,k} Y_{k,j})(\zeta, u'), \tag{4.1}$$

where $u' = u/|u|$,

$$(W_\ell f_{k,j})(t) = \begin{cases} \pi^{d/2}\,(\ell+d-3)!\,(\mathcal{G}_-^{(d-1)/2,\ell} f_{k,j})(t) & \text{if } \ d \ge 2, \\[2ex] \sqrt{\pi}\,(\mathcal{T}_-^{\ell} f_{k,j})(t) & \text{if } \ d = 1, \end{cases} \tag{4.2}$$

$$(\mathcal{H}_{\ell,k} Y_{k,j})(\zeta, u') = \frac{\delta_{d+1}(\ell)}{\sigma_d} \int_{S^{n-1}\cap(\zeta\oplus\mathbb{R}u)} Y_{k,j}(\theta)\,P_\ell^{(d)}(\theta\cdot u')\,d\omega, \tag{4.3}$$

cf. (3.1), (3.2).

We denote by $\mathcal{G}(n,d)$ the set of all pairs $(\zeta, u')$, where $\zeta \in \mathfrak{G}_{n,d}$ and $u' \in S^{n-1} \cap \zeta^\perp$. This set has the topological structure of the fiber bundle with the base $\mathfrak{G}_{n,d}$ and the fibers $u' \in S^{n-1} \cap \zeta^\perp$. It can be equipped with the canonical product measure $d\zeta du'$.

PROBLEM 1. Is it true that the functions

$$\mathfrak{H}_{k,j,\ell}(\zeta, u') = (\mathcal{H}_{\ell,k} Y_{k,j})(\zeta, u')$$

form an orthogonal system in $L^2(\mathcal{G}(n,d))$, that is

$$\int_{\mathcal{G}(n,d)} \mathfrak{H}_{k,j,\ell}(\zeta,u')\,\mathfrak{H}_{k',j',\ell'}(\zeta,u')\,d\zeta du' = c\,\delta_{k,k'}\,\delta_{j,j'}\,\delta_{\ell,\ell'}\ ?$$

**2.** A remarkable result of Armitage and Goldstein [**2**] states that there is a nonconstant harmonic function $h$ on $\mathbb{R}^n$, $n \ge 2$, such that $\int_\tau |h| < \infty$ and $\int_\tau h = 0$ for every hyperplane $\tau$; see also Zalcman [**18**] ($n = 2$) and Armitage [**1**].

PROBLEM 2. Can this statement be extended to the case when $\tau$ is a $d$-dimensional plane with $d < n - 1$?

These challenging problems, that would increase our knowledge of the $d$-plane transform, might be of interest for the reader.

## Acknowledgments

The authors are grateful to Gestur Ólaffson, who shared with us his knowledge of the subject, and to Alexander Katsevich, who brought our attention to his works [**8**–**10**]. The papers [**8**, **9**] employ Gegenbauer-Chebyshev integrals in the study of the range of the hyperplane Radon transform and its dual. The paper [**10**, Theorem 2.6] contains a description of the kernel of the dual Radon transform on some spaces of distributions. Special thanks go to the referee for his valuable remarks.

## References

[1] David H. Armitage, *A non-constant continuous function on the plane whose integral on every line is zero*, Amer. Math. Monthly **101** (1994), no. 9, 892–894. MR1300495

[2] David H. Armitage and Myron Goldstein, *Nonuniqueness for the Radon transform*, Proc. Amer. Math. Soc. **117** (1993), no. 1, 175–178. MR1106177

[3] Ricardo Estrada and Boris Rubin, *Null spaces of Radon transforms*, Adv. Math. **290** (2016), 1159–1182. MR3451950

[4] Arthur Erdélyi (Editor), *Higher transcendental functions*, Vol. I and II, McGraw-Hill, New York, 1953.

[5] Fulton B. Gonzalez, *Notes on integral geometry and harmonic analysis*, COE Lecture Note, vol. 24, Kyushu University, Faculty of Mathematics, Fukuoka, 2010. Math-for-Industry (MI) Lecture Note Series. MR2606155

[6] Sigurdur Helgason, *Geometric analysis on symmetric spaces*, 2nd ed., Mathematical Surveys and Monographs, vol. 39, American Mathematical Society, Providence, RI, 2008. MR2463854

[7] Sigurdur Helgason, *Integral geometry and Radon transforms*, Springer, New York, 2011. MR2743116

[8] Alexander I. Katsevich, *Range of the Radon transform on functions which do not decay fast at infinity*, SIAM J. Math. Anal. **28** (1997), no. 4, 852–866. MR1453309

[9] Alexander Katsevich, *New range theorems for the dual Radon transform*, Trans. Amer. Math. Soc. **353** (2001), no. 3, 1089–1102. MR1804413

[10] Alexander Katsevich, *Range theorems for the Radon transform and its dual*, Radon transforms and tomography (South Hadley, MA, 2000), Contemp. Math., vol. 278, Amer. Math. Soc., Providence, RI, 2001, pp. 167–191. MR1851487

[11] Andrew Markoe, *Analytic tomography*, Encyclopedia of Mathematics and its Applications, vol. 106, Cambridge University Press, Cambridge, 2006. MR2220852

[12] Boris Rubin, *Reconstruction of functions from their integrals over k-planes*, Israel J. Math. **141** (2004), 93–117. MR2063027

[13] Boris Rubin, *On the Funk-Radon-Helgason inversion method in integral geometry*, Geometric analysis, mathematical relativity, and nonlinear partial differential equations, Contemp. Math., vol. 599, Amer. Math. Soc., Providence, RI, 2013, pp. 175–198. MR3202479

[14] Boris Rubin, *Weighted norm inequalities for k-plane transforms*, Proc. Amer. Math. Soc. **142** (2014), no. 10, 3455–3467. MR3238421

[15] Boris Rubin, *Introduction to Radon transforms. With elements of fractional calculus and harmonic analysis*, Encyclopedia of Mathematics and its Applications, vol. 160, Cambridge University Press, New York, 2015. MR3410931

[16] Robert Thomas Seeley, *Spherical harmonics*, Amer. Math. Monthly **73** (1966), no. 4, 115–121, DOI 10.2307/2313760. MR0201695

[17] Elias M. Stein and Guido Weiss, *Introduction to Fourier analysis on Euclidean spaces*, Princeton University Press, Princeton, N.J., 1971. Princeton Mathematical Series, No. 32. MR0304972

[18] Lawrence Zalcman, *Uniqueness and nonuniqueness for the Radon transform*, Bull. London Math. Soc. **14** (1982), no. 3, 241–245. MR656606

Department of Mathematics, Louisiana State University, Baton Rouge, Louisiana 70803

*Email address*: restrada@math.lsu.edu

Department of Mathematics, Louisiana State University, Baton Rouge, Louisiana 70803

*Email address*: borisr@math.lsu.edu

Contemporary Mathematics
Volume **714**, 2018
https://doi.org/10.1090/conm/714/14379

# Spherical functions on Riemannian symmetric spaces

Sigurdur Helgason

*Dedicated to Professor Gestur Ólafsson on his 65th birthday.*

ABSTRACT. This paper introduces a new kind of spherical functions on a symmetric space $X = G/K$. We develop the connection with joint eigenfunctions for $\mathbf{D}(G/K)$ and $\mathbf{D}(G/MN)$ and relate these to Poisson transform and Fourier transform on $X$. We also prove some results about the Harish Chandra $C$ functions corresponding to the spherical functions introduced.

## 1. Introduction

Let $X = G/K$ be a symmetric space where $G$ is a connected noncompact semisimple Lie group with finite center and $K$ a maximal compact subgroup. Let $\mathbf{D}(X)$ denote the algebra of $G$-invariant differential operators on $X$. Let $\delta$ be a unitary irreducible representation of $K$ on a vector space $V_\delta$.

DEFINITION 1.1. A spherical function of $K$ type $\delta$ is a $C^\infty$ function $\Phi : X \to \operatorname{Hom}(V_\delta, V_\delta)$ satisfying the following conditions.

$$\Phi \text{ is an eigenfunction of each } D \in \mathbf{D}(X) \tag{1.1}$$

$$\Phi(k \cdot x) = \delta(k)\Phi(x) \quad k \in K, x \in X. \tag{1.2}$$

On the right we have multiplication in $\operatorname{Hom}(V_\delta, V_\delta)$.

REMARK 1.2. This definition has some similarity with those of Godement [**G52**] and Harish-Chandra [**HC72**]. However, the first one is modeled after invariance under $g \to kgk^{-1}$; the second deals with bi-invariant differential operators on $G$ and a double representation of $K \times K$. Our definition by conditions (1.1) and (1.2) is thus rather different.

It is also unrelated to a definition of a spherical function by Tirao [**T76**] as a function characterized by the function equation (3.3) for zonal spherical functions but with the measure $dk$ replaced by $\chi(k)\, dk$ where $\chi$ is the character of a representation of $K$.

Our definition (1.1) stresses spherical functions as functions on $X$ rather than as $K$-right invariant functions on $G$. I remark that Harish-Chandra's major papers

---

2010 *Mathematics Subject Classification.* Primary 43A90, 53C35; Secondary 22E30, 22E46.
*Key words and phrases.* Symmetric spaces, spherical functions, zonal spherical functions, Harish Chandra's $c$ function.

[**HC58**] on zonal spherical functions do not actually mention the space $X = G/K$ nor the algebra $\mathbf{D}(X)$.

This paper deals with some simple results about the functions satisfying (1.1) and (1.2), namely new integral formulas, new results about behavior at infinity and some facts about the related $C_\sigma$ functions.

Appreciation to G. Olafsson, A. Pasquale, D. Vogan, and J. A. Wolf for helpful comments. I am also grateful to the referee for several corrective suggestions.

## 2. Notation and background

As usual $\mathbf{R}, \mathbf{C}$, and $\mathbf{Z}$ denote the sets of real numbers, the complex numbers and integers, respectively. If $c = a + ib,\ a, b \in \mathbf{R}$ we write $a = \operatorname{Re} c, b = \operatorname{Im} c$ and $\bar{c} = a - ib$. If $L$ is a Lie group with Lie algebra $\mathfrak{l}$, $\exp : \mathfrak{l} \to L$ denotes the exponential mapping and ad (resp. Ad) the adjoint representation of $\mathfrak{l}$ (resp. $L$).

Going back to §1, our group $G$ has a Lie algebra $\mathfrak{g}$ with Cartan decomposition $\mathfrak{g} = \mathfrak{k} + \mathfrak{p}$ where $\mathfrak{k}$ is the Lie algebra of $K$ and $\mathfrak{p}$ is the orthocomplement of $\mathfrak{k}$ relative to the Killing form $B$ of $\mathfrak{g}$. We fix a maximal abelian subspace $\mathfrak{a}$ of $\mathfrak{p}$ and fix a Weyl chamber $\mathfrak{a}^+ \subset \mathfrak{a}$. All such choices are conjugate under $\operatorname{Ad}(K)$. The choice of $\mathfrak{a}^+$ induces Iwasawa decompositions $G = NAK$ and $G = KAN$ where $A = \exp \mathfrak{a}$ and $N$ is nilpotent. In these decompositions we write $g = n \exp A(g) k, g = k_1 \exp H(g) n_1$ where $A(g)$ and $H(g)$ are uniquely determined in $\mathfrak{a}$ and $A(g) = -H(g^{-1})$.

If $M$ is the centralizer of $A$ in $K$ the "vector valued" inner product $A(gK, kM) = A(k^{-1}g)$ is well defined and considered analog to the Euclidean $(x, w)$ the distance from 0 to the hyperplane through $x$ with unit normal $w$. We put $B = K/M$. If $\widehat{K}$ denotes the set of irreducible representations of $K$, condition (1.2) implies that $\delta(M)$ has a common fixed point; we denote by $\widehat{K}_M$ the set of these $\delta$.

Let $V_\delta$ denote the space on which $\delta$ operates and $V_\delta^M$ the subspace of fixed points under $\delta(M)$. We also use the notation $\mathcal{E}(X)$ (resp. $\mathcal{D}(X)$) for the space of $C^\infty$ functions on $X$ (resp. those of compact support).

We denote by $\pi$ the natural map of $G$ on $G/K$ and put $\tilde{f} = f \circ \pi$ for a function $f$ on $G/K$. We also denote by $o$ the coset $eK$.

Let $\mathfrak{a}^*$ (resp. $\mathfrak{a}^*_{\mathbf{C}}$) be the space of $\mathbf{R}$-linear maps of $\mathfrak{a}$ into $\mathbf{R}$ (resp. $\mathbf{C}$). In the bijection of $\mathfrak{a}$ with $\mathfrak{a}^*$ via the Killing form of $\mathfrak{g}$ let $\mathfrak{a}^*_+$ correspond to $\mathfrak{a}^+$. Let $S(\mathfrak{a})$ denote the symmetric algebra over $\mathfrak{a}$ and $I(\mathfrak{a})$ the subspace of $p \in S(\mathfrak{a})$ invariant under the Weyl group $W$. We put $\bar{N} = \theta N$ if $\theta$ is the Cartan involution. Let $M'$ be the normalizer of $A$ in $K$. If $\sigma \in W$ and $m_\sigma$ representing $\sigma$ in $M'$ we put

$$\bar{N}_\sigma = \bar{N} \cap m_\sigma^{-1} N m_\sigma.$$

This group appears later.

## 3. Zonal spherical functions

A **zonal spherical function** $\phi$ on $G$ is a $C^\infty$ function on $G$ satisfying

$$\text{(3.1)} \qquad \phi \text{ is an eigenfunction of each } D \in \mathbf{D}_K(G).$$

$$\text{(3.2)} \qquad \phi \text{ is bi-invariant under } K,\ \phi(e) = 1.$$

Here $\mathbf{D}_K(G)$ is the algebra of differential operators on $G$ which are left invariant under $G$ and right invariant under $K$.

Properties (3.1) and (3.2) are well known to be equivalent to

$$\int_K \phi(xky)dk = \phi(x)\phi(y). \tag{3.3}$$

The zonal spherical functions are all given by Harish-Chandra's formula ([**HC54**])

$$\phi(g) = \int_K e^{(i\lambda-\rho)(H(gk))}dk \tag{3.4}$$

for some $\lambda \in \mathfrak{a}_C^*$. Here $\rho$ is half the sum of the positive restricted roots with multiplicity. Writing $\phi = \phi_\lambda$ this function has from [**HC58**] an expansion

$$\phi_\lambda(a) = \sum_{s\in W} \mathbf{c}(s\lambda)e^{(is\lambda-\rho)(\log a)} \sum_{\mu\in\Lambda} \Gamma(s\lambda)e^{-\mu(\log a)} \tag{3.5}$$

for $a$ in Weyl chamber $\exp(\mathfrak{a}^+)$, $W$ the Weyl group, $\Lambda$ the lattice

$$\Lambda = \{m_1\alpha_1 + \cdots + m_\ell\alpha_\ell \mid m_i \in \mathbf{Z}^+\},$$

the $\alpha_1, \ldots, \alpha_\ell$ being the simple restricted roots. This is where the remarkable $\mathbf{c}$ function first appears.

The $\Gamma$ are rational functions on $\mathfrak{a}_\mathbf{c}^*$ and $\mathbf{c}$ is a meromorphic function on $\mathfrak{a}_\mathbf{c}^*$ given by Harish-Chandra [**HC58**] as the integral

$$\mathbf{c}(\lambda) = \int_{\bar{N}} e^{-(i\lambda+\rho)(H(\bar{n}))}d\bar{n}. \tag{3.6}$$

Through the work of Harish-Chandra [**HC58**], Bhanu-Murthy [**BM60**] and Gindikin-Karpelevic [**GK62**] the $\mathbf{c}$ function is given by

$$\mathbf{c}(\lambda) = \prod_{\alpha\in\sum_0^+} \frac{2^{-\langle i\lambda-\rho,\alpha_0\rangle}\Gamma\left(\frac{1}{2}(m_\alpha + m_{2\alpha} + 1)\right)\Gamma(\langle i\lambda, \alpha_0\rangle)}{\Gamma\left(\frac{1}{2}(\frac{1}{2}m_\alpha + 1 + \langle i\lambda, \alpha_0\rangle)\right)\Gamma\left(\frac{1}{2}(\frac{1}{2}m_\alpha + m_{2\alpha} + \langle i\lambda, \alpha_0\rangle)\right)}. \tag{3.7}$$

Here $\sum_0^+$ denotes the set of positive, indivisible roots, $m_\alpha$ the multiplicity of $\alpha$ and $\alpha_0 = \alpha/\langle\alpha, \alpha\rangle$.

In Harish-Chandra's work, $|\mathbf{c}(\lambda)|^{-2}$ served as the dual measure for the **spherical transform** on $G$. However, formula (3.7) has many other interesting features. See e.g. [**H00**].

## 4. The spaces $X = G/K$ and its Dual $\Xi = G/MN$

As proved in [**H62**], p. 439 and [**H70**], p. 94 the modified integrand in (3.4), that is the function

$$gK \to e^{(i\lambda+\rho)(A(k^{-1}g))}, \tag{4.1}$$

is for each $k$ an eigenfunction of $\mathbf{D}(X)$ and the eigenvalue is $\Gamma(D)(i\lambda)$ where $D \in \mathbf{D}(X)$ and $\Gamma(D) \in I(\mathfrak{a})$. (This is related to Lemma 3 in [**HC58**], I) but not contained in it). The map $\Gamma$ is spelled out in [**H84**], II, Theorem 5.18.

This led in [**H65**] to the definition of a **Fourier transform** $f \to \tilde{f}$ for a function $f$ on $X$,

$$\tilde{f}(\lambda, b) = \int_X f(x)e^{(-i\lambda+\rho)(A(x,b))}\,dx, \;\; b \in B, \lambda \in \mathfrak{a}_\mathbf{c}^* \tag{4.2}$$

in analogy with the polar coordinate expression

$$\tilde{F}(\lambda w) = \int_X F(x) e^{-i\lambda(x,w)}\, dx, \ |w| = 1$$

for the Fourier transform on $\mathbb{R}^n$. Here $dx$ denotes the volume element in both cases.

In addition we consider the **Poisson transform**

$$(4.3) \qquad (\mathcal{P}_\lambda F)(x) = \int_B e^{(i\lambda+\rho)(A(x,b))} F(b)\, db, \ F \text{ a function on } B.$$

These transforms are intimately related to the **c**-function. By [**H70**], p. 120, the map $f \to \tilde{f}$ is an isometry of $L^2(X)$ onto $L^2(\mathfrak{a}^*_+ \times B; |\mathbf{c}(\lambda)|^{-2}\, d\lambda\, db)$. Secondly, $\mathcal{P}_\lambda$ is related to the **denominator** $\Gamma^+_X(\lambda)$ in (3.7), called the Gamma function of $X$.

This $\mathcal{P}_\lambda$ is closely related to the dual Radon transform $\phi \to \overset{\vee}{\phi}$ from $\Xi$, the space of horocycles in $X$, to $X$ which to a function $\phi$ on $\Xi$ associates $\overset{\vee}{\phi}(x)$, the average of $\phi$ over horicycles $\xi \in \Xi$ passing through $x \in X$ ([**H08**], p. 103).

The element $\lambda \in \mathfrak{a}^*_c$ is said to be **simple** if $\mathcal{P}_\lambda$ is injective. The connection with the denominator in (3.7) is ([**H76**]):

THEOREM 4.1. *$\lambda$ is non-simple if and only if* $\Gamma^+_X(\lambda)^{-1} = 0$.

On the other hand, the **numerator** in (3.7) is connected with analysis on the dual of $X$, that is, the space $\Xi = G/MN$ of horocycles in $X$. The counterpart to the zonal spherical functions on $X$ would be the $MN$-invariant eigendistributions of $\mathbf{D}(G/MN)$, the algebra of $G$-invariant differential operators on $\Xi$. These "**conical distributions**" have a construction and theory in [**H70**], using the numerator of (3.7).

While the set of zonal spherical functions is parametrized by $\mathfrak{a}^*_c/W$ via (3.4), the set of conical distributions turned out to be essentially parametrized by $\mathfrak{a}^*_c \times W$.

In the proof of (3.7) the following partial **c**-function $\mathbf{c}_\sigma$ enters for each $\sigma \in W$. In analogy with (3.6), it is defined by

$$\mathbf{c}_\sigma(\lambda) = \int_{\bar{N}_\sigma} e^{-(i\lambda+\rho)(H(\bar{n}))} d\bar{n}, \quad \bar{N}_\sigma = \bar{N} \cap N^{\sigma^{-1}},$$

for a suitable normalization of $d\bar{n}$ on $\bar{N}_\sigma$. It has a formula generalizing (3.7),

$$\mathbf{c}_\sigma(\lambda) = \prod_{\alpha \in \sum_0^+ \cap \sigma^{-1} \sum_0^-} \mathbf{c}_\alpha(\lambda_\alpha), \quad \lambda_\alpha = \lambda|\mathfrak{a}_\alpha,$$

where $\mathbf{c}_\alpha$ is the $c$-function for the group $G_\alpha \subset G$ whose Lie algebra is the subalgebra of $\mathfrak{g}$ generated by $\mathfrak{g}_\alpha$ and $\mathfrak{g}_{-\alpha}$.

The numerous beautiful features of (3.7) are the cause of the title in [**H00**].

## 5. Global descriptions of eigenspaces

Let $V$ be a finite dimensional vector space. A function $\Phi : X \to V$ satisfying (1.1) is called a *joint eigenfuction* of $\mathbf{D}(X)$. The eigenvalue $\chi(D)$ in $D\Phi = \chi(D)\Phi$ is a homomorphism of $I(\mathfrak{a})$ into $\mathbf{C}$ and thus has the form $\chi(D) = \Gamma(D)(i\lambda)$ for some $\lambda \in \mathfrak{a}^*$. Corresponding eigenspaces are *joint eigenspaces*. The scalar version of Proposition 5.1 is from [**H62**], X, §7. We shall use [**H84**], mainly Ch. II, §4 and Ch.IV, §2.

PROPOSITION 5.1. *The joint eigenfunctions* $\Phi : X \to V$ *of* $\mathbf{D}(X)$ *are the continuous functions* $\Phi$ *satisfying.*

$$\int_K \Phi(xky \cdot o)dk = \Phi(x \cdot o)\phi_\lambda(y \cdot o), \quad x, y \in G \tag{5.1}$$

*for some* $\lambda \in \mathfrak{a}_c^*$.

PROOF. First assume (5.1). Integrating (5.1) against a test function $f$ in $y$, $\Phi(x)$ can be written in terms of derivative of $f$ in $x$ so $\Phi$ is smooth. Then applying $D_y$ to (5.1) and putting $y = e$ we find

$$D\Phi = \Gamma(D)(i\lambda)\Phi \tag{5.2}$$

so $\Phi$ is an eigenfunction. For the converse assume (5.2) and put

$$\Psi_x(y) = \int_K \Phi(xky)\, dk$$

and note that

$$D\Psi_x = \Gamma(D)(i\lambda)\Psi_x. \tag{5.3}$$

Let $\mathbf{D}(G)$ denote the algebra of left invariant differential operators on $G$ and as before $\mathbf{D}_K(G)$ the subalgebra of those which are also right $K$-invariant. For $D \in \mathbf{D}(G)$ let

$$D_0 = \int_K \mathrm{Ad}\,(k)D\, dk \tag{5.4}$$

and recall that $D \to D_0$ maps $\mathbf{D}(G)$ onto $\mathbf{D}_K(G)$. Also consider the map $\mu$ given by

$$(\mu(u)f)^\sim = u\tilde{f} \qquad u \in \mathbf{D}_K(G),\ \tilde{f} = f \circ \pi \tag{5.5}$$

which maps $\mathbf{D}_K(G)$ onto $\mathbf{D}(G/K)$ ([**H84**], Ch. II, §4).

Among $D$ in (5.2) is the Laplace-Beltrami operator so $\Phi$ is an analytic function. If $F \in C^\infty(G)$ is bi-invariant under $K$ we have by [**H84**] (3), p. 400,

$$(D_0F)(e) = (DF)(e). \tag{5.6}$$

This applies both to $F(y) = \Psi_x(y \cdot o)$ and $F(y) = \phi_\lambda(y \cdot o)$ and to

$$f(y \cdot o) = \phi_\lambda(e)\Psi_x(y \cdot o) - \Psi_x(e)\phi_\lambda(y \cdot o).$$

For $D \in \mathbf{D}(G)$ arbitrary we take $u = D_0$ in (5.5). Then using (5.6), $(D\tilde{f})(e) = 0$. Since $f$ is analytic and $f(o) = 0$ we have $f \equiv 0$ which is formula (5.1). □

With $V$ as before we consider joint eigenfunctions for the algebra $\mathbf{D}(\Xi) = \mathbf{D}(G/MN)$. If $\mathbf{D}(A)$ denotes the left invariant differential operators on $A$ each $U \in \mathbf{D}(A)$ induces an $D_U \in \mathbf{D}(\Xi)$ by

$$(D_U\phi)(kaMN) = U_a(\phi(kaMN))$$

and $U \to D_U$ is an isomorphism of $\mathbf{D}(A)$ onto $D(\Xi)$ ([**H70**], I, §2). We denote by $\widehat{\Gamma}$ its inverse. Let $\mathcal{E}_\lambda(\Xi)$ denote the joint eigenspace

$$\mathcal{E}_\lambda(\Xi) = \left\{\Psi \in \mathcal{E}(\Xi) : D\Psi = \widehat{\Gamma}(D)(i\lambda - \rho)\Psi\right\}.$$

PROPOSITION 5.2. *The joint eigenfunctions $\Psi$ of $\Xi$ into $V$ are the smooth functions satisfying*

$$\Psi(gaMN) = \Psi(gMN)e^{(i\lambda-\rho)(\log a)}, \quad a \in A, \tag{5.7}$$

*for some $\lambda \in \mathfrak{a}_c^*$.*

For proof see [**H08**], II, §2, also for distributions on $\Xi$.

## 6. Spherical functions of a given $K$-type

Let $\Phi$ be a spherical function of type $\delta$ as in §1. Thus we take $V = \operatorname{Hom}(V_\delta, V_\delta)$ in §5 and assume (1.2). As mentioned in §5 there exists a $\lambda \in \mathfrak{a}_c^*$ such that

$$D\Phi = \Gamma(D)(i\lambda)\Phi, \quad D \in \mathbf{D}(X). \tag{6.1}$$

This $\lambda$ is unique up to conjugacy by $W$ and since each $\lambda \in \mathfrak{a}_{\mathbf{c}}^*$ is $W$ conjugate to one which is simple we can take $\lambda \in \mathfrak{a}_{\mathbf{c}}^*$ to be simple. By definition, $\mathcal{P}_\lambda$ is injective.

Let $v_1, \ldots, v_{d(\delta)}$ be an orthonormal basis of $V_\delta$ such that $v_1, \ldots, v_{\ell(\delta)}$ span $V_\delta^M$. Then

$$\Phi(x)v_j = \sum_{i=1}^{d(\delta)} \phi_{ij}(x)v_i \tag{6.2}$$

Then condition (1.2) implies

$$\phi_{rj}(k \cdot x) = \sum_i \delta_{ri}(k)\phi_{ij}(x),$$

where $\delta_{ri}(k)$ is the expression of $\delta(k)$ in the basis $(v_i)$. Let $\mathcal{E}_\lambda(X)$ denote the space of joint eigenfunctions of $\mathbf{D}(X)$ with eigenvalues $\Gamma(D)(i\lambda)$. Let $\check{\delta}$ denote the contragredient to $\delta$, $d(\check{\delta}) = d(\delta)$ its dimension and $\chi_{\check{\delta}} = \bar{\chi}_\delta$ its character. Let $\pi$ denote the representation of $K$ on $\mathcal{E}_\lambda(X)$ given by $\pi(k) : f(x) \to f(k^{-1} \cdot x)$. By [**H84**], IV, §1, the map

$$d(\check{\delta})\pi(\bar{\chi}_{\check{\delta}})$$

is the projection of $\mathcal{E}_\lambda(X)$ onto the space $\mathcal{E}_{\lambda,\check{\delta}}(X)$ of $K$-finite elements in $\mathcal{E}_\lambda(X)$ of type $\check{\delta}$.

LEMMA 6.1. *Each function $\phi_{ij}$ in (7.2) belongs to $\mathcal{E}_{\lambda,\check{\delta}}(X)$.*

PROOF. We have

$$\begin{aligned} d(\check{\delta})\pi(\bar{\chi}_{\check{\delta}})(\phi_{rj})(x) &= d(\check{\delta}) \int_K \overline{\chi_{\check{\delta}}(k)}\phi_{rj}(k^{-1} \cdot x)dk \\ &= d(\delta) \int_K \overline{\chi_\delta(k)} \sum_i \delta_{ri}(k)\phi_{ij}(x)dk \end{aligned}$$

which by Schur's orthogonality relations reduces to $\phi_{rj}(x)$. Thus $\phi_{ij} \in \mathcal{E}_{\lambda,\check{\delta}}(X)$ for all $i, j$. $\square$

By [**H70**], IV, §1 and [**H76**] §7, invoking the Paley-Wiener theorem for (4.2), each $K$-finite joint eigenfunction of $\mathbf{D}(X)$ is the Poisson transform of a $K$-finite member of $\mathcal{E}(B)$. We apply this to $\phi_{ij}(x)$. As is well known the functions

$$\langle \delta(k)v_j, v_i\rangle \quad 1 \leq i \leq d(\delta),\ 1 \leq j \leq \ell(\delta)$$

form a basis of $\mathcal{E}_{\check{\delta}}(B)$, the space of $K$-finite functions in $\mathcal{E}(B)$ of type $\check{\delta}$. The corresponding images under $\mathcal{P}_\lambda$

$$\int_{K/M} e^{(i\lambda+\rho)(A(x,kM))}\langle \delta(k)v_j, v_i\rangle\, dk_M \tag{6.3}$$

will by [**H73, H76**] span the space $\mathcal{E}_{\lambda,\check{\delta}}(X)$. Consider the Eisenstein integral

$$\Phi_{\lambda,\delta}(x) = \int_K e^{(i\lambda+\rho)(A(x,kM))}\delta(k)\, dk, \tag{6.4}$$

whose matrix entries are given by (6.3).

Changing from $\lambda$ to $s\lambda (s \in W)$ only changes $\Phi_{\lambda,\delta}$ by a factor independent of $x$ [**H73**]. This proves following result.

THEOREM 6.2. *Each spherical function of type $\delta$ has each of its matrix entries linear combinations of the functions*

$$\langle \Phi_{\lambda,\delta}(x)v_j, v_i\rangle \quad 1 \leq j \leq \ell(\delta), 1 \leq i \leq d(\delta).$$

These functions all turn out to be suitable derivative of the zonal spherical functions [**H76**].

THEOREM 6.3. *Fix $v, w \in V_\delta$ and assume $-\lambda$ simple. Then there exists a right invariant differential operator $D$ on $G$ such that*

$$\langle v, \Phi_{\lambda,\delta}(gK), w\rangle = (D\phi_\lambda)(g).$$

This in return implies a series expansion of $\Phi_{\lambda,\delta}$ on $\exp \mathfrak{a}^+$, generalizing Harish-Chandra's expansion (3.5) which introduced the $\mathbf{c}$ function.

THEOREM 6.4. *There exist meromorphic functions $\mathbf{C}_\sigma(\sigma \in W)$ and rational functions $\mathbf{\Gamma}_\mu(\mu \in \Lambda)$ all with values in* Hom $(V_\delta^M, V_\delta^M)$ *such that for $H \in \mathfrak{a}^+, v \in V_\delta^M$,*

$$\Phi_{\lambda,\delta}(\exp H \cdot o)v = \sum_{\sigma\in W,\mu\in\Lambda} e^{(i\sigma\lambda-\rho-\mu)(H)}\mathbf{\Gamma}_\mu(\sigma\lambda)\mathbf{C}_\sigma(\lambda)v$$

The $\mathbf{\Gamma}_\mu$ are given by explicit recursion formulas. Note that in contrast to (3.5) the order of the factors in each term of the series is important. Also by [**H70**], [**H73**]

$$\Phi_{\sigma\lambda,\delta}(x) = \Phi_{\lambda,\delta}(x)\Gamma_{\sigma,\lambda},$$

where $\Gamma_{\sigma,\lambda}$ is meromorphic on $\mathfrak{a}_{\mathbf{c}}^*$ with values in Hom $(V_\delta, V_\delta)$ and

$$\Gamma_{\sigma,\lambda}v = \frac{C_{\sigma^{-1}}(\sigma\lambda)}{\mathbf{c}(\lambda)}v \quad \text{for } v \in V_\delta^M.$$

We could also consider the analog of this for the dual space $\Xi = G/MN$. For this consider a representation $\sigma$ of $MN$ on a finite dimension space $V_\sigma$ and then take $V =$ Hom $(V_\sigma, V_\sigma)$ in §5 and replace (1.2) by

$$\Psi(h \cdot \xi) = \sigma(h)\Psi(\xi) \quad h \in MN,\ \xi \in \Xi,$$

extending $\Psi$ to distributions.

This leads to *Whittaker functions* and *Whittaker distributions*, studied for example in Goodman-Wallach [80] where relations with conical distributions is also established.

## 7. The asymptotics of $\Phi_{\lambda,\delta}$

In this section we limit ourselves to the case rank $X = 1$. Then by Kostant [**K69**] $\ell(\delta) = \dim V_\delta^M = 1$. We fix $v \in V_\delta^M$ of norm 1 and take $H \in \mathfrak{a}^+$ such that $\alpha(H) = 1$. We put $a_t = \exp tH$ and

$$\phi_{\lambda,\delta}(x) = \langle \Phi_{\lambda,\delta}(x)v, v\rangle. \tag{7.1}$$

Then by Theorem 6.4

$$\begin{aligned}\phi_{\lambda,\delta}(a_t \cdot o) &= e^{(i\lambda-\rho)(tH)} \sum_{n=0}^{\infty} e^{-nt}\Gamma_n(\lambda)\langle \mathbf{C}_e(\lambda)v, v\rangle \\ &\quad + e^{-(i\lambda+\rho)(tH)} \sum_{n=0}^{\infty} e^{-nt}\Gamma_n(-\lambda)\langle \mathbf{C}_\sigma(\lambda)v, v\rangle\end{aligned}$$

Multiply by $e^{-(i\lambda-\rho)(tH)}$, $\lambda = \xi + i\eta$.

Then if $\eta < 0$ we have by [**H08**], II, Theorem 3.16,

$$\lim_{t\to\infty} e^{-(i\lambda-\rho)(tH)}\phi_{\lambda,\delta}(a_t \cdot o) = \mathbf{c}(\lambda)\langle v, v\rangle.$$

The left hand side is

$$\sum_0^\infty e^{-nt}\Gamma_n(\lambda)\langle \mathbf{C}_e(\lambda)v, v\rangle + e^{-2i\lambda(tH)} \sum_0^\infty e^{-nt}\Gamma_n(-\lambda)\langle \mathbf{C}_\sigma(\lambda)v, v\rangle$$

Since $\Gamma_n$ grows at most exponentially $< \frac{1}{2}n$ (see [**H08**], III, §5) and $-2i\lambda = -2i\xi + 2\eta$ the limit for $t \to \infty$ equals $\mathbf{C}_e(\lambda)$. Since both sides are meromorphic in $\lambda$ we conclude

$$\mathbf{C}_e(\lambda) = \mathbf{c}(\lambda). \tag{7.2}$$

We can also multiply the expansion by $e^{(i\lambda+\rho)(tH)}$ and for suitable $\eta$ deduce

$$\lim_{t\to\infty} e^{(i\lambda+\rho)(tH)}\phi_{\lambda,\delta}(a_t \cdot o) = \mathbf{C}_\sigma(\lambda). \tag{7.3}$$

This in itself does not give much information.

On the other hand, $\phi_{\lambda,\delta}$ is an eigenfunction of the Laplacian and by [**H76**] or [**H08**], p. 328, given by

$$\begin{aligned}\phi_{\lambda,\delta}(a_t \cdot o) &= c_{\lambda,\delta} \tanh^s t \cosh^l t \\ &\times F\left(\tfrac{1}{2}(s+r-l), \tfrac{1}{2}(s-r-l+1-m_{2\alpha}), s+\tfrac{1}{2}(m_\alpha+m_{2\alpha}+1), \tanh^2 t\right)\end{aligned} \tag{7.4}$$

where $l = (i\lambda - \rho)(H)$, $m_\alpha$ and $m_{2\alpha}$ are the multiplicities of $\alpha$ and $2\alpha$, $r$ and $s$ are integers, $r \leq s$ given by

$$r(r + m_{2\alpha} - 1) = -\frac{1}{4}d_{2\alpha}$$

$$s(s + m_\alpha + m_{2\alpha} - 1) = -d_\alpha - \frac{1}{4}d_{2\alpha}$$

and $d_\alpha$ and $d_{2\alpha}$ determined by

$$\delta(\omega_\alpha)|V_\delta^M = d_\alpha(2(m_\alpha+4m_{2\alpha}))^{-1}$$
$$\delta(\omega_{2\alpha})|V_\delta^M = d_{2\alpha}(2(m_\alpha+4m_{2\alpha}))^{-1}.$$

The operators $w_\alpha$ and $w_{2\alpha}$ are defined in [**H76**], §4.
Also by [**H76**]

(7.5) $$c_{\lambda,\delta} = \frac{\Gamma(\frac{1}{2}(\langle i\lambda+\rho,\alpha_0\rangle+s+r))}{\Gamma(\frac{1}{2}(\langle i\lambda+\rho,\alpha_0\rangle))}\,\frac{\Gamma(\frac{1}{2}(\langle i\lambda+\rho,\alpha_0\rangle+1-m_{2\alpha}+s-r))}{\Gamma(\frac{1}{2}(\langle i\lambda+\rho,\alpha_0\rangle+1-m_{2\alpha}))}$$

Consider the factor $(\cosh t)^l$, $l = i\xi(H)-(\eta+\rho)(H)$ and choose $\eta>0$.
Multiply $\phi_{\lambda,\delta}(a_t\cdot o)$ by $(2\cosh t)^{-\ell}$ and let $t\to+\infty$. Then

(7.6) $$\lim_{t\to\infty}(2\cosh t)^{-\ell}\phi_{\lambda,\delta}(a_t\cdot o) = c_{\lambda,\delta}\;\tfrac{1}{2^l}\,\frac{\Gamma(c-a-b)\Gamma(c)}{\Gamma(c-a)\Gamma(c-b)}.$$

by the limit formula $F(a,b,c,\infty)$, where

$$a=\tfrac{1}{2}(s+r-l),\quad b=\tfrac{1}{2}(s-r-l+1-m_{2\alpha}),\quad c=s+\tfrac{1}{2}(m_\alpha+m_{2\alpha}+1).$$

Then

$$c-a-b = i\lambda(H)$$
$$c-a = \tfrac{1}{2}s-\tfrac{1}{2}r+\tfrac{1}{2}i\lambda(H)+\tfrac{1}{4}m_\alpha+\tfrac{1}{2}$$
$$c-b = \tfrac{1}{2}s+\tfrac{1}{2}r+\tfrac{1}{2}i\lambda(H)+\tfrac{1}{4}m_\alpha+\tfrac{1}{2}m_{2\alpha}$$

In (7.5) recall that $\alpha_0=\alpha/\langle\alpha,\alpha\rangle$

$$\langle\rho,\alpha_0\rangle = \langle\tfrac{1}{2}m_\alpha\alpha+m_{2\alpha}\alpha,\alpha_0\rangle = \tfrac{1}{2}m_\alpha+m_{2\alpha}$$
$$\langle i\lambda+\rho,\alpha_0\rangle = i\lambda(H)+\frac{1}{2}m_\alpha+m_{2\alpha}$$

Thus the right hand side of (7.6) equals

$$\frac{\Gamma\left(\frac{1}{2}\left(i\lambda(H)+\frac{1}{2}m_\alpha+m_{2\alpha}+s+r\right)\right)\Gamma\left(\frac{1}{2}\left(i\lambda(H)+\frac{1}{2}m_\alpha+1+s-r\right)\right)}{\Gamma\left(\frac{1}{2}\left(i\lambda(H)+\frac{1}{2}m_\alpha+m_{2\alpha}\right)\right)\Gamma\left(\frac{1}{2}\left(i\lambda(H)+\frac{1}{2}m_\alpha+1\right)\right)}$$
$$\times\frac{1}{2^l}\,\frac{\Gamma\left(i\lambda(H)\right)\Gamma\left(s+\frac{1}{2}\left(m_\alpha+m_{2\alpha}+1\right)\right)}{\Gamma\left(\frac{1}{2}s-\frac{1}{2}r+\frac{1}{2}i\lambda(H)+\frac{1}{4}m_\alpha+\frac{1}{2}\right)\Gamma\left(\frac{s}{2}+\frac{r}{2}+\frac{i\lambda(H)}{2}+\frac{m_\alpha}{4}+\frac{1}{2}m_{2\alpha}\right)}$$

Here top of the first fraction cancels against the bottom of second fraction. Thus (7.6) becomes (considering (3.7)),

$$\frac{1}{2^l}\,\frac{\Gamma(i\lambda(H))\Gamma\left(s+\frac{1}{2}(m_\alpha+m_{2\alpha}+1)\right)}{\Gamma\left(\frac{1}{2}\left(i\lambda(H)+\frac{1}{2}m_\alpha+m_{2\alpha}\right)\right)\Gamma\left(\frac{1}{2}\left(i\lambda(H)+\frac{1}{2}m_\alpha+1\right)\right)}$$
$$=\frac{\Gamma\left(s+\frac{n}{2}\right)}{\Gamma\left(\frac{n}{2}\right)}\mathbf{c}(\lambda) \qquad \text{if } n=\dim X.$$

THEOREM 7.1. *For* $\lambda=\xi+i\eta$, $\eta>0$,

(7.7) $$\lim_{t\to\infty}(2\cosh t)^{-\ell}\phi_{\lambda,\delta}(a_t\cdot o) = \frac{\Gamma\left(s+\frac{n}{2}\right)}{\Gamma\left(\frac{n}{2}\right)}\mathbf{c}(\lambda).$$

So far we have studied the behavior of $\phi_{\lambda,\delta}$ for large $t$. For behavior for small $t$ we just use (7.4) and conclude

$$\lim_{t\to 0}\frac{\phi_{\lambda,\delta}(a_t\cdot o)}{\phi_{-\lambda,\delta}(a_t\cdot o)}=\frac{c_{\lambda,\delta}}{c_{-\lambda,\delta}}.$$

On the other hand as mentioned in §6

$$\Phi_{s\lambda,\delta}(x)=\Phi_{\lambda,\delta}(x)\Gamma_{s,\lambda},$$

where

$$\Gamma_{s,\lambda}v=\frac{\mathbf{C}_{s^{-1}}(s\lambda)}{\mathbf{c}(\lambda)}v,\qquad v\in V_\delta^M.$$

In the rank one case where $\mathbf{C}_\sigma$ is a scalar this implies

$$\mathbf{C}_\sigma(-\lambda)=\frac{c_{-\lambda,\delta}}{c_{\lambda,\delta}}\,\mathbf{c}(\lambda). \tag{7.8}$$

THEOREM 7.2. *Formulas* (7.2) *and* (7.8) *determine the* $\mathbf{C}$ *functions in the rank one case.*

REMARK 7.3. By [**H73**], Lemma 6.1 and Lemma 6.5 we have

$$\mathbf{C}_\sigma(-\lambda)=\int_N e^{-(i\lambda+\rho)(H(\bar n))}\delta(k(\bar n)^{-1}m^*)\,d\bar n \tag{7.9}$$

where $m^*\in W$ is $-I$ on $\mathfrak{a}$. Thus (7.8) gives an evaluation of the unwieldy integral (7.9). In the papers [**J76**] and [**JW72**], [**JW77**] Johnson and Wallach determined (7.9) by using a classification of $\widehat{K}_M$ related to a parametrization from Kostant [**K69**]. In each case they have a formula in the spirit of (7.8). Their models for the $\delta\in\widehat{K}_M$ are spaces of homogeneous harmonic polynomials on $\mathfrak{p}$, restricted by different conditions, according to the multiplicities $m_\alpha$ and $m_{2\alpha}$. This leads to a determinations of the integers s, r for each $\delta$ in (7.4) ([**H74**], p. 336–337). See also [**C74**] for the simplest cases.

## 8. The case of higher rank

Using a method of Schiffman [**S71**] we shall now investigate the $\mathbf{C}$-functions for $X$ of higher rank.

As in [**H73**] we consider the endomorphism $A(\lambda,\sigma)$ of $V_\delta^M$ given by

$$A(\lambda,\sigma)v=\int_{\bar N_\sigma} e^{-(i\lambda+\rho)(H(\bar n))}\delta(m_\sigma k(\bar n))v\,d\bar n, \tag{8.1}$$

where $m_\sigma$ is a representative of $\sigma\in W$ in the normalizer $M'$ of $A$ in $K$. Under conjugation $g\to mgm^{-1}$ by an $m\in M$ we have $(k(\bar n))^m=k(\bar n^m)$ so $A(\lambda,\sigma)$ in (8.1) is independent of the choice of $m_\sigma\in M'$ representing $\sigma\in W$. Let $\sigma=\sigma_1\ldots\sigma_p$ be a reduced expression of $\sigma$, that is each $\sigma_i$ is a reflection in the plane $\alpha_i=0$ where $\alpha_i$ is a simple root and $p$ is as small as possible.

Then by [**H73**], following a method by [**S71**],

$$A(\lambda,\sigma)=A(\sigma^{(1)}\lambda,m_{\sigma_1})\cdots A(\sigma^{(p)}\lambda,m_{\sigma_p}), \tag{8.2}$$

where $\sigma^{(q)}=\sigma_{q+1}\cdots\sigma_p$.

For each simple root $\alpha$ consider the rank-one symmetric space $G_\alpha/K_\alpha$ where $G_\alpha$ is the analytic subgroup of $G$ whose Lie algebra is generated by the root spaces $\mathfrak{g}_\alpha$ and $\mathfrak{g}_{-\alpha}$ and $K_\alpha=G_\alpha\cap K$. If $\mathfrak{a}_\alpha=\mathbf{R}H_\alpha$ and $A_\alpha=\exp\mathfrak{a}_\alpha$, $\bar N_\alpha=\bar N_{s_\alpha}$ then

$G_\alpha = K_\alpha A_\alpha \bar{N}_\alpha$ is an Iwasawa decomposition of $G_\alpha$. If $\mathbf{c}_\alpha$ is the $\mathbf{c}$ function for $G_\alpha/K_\alpha$ we have by (8.2) (for $\delta$ trivial)

$$\mathbf{c}_\sigma(\lambda) = \prod_1^p \mathbf{c}_{\sigma_j}(\lambda_j) \tag{8.3}$$

where $\lambda_j = (\sigma^{(j)}\lambda)|\mathfrak{a}_{a_j}$.

LEMMA 8.1. *With $\delta \in \widehat{K}_M$ arbitrary let $V$ denote the $K_\alpha M$ invariant subspace of $V_\delta$ generated by $V_\delta^M$ and $V = \bigoplus_{i=1}^{l} V_i$ a decomposition into $K_\alpha M$-irreducible subspaces. Then $l = l(\delta)$, the dimension of $V_\delta^M$ and $\dim(V_\delta^M \cap V_i) = 1$ for each $i$.*

PROOF. See [**H73**], p. 469 and [**H08**], Ch.III Lemma 3.10. □

In this lemma take $\alpha = \sigma_j$ and let $\delta_i$ denote the representation of $K_{\sigma_j}M$ on $V_i$ given by $\delta$. We choose a unit vector $v_i$ in $V_\delta^M \cap V_i$. Since $K_{\sigma_j}$ maps $V_\delta^M \cap V_i$ into itself the operator $A(\sigma^{(j)}\lambda, m_{\sigma_j})$ does too and operates by multiplication with the scalar

$$\int_{\bar{N}_{\sigma_j}} \langle \delta_i(m_{\sigma_j} k(\bar{n})v_i, v_i)\rangle e^{-(i\lambda_j+\rho_{\sigma_j})(H(\bar{n}))}\, d\bar{n}, \tag{8.4}$$

where $\lambda_j = (\sigma^{(j)}\lambda)|\mathfrak{a}_{\sigma_j}$ Here we have used the fact that the restriction $\rho|\mathfrak{a}_{\sigma_j}$ equals the $\rho$-function for $G_{\sigma_j}/K_{\sigma_j}$ [**H84**] (34) page 446. By [**H73**], Lemma 6.5 this number is equal to

$$\int_{\bar{N}_{\sigma_j}} e^{-(i\lambda_j+\rho_{\sigma_j})(H(\bar{n}))} \langle \delta_i(k(\bar{n})^{-1} m_{\sigma_j})v_i, v_i\rangle\, d\bar{n}, \tag{8.5}$$

which we have calculated in (7.8), (7.9). Thus the value of (8.5) equals

$$\frac{c_{-\lambda_j,\delta_i}}{c_{\lambda_j,\delta_i}} \mathbf{c}_{\sigma_j}(\lambda_j). \tag{8.6}$$

The root multiplicities are now the ones in $G_{\sigma_j}$ and the integers $r$ and $s$ are the ones which belong to $\delta_i$. Also

$$V_\delta^M = \bigoplus_i \mathbf{C} v_i. \tag{8.7}$$

For a fixed $j$ this represents diagonalization of $A(\sigma^{(j)}\lambda, m_{\sigma_j})$, $1 \le i \le l(\delta)$. The determinant of this endomorphism of $V_\delta^M$ is then

$$\prod_{i=1}^{l(\delta)} \left( \frac{c_{-\lambda_j,\delta_i}}{c_{\lambda_j,\delta_i}} \mathbf{c}_{\sigma_j}(\lambda_j)\right). \tag{8.8}$$

Changing to another $j$ will change the basis $(v_i)$ giving new representations $\delta(i,j)$. Consequently, by (8.2) and (8.3)

$$\det(A(\lambda,\sigma))|_{V_\delta^M} = \mathbf{c}_\sigma(\lambda)^{l(\delta)} \prod_{j=1}^{p} \left( \prod_{i=1}^{l(\delta)} \frac{c_{-\lambda_j,\delta(i,j)}}{c_{\lambda_j,\delta(i,j)}} \right). \tag{8.9}$$

On the other hand, the adjoint of $A(\bar{\lambda},\sigma)$ is given by

$$(A(\bar{\lambda},\sigma))^* v = \int_{\bar{N}_\sigma} e^{-(-i\lambda+\rho)(H(\bar{n}))} \delta(k(\bar{n})^{-1} m_\sigma^{-1})\, d\bar{n}\, v,$$

which by [**H73**] equals

$$\frac{1}{\mathbf{c}(-\lambda)}\mathbf{c}_\sigma(-\lambda)\mathbf{C}_{\sigma^{-1}}(-\sigma\lambda)v.$$

This by (8.9) determines the determinant of $\mathbf{C}_\sigma(\lambda)$.

Formula (8.9) has some resemblance to Theorem 5 in [**C74**] but it is not clear whether there is a connection.

We do not deal with the problem of determining $C_\sigma$ itself but observe that the Hilbert-Schmidt norm is given by [**H73**]:

$$||C_\sigma(\lambda)||^2 = |\mathbf{c}(\lambda)|^2\, l(\delta).$$

## References


[BM60] T. S. Bhanu Murti, *The asymptotic behavior of zonal spherical functions on the Siegel upper half-plane*, Soviet Math. Dokl. **1** (1960), 1325–1329. MR0123642

[C74] Leslie Cohn, *Analytic theory of the Harish-Chandra C-function*, Lecture Notes in Mathematics, Vol. 429, Springer-Verlag, Berlin-New York, 1974. MR0422509

[G52] Roger Godement, *A theory of spherical functions. I*, Trans. Amer. Math. Soc. **73** (1952), 496–556. MR0052444

[GK62] S. G. Gindikin and F. I. Karpelevič, *Plancherel measure for symmetric Riemannian spaces of non-positive curvature* (Russian), Dokl. Akad. Nauk SSSR **145** (1962), 252–255. MR0150239

[GW80] Roe Goodman and Nolan R. Wallach, *Whittaker vectors and conical vectors*, J. Funct. Anal. **39** (1980), no. 2, 199–279. MR597811

[HC54] Harish-Chandra, *Representations of semisimple Lie groups. II*, Trans. Amer. Math. Soc. **76** (1954), 26–65. MR0058604

[HC57] Harish-Chandra, *Differential operators on a semisimple Lie algebra*, Amer. J. Math. **79** (1957), 87–120. MR0084104

[HC58] Harish-Chandra, *Spherical functions on a semisimple Lie group. I*, Amer. J. Math. **80** (1958), 241–310. MR0094407

[HC72] Harish-Chandra, *On the theory of the Eisenstein integral*, Conference on Harmonic Analysis (Univ. Maryland, College Park, Md., 1971), Springer, Berlin, 1972, pp. 123–149. Lecture Notes in Math., Vol. 266. MR0399355

[H62] Sigurdur Helgason, *Differential geometry and symmetric spaces*, Pure and Applied Mathematics, Vol. XII, Academic Press, New York-London, 1962. MR0145455

[H65] S. Helgason, *Radon-Fourier transforms on symmetric spaces and related group representations*, Bull. Amer. Math. Soc. **71** (1965), 757–763. MR0179295

[H70] Sigurdur Helgason, *A duality for symmetric spaces with applications to group representations*, Advances in Math. **5** (1970), 1–154 (1970). MR0263988

[H73] Sigurdur Helgason, *The surjectivity of invariant differential operators on symmetric spaces. I*, Ann. of Math. (2) **98** (1973), 451–479. MR0367562

[H74] Sigurdur Helgason, *Eigenspaces of the Laplacian; integral representations and irreducibility*, J. Functional Analysis **17** (1974), 328–353. MR0367111

[H76] Sigurdur Helgason, *A duality for symmetric spaces with applications to group representations. II. Differential equations and eigenspace representations*, Advances in Math. **22** (1976), no. 2, 187–219. MR0430162

[H84] Sigurdur Helgason, *Groups and geometric analysis*, Pure and Applied Mathematics, vol. 113, Academic Press, Inc., Orlando, FL, 1984. Integral geometry, invariant differential operators, and spherical functions. MR754767

[H00] Sigurdur Helgason, *Harish-Chandra's c-function. A mathematical jewel.* In E. A. Tanner and R. Wilson (eds.), Noncompact Lie groups and some of their applications, pp. 55-67. Kluwer Academic Publishers, 1994.

[H08] Sigurdur Helgason, *Geometric analysis on symmetric spaces*, 2nd ed., Mathematical Surveys and Monographs, vol. 39, American Mathematical Society, Providence, RI, 2008. MR2463854

[J76] Kenneth D. Johnson, *Composition series and intertwining operators for the spherical principal series. II*, Trans. Amer. Math. Soc. **215** (1976), 269–283. MR0385012

[JW72] Kenneth Johnson and Nolan R. Wallach, *Composition series and intertwining operators for the spherical principal series*, Bull. Amer. Math. Soc. **78** (1972), 1053–1059. MR0310136

[JW77] Kenneth D. Johnson and Nolan R. Wallach, *Composition series and intertwining operators for the spherical principal series. I*, Trans. Amer. Math. Soc. **229** (1977), 137–173. MR0447483

[K69] Bertram Kostant, *On the existence and irreducibility of certain series of representations*, Bull. Amer. Math. Soc. **75** (1969), 627–642. MR0245725

[S71] Gérard Schiffmann, *Intégrales d'entrelacement et fonctions de Whittaker* (French), Bull. Soc. Math. France **99** (1971), 3–72. MR0311838

[T76] Juan A. Tirao, *Spherical functions*, Rev. Un. Mat. Argentina **28** (1976/77), no. 2, 75–98. MR0473723

[W75] Nolan R. Wallach, *On Harish-Chandra's generalized C-functions*, Amer. J. Math. **97** (1975), 386–403. MR0399357

Department of Mathematics, Massachusetts Institute of Technology, Cambridge, Massachusetts 02139

*Email address*: `helgason@mit.edu`

Contemporary Mathematics
Volume **714**, 2018
https://doi.org/10.1090/conm/714/14381

# Schrödinger-type equations and unitary highest weight representations of the metaplectic group

Markus Hunziker, Mark R. Sepanski, and Ronald J. Stanke

*Dedicated to Gestur Olafsson*

ABSTRACT. By the work of Kashiwara–Vergne and Enright–Parthasarathy, every unitary highest weight representation of the metaplectic group $\mathrm{Mp}(n,\mathbb{R})$ can be embedded in $\mathcal{L}^2(\mathrm{M}_{n,k})$ for some $k \geq 1$, where $\mathrm{M}_{n,k}$ denotes the space of real $n \times k$ matrices. Furthermore, every unitary highest weight representation can be embedded in a space of sections of a holomorphic vector bundle on the Siegel upper half-space or, via boundary values, in a degenerate principal series representation. In this paper, we give a new realization of unitary highest weight representations in the kernel of a system of Schrödinger-type equations on the space $\mathrm{M}_{n,k} \times \mathrm{Sym}_n$, where $\mathrm{Sym}_n$ denotes the space of symmetric real $n \times n$ matrices. Our realization has simple intertwining maps to the previously known realizations mentioned above.

## 1. Introduction

**1.1.** The metaplectic group, $G := \mathrm{Mp}(n,\mathbb{R})$, plays an important role in many areas of representation theory, physics, and number theory. In 1978, Kashiwara and Vergne [**8**] constructed a large family of irreducible, unitary hightest weight representations for $G$ inside the $k$-fold tensor product of the oscillator representation. They conjectured that their construction, in fact, gave all irreducible, unitary hightest weight representations for $G$. This conjecture was proved in 1981 by Enright and Parthasarathy [**3**].

Kashiwara and Vergne work with two explicit realizations of their representations and, briefly, mention a third. This paper will construct a fourth realization based on generalized Schrödinger operators that has simple intertwining maps to the aforementioned pictures and that ties everything together in a single commutative diagram. In addition to the importance of the Schrödinger equation, the motivation for these results comes out of invariant theory (see §4.2) and provides a framework for naturally generalizing all of these results to the tube-type case. This paper is a natural extension of the cluster of ideas found in [**5**–**7**, **9**, **10**]. More generally, our line of investigation sits within the extensive body of literature devoted to providing explicit realization for unitary representations. There is also some overlap with the work by R. Berndt and R. Schmidt [**1**, **2**].

---

2010 *Mathematics Subject Classification.* Primary 22E46.
*Key words and phrases.* Metaplectic group, Schrödinger equation, unitary highest weight representations.

**1.2.** In order to summarize our result in more detail, we first outline the three constructions of Kashiwara and Vergne. For their first construction, they show there is a linear action of $G$ given by (see §5.5 for the precise lift of this from $\mathrm{Sp}(n,\mathbb{R})$ to $G$) on $\mathcal{L}^2(\mathrm{M}_{n,k})$

$$(\begin{pmatrix} A & 0 \\ 0 & A^{-1,T} \end{pmatrix} \cdot f)(\xi) = (\det A)^{\frac{k}{2}} f(A^T \xi)$$

$$(\begin{pmatrix} I_n & B \\ 0 & I_n \end{pmatrix} \cdot f)(\xi) = e^{-\frac{i}{2}\operatorname{tr}(\xi^T B \xi)} f(\xi)$$

$$(\begin{pmatrix} 0 & -I_n \\ I_n & 0 \end{pmatrix} \cdot f)(\xi) = \left(\frac{i}{2\pi}\right)^{\frac{nk}{2}} \int_{\mathrm{M}_{n,k}} f(x) e^{i \operatorname{tr}(\xi^T x)}\, dx.$$

There is also a linear $\mathrm{O}(k)$-action on $\mathcal{L}^2(\mathrm{M}_{n,k})$ given by $(c \cdot f)(\xi) = f(\xi c)$ which gives rise to the decomposition

$$\mathcal{L}^2(\mathrm{M}_{n,k}) = \bigoplus_{\lambda \in \Sigma \subseteq \widehat{\mathrm{O}}(k)} \mathcal{L}^2(\mathrm{M}_{n,k}, V_\lambda) \otimes V'_\lambda$$

as a $G \times \mathrm{O}(k)$-module (see §5.4 for the definition of $\Sigma$). Here $V'_\lambda$ is the dual of $V_\lambda$ and $\mathcal{L}^2(\mathrm{M}_{n,k}, V_\lambda)$ is the space of square-integrable $V_\lambda$-valued functions $f : \mathrm{M}_{n,k} \to V_\lambda$ satisfying $f(xc) = \lambda(c)^{-1} f(x)$ for all $c \in \mathrm{O}(k)$. The space $\mathcal{L}^2(\mathrm{M}_{n,k}, V_\lambda)$ is shown to be an irreducible, unitary highest weight $G$-module.

**1.3.** For their second construction, they consider the differential operators

$$\Delta_{ij} := \sum_{\nu=1}^{k} \partial_{x_{i\nu}} \partial_{x_{j\nu}}$$

and write $\Delta := (\Delta_{ij})_{1 \le i \le j \le n}$ for the resulting family of differential operators. For $\lambda \in \Sigma$, let $\tau = \tau(\lambda)$ be the corresponding representation of $\mathrm{GL}(n,\mathbb{C})$ given by Kashiwara–Vergne (see §5.4 and §6.1 for a precise definition) that is realized on

$$W_\tau := \mathfrak{H}(\mathrm{M}_{n,k}, V_\lambda),$$

the space of all $V_\lambda$-valued polynomial functions $f : \mathrm{M}_{n,k} \to V_\lambda$ satisfying

$$\Delta f = 0$$

and $f(xh) = \lambda^{-1}(h) f(x)$ for all $x \in \mathrm{M}_{n,k}$, $h \in \mathrm{O}(k)$, which is shown to be a finite dimensional, irreducible $\mathrm{GL}(n,\mathbb{C})$-module. Write $\Omega_n$ for the Siegel upper half-space (all $Z = X + iY$ with $X, Y \in \mathrm{Sym}_n$ and $Y > 0$).

Kashiwara and Vergne show that there is an action of $G$ on

$$\mathcal{O}(\Omega_n, W_\tau),$$

the space of holomorphic $W_\tau$-valued functions on $\Omega_n$, with $G$-action given by

$$(\begin{pmatrix} A & B \\ C & D \end{pmatrix} \cdot f)(z) = \det(A - ZC)^{-\frac{k}{2}} \tau(A - ZC) \cdot f((A - ZC)^{-1}(ZD - B)).$$

They also construct a $G$-intertwining injection

$$\mathrm{KV}_\lambda : \mathcal{L}^2(\mathrm{M}_{n,k}, V_\lambda) \hookrightarrow \mathcal{O}(\Omega_n, W_\tau)$$

given by

$$(\mathrm{KV}_\lambda f)(Z) := \int_{\mathrm{M}_{n,k}} e^{\frac{i}{2}\operatorname{tr}(x^T Z x)} I^*_\lambda(x) f(x)\, dx$$

for a certain $\mathrm{Hom}(V_\lambda, W_\tau)$-valued polynomial in $x \in \mathrm{M}_{n,k}$ , $I_\lambda^*(x)$.

**1.4.** For their third realization (only mentioned in passing in [**8**] on page 3), let $MA\overline{N}$ be the maximal parabolic subgroup of $G$ covering $\{\left(\begin{smallmatrix} A & 0 \\ C & A^{-1,T}\end{smallmatrix}\right)\} \subseteq \mathrm{Sp}(n,\mathbb{R})$ and let $\det^{-1/2} : MA\overline{N} \to \mathbb{C}$ the character whose square is the lift of the inverse of the determinant function on $A \in \mathrm{GL}(n,\mathbb{R})$. Consider the induced representation $\mathrm{Ind}_{MA\overline{N}}^G(W_\tau \otimes \det^{-k/2})$. Using the (lift of the) embedding $\iota : \mathrm{Sym}_n \to G$ where $\iota(t) := \left(\begin{smallmatrix} I_n & t \\ 0 & I_n\end{smallmatrix}\right)$, the noncompact picture for $\mathrm{Ind}_{MA\overline{N}}^G(W_\tau \otimes \det^{-k/2})$ is given by

$$\begin{aligned}\mathcal{E}(\mathrm{Sym}_n, &W_\tau \otimes \det{}^{-k/2}) \\ &:= \{\psi \in \mathcal{C}^\infty(\mathrm{Sym}_n, W_\tau) \mid \exists \phi \in \mathrm{Ind}_{MA\overline{N}}^G(W_\tau \otimes \det{}^{-k/2}) \text{ with } \psi = \iota^*\phi\}.\end{aligned}$$

The action of $G$ on $\psi \in \mathcal{E}(\mathrm{Sym}_n, W_\tau \otimes \det^{-k/2})$ is given by

$$(g \cdot \psi)(t) = \det{}^{-k/2}(A - tC)\, \tau(a - tc) \cdot \psi((A - tC)^{-1}(tD - B))$$

when $\det(A - tC) \neq 0$.

Let

$$\mathrm{BV} : \mathcal{O}(\Omega_n, W_\tau) \longrightarrow \mathcal{E}(\mathrm{Sym}_n, W_\tau \otimes \det{}^{-k/2})$$

by $(\mathrm{BV}\, f)(t) = \lim_{s \to 0^+} f(t + is)$, $(t, s \in \mathrm{Sym}_n,\ s > 0)$. They show that BV is intertwining and injective on the image of $\mathrm{KV}_\lambda$. Putting all three realizations together, there is a diagram of $G$-maps:

$$\begin{array}{ccc} & & \mathcal{E}(\mathrm{Sym}_n, W_\tau \otimes \det^{-k/2}) \\ & & \uparrow{\scriptstyle \mathrm{BV}} \\ \mathcal{L}^2(\mathrm{M}_{n,k}, V_\lambda) & \xrightarrow[\mathrm{KV}_\lambda]{} & \mathcal{O}(\Omega_n, W_\tau) \end{array}$$

**1.5.** Our new realization will fit in the upper left corner of the diagram above. A key ingredient is to look at induction outside of the semisimple category.

Write the matrices $\mathrm{M}_{2n,k}$ in the form $\binom{v}{w}$, where $v, w \in \mathrm{M}_{n,k}$. Let

$$H := \mathrm{M}_{2n,k} \oplus \mathbb{R}$$

be the Heisenberg group with $\mathrm{Sp}(n,\mathbb{R}) \times \mathrm{O}(k)$-action $(g,c) \cdot (\binom{v}{w}, s) = (g\binom{v}{w}c^{-1}, s)$. Define

$$W := \{(\tbinom{0}{w}, s) \mid w \in \mathrm{M}_{n,k}, s \in \mathbb{R}\} \subseteq H.$$

The stabilizer of $W$ in $G$ is the maximal parabolic subgroup with Langlands decomposition $MA\overline{N}$.

We define

$$\overline{P} := MA\overline{N} \ltimes W$$

with character $\chi : \overline{P} \to \mathbb{C}$ by

$$\chi(ma\overline{n}, (\tbinom{0}{w}, s)) := \det(ma)^{-\frac{k}{2}}\, e^{\frac{is}{2}}.$$

Via the embedding $\iota : \mathrm{M}_{n,k} \times \mathrm{Sym}_n \to (G \ltimes H)$ by (the lift of)

$$\iota(x,t) := (\left(\begin{smallmatrix} I_n & t \\ 0 & I_n\end{smallmatrix}\right), (\tbinom{x}{0}, 0)),$$

we work in the "noncompact" picture of

$$\mathcal{E}(\mathrm{M}_{n,k} \times \mathrm{Sym}_n) := \{\psi \in \mathcal{C}^\infty(\mathrm{M}_{n,k} \times \mathrm{Sym}_n, \mathbb{C}) \mid \exists \phi \in \mathrm{Ind}_{\overline{P}}^{G \ltimes H} \chi \text{ so } \psi = \iota^*\phi\}$$

which carries a $(G \times \mathrm{O}(k)) \ltimes H$-action. It can be shown that $\left(\begin{smallmatrix} A & B \\ C & D \end{smallmatrix}\right) \in G$ acts as

$$(g \cdot \psi)(x,t) = \det{}^{-k/2}(A - tC) e^{\frac{i}{2}\operatorname{tr}(x^T C (A-tC)^{-1} x)} \times \psi((A-tC)^{-1}x,\, (A-tC)^{-1}(tD - B))$$

when $\det(A - tC) \neq 0$ (see Theorem 3.1).

**1.6.** Now let $\mathcal{H}(\mathrm{M}_{n,k} \times \mathrm{Sym}_n)_{\mathrm{pre}}$ be the space of all solutions to the Schrödinger equation[1]

$$i\partial_t \psi(x,t) = \Delta\psi(x,t)$$

with $\psi \in \mathcal{E}(\mathrm{M}_{n,k} \times \mathrm{Sym}_n)$ such that the initial condition $\psi|_{t=0} \in \mathscr{S}(\mathrm{M}_{n,k})$. It is a lemma (Lemma 4.2) that $\mathcal{H}(\mathrm{M}_{n,k} \times \mathrm{Sym}_n)_{\mathrm{pre}}$ is invariant under the action of $(G \times \mathrm{O}(k)) \ltimes H$.

The $\mathrm{O}(k)$ action gives rise to the decomposition

$$\mathcal{H}(\mathrm{M}_{n,k} \times \mathrm{Sym}_n)_{\mathrm{pre}} = \bigoplus_{\lambda \in \Sigma \subseteq \widehat{\mathrm{O}}(k)} \mathcal{H}(\mathrm{M}_{n,k} \times \mathrm{Sym}_n, V_\lambda)_{\mathrm{pre}} \otimes V_\lambda'$$

as a $G \times \mathrm{O}(k)$-module. Here $\mathcal{H}(\mathrm{M}_{n,k} \times \mathrm{Sym}_n, V_\lambda)_{\mathrm{pre}}$ is the space of smooth functions $\psi : \mathrm{M}_{n,k} \times \mathrm{Sym}_n \to V_\lambda$ satisfying $\psi(xc,t) = \lambda(c)^{-1}\psi(x,t)$ for all $c \in \mathrm{O}(k)$ and $\langle \psi, f \rangle \in \mathcal{H}(\mathrm{M}_{n,k} \times \mathrm{Sym}_n)_{\mathrm{pre}}$ for all $f \in V_\lambda'$.

Define $\mathrm{ev}_{t=0} : \mathcal{H}(\mathrm{M}_{n,k} \times \mathrm{Sym}_n, V_\lambda)_{\mathrm{pre}} \to \mathcal{L}^2(\mathrm{M}_{n,k}, V_\lambda)$ by $(\mathrm{ev}_{t=0}\,\psi)(x) = \psi(x,0)$ and $\mathscr{F} : \mathcal{L}^2(\mathrm{M}_{n,k}, V_\lambda) \to \mathcal{L}^2(\mathrm{M}_{n,k}, V_\lambda)$ be given by $(\mathscr{F} f)(\xi) = \widehat{f}(\xi)$. In Theorem 5.2 we show that the map

$$\begin{array}{c} \mathcal{H}(\mathrm{M}_{n,k} \times \mathrm{Sym}_n, V_\lambda)_{\mathrm{pre}} \\ \Big\downarrow {\scriptstyle \mathscr{F} \circ \mathrm{ev}_{t=0}} \\ \mathcal{L}^2(\mathrm{M}_{n,k}, V_\lambda) \end{array}$$

is nonzero, injective, $G$-intertwining, and an isomorphism on $K$-finite vectors. It follows that $\mathcal{H}(\mathrm{M}_{n,k} \times \mathrm{Sym}_n, V_\lambda)_{\mathrm{pre}}$ completes to an irreducible unitary highest weight representation of $G$.

**1.7.** Now recall the $\mathrm{Hom}(V_\lambda, W_\tau)$-valued polynomial in $x \in \mathrm{M}_{n,k}$, $I_\lambda^*(x)$ (used in the map $\mathrm{KV}_\lambda : \mathcal{L}^2(\mathrm{M}_{n,k}, V_\lambda) \to \mathcal{O}(\Omega_n, W_\tau)$) and define

$$\nabla_\lambda : \mathcal{C}^\infty(\mathrm{M}_{n,k} \times \mathrm{Sym}_n, V_\lambda) \to \mathcal{C}^\infty(\mathrm{M}_{n,k} \times \mathrm{Sym}_n, W_\tau)$$

by

$$\nabla_\lambda := I_\lambda^*(-i\partial_x).$$

In Theorem 6.1, we show that the map

$$\mathcal{H}(\mathrm{M}_{n,k} \times \mathrm{Sym}_n, V_\lambda)_{\mathrm{pre}} \xhookrightarrow{\nabla_\lambda|_{x=0}} \mathcal{E}(\mathrm{Sym}_n, W_\tau \otimes \det{}^{-k/2})$$

is injective and $G$-intertwining. For $\lambda \in \Sigma \subseteq \widehat{\mathrm{O}}(k)$, we can combine everything into

---

[1] The equation here is really a system of partial differential equations. Furthermore, the sign in front of $\Delta$ is chosen so that we obtain highest weight representations. The choice

$$i\partial_t \psi(x,t) = -\Delta\psi(x,t)$$

would lead to lowest weight representations. See Section 7 for more details.

one commutative diagram of $G$-maps:

$$\begin{array}{ccc}
\mathcal{H}(\mathrm{M}_{n,k} \times \mathrm{Sym}_n, V_\lambda)_{\mathrm{pre}} & \xhookrightarrow{\nabla_\lambda|_{x=0}} & \mathcal{E}(\mathrm{Sym}_n, W_\tau \otimes \det^{-k/2}) \\
\scriptstyle{\mathscr{F} \circ \mathrm{ev}_{t=0}} \big\downarrow & & \big\uparrow \scriptstyle{\mathrm{BV}} \\
\mathcal{L}^2(\mathrm{M}_{n,k}, V_\lambda) & \xhookrightarrow[\mathrm{KV}_\lambda]{} & \mathcal{O}(\Omega_n, W_\tau)
\end{array}$$

**1.8.** This work generalizes our papers [**5, 6**] ($k = 1$) and [**9**] ($n = 1$). We recently also obtained analogous results for $\mathrm{U}(n, n)$. We expect that there is a uniform generalization to all groups associated to Hermitian symmetric spaces of tube type via Jordan algebras. Finally, we anticipate that our work will eventually tie in with Enright–Wallach [**4**].

## 2. The Group

**2.1.** In the following, we write the matrices $\mathrm{M}_{2n,k} = \mathrm{M}_{2n,k}(\mathbb{R})$ in the form $\binom{v}{w}$, where $v, w \in \mathrm{M}_{n,k}$. We then define the Heisenberg group $H := \mathrm{M}_{2n,k} \oplus \mathbb{R}$ with multiplication given by

$$\left(\begin{pmatrix} v_1 \\ w_1 \end{pmatrix}, s_1\right)\left(\begin{pmatrix} v_2 \\ w_2 \end{pmatrix}, s_2\right) = \left(\begin{pmatrix} v_1 + v_2 \\ w_1 + w_2 \end{pmatrix}, s_1 + s_2 + \mathrm{Tr}(w_1^T v_2 - v_1^T w_2)\right).$$

The symplectic group $\mathrm{Sp}(n, \mathbb{R}) \subseteq \mathrm{GL}(2n, \mathbb{R})$ acts on $H$ by $g \cdot (\binom{v}{w}, s) = (g\binom{v}{w}, s)$. Here $\mathrm{Sp}(n, \mathbb{R})$ is realized with respect to the standard symplectic form $\left(\begin{smallmatrix} 0 & -I_n \\ I_n & 0 \end{smallmatrix}\right)$. The orthogonal group $\mathrm{O}(k)$ acts on $H$ by $h \cdot (\binom{v}{w}, s) = (\binom{v}{w} h^{-1}, s)$.

**2.2.** Let $G := \mathrm{Mp}(n, \mathbb{R})$ be the double cover of $\mathrm{Sp}(n, \mathbb{R})$ defined as in [**5**]. We recall that $G$ consists of the set of pairs $(g, \varepsilon)$ with $g \in \mathrm{Sp}(n, \mathbb{R})$ and smooth $\varepsilon : \Omega_n \to \mathbb{C}$ satisfying $\varepsilon(Z)^2 = \det(CZ + D)$ where $g = \left(\begin{smallmatrix} A & B \\ C & D \end{smallmatrix}\right)$ and $\Omega_n$ is the Siegel upper half-space (all $Z = X + iY$ with $X, Y \in \mathrm{Sym}_n := \mathrm{Sym}_n(\mathbb{R})$ and $Y > 0$). We extend $\varepsilon$ almost everywhere to the boundary of $\Omega_n$, $\mathrm{Sym}_n$, by $\varepsilon(x) := \lim_{Y \to 0^+} \varepsilon(X + iY)$ when $\det(CX + D) \neq 0$.

The Siegel upper half-space carries a transitive action by $\mathrm{Sp}(n, \mathbb{R})$ by linear fractional transformations,

$$g \cdot Z = (AZ + B)(CZ + D)^{-1}.$$

This also extends to an almost everywhere defined action on $X \in \mathrm{Sym}_n$ when $\det(CX + D) \neq 0$.

We form the semidirect product

$$G \ltimes H$$

by having $G$ act on $H$ via its projection to $\mathrm{Sp}(n, \mathbb{R})$. Finally, we will also consider the group $(G \times \mathrm{O}(k)) \ltimes H$.

## 3. The Induced Representation and the Noncompact Picture

**3.1.** Define $W := \{(\binom{0}{w}, s) \mid w \in \mathrm{M}_{n,k},\ s \in \mathbb{R}\} \subseteq H$. The stabilizer of $W$ in $G$ is a maximal parabolic subgroup with Langlands decomposition $MA\overline{N}$ as in [**5**]. The Levi factor, $MA$, is a double cover of $\mathrm{GL}(n, \mathbb{R})$ whose projection to $\mathrm{Sp}(n,\mathbb{R})$ is $\{\left(\begin{smallmatrix} A & 0 \\ 0 & A^{-1,T}\end{smallmatrix}\right) \mid A \in \mathrm{GL}(n,\mathbb{R})\}$ while $\overline{N}$ projects isomorphically to $\{\left(\begin{smallmatrix} I & 0 \\ C & I\end{smallmatrix}\right) \mid C \in \mathrm{Sym}_n\}$. We define

$$\overline{P} := MA\overline{N} \ltimes W.$$

In particular, there is a function(s) $\det^{-1/2} : MA \to \mathbb{C}$ whose square is the lift of the inverse of the determinant function on $\mathrm{GL}(n,\mathbb{R})$. Precisely, elements of $MA$ consist of pairs $l_{A,c} := (\left(\begin{smallmatrix} A & 0 \\ 0 & A^{-1,T}\end{smallmatrix}\right), Z \mapsto c)$ with $c \in \mathbb{C}$ satisfying $c^2 = \det A^{-1}$. Define the function $\det^{-1/2} : MA \to \mathbb{C}$ by

$$\det{}^{-1/2}(l_{A,c}) := c\,.$$

We extend $\det^{-1/2}$ to $MA\overline{N}$ by having it act trivially on $\overline{N}$. Finally, we define two characters $\chi : \overline{P} \to \mathbb{C}$ by

$$\chi(ma\overline{n}, (\begin{pmatrix} 0 \\ w \end{pmatrix}, s)) := \det{}^{-k/2}(ma)\, e^{\frac{is}{2}}.$$

where, of course, $\det^{-k/2}(ma) := (\det^{-1/2}(ma))^k$. Note that $MA$ is isomorphic to the double cover $\widetilde{\mathrm{GL}}(n,\mathbb{R})$ and that $(MA)_{\mathbb{C}}$ is isomorphic to the double cover $\widetilde{\mathrm{GL}}(n,\mathbb{C})$. Clearly $\det^{-1/2}$ uniquely extends to a holomorphic character of $(MA)_{\mathbb{C}}$. We also remark that, in [**5**] with $k = 1$, the character $\chi$ was written as $\chi_{1,-1/2,-i/2}$ and eventually led to the the study of unitary highest weight modules.

**3.2.** The induced representation

$$\mathrm{Ind}_{\overline{P}}^{G\ltimes H}\chi := \{\phi \in \mathcal{C}^\infty(G \ltimes H, \mathbb{C}) \mid \phi(g\overline{p}) = \chi(\overline{p})^{-1}\phi(g)\ \forall g \in G \ltimes H,\, \overline{p} \in \overline{P}\}$$

is an analog of a degenerate principal series representation for the group $G \ltimes H$. Our focus will be on the non-compact picture of this representation. Embed $\iota$ : $\mathrm{M}_{n,k} \times \mathrm{Sym}_n \to (G \ltimes H)$ by

$$\iota(x,t) := ((\begin{pmatrix} I_n & t \\ 0 & I_n \end{pmatrix}, Z \mapsto 1), (\begin{pmatrix} x \\ 0 \end{pmatrix}, 0)).$$

The projection of the image of this map to $(G \ltimes H)/\overline{P}$ is open and dense and so the elements of $\mathrm{Ind}_{\overline{P}}^{G\ltimes H}\chi$ are determined by their restriction to the image of $\iota$. Let $\mathcal{E}(\mathrm{M}_{n,k} \times \mathrm{Sym}_n)$ denote the set of these restrictions,

$$\mathcal{E}(\mathrm{M}_{n,k} \times \mathrm{Sym}_n) := \{\psi \in \mathcal{C}^\infty(\mathrm{M}_{n,k} \times \mathrm{Sym}_n, \mathbb{C}) \mid \exists \phi \in \mathrm{Ind}_{\overline{P}}^{G\ltimes H}\chi \text{ so } \psi = \iota^*\phi\},$$

and view it as a $G \ltimes H$-module for which the map $\phi \mapsto \psi = \iota^*\phi$ is an intertwining isomorphism. Observe that $\mathrm{O}(k)$ also acts on $\mathcal{E}(\mathrm{M}_{n,k} \times \mathrm{Sym}_n)$ by

$$(h \cdot \psi)(x,t) := \psi(xh, t).$$

This action commutes with the action of $G$.

**3.3.** In the following result, we write $g = (\left(\begin{smallmatrix} A & B \\ C & D\end{smallmatrix}\right), \varepsilon) \in G$. When $\det(A - tC) \neq 0$ for $t \in \mathrm{Sym}_n$, a calculation shows that $g^{-1} \cdot t = (A - tC)^{-1}(tD - B)$ and that $\varepsilon(g^{-1} \cdot t)^2 = \det(A - tC)^{-1}$. In order to make the formulas below more transparent, we will use the imprecise notation (when $k$ is odd) $\det^{-k/2}(A - tC)$ to denote $\varepsilon(g^{-1} \cdot t)^k$. Alternately, one can write $g^{-1}$ as $(\left(\begin{smallmatrix} A & B \\ C & D\end{smallmatrix}\right)^{-1}, \varepsilon_{g^{-1}})$ and show $\varepsilon(g^{-1} \cdot t) = \varepsilon_{g^{-1}}(t)$.

THEOREM 3.1. *For $\psi \in \mathcal{E}(\mathrm{M}_{n,k} \times \mathrm{Sym}_n)$, the action of $(g, \varepsilon) \in G$ on $\psi$ is given by*

$$\begin{aligned}((g,\varepsilon)\cdot\psi)(x,t) &= \det{}^{-k/2}(A-tC)e^{\frac{i}{2}\operatorname{tr}(x^TC(A-tC)^{-1}x)} \\ &\quad\cdot \psi((A-tC)^{-1}x,\ (A-tC)^{-1}(tD-B))\end{aligned}$$

*when* $\det(A - tC) \neq 0$.

*The action of* $((\begin{smallmatrix} v \\ w \end{smallmatrix}), s) \in H$ *on* $\psi$ *is given by*

$$(((\begin{smallmatrix} v \\ w \end{smallmatrix}), s)\cdot\psi)(x,t) = e^{\frac{i}{2}(s+\operatorname{tr}(2x^Tw-v^Tw+w^Ttw))}\psi(x-v+tw,t).$$

*The action for* $h \in \mathrm{O}(k)$ *is*

$$(h\cdot\psi)(x,t) = \psi(xh,t).$$

PROOF. This result follows by straightforward calculations. For the action of $G$, we give the details in the most involved case and leave the rest to the reader. Full details may be extracted by the patient reader from Theorem 1 of [**5**].

Suppose $\psi = \iota^*\phi$ with $\phi \in \mathrm{Ind}_{\overline{P}}^{G\ltimes H}\chi$ and $\overline{n} = ((\begin{smallmatrix} I_n & 0 \\ C & I_n \end{smallmatrix}), \varepsilon(Z)) \in G$. By definition, we calculate

$$\begin{aligned}
&(\overline{n}\cdot\psi)(x,t)\\
&\quad= \phi(\overline{n}^{-1}\iota(x,t))\\
&\quad= \phi(((\begin{smallmatrix} I_n & 0 \\ -C & I_n \end{smallmatrix}), \det(I_n-CZ)^{1/2})((\begin{smallmatrix} I_n & t \\ 0 & I_n \end{smallmatrix}),1)((\begin{smallmatrix} x \\ 0 \end{smallmatrix}),0))\\
&\quad= \phi((\begin{pmatrix} I_n & (I_n-tC)^{-1}t \\ 0 & I_n \end{pmatrix},1)(\begin{pmatrix} (I_n-tC)^{-1} & 0 \\ -C & (I_n-Ct) \end{pmatrix}, \det(I_n-C(t+Z))^{1/2})\\
&\qquad\times ((\begin{smallmatrix} x \\ 0 \end{smallmatrix}),0))\\
&\quad= \phi((\begin{pmatrix} I_n & (I_n-tC)^{-1}t \\ 0 & I_n \end{pmatrix},1)((\begin{smallmatrix} (I_n-tC)^{-1}x \\ -CX \end{smallmatrix}),0)\\
&\qquad\times (\begin{pmatrix} (I_n-tC)^{-1} & 0 \\ -C & (I_n-Ct) \end{pmatrix}, \det(I_n-C(t+Z))^{1/2})\\
&\quad= \phi((\begin{pmatrix} I_n & (I_n-tC)^{-1}t \\ 0 & I_n \end{pmatrix},1)((\begin{smallmatrix} (I_n-tC)^{-1}x \\ 0 \end{smallmatrix}),0)((\begin{smallmatrix} 0 \\ -Cx \end{smallmatrix}), -\operatorname{tr}(x^T(I_n-Ct)^{-1}Cx))\\
&\qquad\times \det{}^{-k/2}(I_n-tC)\\
&\quad= \det{}^{-k/2}(I_n-tC)e^{\frac{i}{2}\operatorname{tr}(x^T(I_n-Ct)^{-1}Cx)}\psi((I_n-tC)^{-1}x,(I_n-tC)^{-1}t).
\end{aligned}$$

For the action of $H$, we calculate

$$\begin{aligned}
&(((\begin{smallmatrix} v \\ w \end{smallmatrix}),s)\cdot\psi)(x,t)\\
&\quad= \phi(((\begin{smallmatrix} -v \\ -w \end{smallmatrix}),-s)((\begin{smallmatrix} I_n & t \\ 0 & I_n \end{smallmatrix}),1)((\begin{smallmatrix} x \\ 0 \end{smallmatrix}),0))\\
&\quad= \phi(((\begin{smallmatrix} I_n & t \\ 0 & I_n \end{smallmatrix}),1)((\begin{smallmatrix} -v+tw \\ -w \end{smallmatrix}),-s)((\begin{smallmatrix} x \\ 0 \end{smallmatrix}),0))\\
&\quad= \phi(((\begin{smallmatrix} I_n & t \\ 0 & I_n \end{smallmatrix}),1)((\begin{smallmatrix} x-v+tw \\ -w \end{smallmatrix}),-s-\operatorname{tr}(w^Tx)))\\
&\quad= \phi(((\begin{smallmatrix} I_n & t \\ 0 & I_n \end{smallmatrix}),1)((\begin{smallmatrix} x-v+tw \\ 0 \end{smallmatrix}),0)((\begin{smallmatrix} 0 \\ -w \end{smallmatrix}),-s-\operatorname{tr}(2x^Tw-v^Tw+w^Ttw)))\\
&\quad= e^{\frac{i}{2}(s+\operatorname{tr}(2x^Tw-v^Tw+w^Ttw))}\psi(x-v+tw,t).
\end{aligned}$$

□

Exponentiating and differentiating gives the following action of the Lie algebra. Here we write $\partial_M^x := \sum_{ij} M_{ij}\partial_{x_{ij}}$ and $\partial_M^t := \sum_{i\leq j} M_{ij}\partial_{(t)_{ij}}$ and identify $\mathrm{Lie}(G)$ with $\mathfrak{sp}(n,\mathbb{R})$.

THEOREM 3.2. *For* $X = \left(\begin{smallmatrix} a & b \\ c & -a^T \end{smallmatrix}\right) \in \mathfrak{sp}(n,\mathbb{R})$ *and* $\psi \in \mathcal{E}(\mathrm{M}_{n,k} \times \mathrm{Sym}_n)$,

$$\begin{aligned}(X\cdot\psi)(x,t) &= [\frac{k}{2}\operatorname{tr}(tc-a) + \frac{i}{2}\operatorname{tr}(x^Tcx)]\psi(x,t)\\ &\quad + \partial^x_{(tc-a)x}\psi(x,t) + \partial^t_{tct}\psi(x,t) - \partial^t_{(ta^T+at)}\psi(x,t) - \partial^t_b\psi(x,t).\end{aligned}$$

PROOF. In $\mathrm{Sp}(n,\mathbb{R})$, $\exp(s\left(\begin{smallmatrix} a & b \\ c & -a^T \end{smallmatrix}\right)) = \left(\begin{smallmatrix} I_n+sa & sb \\ sc & I_n-sa^T \end{smallmatrix}\right) + \cdots$ so that

$$\begin{aligned}&(\begin{pmatrix} a & b \\ c & -a^T \end{pmatrix}\cdot\psi)(x,t)\\ &= \frac{d}{ds}\bigg|_{s=0} \{\det{}^{-k/2}(I_n - s(tc-a))e^{\frac{i}{2}s\operatorname{tr}(x^Tc(I_n-s(tc-a))^{-1}x)}\\ &\qquad \times \psi((I_n-s(tc-a))^{-1}x,\ (I_n-s(tc-a))^{-1}(t-s(ta^T+b)))\}\\ &= [\frac{k}{2}\operatorname{tr}(tc-a)+\frac{i}{2}\operatorname{tr}(x^Tcx)]\,\psi(x,t)\\ &\qquad + \partial^x_{(tc-a)x}\psi(x,t) + \partial^t_{tct}\psi(x,t) - \partial^t_{(ta^T+at)}\psi(x,t) - \partial^t_b\psi(x,t).\end{aligned}$$

□

## 4. A System of Schrödinger-Type Operators

**4.1.** The set $\{E_{ij}+E_{ji} \mid 1 \leq i \leq j \leq n\}$ forms an orthogonal basis for $\mathrm{Sym}_n$ with respect to the trace form. Let $\{t_{ij} \mid 1\leq i\leq j\leq n\}$ be the dual basis for $\mathrm{Sym}_n^*$ which we use to coordinatize $\mathrm{Sym}_n$ and to identify it with $\mathbb{R}^{n(n+1)/2}$. Note that if we write $(t)_{ij}$ for the $ij$-matrix entry on $\mathrm{Sym}_n$, then $(t)_{ij} = t_{ij}$ when $i<j$ while $(t)_{ii} = 2t_{ii}$ (and $2\partial_{(t)_{ii}} = \partial_{t_{ii}}$). We write $\partial_{t_{ij}}$ for the corresponding differential operator on $\mathcal{C}^\infty(\mathrm{Sym}_n,\mathbb{C})$ and $\partial_t = (\partial_{t_{ij}})_{1\leq i\leq j\leq n}$ for the resulting family of differential operators.

**4.2.** Next define $\pi : \mathrm{M}_{n,k} \to \mathrm{Sym}_n$ by $\pi(x) := xx^T$ and let $q_{ij} := \pi^*((t)_{ij}) \in \mathcal{C}^\infty(\mathrm{M}_{n,k})$. Then $q_{ij}(x) = \sum_{\nu=1}^k x_{i\nu}x_{j\nu}$. Write

$$\Delta_{ij} := q_{ij}(\partial_x) = \sum_{\nu=1}^k \partial_{x_{i\nu}}\partial_{x_{j\nu}}$$

for the corresponding polynomial differential operator with constant coefficients and $\Delta := (\Delta_{ij})_{1\leq i\leq j\leq n}$ for the resulting family of differential operators.

Let

$$\mathcal{H}(\mathrm{M}_{n,k}\times\mathrm{Sym}_n)_{\mathrm{pre}}$$

be the space of all solutions to

$$i\partial_t\psi(x,t) = \Delta\psi(x,t)$$

with $\psi \in \mathcal{E}(\mathrm{M}_{n,k}\times\mathrm{Sym}_n)$ such that the initial condition $\psi|_{t=0} \in \mathscr{S}(\mathrm{M}_{n,k})$, the Schwartz space of rapidly decreasing functions on $\mathrm{M}_{n,k} \cong \mathbb{R}^{nk}$. We will shortly see that $\mathcal{H}(\mathrm{M}_{n,k}\times\mathrm{Sym}_n)_{\mathrm{pre}}$ is a pre-Hilbert space and is invariant under $(G\times\mathrm{O}(k))\ltimes H$.

For $\Psi \in L^1(\mathrm{M}_{n,k})$, we write the Fourier transform as

$$\widehat{\Psi}(\xi) := \int_{\mathrm{M}_{n,k}} \Psi(x) e^{-i\operatorname{tr}(x^T\xi)}\, dx.$$

For $\psi \in \mathcal{H}(\mathrm{M}_{n,k} \times \mathrm{Sym}_n)_{\mathrm{pre}}$, we also write $\widehat{\psi}(\xi, t)$ for the Fourier transform of $\psi$ with respect to $x$ with fixed $t$.

THEOREM 4.1. *Any $\psi \in \mathcal{H}(\mathrm{M}_{n,k} \times \mathrm{Sym}_n)_{\mathrm{pre}}$ may be written in terms of its initial condition by*

$$\psi(x,t) = \frac{1}{(2\pi)^{nk}} \int_{\mathrm{M}_{n,k}} \widehat{\psi}(\xi, 0) e^{\frac{i}{2}\operatorname{tr}(\xi^T t\xi)} e^{i\operatorname{tr}(x^T\xi)}\, d\xi.$$

PROOF. By standard Fourier techniques, it follows that

$$-i\partial_{t_{ij}} \widehat{\psi}(\xi, t) = \sum_{\nu=1}^{k} \xi_{i\nu}\xi_{j\nu}\widehat{\psi}(\xi, t)$$

so that (recall $2t_{ii} = (t)_{ii}$)

$$\widehat{\psi}(\xi, t) = \widehat{\psi}(\xi, 0) e^{i\sum_{i\le j}\sum_{\nu=1}^{k} \xi_{i\nu}\xi_{j\nu}t_{ij}} = \widehat{\psi}(\xi, 0) e^{\frac{i}{2}\operatorname{tr}(\xi^T t\xi)}.$$

The result now follows by applying the inverse Fourier transform. □

Note that it is not automatic that the solution to the Cauchy problem for an arbitrary Schwartz initial condition lies in $\mathcal{E}(\mathrm{M}_{n,k} \times \mathrm{Sym}_n)$.

The next lemma follows from Theorem 3.1 and straightforward calculations as in Theorems 2 and 4 of [**5**].

LEMMA 4.2. *$\mathcal{H}(\mathrm{M}_{n,k} \times \mathrm{Sym}_n)_{\mathrm{pre}}$ is invariant under the action of $(G \times \mathrm{O}(k)) \ltimes H$.*

PROOF. Invariance of the set of solutions to $(-i\partial_t + \Delta)\psi = 0$ follows from Theorem 3.2 and a straightforward calculation showing that

$$\frac{1}{2}[X, (-i\partial_t + \Delta)_{ij}] = ((ct - a^T)(i\partial_t - \Delta))_{ij} + ((ct - a^T)(i\partial_t - \Delta))_{ji}$$

for each $X \in \mathfrak{sp}(n, \mathbb{R})$. See Theorem 2 of [**5**] for explicit analogous calculations. Preservation of the Schwartz initial condition follows from Theorems 3.1 and 4.1 and Fourier theory as in Theorem 4 of [**5**]. □

## 5. O($k$)-decomposition of $\mathcal{H}(\mathrm{M}_{n,k} \times \mathrm{Sym}_n)_{\mathrm{pre}}$

**5.1.** For an irreducible representation $(\lambda, V_\lambda)$ of $\mathrm{O}(k)$, write the dual space as $V_\lambda' = \mathrm{Hom}_{\mathbb{C}}(V_\lambda, \mathbb{C})$ (in this case known to be isomorphic to $V_\lambda$). For $f \in V_\lambda'$ and $\psi \in \mathcal{C}^\infty(\mathrm{M}_{n,k} \times \mathrm{Sym}_n, V_\lambda)$, define $\langle \psi, f\rangle \in \mathcal{C}^\infty(\mathrm{M}_{n,k} \times \mathrm{Sym}_n)$ by $\langle \psi, f\rangle (x,t) := f(\psi(x,t))$. Now define

$$\begin{aligned}\mathcal{E}(\mathrm{M}_{n,k} \times \mathrm{Sym}_n, V_\lambda) := \{\psi \in \mathcal{C}^\infty(\mathrm{M}_{n,k} \times \mathrm{Sym}_n, V_\lambda) \mid \\ \langle \psi, f\rangle \in \mathcal{E}(\mathrm{M}_{n,k} \times \mathrm{Sym}_n)\ \forall f \in V_\lambda' \text{ and} \\ \psi(xh, t) = \lambda(h)^{-1}\psi(x,t)\ \forall h \in \mathrm{O}(k)\}.\end{aligned}$$

It is easy to see that $\mathcal{E}(\mathrm{M}_{n,k} \times \mathrm{Sym}_n, V_\lambda) \cong \mathrm{Hom}_{\mathrm{O}(k)}(V'_\lambda, \mathcal{E}(\mathrm{M}_{n,k} \times \mathrm{Sym}_n))$. Therefore the canonical decomposition of the $\mathrm{O}(k)$-finite vectors of $\mathcal{E}(\mathrm{M}_{n,k} \times \mathrm{Sym}_n)$ can be written as

$$\mathcal{E}(\mathrm{M}_{n,k} \times \mathrm{Sym}_n)_{\mathrm{O}(k)} \cong \bigoplus_{\lambda \in \widehat{\mathrm{O}}(k)} V'_\lambda \otimes \mathcal{E}(\mathrm{M}_{n,k} \times \mathrm{Sym}_n, V_\lambda).$$

Write $\mathcal{E}(\mathrm{M}_{n,k} \times \mathrm{Sym}_n)_{\lambda'}$ for the $V'_\lambda$-isotypic $\mathrm{O}(k)$-component of $\mathcal{E}(\mathrm{M}_{n,k} \times \mathrm{Sym}_n)$. For each $\lambda$, the isomorphism is implemented by the map

$$\mathcal{E}(\mathrm{M}_{n,k} \times \mathrm{Sym}_n, V_\lambda) \otimes V'_\lambda \xrightarrow{\cong} \mathcal{E}(\mathrm{M}_{n,k} \times \mathrm{Sym}_n)_{\lambda'}$$

induced by $\psi \otimes f \mapsto \langle \psi, f \rangle$. As $G$ commutes with $\mathrm{O}(k)$, $\mathcal{E}(\mathrm{M}_{n,k} \times \mathrm{Sym}_n, V_\lambda)$ inherits the structure of a $G$-module. Specifically, by Theorem 3.1 and with the same conventions, the action of $(g, \varepsilon) \in G$ on $\psi$ is given by

$$\begin{aligned} ((g,\varepsilon)\cdot\psi)(x,t) &= \det(A - tC)^{-\frac{k}{2}} e^{\frac{i}{2}\operatorname{tr}(x^T C(A-tC)^{-1}x)} \\ &\quad \cdot \psi((A-tC)^{-1}x,\, (A-tC)^{-1}(tD-B)) \end{aligned} \tag{5.1.1}$$

when $\det(A - tC) \neq 0$.

**5.2.** Now let

$$\begin{aligned} \mathcal{H}(\mathrm{M}_{n,k} \times \mathrm{Sym}_n, V_\lambda)_{\mathrm{pre}} := \{\psi \in \mathcal{E}(\mathrm{M}_{n,k} \times \mathrm{Sym}_n, V_\lambda) \mid \\ \langle \psi, f \rangle \in \mathcal{H}(\mathrm{M}_{n,k} \times \mathrm{Sym}_n)_{\mathrm{pre}}\ \forall f \in V'_\lambda\}. \end{aligned}$$

The above discussion and restriction results in a mapping

$$\mathcal{H}(\mathrm{M}_{n,k} \times \mathrm{Sym}_n, V_\lambda)_{\mathrm{pre}} \otimes V'_\lambda \to \mathcal{H}(\mathrm{M}_{n,k} \times \mathrm{Sym}_n)_{\mathrm{pre}}$$

induced by $\psi \otimes f \to \langle \psi, f \rangle$ that is an isomorphism onto the $V'_\lambda$-isotypic $\mathrm{O}(k)$-component of $\mathcal{H}(\mathrm{M}_{n,k} \times \mathrm{Sym}_n)_{\mathrm{pre}}$. Note that these spaces may be trivial for certain $\lambda$. As before, $\mathcal{H}(\mathrm{M}_{n,k} \times \mathrm{Sym}_n, V_\lambda)_{\mathrm{pre}}$ inherits the structure of a $G$-module with action given by Equation 5.1.1.

**5.3.** Extend the Fourier Transform in Equation 4.2 from scalar valued integration to $V_\lambda$-vector valued integration on $\mathrm{M}_{n,k}$. By a simple modification of Equation 4.1, any $\psi \in \mathcal{H}(\mathrm{M}_{n,k} \times \mathrm{Sym}_n, V_\lambda)_{\mathrm{pre}}$ can be written in the form

$$\psi(x,t) = \frac{1}{(2\pi)^{nk}} \int_{\mathrm{M}_{n,k}} \widehat{\psi}(\xi, 0) e^{\frac{i}{2}\operatorname{tr}(\xi^T t \xi)} e^{i \operatorname{tr}(x^T \xi)}\, d\xi.$$

**5.4.** It still remains to show that $\mathcal{H}(\mathrm{M}_{n,k} \times \mathrm{Sym}_n, V_\lambda)_{\mathrm{pre}} \neq 0$ for certain $\lambda$. The difficulty is that any particular solution to the Cauchy problem may not originate from a function coming from the induced representation, $\mathrm{Ind}_P^{G \ltimes H} \chi$.

To this end, let $r_s := \exp_G\left(\begin{smallmatrix} 0 & -sI_n \\ sI_n & 0 \end{smallmatrix}\right) = \left(\left(\begin{smallmatrix} I_n \cos s & -I_n \sin s \\ I_n \sin s & I_n \cos s \end{smallmatrix}\right), \varepsilon_s\right)$ where $\varepsilon_s : \Omega_n \to \mathbb{C}$ is determined by the conditions $\varepsilon_s(Z)^2 = \det(I_n \cos s + Z \sin s)$ and $\varepsilon_s(iI_n) = e^{\frac{isn}{2}}$. In particular, $r_{\pi/2} = \left(\left(\begin{smallmatrix} 0 & -I_n \\ I_n & 0 \end{smallmatrix}\right), \varepsilon_{\pi/2}\right)$ where $\varepsilon_{\pi/2}$ is determined by the the conditions $\varepsilon_{\pi/2}(Z)^2 = \det Z$ and $\varepsilon_{\pi/2}(iI_n) = e^{\frac{i\pi n}{4}}$. It is easy to see that them map $s \mapsto r_s$ is a 1-parameter subgroup of $G$.

The *Cartan involution* $\theta : G \to G$ is the anti-involution $\theta(g,\varepsilon) := r_{\pi/2}(g,\varepsilon)^{-1} r_{\pi/2}^{-1}$. We also define $\varepsilon^T : \Omega_n \to \mathbb{C}$ so that

$$\theta(g, \varepsilon) = (g^T, \varepsilon^T).$$

By definition, $\varepsilon^T(Z)^2 = \det(B^T Z + D^T)$. Pinning down the exact square root is a bit messy, but can be done with a straightforward calculation found in [**5**].

By definition, $((\begin{smallmatrix} I_n & t \\ 0 & I_n \end{smallmatrix}), 1)^T = ((\begin{smallmatrix} I_n & 0 \\ t & I_n \end{smallmatrix}), 1^T)$. Here, $1^T(Z)^2 = \det(tZ + I_n)$. Since $1^T(Z)^2 \to 1$ as $t \to 0$, it follows that the function $1^T(Z)$ is the analytic continuation of the function $Z \to \sqrt{\det(tZ + I_n)}$ that is initially defined only for sufficiently small $Z$. In the following, we will abuse notation and write $\det(I_n + tZ)^{1/2}$ for the function $1^T(Z)$. We will also write $\det(I_n + t\overline{Z})^{1/2}$ for the function $\overline{\det(I_n + tZ)}^{1/2}$ (especially for the case of $Z = iI_n$).

**5.5.** Let $\mathfrak{H}(\mathrm{M}_{n,k}, V_\lambda)$ be the space of all $V_\lambda$-valued polynomial functions $f : \mathrm{M}_{n,k} \to V_\lambda$ satisfying $\Delta f = 0$ and $f(xh) = \lambda^{-1}(h)f(x)$ for all $x \in \mathrm{M}_{n,k}$, $h \in \mathrm{O}(k)$. The $\lambda$'s for which $\mathfrak{H}(\mathrm{M}_{n,k}, V_\lambda)$ are nonzero are explicitly calculated in [**8**]. We write

$$\Sigma := \{\lambda \in \widehat{\mathrm{O}}(k) \mid \mathfrak{H}(\mathrm{M}_{n,k}, V_\lambda) \neq 0\}.$$

For such $\lambda$, by [**8**], $\mathfrak{H}(\mathrm{M}_{n,k}, V_\lambda)$ is known to be a finite dimensional, irreducible representation of $\mathrm{GL}(n, \mathbb{C})$. Note that here we extend the action of $\mathrm{GL}(n, \mathbb{R})$ on $\mathfrak{H}(\mathrm{M}_{n,k}, V_\lambda)$ holomorphically to $\mathrm{GL}(n, \mathbb{C})$ by identifying complex valued polynomial functions on $\mathrm{M}_{n,k}$ with complex valued polynomial functions on $\mathrm{M}_{n,k}(\mathbb{C})$.

LEMMA 5.1. *Let $\lambda \in \Sigma \subseteq \widehat{\mathrm{O}}(k)$ and $f \in \mathfrak{H}(\mathrm{M}_{n,k}, V_\lambda)$. Then the function $\Psi_f \in \mathcal{C}^\infty(\mathrm{M}_{n,k} \times \mathrm{Sym}_n, V_\lambda)$ given by*

$$\Psi_f(x,t) = \det(I_n - it)^{-k/2} e^{-\frac{1}{2}\operatorname{tr}(x^T(I_n - it)^{-1}x)} f((I_n - it)^{-1}x).$$

*lies in $\mathcal{H}(\mathrm{M}_{n,k} \times \mathrm{Sym}_n, V_\lambda)_{\mathrm{pre}}$.*

PROOF. We begin by calculating $\varepsilon^T_{\overline{p}}$ explicitly for

$$\overline{p} = ((\begin{smallmatrix} A & 0 \\ C & A^{-1,T} \end{smallmatrix}), \varepsilon_{\overline{p}}) \in MA\overline{N}.$$

We know that $\varepsilon^T_{\overline{p}}(Z)^2 = \det(A^{-1}) = (\det^{-1/2}\overline{p})^2$. To determine the precise value, first look at the special case of $C = 0$, $A = \lambda I_n$ with $\lambda > 0$, and $\varepsilon_{\overline{p}} = \det_1^{-1/2}\overline{p} = \epsilon_0 \lambda^{-n/2}$ for a fixed $\epsilon_0 \in \{\pm 1\}$. By continuity, there is a choice of $\epsilon \in \{\pm 1\}$ so that $\varepsilon^T_{\overline{p}} = \epsilon\epsilon_0\lambda^{-n/2}$ for all $\lambda$. But as $\lambda \to 1$, $r_{\pi/2}\overline{p}^{-1}r^{-1}_{\pi/2} \to (\mathrm{I}_{2n}, \epsilon_0)$. Thus $\epsilon = 1$ and $\varepsilon^T_{\overline{p}} = \det_1^{-1/2}\overline{p} = \overline{\det^{-1/2}}\overline{p}$ for this special case. Next, look at the special case of $C = 0$, $A \in \mathrm{O}(n)$, and $\varepsilon_{\overline{p}} = \det_1^{-1/2}\overline{p} \in \{\pm 1, \pm i\}$. In this case, $r_{\pi/2}\overline{p}^{-1}r^{-1}_{\pi/2} = \overline{p}^{-1}$ so that $\varepsilon^T_{\overline{p}} = \varepsilon^{-1}_{\overline{p}} = \overline{\det^{-1/2}}\overline{p}$ here. Since it is easy to show $\varepsilon^T_{\overline{p}} = 1$ when $\overline{p} \in \overline{N}$, it follows that $\varepsilon^T_{\overline{p}} = \overline{\det^{-1/2}}\overline{p}$ for all $\overline{p} \in MA\overline{N}$. In particular, $((g,\varepsilon)\overline{p})^T = (\left(\begin{smallmatrix} A^T & C \\ 0 & A^{-1} \end{smallmatrix}\right) g^T, \overline{\det^{-1/2}}\overline{p}\,\varepsilon^T)$ for any $(g, \varepsilon) \in G$.

Identifying $f$ with a complex valued polynomial on $\mathrm{M}_{n,k}(\mathbb{C})$ when necessary, consider the function $\Phi \in \mathcal{C}^\infty(G \ltimes H, V_\lambda)$ given by

$$\Phi((g,\varepsilon)\,((\begin{smallmatrix} x \\ y \end{smallmatrix}), s)) := \overline{\varepsilon^T}(iI_n)^{-k}\, e^{-\frac{i}{2}(s + \operatorname{tr}(y^T x + x^T(g^T \cdot (iI_n))x))}$$
$$\times f((D_g + iB_g)^{-1,T}x)$$

where $g = \left(\begin{smallmatrix} A_g & B_g \\ C_g & D_g \end{smallmatrix}\right)$. Calculate

$$\begin{aligned}
&\Phi((g,\varepsilon)\,(\left(\begin{smallmatrix} x \\ y \end{smallmatrix}\right), s)\,\overline{p}\,(\left(\begin{smallmatrix} 0 \\ y_0 \end{smallmatrix}\right), s_0)) \\
&= \Phi((g,\varepsilon)\,\overline{p}\,(\left(\begin{smallmatrix} A^{-1}x \\ A^T y - Cx + y_0 \end{smallmatrix}\right), s + s_0 - \operatorname{tr}(y_0^T A^{-1} x))\,) \\
&= [(\det{}^{-k/2}\overline{p})e^{\frac{is_0}{2}}]^{-1}\overline{\varepsilon^T}(iI_n)^{-k}e^{-\frac{i}{2}(s+\operatorname{tr}(y^T x + x^T(g^T\cdot(i\sigma I_n))x))} \\
&\quad \times f((-iB_g + D_g)^{-1,T}x) \\
&= \chi(\overline{p}\,(\left(\begin{smallmatrix} 0 \\ y_0 \end{smallmatrix}\right), s_0))^{-1}\Phi((g,\varepsilon)\,(\left(\begin{smallmatrix} x \\ y \end{smallmatrix}\right), s)).
\end{aligned}$$

From this it easily follows that $\Psi := \iota^*\Phi \in \mathcal{E}(\mathrm{M}_{n,k} \times \mathrm{Sym}_n, V_\lambda)$.

A quick examination shows that

$$\Psi(x,t) = \det(I_n - it)^{-k/2}e^{-\frac{1}{2}\operatorname{tr}(x^T(I_n - it)^{-1}x)}\, f((I_n - it)^{-1}x).$$

It remains to see that $\Psi \in \mathcal{H}(\mathrm{M}_{n,k} \times \mathrm{Sym}_n, V_\lambda)_{\mathrm{pre}}$. Using only the fact that $\Delta f = 0$, it is a straightforward exercise with the chain rule to verify that $\Psi$ satisfies the Schrödinger Equation 1.6. We omit the calculation (see Theorem 5 of [**5**] for a similar calculation in the scalar-valued case). □

**5.6.** Let $\mathcal{L}^2(\mathrm{M}_{n,k}, V_\lambda)$ be the space of square-integrable $V_\lambda$-valued functions $f : \mathrm{M}_{n,k} \to V_\lambda$ satisfying $f(xh) = \lambda(h)^{-1}f(x)$ for all $h \in \mathrm{O}(k)$ equipped with the $G$-action generated by

$$\begin{aligned}
(l_{A,c}\cdot f)(\xi) &= \det{}^{k/2}(A)f(A^T\xi) \\
(\iota(0,B)\cdot f)(\xi) &= e^{-\frac{i}{2}\operatorname{tr}(\xi^T B\xi)}f(\xi) \\
(r_{\pi/2}\cdot f)(\xi) &= (\frac{i}{2\pi})^{\frac{nk}{2}}\int_{\mathrm{M}_{n,k}} f(x)e^{i\operatorname{tr}(\xi^T x)}\,dx
\end{aligned}$$

where $(i)^{nk/2}$ is shorthand for $e^{i\pi nk/4}$. In [**8**], Kashiwara and Vergne showed that $\mathcal{L}^2(\mathrm{M}_{n,k}, V_\lambda)$ is an irreducible unitary highest weight representation of $G$.

Let $\mathrm{ev}_{t=0} : \mathcal{H}(\mathrm{M}_{n,k} \times \mathrm{Sym}_n, V_\lambda)_{\mathrm{pre}} \to \mathcal{L}^2(\mathrm{M}_{n,k}, V_\lambda)$ be given by $(\mathrm{ev}_{t=0}\,\psi)(x) = \psi(x,0)$ and $\mathscr{F} : \mathcal{L}^2(\mathrm{M}_{n,k}, V_\lambda) \to \mathcal{L}^2(\mathrm{M}_{n,k}, V_\lambda)$ be given by $(\mathscr{F}f)(\xi) = (2\pi)^{-nk}\,\widehat{f}(\xi)$.

THEOREM 5.2. *For $\lambda \in \Sigma \subseteq \widehat{\mathrm{O}}(k)$, the map*

$$\mathscr{F}\circ\mathrm{ev}_{t=0}\qquad \begin{array}{c} \mathcal{H}(\mathrm{M}_{n,k} \times \mathrm{Sym}_n, V_\lambda)_{\mathrm{pre}} \\ \downarrow \\ \mathcal{L}^2(\mathrm{M}_{n,k}, V_\lambda) \end{array}$$

*is nonzero, injective, and $G$-intertwining.*

PROOF. This result follows by standard Fourier analysis calculations. For example, by Theorem 3.1,

$$((\mathscr{F}\circ\mathrm{ev}_{t=0})(l_{A,c}\cdot\psi))(\xi) = \frac{\det^{-k/2}A}{(2\pi)^{nk}}\int_{\mathrm{M}_{n,k}} \psi(A^{-1}x,0)e^{-i\operatorname{tr}(x^T\xi)}\,dx$$

while

$$\begin{aligned}
(l_{A,c}\cdot((\mathscr{F}\circ\mathrm{ev}_{t=0})(\psi)))(\xi) &= \frac{\det^{k/2}A}{(2\pi)^{nk}}\int_{\mathrm{M}_{n,k}} \psi(x,0)e^{-i\operatorname{tr}(x^T A^T\xi)}\,dx \\
&= \frac{\det^{k/2}A\,|\det A|^{-k}}{(2\pi)^{nk}}\int_{\mathrm{M}_{n,k}} \psi(A^{-1}x,0)e^{-i\operatorname{tr}(x^T\xi)}\,dx.
\end{aligned}$$

As it is straightforward to show that $\det^{-k/2} A = \det^{k/2} A\, |\det A|^{-k}$ (see Theorem 6 of [**5**] for an explicit calculation), it follows that the action of $MA$ commutes with $\mathscr{F} \circ \mathrm{ev}_{t=0}$.

For $N$, observe that Equation 5.3 may be rewritten as

$$\psi(x,t) = \mathscr{F}^{-1}[e^{\frac{i}{2}\operatorname{tr}(\cdot^T t \cdot)}(\mathscr{F} \circ \mathrm{ev}_{t=0})(\psi)](x)$$

and calculate that

$$\begin{aligned}((\mathscr{F} \circ \mathrm{ev}_{t=0})(\iota(0,B)\cdot\psi))(\xi) &= \frac{1}{(2\pi)^{nk}}\int_{\mathrm{M}_{n,k}} \psi(x,-B)e^{-i\operatorname{tr}(x^T\xi)}\,dx \\ &= \mathscr{F}(\mathscr{F}^{-1}[e^{-\frac{i}{2}\operatorname{tr}(\cdot^T B\cdot)}(\mathscr{F} \circ \mathrm{ev}_{t=0})(\psi)])(\xi) \\ &= \frac{e^{-\frac{i}{2}\operatorname{tr}(\xi^T B\xi)}}{(2\pi)^{nk}}\int_{\mathrm{M}_{n,k}} \psi(x,0)e^{-i\operatorname{tr}(x^T\xi)}\,dx \\ &= (\iota(0,B)\cdot((\mathscr{F} \circ \mathrm{ev}_{t=0})(\psi)))(\xi).\end{aligned}$$

Finally turn to $r_{\pi/2}$. The relation $\psi(x,t) = \mathscr{F}^{-1}[e^{\frac{i}{2}\operatorname{tr}(\cdot^T t\cdot)}(\mathscr{F} \circ \mathrm{ev}_{t=0})(\psi)](x)$ can be written as (for $\lambda \neq 0$)

$$\begin{aligned}&\psi(x,-\lambda^{-1}I_n) \\ &= \frac{1}{(2\pi)^{nk}}\int_{\mathrm{M}_{n,k}} \widehat{\psi}(\xi,0)e^{-\frac{i}{2\lambda}\operatorname{tr}(\xi^T\xi)}e^{i\operatorname{tr}(x^T\xi)}\,d\xi \\ &= \lim_{\epsilon\to 0^+}\frac{1}{(2\pi)^{nk}}\int_{\mathrm{M}_{n,k}} \widehat{\psi}(\xi,0)e^{(-\epsilon-\frac{i}{2\lambda})\operatorname{tr}(\xi^T\xi)}e^{i\operatorname{tr}(x^T\xi)}\,d\xi \\ &= \lim_{\epsilon\to 0^+}\frac{1}{(2\pi)^{nk}}\int_{\mathrm{M}_{n,k}}\int_{\mathrm{M}_{n,k}} \psi(w,0)e^{-i\operatorname{tr}(\xi^T w)}e^{(-\epsilon-\frac{i}{2\lambda})\operatorname{tr}(\xi^T\xi)}e^{i\operatorname{tr}(x^T\xi)}\,dw\,d\xi \\ &= \lim_{\epsilon\to 0^+}\frac{1}{(2\pi)^{nk}}\int_{\mathrm{M}_{n,k}} \psi(w,0)\int_{\mathrm{M}_{n,k}} e^{(-\epsilon-\frac{i}{2\lambda})\operatorname{tr}(\xi^T\xi)}e^{i\operatorname{tr}((x-w)^T\xi)}\,d\xi dw \\ &= \lim_{\epsilon\to 0^+}\frac{\pi^{\frac{nk}{2}}}{(2\pi)^{nk}(\epsilon+\frac{i}{2\lambda})^{\frac{nk}{2}}}\int_{\mathrm{M}_{n,k}} \psi(w,0)e^{\frac{1}{4(-\epsilon-\frac{i}{2\lambda})}\operatorname{tr}((x-w)^T(x-w))}\,dw \\ &= (-\frac{i\lambda}{2\pi})^{\frac{nk}{2}}\int_{\mathrm{M}_{n,k}} \psi(w,0)e^{\frac{i\lambda}{2}\operatorname{tr}((x-w)^T(x-w))}\,dw.\end{aligned}$$

Thus

$$\begin{aligned}&(r_{\pi/2}\cdot\psi)(x,\lambda I_n) \\ &= \det{}^{-k/2}(-\lambda I_n)^{-\frac{k}{2}}e^{-\frac{i}{2\lambda}\operatorname{tr}(x^T x)}\psi(-\lambda^{-1}x,-\lambda^{-1}I_n) \\ &= (-\frac{i\lambda}{2\pi})^{\frac{nk}{2}}\det{}^{-k/2}(-\lambda I_n) \\ &\qquad \times\int_{\mathrm{M}_{n,k}} \psi(w,0)e^{-\frac{i}{2\lambda}\operatorname{tr}(x^T x)}e^{\frac{i\lambda}{2}\operatorname{tr}((\lambda^{-1}x+w)^T(\lambda^{-1}x+w))}\,dw \\ &= (\frac{i}{2\pi})^{\frac{nk}{2}}\int_{\mathrm{M}_{n,k}} \psi(w,0)e^{\frac{i}{2}(\lambda\operatorname{tr}(w^T w)+2\operatorname{tr}(x^T w))}\,dw\end{aligned}$$

so

$$(r_{\pi/2}\cdot\psi)(x,0) = (\frac{i}{2\pi})^{\frac{nk}{2}}(\mathscr{F}^{-1}\psi)(x,0)$$

and

$$((\mathscr{F} \circ \mathrm{ev}_{t=0})(r_{\pi/2} \cdot \psi))(\xi) = (\frac{i}{2\pi})^{\frac{nk}{2}} \psi(\xi, 0).$$

This will finish the proof since

$$\begin{aligned}(r_{\pi/2} \cdot ((\mathscr{F} \circ \mathrm{ev}_{t=0})(\psi)))(\xi) &= (\frac{i}{2\pi})^{\frac{nk}{2}} \int_{\mathrm{M}_{n,k}} ((\mathscr{F} \circ \mathrm{ev}_{t=0})(\psi))(x) e^{i\,\mathrm{tr}(\xi^T x)}\, dx \\ &= (\frac{i}{2\pi})^{\frac{nk}{2}} \psi(\xi, 0).\end{aligned}$$

□

For $\lambda \in \Sigma \subseteq \widehat{\mathrm{O}}(k)$, it follows that $\mathcal{H}(\mathrm{M}_{n,k} \times \mathrm{Sym}_n, V_\lambda)_{\mathrm{pre}}$ completes to a irreducible unitary highest weight representation of $G$.

## 6. The Commutative Diagram

**6.1.** Recall from Subsection 5.5 that $\mathfrak{H}(\mathrm{M}_{n,k}, V_\lambda)$ is the space of all $V_\lambda$-valued polynomial functions $f : \mathrm{M}_{n,k} \to V_\lambda$ satisfying $\Delta f = 0$ and $f(xh) = \lambda^{-1}(h) f(x)$ for all $x \in \mathrm{M}_{n,k}$, $h \in \mathrm{O}(k)$. By the structure of the differential operators formed out of rows and since the ring of $\mathrm{O}(k)$-invariant constant coefficient differential operators on $\mathrm{M}_{n,k}$ is generated by the $\Delta_{ij}$'s, $\mathfrak{H}(\mathrm{M}_{n,k}, V_\lambda)$ has the structure of a $\mathrm{GL}(n,\mathbb{C}) \times \mathrm{O}(k)$ modules under the action $((g,h) \cdot f)(x) = f(g^{-1}xh)$ for $g \in \mathrm{GL}(n,\mathbb{C})$, $h \in \mathrm{O}(k)$. Note that here we extend $\tau$ holomorphically to $\mathrm{GL}(n,\mathbb{C})$ by identifying complex valued polynomial functions on $\mathrm{M}_{n,k}$ with complex valued polynomial functions on $\mathrm{M}_{n,k}(\mathbb{C})$.

**6.2.** By [**8**], $\mathfrak{H}(\mathrm{M}_{n,k}, V_\lambda)$, $\lambda \in \Sigma$, is a finite dimensional, irreducible representation of $\mathrm{GL}(n,\mathbb{C})$. Write $(\tau, W_\tau)$ for this $\mathrm{GL}(n,\mathbb{C})$-representation (suppressing the dependence of $\tau$ on $\lambda$) and, following [**8**], define a map $I_\lambda : \mathrm{M}_{n,k} \to \mathrm{Hom}(W_\tau, V_\lambda)$ by

$$(I_\lambda(x))f := f(x),$$

where $x \in \mathrm{M}_{n,k}$, $f \in W_\tau = \mathfrak{H}(\mathrm{M}_{n,k}, V_\lambda)$.

**6.3.** Fix an inner product $(\cdot,\cdot)_{W_\tau}$ on $W_\tau$ so that the adjoint of the $\mathrm{GL}(n,\mathbb{C})$-action is given by the conjugate transpose and fix an inner product $(\cdot,\cdot)_{V_\lambda}$ on $V_\lambda$ that is $\mathrm{O}(k)$-invariant. Given $I_\lambda(x) \in \mathrm{Hom}(W_\tau, V_\lambda)$, we define $I_\lambda^*(x) \in \mathrm{Hom}(V_\lambda, W_\tau)$ by the condition

$$(I_\lambda(x)w, v)_{V_\lambda} = (w, I_\lambda^*(x)v)_{W_\tau}$$

for any $w \in W_\tau$, $v \in V_\lambda$. This gives a map

$$I_\lambda^* : \mathrm{M}_{n,k} \to \mathrm{Hom}(V_\lambda, W_\tau)$$

by $x \mapsto I_\lambda^*(x)$.

**6.4.** Holomorphically extend the character $\det^{-k/2}$ of $MA$ to a character of $(MA)_\mathbb{C}$, isomorphic to the double cover $\widetilde{\mathrm{GL}}(n,\mathbb{C})$. Writing

$$\mathbf{c} = \frac{1}{\sqrt{2}} \begin{pmatrix} I_n & iI_n \\ iI_n & I_n \end{pmatrix}$$

for the Cayley transform in $\mathrm{Sp}(n,\mathbb{C})$, pull back and restrict the character $\det^{-k/2}$ of $(MA)_\mathbb{C}$ to a character of the usual maximal compact subgroup $K$ of $G$, isomorphic to a double cover $\widetilde{\mathrm{U}}(n)$, by defining $\det^{-k/2} : K \to \mathbb{C}$ as $\det^{-k/2}(k) := \det^{-k/2}(\mathbf{c}^{-1}k\mathbf{c})$. Similarly using the Cayley transform, pull back the representation of $\mathrm{GL}(n, \mathbb{C})$ on $W_\tau$ to a representation of the maximal compact subgroup of $\mathrm{Sp}(n, \mathbb{R})$, isomorphic to $\mathrm{U}(n)$, which may then be lifted to a nongenuine representation of $K$ via the projection map. In this way, we may view $W_\tau \otimes \det^{-k/2}$ as a representation of $K$. Write $\mathcal{O}(W_\tau \otimes \det^{-k/2})$ for the space of holomorphic sections of $G \times_K (W_\tau \otimes \det^{-k/2})$ with the usual $G$-action and write $\mathcal{O}(\Omega_n, W_\tau)$ for the space of holomorphic $W_\tau$-valued functions on $\Omega_n$.

**6.5.** Kashiwara and Vergne showed that $\mathcal{O}(W_\tau \otimes \det^{-k/2})$ is equivalent to $\mathcal{O}(\Omega_n, W_\tau)$ as a $G$-module ([**8**, Prop. 3.3]), where $(g, \varepsilon) \in G$ acts on $f \in \mathcal{O}(\Omega_n, W_\tau)$ by

$$(g \cdot f)(z) = \det{}_1^{-k/2}(a - zc)\,\tau(a - zc) \cdot f((a - zc)^{-1}(zd - b))$$

where we view $a - zc \in \mathrm{GL}(n, \mathbb{C})$ (for use in $\tau$) and as shorthand for $l_{a-zc,\varepsilon'(z)}$ (for use in $\det^{-1/2}$) where $(g, \varepsilon)^{-1} = (g^{-1}, \varepsilon')$. They also showed that the map $\mathrm{KV}_\lambda : \mathcal{L}^2(\mathrm{M}_{n,k}, V_\lambda) \to \mathcal{O}(\Omega_n, W_\tau)$, given by

$$(\mathrm{KV}_\lambda f)(z) := \int_{\mathrm{M}_{n,k}} e^{\frac{i}{2}\operatorname{tr}(x^T zx)} I_\lambda^*(x) f(x)\, dx$$

is $G$-intertwining and injective.

**6.6.** Notice that if elements of $\mathrm{Hom}(W_\tau, V_\lambda)$ and $\mathrm{Hom}(V_\lambda, W_\tau)$ are identified with matrices by a choice of bases for $W_\tau$ and $V_\lambda$, then, by construction, both $I_\lambda(x)$ and $I_\lambda^*(x)$ have polynomial entries in $\mathfrak{H}(\mathrm{M}_{n,k})$, the space of $\mathbb{C}$-valued harmonic polynomials on $\mathrm{M}_{n,k}$. Therefore, we may define

$$\nabla_\lambda : \mathcal{C}^\infty(\mathrm{M}_{n,k} \times \mathrm{Sym}_n, V_\lambda) \to \mathcal{C}^\infty(\mathrm{M}_{n,k} \times \mathrm{Sym}_n, W_\tau)$$

by

$$\nabla_\lambda := I_\lambda^*(-i\partial_x).$$

Here we again recall that we identify complex valued polynomial functions on $\mathrm{M}_{n,k}$ with complex valued polynomial functions on $\mathrm{M}_{n,k}(\mathbb{C})$. Thus

$$\nabla_\lambda|_{x=0}\, e^{i\operatorname{tr}(x\xi^T)}\psi(\xi) = I_\lambda^*(\xi)\psi(\xi).$$

**6.7.** Lift $\tau$ by the projection map to a representation of $MA$ by $\tau(l_{A,c}) := \tau(A)$ and then extend the action trivially to $\overline{N}$. We define

$$\mathcal{E}(\mathrm{Sym}_n, W_\tau \otimes \det^{-k/2}) := \{\psi \in \mathcal{C}^\infty(\mathrm{Sym}_n, W_\tau) \mid \exists \phi \in \mathrm{Ind}_{MA\overline{N}}^G(\tau \otimes \chi) \text{ with } \psi = \iota^*\phi\}$$

where here $\iota(t) := (\left(\begin{smallmatrix} I_n & t \\ 0 & I_n \end{smallmatrix}\right), Z \mapsto 1)$ and we identify $W_\tau \otimes \det^{-k/2}$ with $W_\tau$ as vector spaces. It is easy to see from the proof of Theorem 3.1 that the action of $(g, \varepsilon) \in G$ on $\psi \in \mathcal{E}(\mathrm{Sym}_n, W_\tau \otimes \det^{-k/2})$ is given by

$$((g, \varepsilon) \cdot \psi)(t) = \det^{-k/2}(A - tC)\,\tau(a - tc) \cdot \psi((A - tC)^{-1}(tD - B))$$

when $\det(A - tC) \neq 0$.

**6.8.** We will see below that $\nabla_\lambda|_{x=0} : \mathcal{H}(\mathrm{M}_{n,k} \times \mathrm{Sym}_n, V_\lambda)_{\mathrm{pre}} \hookrightarrow \mathcal{E}(\mathrm{Sym}_n, W_\tau \otimes \det^{-k/2})$. We also define $\mathrm{BV} : \mathcal{O}(\Omega_n, W_\tau) \to \mathcal{E}(\mathrm{Sym}_n, W_\tau \otimes \det^{-k/2})$ by $(\mathrm{BV}\, f)(t) = \lim_{s\to 0^+} f(t+is)$, $(t, s \in \mathrm{Sym}_n,\ s >> 0)$. This will certainly be well-defined on the image of $\mathcal{S}(\mathrm{M}_{n,k}, V_\lambda)$ under $\mathrm{KV}_\lambda$ which will suffice for our purposes.

THEOREM 6.1. *For $\lambda \in \Sigma \subseteq \widehat{\mathrm{O}}(k)$, there is a commutative diagram of $G$-maps*

$$\begin{array}{ccc}
\mathcal{H}(\mathrm{M}_{n,k} \times \mathrm{Sym}_n, V_\lambda)_{\mathrm{pre}} & \xrightarrow{\ \nabla_\lambda|_{x=0}\ } & \mathcal{E}(\mathrm{Sym}_n, W_\tau \otimes \det^{-k/2}) \\
\Big\downarrow {\scriptstyle \mathscr{F} \circ \mathrm{ev}_{t=0}} & & \Big\uparrow {\scriptstyle \mathrm{BV}} \\
\mathcal{L}^2(\mathrm{M}_{n,k}, V_\lambda) & \xrightarrow[\mathrm{KV}_\lambda]{} & \mathcal{O}(\Omega_n, W_\tau)
\end{array}$$

*where $\mathscr{F} \circ \mathrm{ev}_{t=0}$ is an isomorphism on $K$-finite vectors and $\mathrm{KV}_\lambda$ and $\nabla_\lambda|_{x=0}$ are injective.*

PROOF. We have already seen that $\mathscr{F} \circ \mathrm{ev}_{t=0}$ is $G$-intertwining in Theorem 5.2 and injective by Equation 5.3. It is shown in [**8**] that $\mathrm{KV}_{\lambda,\sigma}$ is injective and $G$-intertwining and that BV is injective. Since it is obvious that BV is also a $G$-map, it only remains to show that the diagram is commutative. For this we simply compute. Let $\psi \in \mathcal{H}(\mathrm{M}_{n,k} \times \mathrm{Sym}_n, V_\lambda)_{\mathrm{pre}}$. Then

$$((\mathrm{KV}_\lambda \circ \mathscr{F} \circ \mathrm{ev}_{t=0})\psi)(z) = \frac{1}{(2\pi)^{nk}} \int_{\mathrm{M}_{n,k}} e^{\frac{i}{2}\operatorname{tr}(\xi^T z \xi)} I_\lambda^*(\xi) \widehat{\psi}(\xi, 0)\, d\xi$$

so

$$((\mathrm{BV} \circ \mathrm{KV}_\lambda \circ \mathscr{F} \circ \mathrm{ev}_{t=0})\psi)(t) = \frac{1}{(2\pi)^{nk}} \int_{\mathrm{M}_{n,k}} e^{\frac{i}{2}\operatorname{tr}(\xi^T t \xi)} I_\lambda^*(\xi) \widehat{\psi}(\xi, 0)\, d\xi.$$

On the other hand, by Equation 5.3,

$$\psi(x,t) = \frac{1}{(2\pi)^{nk}} \int_{\mathrm{M}_{n,k}} \widehat{\psi}(\xi, 0) e^{\frac{i}{2}\operatorname{tr}(\xi^T t \xi)} e^{i \operatorname{tr}(x\xi^T)}\, d\xi$$

so the proof is finished by using Equation 6.6 to calculate

$$\begin{aligned}
(\nabla_\lambda|_{x=0}\psi)(t) &= \frac{1}{(2\pi)^{nk}} \int_{\mathrm{M}_{n,k}} e^{\frac{i}{2}\operatorname{tr}(\xi^T t \xi)} I_\lambda^*(-i\partial_x)|_{x=0} e^{i\operatorname{tr}(x\xi^T)} \widehat{\psi}(\xi, 0)\, d\xi \\
&= \frac{1}{(2\pi)^{nk}} \int_{\mathrm{M}_{n,k}} e^{\frac{i}{2}\operatorname{tr}(\xi^T t \xi)} I_\lambda^*(\xi) \widehat{\psi}(\xi, 0)\, d\xi.
\end{aligned}$$

□

**6.9.** It is shown in [**8**] that the highest weight vector in $\mathcal{L}^2(\mathrm{M}_{n,k}, V_\lambda)$ is the function

$$e^{-\frac{1}{2}\operatorname{tr}(x^T x)} f_\tau(x)$$

where $f_\tau \in W_\tau$ is the highest weight vector of $W_\tau$. It is also shown in [**8**] that the image of $e^{-\frac{1}{2}\operatorname{tr}(x^T x)} f_\tau(x)$ in $\mathcal{O}(\Omega_n, W_\tau)$ under $\mathrm{KV}_\lambda$ is the function

$$\det(I_n - iz)^{-\frac{k}{2}}\, \tau(I_n - iz) \cdot f_\tau$$

where $I_n - iz \in \mathrm{GL}(n, \mathbb{C})$ acts via $\tau$ on $f_\tau \in W_\tau$. It is obvious that the image of this function in $\mathcal{E}(\mathrm{Sym}_n, W_\tau \otimes \det^{-k/2})$ under BV is the function

$$\det(I_n - it)^{-\frac{k}{2}}\, \tau(I_n - it) \cdot f_\tau.$$

**6.10.** Finally, for $f \in W_\tau$, use [8] 4.5 to calculate that

$$\begin{aligned}((\mathscr{F} \circ \mathrm{ev}_{t=0})(\Psi_f))(\xi) &= \frac{1}{(2\pi)^{nk}} \widehat{\Psi}_f(\xi, 0) \\ &= \frac{1}{(2\pi)^{nk}} \int_{\mathrm{M}_{n,k}} f(x) e^{-\frac{1}{2}\operatorname{tr}(x^T x)} e^{-i\operatorname{tr}(x^T \xi)}\, dx \\ &= (2\pi)^{-\frac{nk}{2}} e^{-\frac{1}{2}\operatorname{tr}(\xi^T \xi)} f(-i\xi).\end{aligned}$$

Since elements of $W_\tau$ are homogeneous, it follows that, up to a scalar, the function in $\mathcal{H}(\mathrm{M}_{n,k} \times \mathrm{Sym}_n, V_\lambda)_{\mathrm{pre}}$ corresponding to $e^{-\frac{1}{2}\operatorname{tr}(x^T x)} f_\tau(x)$ is

$$\det(I_n - it)^{-k/2} e^{-\frac{1}{2}\operatorname{tr}(x^T (I_n - it)^{-1} x)} f_\tau((I_n - it)^{-1} x).$$

## 7. Lowest Weights

**7.1.** We remark that if one instead begins with the character $\overline{\chi} : \overline{P} \to \mathbb{C}$ given by

$$\overline{\chi}(ma\overline{n}, (\binom{0}{w}, s)) := \overline{\det^{-k/2}(ma)}\, e^{-\frac{is}{2}},$$

constructs $\operatorname{Ind}_{\overline{P}}^{G \ltimes H} \overline{\chi}$ and its noncompact picture, $\mathcal{E}_{\overline{\chi}}(\mathrm{M}_{n,k} \times \mathrm{Sym}_n)$, as usual, and then looks at $\mathcal{H}_{\overline{\chi}}(\mathrm{M}_{n,k} \times \mathrm{Sym}_n)_{\mathrm{pre}}$ defined as the space of all solutions to

$$i\partial_t \psi(x,t) = -\Delta \psi(x,t)$$

with $\psi \in \mathcal{E}_{\overline{\chi}}(\mathrm{M}_{n,k} \times \mathrm{Sym}_n)$ such that the initial condition $\psi|_{t=0} \in \mathscr{S}(\mathrm{M}_{n,k})$, it is possible to show $\mathcal{H}_{\overline{\chi}}(\mathrm{M}_{n,k} \times \mathrm{Sym}_n)_{\mathrm{pre}}$ is a pre-Hilbert space and is invariant under $(G \times \mathrm{O}(k)) \ltimes H$ that is dual to $\mathcal{H}(\mathrm{M}_{n,k} \times \mathrm{Sym}_n)_{\mathrm{pre}}$. By extending to antiholomorphic functions, there is an analogous result to Theorem 6.1 except with lowest weight representations.

## References


[1] Rolf Berndt, *The heat equation and representations of the Jacobi group*, The ubiquitous heat kernel, Contemp. Math., vol. 398, Amer. Math. Soc., Providence, RI, 2006, pp. 47–68. MR2218013

[2] Rolf Berndt and Ralf Schmidt, *Elements of the representation theory of the Jacobi group*, Modern Birkhäuser Classics, Birkhäuser/Springer Basel AG, Basel, 1998. [2011 reprint of the 1998 original] [MR1634977]. MR3013729

[3] T. J. Enright and R. Parthasarathy, *A proof of a conjecture of Kashiwara and Vergne*, Noncommutative harmonic analysis and Lie groups (Marseille, 1980), Lecture Notes in Math., vol. 880, Springer, Berlin-New York, 1981, pp. 74–90. MR644829

[4] Thomas J. Enright and Nolan R. Wallach, *Embeddings of unitary highest weight representations and generalized Dirac operators*, Math. Ann. **307** (1997), no. 4, 627–646. MR1464134

[5] Markus Hunziker, Mark R. Sepanski, and Ronald J. Stanke, *A system of Schrödinger equations and the oscillator representation*, Electron. J. Differential Equations (2015), No. 260, 28. MR3414114

[6] Markus Hunziker, Mark R. Sepanski, and Ronald J. Stanke, *Global Lie symmetries of a system of Schrödinger equations and the oscillator representation*, Miskolc Math. Notes **14** (2013), no. 2, 647–657. MR3144103

[7] Markus Hunziker, Mark R. Sepanski, and Ronald J. Stanke, *The minimal representation of the conformal group and classical solutions to the wave equation*, J. Lie Theory **22** (2012), no. 2, 301–360. MR2976923

[8] M. Kashiwara and M. Vergne, *On the Segal-Shale-Weil representations and harmonic polynomials*, Invent. Math. **44** (1978), no. 1, 1–47. MR0463359

[9] Mark R. Sepanski and Ronald J. Stanke, *On global* SL(2, ℝ) *symmetries of differential operators*, J. Funct. Anal. **224** (2005), no. 1, 1–21. MR2139102

[10] Mark R. Sepanski and Ronald J. Stanke, *Global Lie symmetries of the heat and Schrödinger equation*, J. Lie Theory **20** (2010), no. 3, 543–580. MR2743104

Department of Mathematics, Baylor University, Waco, Texas 76798
*Email address*: markus_hunziker@baylor.edu

Department of Mathematics, Baylor University, Waco, Texas 76798
*Email address*: mark_sepanski@baylor.edu

Department of Mathematics, Baylor University, Waco, Texas 76798
*Email address*: ronald_stanke@baylor.edu

Contemporary Mathematics
Volume **714**, 2018
https://doi.org/10.1090/conm/714/14380

# Residue formula for regular symmetry breaking operators

Toshiyuki Kobayashi

*To Gestur Ólafsson on his 65th birthday.*

ABSTRACT. We prove an explicit residue formula for a meromorphic continuation of conformally covariant integral operators between differential forms on $\mathbb{R}^n$ and on its hyperplane. The results provide a simple and new construction of the conformally covariant differential symmetry breaking operators between differential forms on the sphere and those on its totally geodesic hypersurface that were introduced in [Kobayashi–Kubo–Pevzner, Lect. Notes Math. (2016)]. Moreover, we determine the zeros of the matrix-valued regular symmetry breaking operators between principal series representations of $O(n+1,1)$ and $O(n,1)$.

CONTENTS



## 1. Statement of the main results

Let $\mathcal{E}^i(\mathbb{R}^n)$ be the space of (complex-valued) differential $i$-forms on $\mathbb{R}^n$, and $\mathcal{E}^i_c(\mathbb{R}^n)$ the subspace of those having compact support. The object of study in this article is a meromorphic family of operators

$$\mathbb{A}^{i,j}_{\lambda,\nu,\pm} : \mathcal{E}^i_c(\mathbb{R}^n) \to \mathcal{E}^j(\mathbb{R}^{n-1}),$$

2010 *Mathematics Subject Classification.* Primary 22E30, 45P05; Secondary 14F10, 22E46, 26A33, 53A30.

*Key words and phrases.* Riesz potential, conformal geometry, branching law, symmetry breaking operator.

The author was partially supported by Grant-in-Aid for Scientific Research (A) (25247006), Japan Society for the Promotion of Science.

which are initially defined as integral operators when $\mathrm{Re}\lambda \gg |\mathrm{Re}\nu|$.

The operators $\mathbb{A}^{i,j}_{\lambda,\nu,\pm}$ arise as

- matrix-valued regular *symmetry breaking operators* for principal series representations for strong Gelfand pair $(O(n+1,1), O(n,1))$;
- conformally covariant operators on differential forms on the model space $(S^n, S^{n-1})$

among others.

By choosing appropriate Gamma factors $a_\pm(\lambda,\nu)$ (see (1.5)), we renormalize $\mathbb{A}^{i,j}_{\lambda,\nu,\pm}$ by

$$\widetilde{\mathbb{A}}^{i,j}_{\lambda,\nu,\pm} := a_\pm(\lambda,\nu)\mathbb{A}^{i,j}_{\lambda,\nu,\pm}$$

so that $\widetilde{\mathbb{A}}^{i,j}_{\lambda,\nu,\pm}$ depend holomorphically on $(\lambda,\nu)$ in the entire plane $\mathbb{C}^2$, see Fact 1.1 below.

The goal of this paper is in twofold:

- to find the residue formula of the matrix-valued operators $\mathbb{A}^{i,j}_{\lambda,\nu,\pm}$ along $\nu-\lambda \in \mathbb{N}$ (see Theorem 1.3);
- to determine all the (isolated) zeros of the normalized operators $\widetilde{\mathbb{A}}^{i,j}_{\lambda,\nu,\pm}$ (see Theorem 8.1 and Remark 8.3).

**1.1. Integral operators $\mathbb{A}^{i,j}_{\lambda,\nu,\pm}\colon \mathcal{E}^i_c(\mathbb{R}^n) \to \mathcal{E}^j(\mathbb{R}^{n-1})$.** To state our main results, we fix some notation.

Let $|x| := (x_1^2+\cdots+x_n^2)^{\frac{1}{2}}$ for $x \in \mathbb{R}^n$. We define

$$\psi_n\colon \mathbb{R}^n \setminus \{0\} \to O(n), \quad x \mapsto I_n - \frac{2x\,{}^t x}{|x|^2}, \tag{1.1}$$

as a matrix expression of the reflection with respect to normal vector $x$. Let $\sigma^{(i)}\colon O(n) \to GL_{\mathbb{C}}(\bigwedge^i(\mathbb{C}^n))$ be the $i$th exterior representation of the natural representation of $O(n)$ on $\mathbb{C}^n$.

We identify $\mathcal{E}^i(\mathbb{R}^n)$ with $C^\infty(\mathbb{R}^n)\otimes \bigwedge^i(\mathbb{C}^n)$, and similarly, $\mathcal{E}^j(\mathbb{R}^{n-1})$ with $C^\infty(\mathbb{R}^{n-1})\otimes \bigwedge^j(\mathbb{C}^{n-1})$. We define $\mathrm{pr}_{i\to j}\colon \bigwedge^i(\mathbb{C}^n) \to \bigwedge^j(\mathbb{C}^{n-1})$ for $j=i-1,i$ to be the first and second projections of the decomposition $\bigwedge^i(\mathbb{C}^n) = \bigwedge^{i-1}(\mathbb{C}^{n-1}) \oplus \bigwedge^i(\mathbb{C}^{n-1})$, respectively, so that the following linear map

$$\mathrm{Rest}_{x_n=0}\circ(\mathrm{id}\oplus \iota_{\frac{\partial}{\partial x_n}})\colon \mathcal{E}^i(\mathbb{R}^n) \to \mathcal{E}^i(\mathbb{R}^{n-1})\oplus\mathcal{E}^{i-1}(\mathbb{R}^{n-1}) \tag{1.2}$$

is identified with $\mathrm{id}\otimes(\mathrm{pr}_{i\to i}\oplus \mathrm{pr}_{i\to i-1})$. Here $\iota_{\frac{\partial}{\partial x_n}}$ denotes the inner multiplication of the vector field $\frac{\partial}{\partial x_n}$.

For $j\in\{i-1,i\}$ and $\mathrm{Re}\lambda \gg |\mathrm{Re}\nu|$, we define $\mathrm{Hom}_{\mathbb{C}}(\bigwedge^i(\mathbb{C}^n), \bigwedge^j(\mathbb{C}^{n-1}))$-valued, locally integrable functions $\mathcal{A}^{i,j}_{\lambda,\nu,\pm}$ on $\mathbb{R}^n$ by

$$\mathcal{A}^{i,j}_{\lambda,\nu,+}(x) := (x_1^2+\cdots+x_n^2)^{-\nu}|x_n|^{\lambda+\nu-n}\,\mathrm{pr}_{i\to j}\circ\sigma^{(i)}(\psi_n(x)), \tag{1.3}$$

$$\mathcal{A}^{i,j}_{\lambda,\nu,-}(x) := \mathcal{A}^{i,j}_{\lambda,\nu,+}(x)\mathrm{sgn}(x_n). \tag{1.4}$$

We introduce a normalizing factor $a_\pm(\lambda,\nu)$ by

$$a_{\varepsilon(\kappa)}(\lambda,\nu)^{-1} := \Gamma(\frac{\lambda+\nu-n+1+\kappa}{2})\Gamma(\frac{\lambda-\nu+\kappa}{2}) \tag{1.5}$$

for $\kappa=0,1$, where we set

$$\varepsilon\colon \{0,1\}\to\{\pm\}$$

by $\varepsilon(0) = +$ and $\varepsilon(1) = -$. Then $a_\pm(\lambda,\nu)$ are holomorphic functions of $(\lambda,\nu)$ in $\mathbb{C}^2$. We set

$$\widetilde{\mathcal{A}}^{i,j}_{\lambda,\nu,\pm} := a_\pm(\lambda,\nu)\mathcal{A}^{i,j}_{\lambda,\nu,\pm}. \tag{1.6}$$

FACT 1.1 (see [**19**, **21**]). Let $j = i-1$ or $i$. Then the integral operators

$$\widetilde{\mathbb{A}}^{i,j}_{\lambda,\nu,\pm} : \mathcal{E}^i_c(\mathbb{R}^n) \to \mathcal{E}^j(\mathbb{R}^{n-1}), f \mapsto \mathrm{Rest}_{x_n=0} \circ (\widetilde{\mathcal{A}}^{i,j}_{\lambda,\nu,\pm} * f),$$

originally defined when $\mathrm{Re}\lambda \gg |\mathrm{Re}\nu|$, extend to continuous operators that depend holomorphically on $(\lambda,\nu)$ in the entire complex plane $\mathbb{C}^2$. Moreover, $\{(\lambda,\nu) \in \mathbb{C}^2 : \widetilde{\mathbb{A}}^{i,j}_{\lambda,\nu,\pm} = 0\}$ is a discrete subset of $\mathbb{C}^2$.

REMARK 1.2. A key point of (1.6) is that the Gamma factors $a_\pm(\lambda,\nu)$ are chosen in an optimal way that there is no pole of $\widetilde{\mathbb{A}}^{i,j}_{\lambda,\nu,\pm}$ and that the zeros are of codimension two in $\mathbb{C}^2$. We note that $a_+(\lambda,\nu)$ coincides with normalizing factor of the scalar-valued regular symmetry breaking operator $\mathbb{A}_{\lambda,\nu}$ in [**19**, (7.8)] when $i = j = 0$ and $\kappa = 0$.

**1.2. Residue formula of matrix-valued operators $\mathbb{A}^{i,j}_{\lambda,\nu,\pm}$.** The first factor $\Gamma(\frac{\lambda+\nu-n+1+\kappa}{2})$ of $a_{\varepsilon(\kappa)}(\lambda,\nu)$ in (1.5) arises from the normalization of the distribution $|x_n|^{\lambda+\nu-n}(\mathrm{sgn}\, x_n)^\kappa$ of one-variable, and the corresponding residue of $\mathbb{A}^{i,j}_{\lambda,\nu,\pm}$ is easily obtained. On the other hand, the second factor $\Gamma(\frac{\lambda-\nu+\kappa}{2})$ is more involved because it arises not only from the normalization of $(x_1^2 + \cdots + x_n^2)^{-\nu}$ but from the whole $\mathcal{A}^{i,j}_{\lambda,\nu,\pm}$. The main result of this article is to give a closed formula for the residues of the operators $\mathbb{A}^{i,j}_{\lambda,\nu,\pm}$ at the places of the poles of $\Gamma(\frac{\lambda-\nu+\kappa}{2})$ as follows:

THEOREM 1.3 (residue formula of $\mathbb{A}^{i,j}_{\lambda,\nu,\pm}$). *Let $j = i-1$ or $i$, and $\kappa \in \{0,1\}$. Suppose $\nu - \lambda = 2m + \kappa$ with $m \in \mathbb{N}$. Then we have*

$$\widetilde{\mathbb{A}}^{i,j}_{\lambda,\nu,\varepsilon(\kappa)} = \frac{(-1)^{i-j+m+\kappa}\pi^{\frac{n-1}{2}} m!}{2^{2m-1+3\kappa}\Gamma(\nu+1)} \mathbb{C}^{i,j}_{\lambda,\nu}. \tag{1.7}$$

Here $\mathbb{C}^{i,j}_{\lambda,\nu} : \mathcal{E}^i(\mathbb{R}^n) \to \mathcal{E}^j(\mathbb{R}^{n-1})$ is a matrix-valued differential operator which we introduced in [**12**, **13**]. See [**18**, Def. 2.1] for instance, for the definition of differential operators between two manifolds. To review the differential operator $\mathbb{C}^{i,j}_{\lambda,\nu}$, we begin with a scalar-valued differential operator $\widetilde{\mathbb{C}}_{\lambda,\nu}$ which we call *Juhl's operator* from [**6**]. For $l := \nu - \lambda \in \mathbb{N}$, we set $m := [\frac{l}{2}]$, the largest integer that does not exceed $\frac{l}{2}$, and define $\widetilde{\mathbb{C}}_{\lambda,\nu} : C^\infty(\mathbb{R}^n) \to C^\infty(\mathbb{R}^{n-1})$ by

$$\widetilde{\mathbb{C}}_{\lambda,\nu} := \mathrm{Rest}_{x_n=0} \circ \sum_{k=0}^{m} \frac{\prod_{j=1}^{m-k+1} (\nu - \frac{n-1}{2} - m + j)}{2^{2k-l} k!(l-2k)!} (\Delta_{\mathbb{R}^{n-1}})^k (\frac{\partial}{\partial x_n})^{l-2k}. \tag{1.8}$$

More generally, we introduced in [**12**] the following matrix-valued differential operator $\mathbb{C}^{i,j}_{\lambda,\nu} : \mathcal{E}^i(\mathbb{R}^n) \to \mathcal{E}^j(\mathbb{R}^{n-1})$ by the formula (see also [**13**, (2.24) and (2.26)]):

$$\mathbb{C}^{i,i}_{\lambda,\nu} := \widetilde{\mathbb{C}}_{\lambda+1,\nu-1} dd^* - \gamma(\lambda - \frac{n}{2}, \nu - \lambda)\widetilde{\mathbb{C}}_{\lambda,\nu-1} d\iota_{\frac{\partial}{\partial x_n}} + \frac{\nu - i}{2}\widetilde{\mathbb{C}}_{\lambda,\nu},$$

$$\mathbb{C}^{i,i-1}_{\lambda,\nu} := -\widetilde{\mathbb{C}}_{\lambda+1,\nu-1} dd^* \iota_{\frac{\partial}{\partial x_n}} - \gamma(\lambda - \frac{n-1}{2}, \nu - \lambda)\widetilde{\mathbb{C}}_{\lambda+1,\nu} d^* + \frac{\lambda + i - n}{2}\widetilde{\mathbb{C}}_{\lambda,\nu}\iota_{\frac{\partial}{\partial x_n}},$$

where the codifferential $d^*: \mathcal{E}^{j+1}(\mathbb{R}^n) \to \mathcal{E}^j(\mathbb{R}^n)$ is the formal adjoint of the exterior derivative $d: \mathcal{E}^j(\mathbb{R}^n) \to \mathcal{E}^{j+1}(\mathbb{R}^n)$. The constant $\gamma(\mu, a) \in \mathbb{C}$ is given for $\mu \in \mathbb{C}$ and $a \in \mathbb{N}$ by

$$\gamma(\mu, a) := \begin{cases} 1 & \text{if } a \text{ is odd,} \\ \mu + \frac{a}{2} & \text{if } a \text{ is even.} \end{cases} \tag{1.9}$$

REMARK 1.4. (1) Theorem 1.3 gives a method to find an explicit formula of the conformally covariant differential symmetry breaking operators $\mathbb{C}^{i,j}_{\lambda,\nu}$ for $j = i-1, i$, for which we gave a different proof in [**13**] based on the F-method [**8**]. In fact Theorem 1.3 was obtained by 2015, however, we did not adopt this method in the article [**13**] and have postponed its publication to this article, because the residue method gives only a "construction" of differential symmetry breaking operators, whereas the F-method gives both "construction" and "exhaustion".
(2) In contrast, applications of Theorem 1.3 include a proof of the "exhaustion" of *nonlocal* symmetry breaking operators with singular parametes, and then contribute to the multiplicity-formula of confomally covariant symmetry breaking operators for differential forms on $(S^n, S^{n-1})$ given in [**20**], see [**21**] for detailed proof.

**Notation.**

$$\mathbb{N} = \{0, 1, 2, \cdots\}, \quad \mathbb{N}_+ = \{1, 2, \cdots\}.$$

**Acknowledgments.** This work was partially supported by Grant-in-Aid for Scientific Research (A) (25247006), Japan Society for the Promotion of Science. The author would like to express his gratitude to the organizers, Jens Christensen, Susanna Dann, and Matthew Dawson, of the conference in honor of Gestur Ólafsson's 65th birthday that was held on January 4, 2017 in Atlanta, USA.

## 2. Symmetry breaking in conformal geometry

We discuss the operators $\widetilde{\mathbb{A}}^{i,j}_{\lambda,\nu,\pm}$ from two different viewpoints: representation theory of real reductive groups (Section 2.1) and conformal geometry (Section 2.2). With these perspectives, we shall explain two applications of Theorem 1.3 in Section 1. Logically, the results of this section are not used for the proof of Theorem 1.3.

**2.1. Symmetry breaking for reductive groups.** In this subsection, we discuss the operators $\widetilde{\mathbb{A}}^{i,j}_{\lambda,\nu,\pm}$ from the viewpoint of representation theory of real reductive Lie groups.

Let $\Pi$ be a continuous representation of a group $G$ on a topological vector space. If $G'$ is a subgroup of $G$, we may think of $\Pi$ as a representation of the subgroup $G'$, which is called the *restriction* of $\Pi$, to be denoted by $\Pi|_{G'}$. Let $\pi$ be another representation of the subgroup $G'$. A *symmetry breaking operator* is a continuous linear map $\Pi \to \pi$ that intertwines the actions of the subgroup $G'$.

The operators $\widetilde{\mathbb{A}}^{i,j}_{\lambda,\nu,\pm}$ arise as symmetry breaking operators for the pair $(G, G') = (O(n+1, 1), O(n, 1))$ as follows. Let $P = MAN$ be a minimal parabolic subgroup of $G$. For $0 \leq i \leq n$, $\delta \in \{0, 1\} \simeq \mathbb{Z}/2\mathbb{Z}$, and $\lambda \in \mathbb{C}$, we take $\Pi$ to be the principal series representation

$$I_\delta(i, \lambda) = \operatorname{Ind}_P^G(\textstyle\bigwedge^i(\mathbb{C}^n) \otimes \operatorname{sgn}^\delta \otimes \mathbb{C}_\lambda),$$

given by (unnormalized) parabolic induction. Here $\bigwedge^i(\mathbb{C}^n)\otimes\mathrm{sgn}^\delta\otimes\mathbb{C}_\lambda$ stands for an irreducible representation of $P$ that extends the outer tensor product representation of $MA \simeq O(n) \times O(1) \times \mathbb{R}$ with trivial $N$-action. Likewise, we take $\pi$ to be the parabolic induction

$$J_\varepsilon(j,\nu) = \mathrm{Ind}_{P'}^{G'}(\textstyle\bigwedge^j(\mathbb{C}^{n-1}) \otimes \mathrm{sgn}^\varepsilon \otimes \mathbb{C}_\nu)$$

from a minimal parabolic subgroup $P'$ of $G'$. By identifying the open Bruhat cell of $G/P$ with $\mathbb{R}^n$ and that of $G'/P'$ with $\mathbb{R}^{n-1}$, we can realize $\Pi$ and $\pi$ in $\mathcal{E}^i(\mathbb{R}^n)$ and $\mathcal{E}^j(\mathbb{R}^{n-1})$ as the "$N$-picture" of principal series representations, respectively. Then the operators $\widetilde{\mathbb{A}}^{i,j}_{\lambda,\nu,\mu}\colon \mathcal{E}^i_c(\mathbb{R}^n) \to \mathcal{E}^j(\mathbb{R}^{n-1})$ in Fact 1.1 gives rise to symmetry breaking operators, to be denoted by the same symbol,

$$\widetilde{\mathbb{A}}^{i,j}_{\lambda,\nu,\mu}\colon I_\delta(i,\lambda) \to J_\varepsilon(j,\nu)$$

for $\mu = +$ $(\delta \equiv \varepsilon \mod 2)$ and $\mu = -$ $(\delta \not\equiv \varepsilon \mod 2)$. If

$$\nu - \lambda \notin 2\mathbb{N} \text{ for } \delta \equiv \varepsilon \quad \text{or} \quad \nu - \lambda \notin 2\mathbb{N}+1 \text{ for } \delta \not\equiv \varepsilon, \tag{2.1}$$

then the support of the distribution kernel $\widetilde{\mathcal{A}}^{i,j}_{\lambda,\nu,\mu}$ has an interior point, and the operator $\widetilde{\mathbb{A}}^{i,j}_{\lambda,\nu,\mu}$ is called a *regular* symmetry breaking operator ([**19**, Def. 3.3]).

The dimension of the space $\mathrm{Hom}_{G'}(I_\delta(i,\lambda)|_{G'}, J_\varepsilon(j,\nu))$ of symmetry breaking operators is uniformly bounded with respect to the parameters $(\lambda,\nu,\delta,\varepsilon)$ by the general theory ([**17**, **21**]). Moreover, it is one-dimensional and spanned by $\widetilde{\mathbb{A}}^{i,j}_{\lambda,\nu,\mu}$ if the generic condition (2.1) on parameters is satisfied, see [**20**].

REMARK 2.1. From the representation theory of reductive groups, an important property of the family $\{I_\delta(i,\lambda)\}$ of the aforementioned representations is that *any* irreducible admissible smooth representation $\varpi$ of $G$ with trivial infinitesimal character can be obtained as an irreducible subquotient of the representation $I_\delta(i,\lambda)$ for some $\delta, i$ and $\lambda$. Furthermore, any irreducible unitary representation with nonzero $(\mathfrak{g},K)$-cohomologies arises as the Hilbert completion of such $\varpi$, and *vice versa*, see [**21**, Thm. 2.16].

The dimension formula of $\mathrm{Hom}_{G'}(I_\delta(i,\lambda)|_{G'}, J_\varepsilon(j,\nu))$ was given in [**20**] where a part of the proof utilized the results of this article, namely, the vanishing criterion of $\widetilde{\mathbb{A}}^{i,j}_{\lambda,\nu,\pm}$ which will be stated in Theorem 8.1. The complete classification of $\mathrm{Hom}_{G'}(I_\delta(i,\lambda)|_{G'}, J_\varepsilon(j,\nu))$ with explicit generators for general $0 \le i \le n$, $0 \le j \le n-1$, $\lambda,\nu \in \mathbb{C}$ and $\delta,\varepsilon \in \{0,1\}$ is accomplished in [**21**], which may be thought of as "Stage C" of the branching problem (see [**10**]). Among these symmetry breaking operators, the classification and construction of differential operators such as $\mathbb{C}^{i,i-1}_{\lambda,\nu}$ and $\mathbb{C}^{i,i}_{\lambda,\nu}$ was established in [**13**] by using the F-method which was initiated and developed in [**8**, **9**, **16**, **18**]. A preprint [**3**] treats a similar problem by using also the F-method (though the formula for $\mathbb{C}^{i,i-1}_{\lambda,\nu}$ and $\mathbb{C}^{i,i}_{\lambda,\nu}$ *loc.cit.* is not of the present form). A generalization to the pseudo-Riemannian case and to any domains is proved in [**14**].

**2.2. Conformally covariant symmetry breaking operators.** In this section, we explain from the viewpoint of conformal geometry a general question on symmetry breaking with respect to a pair of conformal manifolds $X \supset Y$.

Let $(X,g)$ be a Riemannian manifold. Suppose that a Lie group $G$ acts conformally on $X$. This means that there exists a positive-valued function $\Omega \in C^\infty(G\times X)$

(*conformal factor*) such that

$$L_h^* g_{h\cdot x} = \Omega(h,x)^2 g_x \quad \text{for all } h \in G,\ x \in X,$$

where we write $L_h \colon X \to X, x \mapsto h \cdot x$ for the action of $G$ on $X$. When $X$ is oriented, we define a locally constant function

$$\mathit{or} \colon G \times X \longrightarrow \{\pm 1\}$$

by $\mathit{or}(h)(x) = 1$ if $(L_h)_{*x} \colon T_x X \longrightarrow T_{L_h x} X$ is orientation-preserving, and $\mathit{or}(h)(x) = -1$ if it is orientation-reversing.

Since both the conformal factor $\Omega$ and the orientation map $\mathit{or}$ satisfy cocycle conditions, we can form a family of representations $\varpi_{u,\delta}^{(i)}$ of $G$ with parameters $u \in \mathbb{C}$ and $\delta \in \mathbb{Z}/2\mathbb{Z}$ on the space $\mathcal{E}^i(X)$ of differential $i$-forms on $X$ $(0 \le i \le \dim X)$ defined by

$$\varpi_{u,\delta}^{(i)}(h)\alpha := \mathit{or}(h)^\delta \Omega(h^{-1}, \cdot)^u L_{h^{-1}}^* \alpha, \quad (h \in G). \tag{2.2}$$

The representation $\varpi_{u,\delta}^{(i)}$ of the conformal group $G$ on $\mathcal{E}^i(X)$ will be referred to as the *conformal representation* on differential $i$-forms ([**13**, (1.1)], see also [**15**] in the $i = 0$ and $\delta \equiv 0$ case).

Suppose that $Y$ is an orientable submanifold. The submanifold $Y$ is endowed with a Riemannian structure $g|_Y$ by restriction, and we can define in a similar way a family of representations $\varpi_{v,\varepsilon}^{(j)}$ on $\mathcal{E}^j(Y)$ $(v \in \mathbb{C}, \varepsilon \in \mathbb{Z}/2\mathbb{Z}, 0 \le j \le \dim Y)$ of the conformal group of $(Y, g|_Y)$.

We consider the full group of conformal diffeomorphisms and its subgroup defined as

$$\begin{aligned} \operatorname{Conf}(X) &:= \{\text{conformal diffeomorphisms of } (X,g)\}, \\ \operatorname{Conf}(X;Y) &:= \{\varphi \in \operatorname{Conf}(X) : \varphi(Y) = Y\}. \end{aligned} \tag{2.3}$$

Then there is a natural group homomorphism

$$\operatorname{Conf}(X;Y) \to \operatorname{Conf}(Y), \quad \varphi \mapsto \varphi|_Y. \tag{2.4}$$

The general question is to construct and classify continuous linear maps (*conformally covariant symmetry breaking operators*) $\mathcal{E}^i(X) \to \mathcal{E}^j(Y)$ that intertwine the restriction $\varpi_{u,\delta}^{(i)}|_{\operatorname{Conf}(X;Y)}$ and the representation $\varpi_{v,\varepsilon}^{(j)}$ of the subgroup $\operatorname{Conf}(X;Y)$ of $\operatorname{Conf}(X)$. The integral operators $\widetilde{\mathbb{A}}_{\lambda,\nu,\pm}^{i,j}$ and the differential operators $\mathbb{C}_{\lambda,\nu}^{i,j}$ are such operators for the model space $(X,Y) = (S^n, S^{n-1})$.

In fact, for the model space

$$(X,Y) = (S^n, S^{n-1}), \tag{2.5}$$

where $Y$ is embedded as a totally geodesic hypersurface of the standard sphere $X$, the pair $(\operatorname{Conf}(X), \operatorname{Conf}(X;Y))$ of conformal groups is locally isomorphic to the pair of the Lorentz groups:

$$(G, G') = (O(n+1,1), O(n,1)). \tag{2.6}$$

In turn, the geometry $(X,Y)$ can be recovered as the real flag manifolds of the real reductive Lie groups $G$ and $G'$. Furthermore, the conformal representations are naturally isomorphic to the principal series representations given in Section 2.1 as follows:

LEMMA 2.2 (see [**13**, Prop. 2.3]). *Let* $0 \le i \le n$, $\delta \in \{\pm\}$, $u \in \mathbb{C}$. *Then there is a natural isomorphism*

$$\mathcal{E}^i(S^n)_{u,\delta} \simeq I_{i+\delta}(i, u+i)$$

*as representations of* $G = O(n+1,1)$. *Here* $i+\delta$ *is understood as an element of* $\mathbb{Z}/2\mathbb{Z}$.

Thus the conformal representation $\varpi_{u,\delta}^{(i)}$ of the group $\mathrm{Conf}(S^n)$ on the space $\mathcal{E}^i(S^n)$ of differential $i$-forms may be identified with the so-called "$K$-picture" of the principal series representation $I_\delta(i,\lambda)$ of $G = O(n+1,1)$ after some shift of parameters. Similarly, the conformal representation $(\varpi_{v,\varepsilon}^{(j)}, \mathcal{E}^j(S^{n-1}))$ of the subgroup $\mathrm{Conf}(S^n; S^{n-1})$ is identified with $J_\varepsilon(j,\nu)$. Hence the integral operator $\widetilde{\mathbb{A}}_{\lambda,\nu,\pm}^{i,j}$ and its holomorphic continuation give rise to conformally covariant, symmetry breaking operators $\mathcal{E}^i(S^n) \to \mathcal{E}^j(S^{n-1})$.

**2.3. Applications.** As mentioned briefly in Remark 1.4, Theorem 1.3 in Section 1 leads us to two applications:

(1) (a necessary and sufficient condition for $\widetilde{\mathbb{A}}_{\lambda,\nu,\pm}^{i,j}$ to vanish) The matrix-valued symmetry breaking operators $\widetilde{\mathbb{A}}_{\lambda,\nu,\pm}^{i,j}$ are defined as the holomorphic continuation of integral operators, and it is nontrivial to find the precise location of the zeros. By the residue formula (Theorem 1.3), we can determine the zeros of $\widetilde{\mathbb{A}}_{\lambda,\nu,\pm}^{i,j}$ (see Theorem 8.1 and Remark 8.3). This plays a crucial role in the classification problem of symmetry breaking operators ([**21**]).
(2) (another approach to construct conformally covariant differential operators) It is easy to see that the integral transforms (and its analytic continuation) $\widetilde{\mathbb{A}}_{\lambda,\nu,\pm}^{i,j} \colon I_\delta(i,\lambda) \to J_\varepsilon(j,\nu)$ respect the actions of the subgroup $G' = O(n,1)$ of $G = O(n+1,1)$ ([**19**, **21**]). Equivalently, $\widetilde{\mathbb{A}}_{\lambda,\nu,\pm}^{i,j}$ give conformally covariant, symmetry breaking operators from $\mathcal{E}^i(S^n)$ to $\mathcal{E}^j(S^{n-1})$, and also from $\mathcal{E}_c^i(\mathbb{R}^n)$ to $\mathcal{E}^j(\mathbb{R}^{n-1})$, as is seen in Section 2.2, and in turn, so do the residues. Thus the residue formula (Theorem 1.3) gives a proof that the differential operator $\mathbb{C}_{\lambda,\nu}^{i,j}$ ($j = i, i-1$) is a conformally covariant symmetry breaking operator from $\mathcal{E}^i(\mathbb{R}^n)$ to $\mathcal{E}^j(\mathbb{R}^{n-1})$. In [**13**] we gave another proof for this result based on the F-method ([**8**, **9**, **16**, **18**]). In contrast, the argument of the present article does not use the F-method. Since the operator $\mathbb{C}_{\lambda,\nu}^{i,j}$ is recovered from its matrix coefficients by an elementary computation in differential geometry (*cf.* Facts 7.2 and 7.3), we may think of the residue formula (1.7) in Theorem 1.3 as yet another method to find an explicit formula for the conformally covariant, differential symmetry breaking operators $\mathbb{C}_{\lambda,\nu}^{i,j}$ when $j = i, i-1$.

REMARK 2.3. (scalar-valued case) In the case where $i = j = 0$, the matrix-valued symmetry breaking operator $\mathbb{C}_{\lambda,\nu}^{i,j}$ reduces to a scalar-valued one

(Juhl's operator), and we have

$$\mathbb{C}^{0,0}_{\lambda,\nu} = \frac{1}{2}\nu\widetilde{\mathbb{C}}_{\lambda,\nu},$$

see [**13**, p. 23]. Thus Theorem 1.3 in this case coincides with [**19**, Thm. 12.2 (2)], see Fact 6.3. Actually, our proof of Theorem 1.3 uses the results in the scalar case.

## 3. Some identities in the Weyl algebra

A key technique in our proof of the matrix-valued residue formula (Theorem 1.3) relies on an algebraic manipulation in the Weyl algebra, for which we give a basic set-up in this section. We shall develop it for Juhl's operators in Section 5.

Let $\mathcal{D}'(\mathbb{R}^n)$ be the space of distributions on $\mathbb{R}^n$, and $\mathcal{D}'_{\{0\}}(\mathbb{R}^n)$ the subspace consisting of distributions supported at the origin. Then the Weyl algebra

$$\mathbb{C}[x, \frac{\partial}{\partial x}] \equiv \mathbb{C}[x_1, \cdots, x_n, \frac{\partial}{\partial x_1}, \cdots, \frac{\partial}{\partial x_n}]$$

acts naturally on $\mathcal{D}'(\mathbb{R}^n)$ and leaves the subspace $\mathcal{D}'_{\{0\}}(\mathbb{R}^n)$ invariant. Let $\mathcal{J}$ be the annihilator of the Dirac delta function $\delta(x) = \delta(x_1, \cdots, x_n)$, namely, the kernel of the following $\mathbb{C}[x, \frac{\partial}{\partial x}]$-homomorphism:

$$\Psi : \mathbb{C}[x, \frac{\partial}{\partial x}] \to \mathcal{D}'(\mathbb{R}^n), \qquad P \mapsto P\delta. \tag{3.1}$$

Then $\mathcal{J}$ is the left ideal generated by the coordinate functions $x_1, \cdots, x_n$, and $\Psi$ induces an isomorphism of $\mathbb{C}[x, \frac{\partial}{\partial x}]$-modules.

$$\overline{\Psi} : \mathbb{C}[x, \frac{\partial}{\partial x}]/\mathcal{J} \xrightarrow{\sim} \mathcal{D}'_{\{0\}}(\mathbb{R}^n). \tag{3.2}$$

Our strategy is to reduce (rather complicated) computations in $\mathcal{D}'_{\{0\}}(\mathbb{R}^n)$ to simpler algebraic ones via the isomorphism (3.2) by preparing systematically certain identities in the Weyl algebra $\mathbb{C}[x, \frac{\partial}{\partial x}]$ modulo $\mathcal{J}$ (see Lemmas 3.2, 4.4 and 5.2).

Before entering this part, we give the following observation:

LEMMA 3.1. *If $P$ is a differential operator on $\mathbb{R}^n$ with constant coefficients, then*

$$\Psi(P) * f = Pf \qquad \textit{for all } f \in C^\infty(\mathbb{R}^n). \tag{3.3}$$

PROOF. Let $\alpha = (\alpha_1, \cdots, \alpha_n)$ be a multi-index, and we write $P = \sum_\alpha a_\alpha \frac{\partial^{|\alpha|}}{\partial x^\alpha}$. Then

$$\begin{aligned}\Psi(P) * f(x) &= \int \sum_\alpha a_\alpha \frac{\partial^{|\alpha|}\delta}{\partial y^\alpha}(y) f(x-y)dy \\ &= \sum_\alpha a_\alpha \int \delta(y)(-1)^{|\alpha|}\frac{\partial^{|\alpha|}}{\partial y^\alpha} f(x-y)dy \\ &= \sum_\alpha a_\alpha \frac{\partial^{|\alpha|}}{\partial x^\alpha} f(x).\end{aligned}$$

Hence $\Psi(P) * f(x) = Pf$. □

We set

$$\Delta \equiv \Delta_{\mathbb{R}^n} = \frac{\partial^2}{\partial x_1^2} + \cdots + \frac{\partial^2}{\partial x_n^2}.$$

LEMMA 3.2. *Let $k \in \mathbb{N}$, and $1 \leq p, q \leq n$ with $p \neq q$. Then the following identities hold in the Weyl algebra $\mathbb{C}[x, \frac{\partial}{\partial x}]$ modulo $\mathcal{J}$.*

(1) $x_p \Delta^k \equiv -2k \frac{\partial}{\partial x_p} \Delta^{k-1} \mod \mathcal{J}$.

(2) $x_p x_q \Delta^k \equiv 4k(k-1) \frac{\partial^2}{\partial x_p \partial x_q} \Delta^{k-2} \mod \mathcal{J}$.

(3) $x_p^2 \Delta^k \equiv 4k(k-1) \frac{\partial^2}{\partial x_p^2} \Delta^{k-2} + 2k\Delta^{k-1} \mod \mathcal{J}$.

PROOF. We denote by $[P, Q] := PQ - QP$ the bracket of $P, Q \in \mathbb{C}[x, \frac{\partial}{\partial x}]$ as usual. Then the assertions are derived from the following commutation relations:

$$[x_p, \Delta^k] = -2k \frac{\partial}{\partial x_p} \Delta^{k-1}, \tag{3.4}$$

$$[x_p x_q, \Delta^k] = 4k(k-1) \frac{\partial^2}{\partial x_p \partial x_q} \Delta^{k-2} - 2k\left(\frac{\partial}{\partial x_p} \Delta^{k-1} x_q + \frac{\partial}{\partial x_q} \Delta^{k-1} x_p\right), \tag{3.5}$$

$$[x_p^2, \Delta^k] = 4k(k-1) \frac{\partial^2}{\partial x_p^2} \Delta^{k-2} + 2k\Delta^{k-1} - 4k \frac{\partial}{\partial x_p} \Delta^{k-1} x_p. \tag{3.6}$$

The first equation (3.4) is verified easily by induction on $k$. In turn, the second and third ones (3.5) and (3.6) follow from the iterated use of (3.4) and from the identity $[AB, C] = A[B, C] + [A, C]B$. □

## 4. Residue formulæ of the matrix-valued Knapp–Stein intertwining operators

In this section we consider a baby-case (*i.e.* $G = G'$ case), and apply the machinery in the previous section to find a residue formula for the matrix-valued Knapp–Stein intertwining operator $\widetilde{\mathbb{T}}^i_{\lambda, n-\lambda}$. Since the principal series representation $I_\delta(i, \lambda)$ is realized in the space $\mathcal{E}^i(\mathbb{R}^n)$ of differential forms on $\mathbb{R}^n$, the residue formula should be given by some familiar operators known in differential geometry. Actually, we shall see that the residue formula is proportional to Branson's conformal covariant differential operator [**2**].

Our proof here illustrates an idea of the more complicated argument in later sections, where we give a proof of our main results on symmetry breaking operators $\widetilde{\mathbb{A}}^{i,j}_{\lambda,\nu,\pm}$.

**4.1. Matrix-valued Knapp–Stein intertwining operators.** Suppose $0 \leq i \leq n$. For $\mathrm{Re}\lambda \gg 0$, we define $\widetilde{\mathcal{T}}^i_\lambda(x)$ to be an $\mathrm{End}_{\mathbb{C}}(\bigwedge^i(\mathbb{C}^n))$-valued, locally integrable function on $\mathbb{R}^n$ by the following formula

$$\widetilde{\mathcal{T}}^i_\lambda(x) := \frac{1}{\Gamma(\lambda - \frac{n}{2})} |x|^{2(\lambda-n)} \sigma^{(i)}(\psi_n(x)). \tag{4.1}$$

The Knapp–Stein intertwining operator [**7**] between principal series representations of $G = O(n+1, 1)$,

$$\widetilde{\mathbb{T}}^i_{\lambda,n-\lambda} \colon I_\delta(i, \lambda) \to I_\delta(i, n-\lambda) \qquad \text{for } \delta \in \{0, 1\},$$

is defined in the $N$-picture as the analytic continuation of the convolution map

$$\mathcal{E}^i_c(\mathbb{R}^n) \to \mathcal{E}^i(\mathbb{R}^n), \quad f \mapsto \widetilde{\mathcal{T}}^i_\lambda * f.$$

Next, we recall that Branson–Gover's conformally covariant differential operator (see [**2**])

$$\mathcal{D}_{2l}^{(i)} : \mathcal{E}^i(\mathbb{R}^n) \to \mathcal{E}^i(\mathbb{R}^n)$$

is given by

$$\mathcal{D}_{2l}^{(i)} := \begin{cases} -(\frac{n}{2} - i)\Delta^l + l(d^*d - dd^*)\Delta^{l-1} & \text{for } l \in \mathbb{N}_+, \\ -(\frac{n}{2} - i)\mathrm{id} & \text{for } l = 0, \end{cases} \tag{4.2}$$

where $\Delta = -(dd^* + d^*d)$ is the Laplace–Bertrami operator acting on differential forms. We adopt the normalization of $\mathcal{D}_{2l}^{(i)}$ given by [**13**, (12.1)]. In particular, $\mathcal{D}_{2l}^{(i)}$ vanishes when $i = \frac{n}{2}$ and $l = 0$. The conformally covariant property of Branson's operator is reformulated as the intertwining property between two principal series representations in their $N$-picture:

$$\mathcal{D}_{2l}^{(i)} : I_\delta(i, \frac{n}{2} - l) \to I_\delta(i, \frac{n}{2} + l) \quad \text{for } \delta \in \{0, 1\}.$$

See [**13**, Thm. 12.2] for instance, for the classification of such operators.

Here is a relationship between $\mathrm{End}_{\mathbb{C}}(\bigwedge^i(\mathbb{C}^n))$-valued Knapp–Stein intertwining operators $\widetilde{\mathbb{T}}^i_{\lambda,n-\lambda}$ and Branson's conformally covariant operators $\mathcal{D}_{2l}^{(i)}$:

THEOREM 4.1. *Let $0 \le i \le n$ and $\lambda \in \mathbb{C}$.*

(1) *The matrix-valued Knapp–Stein intertwining operator $\widetilde{\mathbb{T}}^i_{\lambda,n-\lambda}$ reduces to a differential operator if and only if $n - 2\lambda \in 2\mathbb{N}$.*
(2) *Suppose $l \in \mathbb{N}_+$. Then we have*

$$\widetilde{\mathbb{T}}^i_{\lambda,n-\lambda}|_{\lambda=\frac{n}{2}-l} = \frac{(-1)^{l+1}\pi^{\frac{n}{2}}}{2^{2l}\Gamma(\frac{n}{2}+l+1)}\mathcal{D}_{2l}^{(i)}. \tag{4.3}$$

Theorem 4.1 will be proved in Section 4.4 after preparing some basic results.

A different approach to prove Theorem 4.1 may be seen in a recent preprint [**4**, Cor. 4.4], which utilizes the Fourier transform of the distribution kernel (*cf.*, the F-method [**9**] for finding a residue formula).

**4.2. Residue formula of the Riesz potential $|x|^\mu$.** We review a classical result on the Riesz potential

$$|x|^\mu = (x_1^2 + \cdots + x_n^2)^{\frac{\mu}{2}}.$$

This is a meromorphic family of distributions on $\mathbb{R}^n$, and has simple poles at $\mu = -n - 2l$ ($l \in \mathbb{N}$). Thus the normalized Riesz potential on $\mathbb{R}^n$ defined by

$$\widetilde{\mathcal{T}}_\lambda(x) := \frac{1}{\Gamma(\lambda - \frac{n}{2})}|x|^{2(\lambda-n)}$$

depends holomorphically on $\lambda$ in the entire plane $\mathbb{C}$. The residue formula is classically known (see [**5**, Chap.$\tilde{(}$2.2)], [**7**] for example):

FACT 4.2 (residue of $|x|^\mu$). Suppose $l \in \mathbb{N}$. Then we have:

$$\widetilde{\mathcal{T}}_{\frac{n}{2}-l}(x) = C(l,n)\Delta^l\delta(x),$$

where we set

$$C(l,n) := \frac{(-1)^l\pi^{\frac{n}{2}}}{2^{2l}\Gamma(\frac{n}{2}+l)}.$$

**4.3. Index set $\mathfrak{I}_{n,i}$.** In what follows, we use the convention of index sets as below. For $0 \leq i \leq n$, we define

$$\mathfrak{I}_{n,i} := \{I \subset \{1, \cdots, n\} : \# I = i\}.$$

For $I = \{k_1, \cdots, k_i\} \in \mathfrak{I}_{n,i}$ with $k_1 < \cdots < k_i$, we set

$$\begin{aligned} e_I :=& e_{k_1} \wedge e_{k_2} \wedge \cdots \wedge e_{k_i}, \\ dx_I :=& dx_{k_1} \wedge dx_{k_2} \wedge \cdots \wedge dx_{k_i}. \end{aligned}$$

Then $\{e_I\}$ forms a basis of the vector space $\bigwedge^i(\mathbb{C}^n)$, and $\{dx_I\}$ forms a basis of $\mathcal{E}^i(\mathbb{R}^n)$ as a $C^\infty(\mathbb{R}^n)$-module. We then have a natural isomorphism as $C^\infty(\mathbb{R}^n)$-modules:

$$C^\infty(\mathbb{R}^n) \otimes \textstyle\bigwedge^i(\mathbb{C}^n) \xrightarrow{\sim} \mathcal{E}^i(\mathbb{R}^n), \quad \sum f_I \otimes e_I \mapsto \sum f_I dx_I. \tag{4.4}$$

We introduce a family of quadratic polynomials $S_{IJ}(x)$ indexed by $I$, $J \in \mathfrak{I}_{n,i}$ of $x = (x_1, \cdots, x_n)$ as follows:

$$S_{IJ}(x) = \begin{cases} \sum_{k=1}^n \varepsilon_I(k) x_k^2 & \text{if } I = J, \\ \operatorname{sgn}(I, I') x_p x_q & \text{if } \#(I \setminus J) = 1, \\ 0 & \text{if } \#(I \setminus J) \geq 2. \end{cases} \tag{4.5}$$

Here we set

$$\varepsilon_I(k) := \begin{cases} 1 & \text{for } k \in I, \\ -1 & \text{for } k \notin I. \end{cases}$$

For $K \subset \{1, \cdots, n\}$ and $1 \leq p, q \leq n$, $\operatorname{sgn}(K; p, q) \in \{\pm 1\}$ is defined by

$$\operatorname{sgn}(K; p, q) := (-1)^{\#\{r \in K : \min(p,q) < r < \max(p,q)\}}.$$

For $I, I' \in \mathfrak{I}_{n,i}$ with $\#(I \setminus I') = 1$, we set

$$\operatorname{sgn}(I, I') := \operatorname{sgn}(I \cap I'; p, q), \tag{4.6}$$

where $p \in I \setminus I'$ and $q \in I' \setminus I$.

We have the following.

LEMMA 4.3. *Suppose $0 \leq i \leq n$. Then the minor determinant of $\psi_n(x) \in O(n)$ for $I, J \in \mathfrak{I}_{n,i}$ is given by*

$$(\det \psi_n(x))_{IJ} = -\frac{1}{|x|^2} S_{IJ}(x). \tag{4.7}$$

**4.4. Proof of Theorem 4.1.** We complete the proof of Theorem 4.1 by comparing the matrix components of the both sides of the equation (4.3).

For $I, J \in \mathfrak{I}_{n,i}$, we define the $(I, J)$-component of the matrix-valued distribution $\widetilde{\mathcal{T}}_\lambda^i$ by

$$(\widetilde{\mathcal{T}}_\lambda^i)_{IJ} := \langle \widetilde{\mathcal{T}}_\lambda^i(e_I), e_J^\vee \rangle,$$

where $\{e_J^\vee\}$ is the dual basis. By (4.1) and (4.7), we have

$$\begin{aligned} (\widetilde{\mathcal{T}}_\lambda^i)_{IJ} =& \frac{1}{\Gamma(\lambda - \frac{n}{2})} |x|^{2(\lambda-n)} (\det \psi_n(x))_{JI} \\ =& \frac{-1}{\Gamma(\lambda - \frac{n}{2})} |x|^{2(\lambda-n-1)} S_{JI}(x) \\ =& \frac{1}{\frac{n}{2} - \lambda + 1} \widetilde{\mathcal{T}}_{\lambda-1}(x) S_{IJ}(x). \end{aligned}$$

Applying Fact 4.2, we get the first statement of Theorem 4.1, and the following equality in $\mathcal{D}'_{\{0\}}(\mathbb{R}^n)$:

$$(\widetilde{\mathcal{T}}^i_\lambda)_{IJ}|_{\lambda=\frac{n}{2}-l} = \frac{1}{l+1}C(l+1,n)S_{IJ}(x)\Delta^{l+1}\delta(x). \tag{4.8}$$

In order to compute the right-hand side of (4.8), we work with the Weyl algebra by using Lemma 3.2.

LEMMA 4.4. *Suppose $l \in \mathbb{N}_+$ and $I, J \in \mathfrak{I}_{n,i}$. Then the following equalities hold in $\mathbb{C}[x, \frac{\partial}{\partial x}]/\mathcal{J}$.*

(1) *Suppose $I = J$.*

$$S_{IJ}(x)\Delta^{l+1} \equiv 4(l+1)l(\sum_{p\in I}\varepsilon_I(p)\frac{\partial^2}{\partial x_p^2})\Delta^{l-1} + 2(l+1)(2i-n)\Delta^l.$$

(2) *Suppose $\#(I-J) = 1$. Let $\{p,q\} := (I\cup J)\setminus(I\cap J)$.*

$$S_{IJ}(x)\Delta^{l+1} \equiv 8l(l+1)\operatorname{sgn}(I,J)\frac{\partial^2}{\partial x_p\partial x_q}\Delta^{l-1}.$$

PROOF. Direct from the definition (4.5) of $S_{IJ}(x)$ and from Lemma 3.2. □

By Lemma 4.4, the equation (4.8) gives the following.

LEMMA 4.5. *Suppose $l \in \mathbb{N}_+$.*

$$\begin{aligned}&(\widetilde{\mathcal{T}}^{(i)}_\lambda)_{IJ}|_{\lambda=\frac{n}{2}-l}\\ =&4C(l+1,n)\times\begin{cases} l(\sum_{p\in I}\varepsilon_I(p)\frac{\partial^2}{\partial x_p^2})\Delta^{l-1} + (i-\frac{n}{2})\Delta^l & \text{if } I=J,\\ 2l\operatorname{sgn}(I,J) & \text{if } \#(I\setminus J)=1,\\ 0 & \text{otherwise.}\end{cases}\end{aligned}$$

On the other hand, Branson's conformally covariant differential operators $\mathcal{D}^{(i)}_{2l}$ take the following form:

LEMMA 4.6. *Suppose $0 \le i \le n$ and $l \in \mathbb{N}_+$. For any $\alpha = f dx_I \in \mathcal{E}^i(\mathbb{R}^n)$ with $f \in C^\infty(\mathbb{R}^n)$ and $I \in \mathfrak{I}_{n,i}$, we have*

$$\begin{aligned}\mathcal{D}^{(i)}_{2l}\alpha =&((i-\frac{n}{2})\Delta^l f + l(\sum_{p\in I}\varepsilon_I(p)\frac{\partial^2}{\partial x_p^2})\Delta^{l-1}f)dx_I\\ &+2l\sum_{\substack{p\in I\\ q\notin I}}\operatorname{sgn}(I;p,q)\frac{\partial^2}{\partial x_p\partial x_q}(\Delta^{l-1}f)dx_{I\setminus\{p\}\cup\{q\}}.\end{aligned}$$

PROOF. The formula is direct from the definition (4.2) of $\mathcal{D}^{(i)}_{2l}$ and from the following elementary computations:

$$dd^*(f dx_I) = -\sum_{p\in I}\frac{\partial^2 f}{\partial x_p^2}dx_I - \sum_{\substack{p\in I\\ q\notin I}}\operatorname{sgn}(I;p,q)\frac{\partial^2 f}{\partial x_p\partial x_q}dx_{I\setminus\{p\}\cup\{q\}},$$

$$d^*d(f dx_I) = -\sum_{q\notin I}\frac{\partial^2 f}{\partial x_q^2}dx_I + \sum_{\substack{p\in I\\ q\notin I}}\operatorname{sgn}(I;p,q)\frac{\partial^2 f}{\partial x_p\partial x_q}dx_{I\setminus\{p\}\cup\{q\}}.$$

□

PROOF OF THEOREM 4.1. We identify $\mathcal{E}^i(\mathbb{R}^n)$ with $C^\infty(\mathbb{R}^n) \otimes \bigwedge^i(\mathbb{C}^n)$ via (4.4). Comparing the $(I, J)$-component of Branson's conformally covariant operators with that of the matrix-valued Knapp–Stein intertwining operator $\widetilde{\mathbb{T}}^i_{\lambda,n-\lambda}$ in Lemmas 4.5 and 4.6, we get Theorem 4.1. $\square$

**4.5. Vanishing condition of $\widetilde{\mathbb{T}}^i_{\lambda,n-\lambda}$.** As a biproduct of the residue formula (4.3), we obtain the vanishing condition of the matrix-valued Knapp–Stein intertwining operator $\widetilde{\mathbb{T}}^i_{\lambda,n-\lambda}$ as below.

COROLLARY 4.7. *$\widetilde{\mathbb{T}}^i_{\lambda,n-\lambda} = 0$ if and only if $n$ is even and $i = \lambda = \frac{n}{2}$.*

PROOF. Since both $\widetilde{\mathcal{T}}_\lambda(x)$ and $\sigma^{(i)}(\psi_n(x))$ are smooth in $\mathbb{R}^n \setminus \{0\}$, the distribution kernel $\widetilde{\mathcal{T}}^i_\lambda$ vanishes only when $\operatorname{Supp}\widetilde{\mathcal{T}}_\lambda \subset \{0\}$, or equivalently, $n - 2\lambda \in 2\mathbb{N}$. Suppose now $n - 2\lambda = 2l$ for some $l \in \mathbb{N}$. Then it follows from Theorem 4.1 that $\widetilde{\mathcal{T}}^{(i)}_{\lambda,n-\lambda}$ vanishes if and only if $\mathcal{D}^{(i)}_{2l} = 0$. In turn, this happens exactly when $i = \frac{n}{2}$ and $l = 0$ by (4.2). Thus the corollary is proved. $\square$

REMARK 4.8. When $n$ is even, a special attention is required at the middle degree $i = \frac{n}{2}$ with $\lambda = i$. See [**21**, Chap. Sec. 4] for representation theoretic properties of the *renormalization* of $\widetilde{\mathbb{T}}^i_{\lambda,n-\lambda}$ in this case.

## 5. Juhl's operator in the Weyl algebra

In order to prove our main results (Theorem 1.3), we need some further identities in $\mathbb{C}[x, \frac{\partial}{\partial x}]/\mathcal{J}$ where $\mathcal{J}$ is the left ideal generated by $x_1, \cdots, x_n$ as in Section 3.

For $l \in \mathbb{N}$, we define a finite-dimensional vector space of polynomials of one variable by

$$\mathrm{Pol}_l[z]_{\mathrm{even}} := \mathbb{C}\text{-span}\langle z^{l-2j} : 0 \leq 2j \leq l\rangle.$$

We inflate a polynomial $g(z) \in \mathrm{Pol}_l[z]_{\mathrm{even}}$ to a polynomial of two variables $s$ and $t$ by

$$(I_l g)(s,t) := s^{\frac{l}{2}} g\left(\frac{t}{\sqrt{s}}\right). \tag{5.1}$$

If $g(z)$ is given as $g(z) = \sum_{j=0}^{[\frac{l}{2}]} a_j z^{l-2j}$, then the definition (5.1) yields

$$(I_l g)(s,t) = s^{\frac{l}{2}} g\left(\frac{t}{\sqrt{s}}\right) = \sum_{j=0}^{[\frac{l}{2}]} a_j s^j t^{l-2j}. \tag{5.2}$$

We note that $(I_l g)(s^2, t)$ is a homogeneous polynomial of $s$ and $t$ of degree $l$. The following example reveals how Juhl's conformally covariant differential operators [**6**, **16**] arise.

EXAMPLE 5.1. Let $\widetilde{C}^\alpha_l(z)$ be the renormalized Gegenbauer polynomial defined by

$$\widetilde{C}^\alpha_l(z) = \frac{1}{\Gamma(\alpha + [\frac{l+1}{2}])} \sum_{k=0}^{[\frac{l}{2}]} (-1)^k \frac{\Gamma(l-k+\alpha)}{k!(l-2k)!} (2z)^{l-2k}.$$

We adopt the normalization given in [**13**, (14.3)] so that $\widetilde{C}_l^\alpha(z)$ is a nonzero element in $\mathrm{Pol}_l[z]_{\mathrm{even}}$ for any $\alpha \in \mathbb{C}$. Suppose $\nu - \lambda \in \mathbb{N}$. Then Juhl's conformally covariant operator $\widetilde{\mathbb{C}}_{\lambda,\nu}$ (see (1.8)) is expressed as

$$\widetilde{\mathbb{C}}_{\lambda,\nu} = \mathrm{Rest}_{x_n=0} \circ P,$$

where $P$ is a homogeneous differential operator of order $\nu - \lambda$ on $\mathbb{R}^n$ given by

$$P := (I_{\nu-\lambda}\widetilde{C}_{\nu-\lambda}^{\lambda-\frac{n-1}{2}})(-\Delta_{\mathbb{R}^{n-1}}, \frac{\partial}{\partial x_n}).$$

We also define a distribution on $\mathbb{R}^n$ supported at the origin by

$$\widetilde{\mathcal{C}}_{\lambda,\nu} := (I_{\nu-\lambda}\widetilde{C}_{\nu-\lambda}^{\lambda-\frac{n-1}{2}})(-\Delta_{\mathbb{R}^{n-1}}, \frac{\partial}{\partial x_n})\delta(x_1, \cdots, x_n). \tag{5.3}$$

This is the distribution kernel of the scalar-valued, differential symmetry breaking operator $\widetilde{\mathbb{C}}_{\lambda,\nu}$ (see (1.8)), and was denoted by $\widetilde{K}_{\lambda,\nu}^{\mathbb{C}}$ in [**19**, (10.3)] when $\nu - \lambda \in 2\mathbb{N}$. In terms of the map $\Psi\colon \mathbb{C}[x, \frac{\partial}{\partial x}] \to \mathcal{D}_{\{0\}}(\mathbb{R}^n)$ defined in (3.1), Lemma 3.1 shows

$$\begin{aligned}\widetilde{\mathcal{C}}_{\lambda,\nu} &= \Psi(P),\\ \widetilde{\mathbb{C}}_{\lambda,\nu} &= \mathrm{Rest}_{x_n=0} \circ \widetilde{\mathcal{C}}_{\lambda,\nu} \, * \, .\end{aligned}$$

We begin with some basic computations in the Weyl algebra $\mathbb{C}[x, \frac{\partial}{\partial x}]$ modulo the left ideal $\mathcal{J}$.

LEMMA 5.2. *Suppose* $g(z) \in \mathrm{Pol}_l[z]_{\mathrm{even}}$. *Let* $\theta = z\frac{d}{dz}$.

(1) *For any* $p$ *with* $1 \le p \le n-1$, *we have*

$$x_p(I_l g)(-\Delta_{\mathbb{R}^{n-1}}, \frac{\partial}{\partial x_n}) \equiv \frac{\partial}{\partial x_p} I_{l-2}((l-\theta)g)(-\Delta_{\mathbb{R}^{n-1}}, \frac{\partial}{\partial x_n}) \mod \mathcal{J}.$$

(2) $x_n(I_l g)(-\Delta_{\mathbb{R}^{n-1}}, \frac{\partial}{\partial x_n}) \equiv -(I_{l-1}g')(-\Delta_{\mathbb{R}^{n-1}}, \frac{\partial}{\partial x_n}) \mod \mathcal{J}$.

PROOF. Applying $\frac{\partial}{\partial s}$ and $\frac{\partial}{\partial t}$ to the equation (5.2), we get

$$\frac{\partial}{\partial s}(I_l g)(s,t) = \frac{1}{2} I_{l-2}((l-\theta)g)(s,t) = \sum_j a_j j s^{j-1} t^{l-2j}, \tag{5.4}$$

$$\frac{\partial}{\partial t}(I_l g)(s,t) = (I_{l-1}g')(s,t) = \sum_j a_j(l-2j)s^j t^{l-2j-1}. \tag{5.5}$$

(1) By Lemma 3.2,

$$x_p(I_l g)(-\Delta_{\mathbb{R}^{n-1}}, \frac{\partial}{\partial x_n}) \equiv 2\frac{\partial}{\partial x_p}\sum_j a_j j(-\Delta_{\mathbb{R}^{n-1}})^{j-1}(\frac{\partial}{\partial x_n})^{l-2j} \mod \mathcal{J}. \tag{5.6}$$

By (5.4), the right-hand side of (5.6) amounts to

$$\frac{\partial}{\partial x_p} I_{l-2}((l-\theta)g)(-\Delta_{\mathbb{R}^{n-1}}, \frac{\partial}{\partial x_n}) \mod \mathcal{J}.$$

(2) Since $x_n(\frac{\partial}{\partial x_n})^{l-2j} \equiv -(l-2j)(\frac{\partial}{\partial x_n})^{l-2j-1} \mod \mathcal{J}$, we have

$$x_n(I_l g)(-\Delta_{\mathbb{R}^{n-1}}, \frac{\partial}{\partial x_n}) \equiv -\sum_j a_j(l-2j)(-\Delta_{\mathbb{R}^{n-1}})^j(\frac{\partial}{\partial x_n})^{l-2j} \mod \mathcal{J}. \tag{5.7}$$

By (5.5), we get the desired formula. □

LEMMA 5.3. *Suppose $\lambda, \nu \in \mathbb{C}$ with $\nu - \lambda \in \mathbb{N}$. Let $\widetilde{C}_{\lambda,\nu}$ be the distribution on $\mathbb{R}^{n-1}$ defined in* (5.3). *Then the following equations hold.*

$$x_p \widetilde{C}_{\lambda,\nu} = -2\frac{\partial}{\partial x_p}\widetilde{C}_{\lambda+1,\nu-1} \quad \text{for } 1 \le p \le n-1, \tag{5.8}$$

$$x_n \widetilde{C}_{\lambda,\nu} = -2\gamma(\lambda - \frac{n-1}{2}, \nu - \lambda)\widetilde{C}_{\lambda+1,\nu}. \tag{5.9}$$

PROOF. We recall from [**1**], or [**13**, (14.8)] and [**16**, (A.13), (A.14)] that the normalized Gegenbauer polynomial $\widetilde{C}_l^\alpha(z)$ satisfies the following relations

$$(l - z\frac{d}{dz})\widetilde{C}_l^\alpha(z) = -2\widetilde{C}_{l-2}^{\alpha+1}(z), \tag{5.10}$$

$$\frac{d}{dz}\widetilde{C}_l^\alpha(z) = 2\gamma(\alpha, l)\widetilde{C}_{l-1}^{\alpha+1}(z), \tag{5.11}$$

for $\alpha \in \mathbb{C}$ and $l \in \mathbb{N}_+$.

Applying Lemma 5.2 (1) to $g(z) = \widetilde{C}_l^\alpha(z)$, we get from (5.10) the following identity in $\mathbb{C}[x, \frac{\partial}{\partial x}]/\mathcal{J}$:

$$x_p(I_l\widetilde{C}_l^\alpha)(-\Delta_{\mathbb{R}^{n-1}}, \frac{\partial}{\partial x_n}) \equiv -2\frac{\partial}{\partial x_p}(I_{l-2}\widetilde{C}_{l-2}^{\alpha+1})(-\Delta_{\mathbb{R}^{n-1}}, \frac{\partial}{\partial x_n}) \mod \mathcal{J}.$$

Hence (5.8) is verified by putting $\alpha = \lambda - \frac{n-1}{2}$ and $l = \nu - \lambda$. Likewise, applying Lemma 5.2 (2) to $g(z) = \widetilde{C}_{\nu-\lambda}^{\lambda - \frac{n-1}{2}}(z)$, we get (5.9) from (5.11). □

PROPOSITION 5.4. *Suppose $1 \le p, q \le n-1$, and $\lambda, \nu \in \mathbb{C}$. Then we have the following equations in $\mathcal{D}'(\mathbb{R}^n)$.*

$$x_p x_n \widetilde{C}_{\lambda-1,\nu+1} = 4\gamma(\lambda - \frac{n-1}{2}, \nu - \lambda)\frac{\partial}{\partial x_p}\widetilde{C}_{\lambda+1,\nu}. \tag{5.12}$$

$$x_p x_q \widetilde{C}_{\lambda-1,\nu+1} = 4\frac{\partial^2}{\partial x_p \partial x_q}\widetilde{C}_{\lambda+1,\nu-1}. \tag{5.13}$$

$$x_p^2 \widetilde{C}_{\lambda-1,\nu+1} = 4\frac{\partial^2}{\partial x_p^2}\widetilde{C}_{\lambda+1,\nu-1} + 2\widetilde{C}_{\lambda,\nu}. \tag{5.14}$$

$$\begin{aligned} x_n^2 \widetilde{C}_{\lambda-1,\nu+1} &= 4(\lambda - \frac{n-1}{2} + [\frac{\nu-\lambda+1}{2}])\widetilde{C}_{\lambda+1,\nu+1} \\ &= 2(2\nu - n + 1)\widetilde{C}_{\lambda,\nu} - 4\Delta_{\mathbb{R}^{n-1}}\widetilde{C}_{\lambda+1,\nu-1}. \end{aligned} \tag{5.15}$$

$$Q_{n-1}(x)\widetilde{C}_{\lambda-1,\nu+1} = 4\nu\widetilde{C}_{\lambda,\nu} - 4(\lambda - \frac{n-1}{2} + [\frac{\nu-\lambda+1}{2}])\widetilde{C}_{\lambda+1,\nu+1}. \tag{5.16}$$

*Here $Q_{n-1}(x) = x_1^2 + \cdots + x_{n-1}^2$ is a quadratic polynomial of $x = (x_1, \ldots, x_n)$.*

PROOF. The iterated use of Lemma 5.3 yields these identities except for the second equality in (5.15). The second equality is nothing but the three-term relation [**13**, (9.10)]

$$(\lambda - \frac{n-1}{2} + [\frac{\nu-\lambda+1}{2}])\widetilde{C}_{\lambda+1,\nu+1} = (\nu - \frac{n-1}{2})\widetilde{C}_{\lambda,\nu} - \Delta_{\mathbb{R}^{n-1}}\widetilde{C}_{\lambda+1,\nu-1},$$

which is derived from the following three-term relation of the Gegenbauer polynomials:

$$(5.17)\qquad (\mu+a)\widetilde{C}_a^{\mu}(t)+\widetilde{C}_{a-1}^{\mu+1}(t)=(\mu+[\frac{a+1}{2}])\widetilde{C}_a^{\mu+1}(t).$$

□

## 6. Reduction to the scalar-valued case

Our strategy to find the residue of the matrix-valued operators $\mathbb{A}_{\lambda,\nu,\pm}^{i,j}$ is to reduce the problem to the one in the scalar-valued case by proving the following key decomposition formula:

$$(6.1)\qquad \mathbb{A}_{\lambda,\nu,\varepsilon}^{i,j}=Q(\lambda,\nu)^{-1}H(x)\widetilde{\mathbb{A}}_{\lambda+a,\nu-b,+},$$

see Lemma 6.2 below. The primary features of the three factors in the right-hand side are as follows:

- $Q(\lambda,\nu)$ is a polynomial of $(\lambda,\nu)$;
- $H(x)$ is a $\mathrm{Hom}_{\mathbb{C}}(\bigwedge^i(\mathbb{C}^n),\bigwedge^j(\mathbb{C}^{n-1}))$-valued polynomial of $x$;
- $\widetilde{\mathbb{A}}_{\lambda+a,\nu-b,+}$ is the *scalar-valued* regular symmetry breaking operator with parameter *shifted* by $(a,b)\in\mathbb{N}^2$.

The decomposition (6.1) is a technical heart of our trick to overcome some difficulties arising from representation theory.

**6.1. Polynomial $g_{IJ}(x)$.** For $x=(x_1,\cdots,x_n)\neq 0$, we recall $\psi_n(x)=I_n-2|x|^{-2}x\,{}^tx$ where $|x|^2=x_1^2+\cdots+x_n^2$. To find the matrix-valued polynomial $H(x)$ in (6.1), we introduce the following polynomials $g_{IJ}(x)$ for $I\in\mathfrak{I}_{n,i}$, $J\in\mathfrak{I}_{n-1,j}$ with $j\in\{i-1,i\}$:

$$(6.2)\qquad g_{IJ}(x):=|x|^2\langle \mathrm{pr}_{i\to j}\circ\sigma^{(i)}(\psi(x))e_I,e_J^{\vee}\rangle.$$

LEMMA 6.1. *$g_{IJ}(x)$ is a polynomial of $x=(x_1,\cdots,x_{n-1},x_n)$ given by*

$$(6.3)\qquad g_{IJ}=\begin{cases}-S_{JI} & \text{for } j=i,\\ (-1)^iS_{J\cup\{n\},I} & \text{for } j=i-1.\end{cases}$$

PROOF. The right-hand side of (6.2) amounts to

$$|x|^2\sum_{K\in\mathfrak{I}_{n,i}}(\det\psi_n(x))_{KI}\langle \mathrm{pr}_{i\to j}(e_K),e_J^{\vee}\rangle.$$

The projection $\mathrm{pr}_{i\to j}\colon\bigwedge^i(\mathbb{C}^n)\to\bigwedge^j(\mathbb{C}^{n-1})$ in Section 1.1 sends the basis $\{e_I\}$ as follows.

$$\mathrm{pr}_{i\to i}(e_I)=\begin{cases}e_I & \text{for } n\notin I,\\ 0 & \text{for } n\in I,\end{cases}$$

$$\mathrm{pr}_{i\to i-1}(e_I)=\begin{cases}0 & \text{for } n\notin I,\\ (-1)^{i-1}e_{I\setminus\{n\}} & \text{for } n\in I.\end{cases}$$

Therefore, we get the desired result by (4.7). □

**6.2. Matrix components of $\widetilde{\mathbb{A}}^{i,j}_{\lambda,\nu,\pm}$.** For $I \in \mathfrak{I}_{n,i}$ and $J \in \mathfrak{I}_{n-1,j}$, we define $(\widetilde{\mathcal{A}}^{i,j}_{\lambda,\nu,\pm})_{IJ} \in \mathcal{D}'(\mathbb{R}^n)$ as the $(I,J)$-component of the distribution kernel $\widetilde{\mathcal{A}}^{i,j}_{\lambda,\nu,\pm}$ of the matrix-valued symmetry breaking operator $\widetilde{\mathbb{A}}^{i,j}_{\lambda,\nu,\pm}$ by

$$(\widetilde{\mathcal{A}}^{i,j}_{\lambda,\nu,\pm})_{IJ} := \langle \widetilde{\mathcal{A}}^{i,j}_{\lambda,\nu,\pm}(e_I), e_J^{\vee}\rangle.$$

We compare the matrix-valued symmetry breaking operator with the *shifted* scalar-valued one. We set

$$\widetilde{\mathcal{A}}_{\lambda,\nu,+} := a_+(\lambda,\nu)|x|^{-2\nu}|x_n|^{\lambda+\nu-n}. \tag{6.4}$$

This is a distribution with holomorphic parameter $(\lambda,\nu) \in \mathbb{C}^2$, and is identified with $\widetilde{\mathcal{A}}^{0,0}_{\lambda,\nu,+}$.

LEMMA 6.2. *The matrix component $(\widetilde{\mathcal{A}}^{i,j}_{\lambda,\nu,\pm})_{IJ}$ takes the following form:*

$$(\widetilde{\mathcal{A}}^{i,j}_{\lambda,\nu,+})_{IJ} =\frac{2}{\lambda-\nu-2} g_{IJ}\widetilde{\mathcal{A}}_{\lambda-1,\nu+1,+}, \tag{6.5}$$

$$(\widetilde{\mathcal{A}}^{i,j}_{\lambda,\nu,-})_{IJ} =\frac{2}{(\lambda+\nu-n)(\lambda-\nu-1)(\lambda-\nu-3)} x_n g_{IJ}\widetilde{\mathcal{A}}_{\lambda-2,\nu+1,+}. \tag{6.6}$$

PROOF. If $\mathrm{Re}\lambda \gg |\mathrm{Re}\nu|$, then the definition (1.6) of $\widetilde{\mathcal{A}}^{i,j}_{\lambda,\nu,+}$ shows that

$$\begin{aligned}(\widetilde{\mathcal{A}}^{i,j}_{\lambda,\nu,+})_{IJ} =& a_+(\lambda,\nu)|x|^{-2\nu}|x_n|^{\lambda+\nu-n}\langle \mathrm{pr}_{i\to j}\circ\sigma^{(i)}(\psi(x))(e_I), e_J^{\vee}\rangle\\ =& a_+(\lambda,\nu)|x|^{-2\nu-2}|x_n|^{\lambda+\nu-n} g_{IJ}(x),\end{aligned}$$

where the second equality follows from the definition (6.2) of $g_{IJ}(x)$. In view of the definition (6.4) of the scalar-valued distribution $\widetilde{\mathcal{A}}_{\lambda,\nu,+}$, the identity (6.5) holds for $\mathrm{Re}\lambda \gg |\mathrm{Re}\nu|$. A similar computation tells that the identity (6.6) holds for $\mathrm{Re}\lambda \gg |\mathrm{Re}\nu|$. Since $g_{IJ}$ is a polynomial, the right-hand sides of (6.5) and (6.6) are well-defined distributions on $\mathbb{R}^n$ that depend meromorphically on $(\lambda,\nu) \in \mathbb{C}^2$. On the other hand, the left-hand sides of (6.5) and (6.6) depend holomorphically on $(\lambda,\nu) \in \mathbb{C}^2$ by Fact 1.1. Therefore, the identities (6.5) and (6.6) are proved for all $(\lambda,\nu) \in \mathbb{C}^2$ by analytic continuation. □

**6.3. Review on the scalar-valued case.** In the case $i = j = 0$, the operator $\widetilde{\mathbb{A}}^{0,0}_{\lambda,\nu,+}$ is identified with scalar-valued one

$$\widetilde{\mathbb{A}}_{\lambda,\nu,+} \colon C_c^{\infty}(\mathbb{R}^n) \to C^{\infty}(\mathbb{R}^{n-1})$$

with the distribution kernel $\widetilde{\mathcal{A}}_{\lambda,\nu,+} \equiv \widetilde{\mathcal{A}}^{0,0}_{\lambda,\nu,+}$, which was thoroughly studied in [**9, 19**]. In particular, we determined the residue formula for the *scalar-valued* symmetry breaking operators $\widetilde{\mathbb{A}}_{\lambda,\nu,+}$ in [**9**] (see also [**19**, Thm. 12.2 (2)]) as follows. For $(\lambda,\nu) \in \mathbb{C}^2$ with $\nu-\lambda \in 2\mathbb{N}$, we set

$$q_C^A(\lambda,\nu) := \frac{(-1)^{\frac{\nu-\lambda}{2}}(\frac{\nu-\lambda}{2})!\pi^{\frac{n-1}{2}}}{2^{\nu-\lambda}\,\Gamma(\nu)}. \tag{6.7}$$

FACT 6.3 (residue formula in the scalar-valued case). Suppose $(\lambda,\nu) \in \mathbb{C}^2$ satisfies $\nu-\lambda \in 2\mathbb{N}$. Then we have

$$\widetilde{\mathbb{A}}_{\lambda,\nu,+} = q_C^A(\lambda,\nu)\widetilde{\mathbb{C}}_{\lambda,\nu}.$$

There are two approaches to the proof of the above residue formula. One is by using the Fourier transform of the distribution kernel (F-method) given in [**9**], and the other is by using the $(K, K')$-spectrum given in [**19**, Thm. 12.2 (2)].

**6.4. Reduction to the scalar-valued case.** We are ready to formulate an intermediate step for the proof of Theorem 1.3 on the residue formula of the matrix-valued regular symmetry breaking operators $\widetilde{\mathbb{A}}^{i,j}_{\lambda,\nu,\pm}$. We recall that $g_{IJ}(x)$ is a quadratic polynomial of $x$ (see Lemma 6.1) and $\widetilde{\mathcal{C}}_{\lambda,\nu}$ is a distribution on $\mathbb{R}^n$ supported at the origin (see (5.3)). Then we have the following.

PROPOSITION 6.4. (1) *If $\nu - \lambda = 2m$ with $m \in \mathbb{N}$, then*

$$(\widetilde{\mathcal{A}}^{i,j}_{\lambda,\nu,+})_{IJ} = \frac{(-1)^m m! \pi^{\frac{n-1}{2}}}{2^{2m+2}\Gamma(\nu+1)} g_{IJ}(x)\widetilde{\mathcal{C}}_{\lambda-1,\nu+1}.$$

(2) *If $\nu - \lambda = 2m+1$ with $m \in \mathbb{N}$, then*

$$(\widetilde{\mathcal{A}}^{i,j}_{\lambda,\nu,-})_{IJ} = \frac{(-1)^m m! \pi^{\frac{n-1}{2}}}{2^{2m+5}(\lambda+\nu-n)\Gamma(\nu+1)} x_n g_{IJ}(x)\widetilde{\mathcal{C}}_{\lambda-2,\nu+1}.$$

PROOF OF PROPOSITION 6.4. (1) By Fact 6.3, we have

$$\widetilde{\mathcal{A}}_{\lambda-1,\nu+1,+} = q_C^A(\lambda-1,\nu+1)\widetilde{\mathcal{C}}_{\lambda-1,\nu+1}$$

if $\nu - \lambda \in \{-2, 0, 2, 4, \cdots\}$. Combining this with (6.5), we obtain

$$(\widetilde{\mathcal{A}}^{i,j}_{\lambda,\nu,+})_{IJ} = \frac{2}{\lambda-\nu-2} q_C^A(\lambda-1,\nu+1)\widetilde{\mathcal{C}}_{\lambda-1,\nu+1} g_{IJ}.$$

Now a simple computation shows the first statement.
(2) Suppose $\nu - \lambda = 2m+1$ with $m \in \mathbb{N}$. By Fact 6.3, we have

$$\widetilde{\mathcal{A}}_{\lambda-2,\nu+1,+} = q_C^A(\lambda-2,\nu+1)\widetilde{\mathcal{C}}_{\lambda-2,\nu+1}.$$

Then (6.6) and Fact 6.3 tell that

$$\begin{aligned}(\widetilde{\mathcal{A}}^{i,j}_{\lambda,\nu,-})_{IJ} =& \frac{x_n g_{IJ}\widetilde{\mathcal{A}}_{\lambda-2,\nu+1,+}}{2(\lambda+\nu-n)(m+1)(m+2)} \\ =& \frac{q_C^A(\lambda-2,\nu+1)}{2(\lambda+\nu-n)(m+1)(m+2)} x_n g_{IJ}\widetilde{\mathcal{C}}_{\lambda-2,\nu+1}.\end{aligned}$$

Thus the second statement is shown. □

In order to find a closed expression of the right-hand sides of the formulæ in Proposition 6.4, we shall apply in the next section the identities in $\mathbb{C}[x, \frac{\partial}{\partial x}]/\mathcal{J}$ proved in Section 5. In particular, we shall see in Proposition 7.1 that a $\operatorname{Hom}_{\mathbb{C}}(\bigwedge^i(\mathbb{C}^n), \bigwedge^j(\mathbb{C}^{n-1}))$-valued distribution on $\mathbb{R}^n$ $(j = i-1, i)$ whose $(I, J)$-component is equal to $g_{IJ}(x)\widetilde{\mathcal{C}}_{\lambda-1,\nu+1}$ recovers the differential symmetry breaking operator

$$\mathbb{C}^{i,j}_{\lambda,\nu} : \mathcal{E}^i(\mathbb{R}^n) \to \mathcal{E}^j(\mathbb{R}^{n-1})$$

given in Section 1.

## 7. Proof of Theorem 1.3

In this section we complete the proof of the residue formula of the *matrix-valued* symmetry breaking operator $\widetilde{\mathbb{A}}^{i,j}_{\lambda,\nu,\pm}$ by comparing the $(I, J)$-component of the equation (1.7).

**7.1. Matrix components of matrix-valued differential symmetry breaking operators $\mathbb{C}^{i,j}_{\lambda,\nu}$.** We recall the matrix-valued differential operator from Section 1:

$$\mathbb{C}^{i,j}_{\lambda,\nu}\colon \mathcal{E}^i(\mathbb{R}^n) \to \mathcal{E}^j(\mathbb{R}^{n-1}) \quad (j = i-1, i).$$

For $I \in \mathfrak{I}_{n,i}$ and $J \in \mathfrak{I}_{n-1,j}$, we define a linear map

$$(\mathbb{C}^{i,j}_{\lambda,\nu})_{IJ}\colon C^\infty(\mathbb{R}^n) \to C^\infty(\mathbb{R}^{n-1})$$

as the $(I, J)$-component of the differential operator $\mathbb{C}^{i,j}_{\lambda,\nu}$, which is characterized by the formula

$$\mathbb{C}^{i,j}_{\lambda,\nu}(f(x)dx_I) = \sum_{J\in\mathfrak{I}_{n-1,j}} ((\mathbb{C}^{i,j}_{\lambda,\nu})_{IJ} f)dx_J.$$

Then there exist uniquely distributions $(\mathcal{C}^{i,j}_{\lambda,\nu})_{IJ}$ on $\mathbb{R}^n$ supported at the origin such that

$$(\mathbb{C}^{i,j}_{\lambda,\nu})_{IJ} = \operatorname{Rest}_{x_n=0} \circ (\mathcal{C}^{i,j}_{\lambda,\nu})_{IJ} * .$$

The explicit formulæ of $(\mathbb{C}^{i,j}_{\lambda,\nu})_{IJ}$ (or equivalently, of $(\mathcal{C}^{i,j}_{\lambda,\nu})_{IJ}$) are given in Facts 7.2 and 7.3 below.

Both the distributions $(\mathcal{C}^{i,j}_{\lambda,\nu})_{IJ}$ and $g_{IJ}(x)\widetilde{\mathcal{C}}_{\lambda-1,\nu+1}$ are distributions on $\mathbb{R}^n$ supported at the origin. We give its relationship as follows.

PROPOSITION 7.1. *Let $j = i$ or $i-1$. For any $I \in \mathcal{I}_{n,i}$ and $J \in \mathcal{I}_{n-1,j}$, we have*

$$g_{IJ}(x)\widetilde{\mathcal{C}}_{\lambda-1,\nu+1} = 8(-1)^{i-j}(\mathcal{C}^{i,j}_{\lambda,\nu})_{IJ}. \tag{7.1}$$

The next two subsections will be devoted to the proof of Proposition 7.1 by using the identities in Section 5. The cases $j = i$ and $j = i-1$ are treated separately in Sections 7.2 and 7.3. In Proposition 6.4, we have related the left-hand side of (7.1) with the $(I, J)$-component of the regular symmetry breaking operator $\widetilde{\mathbb{A}}^{i,j}_{\lambda,\nu,\pm}$. Thus we shall complete the proof of Theorem 1.3 based on Proposition 7.1. The last step will be given in Section 7.4.

**7.2. Proof of Proposition 7.1 for $j = i$.** Let $j = i$. Suppose $I \in \mathcal{I}_{n,i}$ and $J \in \mathcal{I}_{n-1,i}$. The main cases will be the following.
**Case 1.** $n \notin I$, $J = I$.
**Case 2.** $n \notin I$, $\#(J \setminus I) = 1$. We may write $I = K \cup \{p\}$, $J = K \cup \{q\}$.
**Case 3.** $n \in I$, $\#(J \setminus I) = 1$. We may write $I = K \cup \{n\}$, $J = K \cup \{q\}$.

FACT 7.2 ([**13**, Lem. 9.6]). The $(I, J)$-component $(\mathbb{C}^{i,i}_{\lambda,\nu})_{IJ}$ of the differential operator $\mathbb{C}^{i,i}_{\lambda,\nu}\colon \mathcal{E}^i(\mathbb{R}^n) \to \mathcal{E}^i(\mathbb{R}^{n-1})$ is equal to

**Case 1.** $-\widetilde{\mathbb{C}}_{\lambda+1,\nu-1}(\sum_{p\in I} \frac{\partial^2}{\partial x_p^2}) + \frac{1}{2}(\nu - i)\widetilde{\mathbb{C}}_{\lambda,\nu}$,

**Case 2.** $-\operatorname{sgn}(I; p, q)\widetilde{\mathbb{C}}_{\lambda+1,\nu-1}\frac{\partial^2}{\partial x_p \partial x_q}$,

**Case 3.** $-\operatorname{sgn}(I; q, n)\gamma(\lambda - \frac{n-1}{2}, \nu - \lambda)\widetilde{\mathbb{C}}_{\lambda+1,\nu}\frac{\partial}{\partial x_q}$,

and is zero otherwise.

PROOF OF PROPOSITION 7.1 FOR $j = i$. It is readily seen that the both sides of (7.1) vanish unless $(I, J)$ belongs to Cases 1–3. From now, we focus on Cases 1–3.

**Case 1.** By (4.5), the polynomial $g_{IJ}(x)$ in (6.3) amounts to $g_{IJ}(x) = x_n^2 - \sum_{k=1}^{n-1} \varepsilon_I(k) x_k^2$. Then a small computation using (5.14) and (5.15) shows the following equalities in $\mathcal{D}'_{\{0\}}(\mathbb{R}^n)$:

$$g_{IJ}(x)\widetilde{\mathcal{C}}_{\lambda-1,\nu+1} = 4(\nu - i)\widetilde{\mathcal{C}}_{\lambda,\nu} - 8(\sum_{k\in I} \frac{\partial^2}{\partial x_k^2})\widetilde{\mathcal{C}}_{\lambda+1,\nu-1}.$$

By Fact 7.2 in Case 1 and (3.3), we get

$$\operatorname{Rest}_{x_n=0} \circ (g_{IJ}\widetilde{\mathcal{C}}_{\lambda-1,\nu+1})* = 8(\mathbb{C}_{\lambda,\nu}^{i,i})_{IJ}. \tag{7.2}$$

**Case 2.** In this case, $g_{IJ}(x) = -2\operatorname{sgn}(K;p,q)x_p x_q$. By (5.13),

$$\begin{aligned} g_{IJ}(x)\widetilde{\mathcal{C}}_{\lambda-1,\nu+1} &= -2\operatorname{sgn}(K;p,q)x_p x_q \widetilde{\mathcal{C}}_{\lambda-1,\nu+1} \\ &= -8\operatorname{sgn}(K;p,q)\frac{\partial^2}{\partial x_p \partial x_q}\widetilde{\mathcal{C}}_{\lambda+1,\nu-1}. \end{aligned}$$

Comparing this with Fact 7.2 in Case 2, we get (7.2) in this case.

**Case 3.** Suppose $n \in I$ and $I = K \cup \{n\}$, $J = K \cup \{q\}$. Then

$$g_{IJ}(x) = 2\operatorname{sgn}(K;q,n)x_q x_n.$$

By (5.12), we have

$$\begin{aligned} g_{IJ}(x)\widetilde{\mathcal{C}}_{\lambda-1,\nu+1} &= 2\operatorname{sgn}(K;q,n)x_q x_n \widetilde{\mathcal{C}}_{\lambda-1,\nu+1} \\ &= -8\gamma(\lambda - \frac{n-1}{2}, \nu - \lambda)\operatorname{sgn}(K;q,n)\frac{\partial}{\partial x_q}\widetilde{\mathcal{C}}_{\lambda+1,\nu}. \end{aligned}$$

Again by Fact 7.2, we get (7.2) in this case. □

**7.3. Proof of Proposition 7.1 for $j = i - 1$.** In this section, we give a proof of Proposition 7.1 for $j = i - 1$. Suppose $I \in \mathcal{I}_{n,i}$ and $J \in \mathcal{I}_{n-1,i-1}$. The main cases will be the following.

**Case 1.** $n \in I$, $J = I \setminus \{n\}$.

**Case 2.** $n \in I$, $\#(J \setminus I) = 1$. We may write $I = K \cup \{p, n\}$, $J = K \cup \{q\}$.

**Case 3.** $n \notin I$, $J \subset I$. We may write $I = J \cup \{p\}$.

FACT 7.3 ([**13**, Lem. 9.5]). *The $(I,J)$-component $(\mathbb{C}_{\lambda,\nu}^{i,i-1})_{IJ}$ of the differential operator $\mathbb{C}_{\lambda,\nu}^{i,i-1} : \mathcal{E}^i(\mathbb{R}^n) \to \mathcal{E}^{i-1}(\mathbb{R}^{n-1})$ is equal to*

**Case 1.** $(-1)^{i-1}(-\widetilde{\mathbb{C}}_{\lambda+1,\nu-1}(\sum_{p\notin I} \frac{\partial^2}{\partial x_p^2}) + \frac{\nu+i-n}{2}\widetilde{\mathbb{C}}_{\lambda,\nu})$,

**Case 2.** $(-1)^{i-1}\operatorname{sgn}(I;p,q)\widetilde{\mathbb{C}}_{\lambda+1,\nu-1}\frac{\partial^2}{\partial x_p \partial x_q}$,

**Case 3.** $\operatorname{sgn}(I;p)\gamma(\lambda - \frac{n-1}{2}, \nu - \lambda)\widetilde{\mathbb{C}}_{\lambda+1,\nu}\frac{\partial}{\partial x_p}$,

*and is zero otherwise.*

PROOF OF PROPOSITION 7.1 FOR $j = i - 1$.

**Case 1.** By (4.5), we have $g_{IJ}(x) = (-1)^i(x_n^2 + \sum_{k=1}^{n-1}\varepsilon_I(k)x_k^2)$. Then (5.14) and (5.15) tell that

$$(-1)^i g_{IJ}(x)\widetilde{\mathcal{C}}_{\lambda-1,\nu+1} = 4(\nu - n + i)\widetilde{\mathcal{C}}_{\lambda,\nu} - 8\sum_{k\notin I}\frac{\partial^2}{\partial x_k^2}\widetilde{\mathcal{C}}_{\lambda+1,\nu-1}.$$

Comparing this with Fact 7.3 in Case 1, we get

$$\operatorname{Rest}_{x_n=0} \circ (g_{IJ}\widetilde{\mathcal{C}}_{\lambda-1,\nu+1})* = -8(\mathbb{C}_{\lambda,\nu}^{i,i-1})_{IJ}. \tag{7.3}$$

**Case 2.** In this case, $g_{IJ}(x) = 2(-1)^i \operatorname{sgn}(K; p, q) x_p x_q$. By (5.13), we have

$$g_{IJ}(x)\widetilde{\mathcal{C}}_{\lambda-1,\nu+1} = 8(-1)^i \operatorname{sgn}(K; p, q) \frac{\partial^2}{\partial x_p \partial x_q} \widetilde{\mathcal{C}}_{\lambda+1,\nu-1}.$$

Hence, Fact 7.3 in Case 2 implies (7.3).

**Case 3.** In this case, $g_{IJ}(x) = -2 \operatorname{sgn}(I; p) x_p x_n$. Then (5.12) implies

$$g_{IJ}(x)\widetilde{\mathcal{C}}_{\lambda-1,\nu+1} = -8\gamma(\lambda - \frac{n-1}{2}, \nu - \lambda) \operatorname{sgn}(I; p) \frac{\partial}{\partial x_p} \widetilde{\mathcal{C}}_{\lambda+1,\nu}.$$

Hence Fact 7.3 in Case 3 shows (7.3). □

**7.4. Proof of Theorem 1.3.** We are ready to complete the proof of Theorem 1.3.

Proof of Theorem 1.3. Let $\gamma = 0$. Theorem 1.3 in this case follows from Propositions 6.4 and 7.1.

Next, let $\gamma = 1$. By (5.9) and Proposition 7.1,

$$\begin{aligned} x_n g_{IJ} \widetilde{\mathcal{C}}_{\lambda-2,\nu+1} &= (n - \lambda - \nu) g_{IJ} \widetilde{\mathcal{C}}_{\lambda-1,\nu+1} \\ &= 8(-1)^{i-j+1}(\lambda + \nu - n)(\mathcal{C}^{i,j}_{\lambda,\nu})_{IJ}. \end{aligned}$$

Hence Theorem 1.3 for $\gamma = 1$ is shown. □

## 8. Vanishing condition of the symmetry breaking operator $\widetilde{\mathbb{A}}^{i,j}_{\lambda,\nu,\pm}$

As a corollary of Theorem 1.3, we can determine the (isolated) zeros of the analytic continuation $\widetilde{\mathbb{A}}^{i,j}_{\lambda,\nu,\pm}$ of the integral operators. This generalizes one of the most important discoveries of [**19**] about the isolated zeros of the normalized operator $\widetilde{\mathbb{A}}^{i,j}_{\lambda,\nu,\pm}$ in the special case $i = j = 0$ ([**19**, Thm. 1.5]). However, our method is based on the residue formula, and is different from the previous approach based on the analysis of $(K, K')$-spectrum. Keeping the same notation as in [**19**, Chap. 1], we define two subsets in $\mathbb{Z}^2$ as below:

$$\begin{aligned} L_{\text{even}} &:= \{(-i, -j) : 0 \le j \le i \text{ and } i \equiv j \mod 2\}, \\ L_{\text{odd}} &:= \{(-i, -j) : 0 \le j \le i \text{ and } i \equiv j + 1 \mod 2\}. \end{aligned}$$

Theorem 8.1. (1) *Suppose* $\nu - \lambda \in 2\mathbb{N}$.

$\widetilde{\mathbb{A}}^{i,i}_{\lambda,\nu,+} = 0$ *if and only if*

$$(\lambda, \nu) \in \begin{cases} L_{\text{even}} & \text{for } i = 0, \\ (L_{\text{even}} \setminus \{\nu = 0\}) \cup \{(i, i)\} & \text{for } 1 \le i \le n-1. \end{cases}$$

$\widetilde{\mathbb{A}}^{i,i-1}_{\lambda,\nu,+} = 0$ *if and only if*

$$(\lambda, \nu) \in \begin{cases} (L_{\text{even}} \setminus \{\nu = 0\}) \cup \{(n-i, n-i)\} & \text{for } 1 \le i \le n-1, \\ L_{\text{even}} & \text{for } i = n. \end{cases}$$

(2) *Suppose* $\nu - \lambda \in 2\mathbb{N}+1$.
$\widetilde{\mathbb{A}}^{i,i}_{\lambda,\nu,-} = 0$ *if and only if*

$$(\lambda,\nu) \in \begin{cases} L_{\mathrm{odd}} & \text{for } i = 0, \\ L_{\mathrm{odd}} \setminus \{\nu = 0\} & \text{for } 1 \le i \le n-1. \end{cases}$$

$\widetilde{\mathbb{A}}^{i,i-1}_{\lambda,\nu,-} = 0$ *if and only if*

$$(\lambda,\nu) \in \begin{cases} L_{\mathrm{odd}} \setminus \{\nu = 0\} & \text{for } 1 \le i \le n-1, \\ L_{\mathrm{odd}} & \text{for } i = n. \end{cases}$$

As we have already mentioned, the strategy of our proof is to apply the residue formula (Theorem 1.3). Then we can reduce the proof of Theorem 8.1 to an easier question, that is, to find a necessary and sufficient condition for the matrix-valued differential symmetry breaking operators $\mathbb{C}^{i,j}_{\lambda,\nu}$ to vanish. The latter condition is verified immediately by the formula for the $(I,J)$-component of $\mathbb{C}^{i,j}_{\lambda,\nu}$ (see Facts 7.2 and 7.3), and is described as follows:

LEMMA 8.2 ([**13**, Prop. 1.4 and p. 23]). *Suppose* $(\lambda,\nu) \in \mathbb{C}^2$ *with* $\nu - \lambda \in \mathbb{N}$.

(1) *Let* $1 \le i \le n$. *Then* $\mathbb{C}^{i,i}_{\lambda,\nu}$ *vanishes if and only if* $\lambda = \nu = i$ *or* $\nu = i = 0$.

(2) *Let* $1 \le i \le n-1$. *Then* $\mathbb{C}^{i,i-1}_{\lambda,\nu}$ *vanishes if and only if* $\lambda = \nu = n-i$ *or* $\nu = n - i = 0$.

PROOF OF THEOREM 8.1. Suppose

$$\nu - \lambda \in 2\mathbb{N} \quad (\varepsilon = +) \quad \text{or} \quad \nu - \lambda \in 2\mathbb{N}+1 \quad (\varepsilon = -).$$

In either case, the residue formula in Theorem 1.3 asserts that

$$\widetilde{\mathbb{A}}^{i,j}_{\lambda,\nu,\varepsilon} = \frac{c}{\Gamma(\nu+1)}\mathbb{C}^{i,j}_{\lambda,\nu},$$

for some $c \neq 0$. Therefore we have

$$\widetilde{\mathbb{A}}^{i,j}_{\lambda,\nu,\varepsilon} = 0 \quad \text{if and only if} \quad \nu \in -\mathbb{N}_+ \text{ or } \mathbb{C}^{i,j}_{\lambda,\nu} = 0.$$

In light of Lemma 8.2, we conclude Theorem 8.1. □

REMARK 8.3. By the general results (see [**21**]), $\widetilde{\mathbb{A}}^{i,j}_{\lambda,\nu,\varepsilon}$ vanishes only if

$$\nu - \lambda \in 2\mathbb{N} \quad (\varepsilon = +) \quad \text{or} \quad \nu - \lambda \in 2\mathbb{N}+1 \quad (\varepsilon = -).$$

Hence Theorem 8.1 determines precisely when the regular symmetry breaking operator $\widetilde{\mathbb{A}}^{i,j}_{\lambda,\nu,\pm}$ vanishes.

## References


[1] G. E. Andrews, R. Askey, and R. Roy, *Special functions*, Encyclopedia of Mathematics and its Applications, vol. 71, Cambridge University Press, Cambridge, 1999. MR1688958

[2] T. P. Branson, *Conformally covariant equations on differential forms*, Comm. Partial Differential Equations **7** (1982), no. 4, 393–431. MR652815

[3] M. Fischmann, A. Juhl, P. Somberg, Conformal symmetry breaking differential operators on differential forms, preprint, arXiv:1605.04517.

[4] M. Fischmann, B.Ørsted, A family of Riesz distributions for differential forms on Euclidean space, arXiv:1702.00930.

[5] I. M. Gel'fand and G. E. Shilov, *Generalized functions. Vol. I: Properties and operations*, Translated by Eugene Saletan, Academic Press, New York-London, 1964. MR0166596

[6] A. Juhl, *Families of conformally covariant differential operators, Q-curvature and holography*, Progress in Mathematics, vol. 275, Birkhäuser Verlag, Basel, 2009. MR2521913

[7] A. W. Knapp and E. M. Stein, *Intertwining operators for semisimple groups*, Ann. of Math. (2) **93** (1971), 489–578. MR0460543

[8] Toshiyuki Kobayashi, *F-method for constructing equivariant differential operators*, Geometric analysis and integral geometry, Contemp. Math., vol. 598, Amer. Math. Soc., Providence, RI, 2013, pp. 139–146. MR3156443

[9] Toshiyuki Kobayashi, *F-method for symmetry breaking operators*, Differential Geom. Appl. **33** (2014), no. suppl., 272–289. MR3159963

[10] Toshiyuki Kobayashi, *A program for branching problems in the representation theory of real reductive groups*, Representations of reductive groups, Progr. Math., vol. 312, Birkhäuser/Springer, Cham, 2015, pp. 277–322. MR3495800

[11] Toshiyuki Kobayashi, Toshihisa Kubo, and Michael Pevzner, *Vector-valued covariant differential operators for the Möbius transformation*, Lie theory and its applications in physics, Springer Proc. Math. Stat., vol. 111, Springer, Tokyo, 2014, pp. 67–85. MR3619943

[12] Toshiyuki Kobayashi, Toshihisa Kubo, and Michael Pevzner, *Classification of differential symmetry breaking operators for differential forms* (English, with English and French summaries), C. R. Math. Acad. Sci. Paris **354** (2016), no. 7, 671–676. MR3508562

[13] Toshiyuki Kobayashi, Toshihisa Kubo, and Michael Pevzner, *Conformal symmetry breaking operators for differential forms on spheres*, Lecture Notes in Mathematics, vol. 2170, Springer, Singapore, 2016. MR3559958

[14] Toshiyuki Kobayashi, Toshihisa Kubo, and Michael Pevzner, *Conformal symmetry breaking operators for anti-de Sitter spaces.* in press, Trends in Math., pp. 69–85, Birkhäuser-Springer; available also at arXiv:1610.09475.

[15] Toshiyuki Kobayashi and Bent Ørsted, *Analysis on the minimal representation of* $\mathrm{O}(p,q)$. *I. Realization via conformal geometry*, Adv. Math. **180** (2003), no. 2, 486–512. MR2020550

[16] Toshiyuki Kobayashi, Bent Ørsted, Petr Somberg, and Vladimír Souček, *Branching laws for Verma modules and applications in parabolic geometry. I*, Adv. Math. **285** (2015), 1796–1852. MR3406542

[17] Toshiyuki Kobayashi and Toshio Oshima, *Finite multiplicity theorems for induction and restriction*, Adv. Math. **248** (2013), 921–944. MR3107532

[18] Toshiyuki Kobayashi and Michael Pevzner, *Differential symmetry breaking operators: I. General theory and F-method*, Selecta Math. (N.S.) **22** (2016), no. 2, 801–845. MR3477336

[19] Toshiyuki Kobayashi and Birgit Speh, *Symmetry breaking for representations of rank one orthogonal groups*, Mem. Amer. Math. Soc. **238** (2015), no. 1126, v+110. MR3400768

[20] Toshiyuki Kobayashi and Birgit Speh, Symmetry breaking for orthogonal groups and a conjecture by B. Gross and D. Prasad. In: Geometric Aspects of the Trace Formula. Simons Symp., Springer, Cham, (in press), available also at arXiv:1702.00263.

[21] Toshiyuki Kobayashi and Birgit Speh, *Symmetry breaking for representations of rank one orthogonal groups*, Mem. Amer. Math. Soc. **238** (2015), no. 1126, v+110. MR3400768

Graduate School of Mathematical Sciences, The University of Tokyo, Komaba, Meguro-ku, Tokyo, Japan, 153-8914, and Kavli IPMU

*Email address*: toshi@ms.u-tokyo.ac.jp

Contemporary Mathematics
Volume **714**, 2018
https://doi.org/10.1090/conm/714/14330

# On the geometry of standard subspaces

Karl-Hermann Neeb

Abstract. A closed real subspace $V$ of a complex Hilbert space $\mathcal{H}$ is called standard if $V \cap iV = \{0\}$ and $V + iV$ is dense in $\mathcal{H}$. In this note we study several aspects of the geometry of the space $\mathrm{Stand}(\mathcal{H})$ of standard subspaces. In particular, we show that modular conjugations define the structure of a reflection space and that the modular automorphism groups extend this to the structure of a dilation space. Every antiunitary representation of a graded Lie group $G$ leads to a morphism of dilation spaces $\mathrm{Hom}_{\mathrm{gr}}(\mathbb{R}^\times, G) \to \mathrm{Stand}(\mathcal{H})$. Here dilation invariant geodesics (with respect to the reflection space structure) correspond to antiunitary representations $U$ of $\mathrm{Aff}(\mathbb{R})$ and they are decreasing if and only if $U$ is a positive energy representation. We also show that the ordered symmetric spaces corresponding to euclidean Jordan algebras have natural order embeddings into $\mathrm{Stand}(\mathcal{H})$ obtained from any antiunitary positive energy representation of the conformal group.

Contents



## Introduction

A closed real subspace $V$ of a complex Hilbert space $\mathcal{H}$ is called *standard* if $V \cap iV = \{0\}$ and $V + iV$ is dense in $\mathcal{H}$ ([**Lo08**]). We write $\mathrm{Stand}(\mathcal{H})$ for the set of standard subspaces of $\mathcal{H}$. The main goal of this note is to shed some light on the geometric structure of this space and how it can be related to geometric structures on manifolds on which Lie groups $G$ act via antiunitary representations on $\mathcal{H}$.

Standard subspaces arise naturally in the modular theory of von Neumann algebras. If $\mathcal{M} \subseteq B(\mathcal{H})$ is a von Neumann algebra and $\xi \in \mathcal{H}$ is a cyclic separating

2010 *Mathematics Subject Classification.* Primary 22E45; Secondary 81R05, 81T05.
K.-H. Neeb acknowledges support by DFG-grant NE 413/9-1.

vector for $\mathcal{M}$, i.e., $\mathcal{M}\xi$ is dense in $\mathcal{H}$ and the map $\mathcal{M} \to \mathcal{H}, M \mapsto M\xi$ is injective, then

$$V_{\mathcal{M}} := \overline{\{M\xi \colon M^* = M, M \in \mathcal{M}\}}$$

is a standard subspace of $\mathcal{H}$. Conversely, one can associate to every standard subspace $V \subseteq \mathcal{H}$ in a natural way a von Neumann algebra in the bosonic and fermionic Fock space of $\mathcal{H}$, and this assignment has many nice properties (see [**NÓ17**, §§4,6] and [**Lo08**] for details). This establishes a direct connection between standard subspaces and pairs $(\mathcal{M}, \xi)$ of von Neumann algebras with cyclic separating vectors. Since the latter objects play a key role in Algebraic Quantum Field Theory in the context of Haag–Kastler nets ([**Ar99**, **Ha96**, **BW92**]), it is important to understand the geometric structure of the space $\mathrm{Stand}(\mathcal{H})$. Here a key point is that it reflects many important properties of von Neumann algebras related to modular inclusions and symmetry groups quite faithfully in a much simpler environment ([**NÓ17**, §4.2]). We refer to [**Lo08**] for an excellent survey on this correspondence. In QFT, standard subspaces provide the basis for the technique of modular localization, developed by Brunetti, Guido and Longo in [**BGL02**].

Every standard subspace $V$ determines by the polar decomposition of the closed operator $S$, defined on $V + iV$ by $S(x + iy) = x - iy$, a pair $(\Delta_V, J_V)$ of so-called modular objects, i.e., $\Delta_V$ is a positive selfadjoint operator and $J_V$ is a conjugation (an antiunitary involution) satisfying $J_V \Delta_V J_V = \Delta_V^{-1}$. This correspondence leads to a bijection

$$\Psi \colon \mathrm{Mod}(\mathcal{H}) \to \mathrm{Stand}(\mathcal{H}), \quad (\Delta, J) \mapsto \mathrm{Fix}(J\Delta^{1/2})$$

between the set $\mathrm{Mod}(\mathcal{H})$ of pairs of modular objects and $\mathrm{Stand}(\mathcal{H})$.

There actually is a third model of $\mathrm{Stand}(\mathcal{H})$ that comes from the fact that each pair $(\Delta, J)$ defines a homomorphism

$$\gamma \colon \mathbb{R}^\times \to \mathrm{AU}(\mathcal{H}) \quad \text{by} \quad \gamma(e^t) := \Delta^{-it/2\pi}, \quad \gamma(-1) := J,$$

where $\mathrm{AU}(\mathcal{H})$ denotes the group of unitary and antiunitary operators on $\mathcal{H}$. This perspective will play a crucial role in our analysis of $\mathrm{Stand}(\mathcal{H})$.

In Section 1 we discuss Loos' concept of a reflection space, which is a generalization of the concept of a symmetric space. Although symmetric spaces play a central role in differential geometry and harmonic analysis for more than a century, reflection spaces never received much attention. As we shall see below, they provide exactly the right framework to study the geometry of $\mathrm{Stand}(\mathcal{H})$. Reflection spaces are specified in terms of a system $(s_x)_{x\in M}$ of involutions satisfying

$$s_x(x) = x \quad \text{and} \quad s_x s_y s_x = s_{s_x y} \quad \text{for} \quad x, y \in M.$$

One sometimes has even more structure encoded in a family $(r_x)_{r\in\mathbb{R}^\times, x\in M}$ of $\mathbb{R}^\times$-actions on $M$ satisfying

$$r_x(x) = x, \quad r_x s_x = (rs)_x \quad \text{and} \quad r_x s_y r_x^{-1} = s_{r_x y} \quad \text{for} \quad x, y \in M, \ r, s \in \mathbb{R}^\times.$$

This defines the structure of a *dilation space*, a concept studied in the more general context of $\Sigma$-spaces by Loos in [**Lo72**]. For $r = -1$, we obtain a reflection space, so that a dilation space is a reflection space with additional structure. Other important classes of dilation spaces are the ruled spaces discussed in [**Be00**, Ch. VI] that arise naturally in Jordan theory. Topological reflection spaces also appeared recently in the context of Kac–Moody symmetric space (see [**FHHK17**]). Discrete reflection spaces also arise as quandles whose structure maps are involutions in knot theory ([**EN15**, **Nel16**]).

In Section 2 we turn to the space $\mathrm{Stand}(\mathcal{H})$ of standard subspaces and show that it carries a natural dilation space structure. This corresponds naturally to dilation space structures on the sets $\mathrm{Mod}(\mathcal{H})$ and $\mathrm{Hom}(\mathbb{R}^\times, \mathrm{AU}(\mathcal{H}))$. The underlying reflection space structure on $\mathrm{Stand}(\mathcal{H})$ is given by

$$V_1 \bullet V_2 := s_{V_1} V_2 = J_1 J_2 V_2$$

and the map

$$q\colon \mathrm{Stand}(\mathcal{H}) \to \mathrm{Conj}(\mathcal{H}), \quad V \mapsto J_V$$

onto the symmetric space $\mathrm{Conj}(\mathcal{H})$ of antiunitary involutions on $\mathcal{H}$ is a morphism of reflection spaces. Here an interesting point is that the symmetric space $\mathrm{Conj}(\mathcal{H})$ does not carry a non-trivial dilation space structure, but that the weaker structure of a reflection space on $\mathrm{Stand}(\mathcal{H})$ actually extends to the much richer dilation space structure. If $(G, \varepsilon_G)$ is a graded topological group, i.e., $\varepsilon_G \to \{\pm 1\}$ is a continuous homomorphism, then, for every antiunitary representation $U\colon G \to \mathrm{AU}(\mathcal{H})$, the natural map $U_*\colon \mathrm{Hom}_{\mathrm{gr}}(\mathbb{R}^\times, G) \to \mathrm{Hom}_{\mathrm{gr}}(\mathbb{R}^\times, \mathrm{AU}(\mathcal{H}))$ defines a morphism of dilation spaces $\mathcal{V}_U\colon \mathrm{Hom}_{\mathrm{gr}}(\mathbb{R}^\times, G) \to \mathrm{Stand}(\mathcal{H})$ which is the Brunetti–Guido–Longo (BGL) map from [**NÓ17**, Prop. 5.6] and [**BGL02**, Thm. 2.5].

A morphism of reflection spaces $\gamma\colon \mathbb{R} \to \mathrm{Stand}(\mathcal{H})$ is called a geodesic. In Proposition 2.9 we describe the geodesics for which $q \circ \gamma$ is continuous in terms of unitary one-parameter groups $(U_t)_{t\in\mathbb{R}}$. They are of the form

$$\gamma(t) = U_t V, \quad \text{where} \quad J_V U_t J_V = U_{-t} \quad \text{for} \quad t \in \mathbb{R}.$$

On the other hand, we have for each $V \in \mathrm{Stand}(\mathcal{H})$ the corresponding dilation group implemented by the unitary operators $(\Delta_V^{it})_{t\in\mathbb{R}}$. Both structures interact nicely for geodesics invariant under the dilation group. In Proposition 2.11 we show that, if $(U_t)_{t\in\mathbb{R}}$ does not commute with the dilations $(\Delta_V^{is})_{s\in\mathbb{R}}$, the geodesic is an orbit of $\mathrm{Aff}(\mathbb{R})_0$ in $\mathrm{Stand}(\mathcal{H})$, where the action is given by an antiunitary representation.

A particularly intriguing structure on $\mathrm{Stand}(\mathcal{H})$ is the order structure defined by set inclusion to which we turn in Section 3. This structure is trivial if $\mathcal{H}$ is finite dimensional and it is also trivial on the subspace

$$\mathrm{Stand}_0(\mathcal{H}) := \{V \in \mathrm{Stand}(\mathcal{H})\colon V + iV = \mathcal{H}\}.$$

But if $\mathcal{H}$ is infinite dimensional non-trivial inclusions can be obtained from antiunitary positive energy representations of $\mathrm{Aff}(\mathbb{R})$, which actually lead to monotone dilation invariant geodesics (Theorem 3.3). This is a direct consequence of the theorems of Borchers and Wiesbrock (cf. [**Lo08**], [**NÓ17**]) and the dilation space structure thus provides a new geometric perspective on these results that were originally formulated in terms of inclusions of von Neumann algebras ([**Bo92**], [**Wi93**]). In view of this characterization of the monotone dilation invariant geodesics, it is an interesting open problem to characterize all monotone geodesics in $\mathrm{Stand}(\mathcal{H})$. To get some more information on the ordered space $\mathrm{Stand}(\mathcal{H})$, one natural strategy is to consider finite dimensional submanifolds, resp., orbits $\mathcal{O}_V := U_{G_1}.V \cong G_1/G_{1,V}$, where $U$ is an antiunitary representation. Then

$$S_V := \{g \in G_1 \colon U_g V \subseteq V\} \tag{0.1}$$

is a closed subsemigroup of $G_1$ with $G_{1,V} = S_V \cap S_V^{-1}$ and $S_V$ determines an order structure on $G_1/G_{1,V}$ by $gG_{1,V} \leq g'G_{1,V}$ if $g \in g'S_V$ for which the inclusion

$G_1/G_{1,V} \hookrightarrow \mathrm{Stand}(\mathcal{H})$ is an equivariant order embedding. Of course, the most natural cases arise if $V$ corresponds to some $\gamma \in \mathrm{Hom}_{\mathrm{gr}}(\mathbb{R}^\times, G)$ under the BGL construction and then $G_{1,\gamma} \subseteq G_{1,V}$, so that $\mathcal{O}_V$ is a $G_1$-equivariant quotient of $G_1/G_{1,\gamma} \cong G_1.\gamma \subseteq \mathrm{Hom}_{\mathrm{gr}}(\mathbb{R}^\times, G)$. We conclude this note by showing that, if $G$ is the conformal group of a euclidean Jordan algebra $E$ and $\gamma\colon \mathbb{R}^\times \to G$ corresponds to scalar multiplication on $E$, the ordered homogeneous spaces $U_{G_1}V$, $V := \mathcal{V}_U(\gamma)$, obtained from an antiunitary positive energy representations $(U, \mathcal{H})$ of $G$, are mutually isomorphic and the order structure can be described by showing that the semigroup $S_V$ coincides with the well-known Olshanski semigroup $S_{E_+}$ of conformal compressions of the open positive cone $E_+$ ([**HN93**], [**Ko95**]). This result is based on the maximality of the subsemigroup $S_{E_+}$ in $G_1$ which is proved in an appendix.

## 1. Reflection and dilation spaces

In this first section we first review some generalities on reflection spaces ([**Lo67**]) and introduce the notion of a dilation space by specialization of Loos' more general concept of a $\Sigma$-space ([**Lo72**]). A key feature of these abstract concepts is that they work well in many categories, in particular in the category of sets and the category of topological spaces and not only in the category of smooth manifolds. Only when it comes the finer geometric points related to the concept of a symmetric space, a smooth structure is required. In Section 2 this will be crucial for the space $\mathrm{Stand}(\mathcal{H})$ which carries no natural smooth structure but which is fibered over the topological space $\mathrm{Conj}(\mathcal{H})$, endowed with the strong operator topology.

### 1.1. Reflection spaces.

DEFINITION 1.1. (a) Let $M$ be a set and

$$\mu\colon M \times M \to M, \quad (x, y) \mapsto x \bullet y =: s_x(y)$$

be a map with the following properties:

(S1) $x \bullet x = x$ for all $x \in M$, i.e., $s_x(x) = x$.
(S2) $x \bullet (x \bullet y) = y$ for all $x, y \in M$, i.e., $s_x^2 = \mathrm{id}_M$.
(S3) $s_x(y \bullet z) = s_x(y) \bullet s_x(z)$ for all $x, y \in M$, i.e., $s_x \in \mathrm{Aut}(M, \bullet)$.

Then we call $(M, \mu)$ a *reflection space* ([**Lo67, Lo67b**]).

(b) If $M$ be a smooth manifold and $\mu\colon M \times M \to M$ is a smooth map turning $(M, \mu)$ into a reflection space, then it is called a smooth reflection space. If, in addition, each $x$ is an isolated fixed point of $s_x$, then it is called a *symmetric space* (in the sense of Loos).

If $M$ is a topological space and $\mu$ is continuous, we call it a *topological reflection space*.

(c) If $(M, \bullet)$ and $(N, \bullet)$ are reflection spaces, then a map $f\colon M \to N$ is called a *morphism of reflection spaces* if

$$f(m \bullet m') = f(m) \bullet f(m') \quad \text{for} \quad m, m' \in M.$$

EXAMPLE 1.2. (a) Any group $G$ is a reflection space with respect to the product

$$g \bullet h := s_g(h) := gh^{-1}g. \tag{1.1}$$

Note that left and right translations

$$\lambda_g(x) = gx \quad \text{and} \quad \rho_g(x) = xg$$

are automorphisms of the reflection space $(G, \bullet)$.

The subset $\mathrm{Inv}(G)$ of involutions in $G$ is a reflection subspace on which the product takes the form $s_g(h) := ghg = ghg^{-1}$.

(b) Suppose that $G$ is a group and $\tau \in \mathrm{Aut}(G)$ is an involution. For any subgroup $H \subseteq G^\tau := \mathrm{Fix}(\tau)$, we obtain on the coset space $M := G/H$ the structure of a reflection space by

$$xH \bullet yH := x\tau(x)^{-1}\tau(y)H. \tag{1.2}$$

For this reflection space structure all left translations $\mu_g \colon G/H \to G/H, xH \mapsto gxH$ are automorphisms.

If, in addition, $G$ is a Banach–Lie group and $H$ is a complemented Lie subgroup, then $G/H$ is a smooth reflection space. It is a symmetric space if and only if $H$ is an open subgroup of the group $G^\tau$. In fact, if $e_M := eH$ is the base point of $G/H$, then $T_{e_M}(G/H) \cong \mathfrak{g}/\mathfrak{h}$ and the tangent map of $s_{e_M}(gH) = \tau(g)H$ is the involution on $\mathfrak{g}/\mathfrak{h}$ induced by $T_e(\tau)$. This equals $-\mathrm{id}_{\mathfrak{g}/\mathfrak{h}}$ if and only if $\mathfrak{h} = \mathfrak{g}^\tau$, which is equivalent to $H$ being an open subgroup of $G^\tau$. For any open subgroup $H \subseteq G^\tau = \mathrm{Fix}(\tau)$, we thus obtain by (1.2) on $G/H$ the structure of a symmetric space.

Note that $H = \{e\}$ is also allowed, showing that

$$x \bullet y := x\tau(x)^{-1}\tau(y) \tag{1.3}$$

also defines on $G$ the structure of a smooth reflection space.

(c) Every manifold $M$ is a smooth reflection space with respect to $x \bullet y := y$ for $x, y \in M$.

(d) If $(M, \bullet)$ and $(N, \bullet)$ are smooth reflection spaces, then so is their product $M \times N$ with respect to

$$(m, n) \bullet (m', n') := (m \bullet m', n \bullet n').$$

(e) If $(M, \bullet)$ is a reflection space and $q \colon M \to N$ is a surjective submersion whose kernel relation is a congruence relation with respect to $\bullet$, i.e., $q(x) = q(x')$ and $q(y) = q(y')$ implies $q(x \bullet y) = q(x' \bullet y')$, then

$$q(x) \bullet q(y) := q(x \bullet y)$$

defines on $N$ the structure of a reflection space.

In fact, that the product on $N$ is well-defined is our assumption. That it is smooth follows from the smoothness of the map $M \times M \to M/N, (x, y) \mapsto q(x \bullet y)$ and the fact that $q \times q \colon M \times M \to N \times N$ is a submersion. Now the relations (S1-3) for $N$ follow immediately from the corresponding relations on $M$.

(f) In addition to (b), we consider a smooth action $\alpha \colon H \to \mathrm{Diff}(F)$ of $H$ on the manifold $F$ and consider the space

$$M := G \times_H F = (G \times F)/H,$$

where $H$ acts on $G \times F$ by $h.(g, f) := (gh^{-1}, \alpha_h(f))$. We write $[g, f]$ for the $H$-orbit of $(g, f)$ in $M$. We claim that

$$[g_1, f_1] \bullet [g_2, f_2] := [g_1\tau(g_1)^{-1}\tau(g_2), f_2] \tag{1.4}$$

defines on $M$ the structure of a smooth reflection space on which $G$ acts by automorphisms via $\mu_g[g', f] = [gg', f]$.

That (1.4) is a well-defined smooth binary operation is clear. That we obtain a reflection space is most naturally derived from (b), (c), (d) and (e). First we note that the product manifold $G \times F$ carries a natural reflection space structure given by

$$(gH, f) \bullet (g'H, f') := (g\tau(g)^{-1}\tau(g'), f'),$$

where we use the reflection space structure from (a) on $G$, and the trivial one from (c) on $F$. Next we note that the quotient map $q(g, f) := [g, f]$ is a submersion and that, for $g, g' \in G, h, h' \in H$, and $f, f' \in F$, the image

$$q((g, f) \bullet (g', f')) = [g\tau(g)^{-1}\tau(g'), f']$$

of the product in $G/H \times F$ does not change on the $H$-orbits:

$$q((gh^{-1}, \alpha_h(f)) \bullet (g'h', \alpha_{h'}(f'))) = [g\tau(g)^{-1}\tau(g')(h')^{-1}, \alpha_{h'}f'] = [g\tau(g)^{-1}\tau(g'), f'].$$

Therefore our claim follows from (e).

One of the main results of [**Lo67, Lo67b**] asserts that every finite dimensional connected reflection space $(M, \bullet)$ is of this form, where

- $G := \langle s_x s_y : x, y \in M\rangle_{\rm grp} \subseteq \mathrm{Diff}(M)$ carries a finite dimensional Lie group structure.
- $H = G^\tau$ for $\tau(g) = s_e g s_e$, where $e \in M$ is a base point.
- $F := \{m \in M : s_m = s_e\}$.

Typical examples with discrete spaces $F$ arise for $F := \pi_0(H)$ on which $H$ acts through the quotient homomorphism $H \to \pi_0(H)$ by translations.

We also note that $G$ acts transitively on $G \times_H F$ if and only if $H$ acts transitively on $F$. For any $f \in F$ we then have $G \times_H F \cong G/H_f$ as a homogeneous space of $G$.

(g) If $(V, \beta)$ is a $\mathbb{K}$-vector space ($\mathrm{char}(\mathbb{K}) \neq 2$) and $\beta : V \times V \to \mathbb{K}$ is a symmetric bilinear form, then the subset $V^\times := \{v \in V : \beta(v, v) \neq 0\}$ is a reflection space with respect to

$$x \bullet y = s_x(y) := -y + 2\frac{\beta(x, y)}{\beta(x, x)}x.$$

Note that $s_{\lambda x} = s_x$ for every $\lambda \in \mathbb{K}^\times$ and, conversely, that $s_x = s_z$ implies $z \in \mathbb{K}^\times x$ because $z \in \ker(s_x + \mathbf{1}) = \mathbb{K}x$.

For $\mathbb{K} = \mathbb{R}$ and $V$ a locally convex space, we thus obtain on $V^\times$ the structure of a real smooth reflection space and each level set

$$V_m^\times = \{x \in V : \beta(x, x) = m\}$$

becomes a symmetric space. The same holds for the image $V^\times/\mathbb{K}^\times$ in the projective space $\mathbb{P}(V)$.

REMARK 1.3. (S1-3) can also be formulated as conditions on the map $s_x : M \to M$, namely that $s_x$ is an involution fixing $x$, and (S3) takes the form

$$s_{x \bullet y} = s_x s_y s_x = s_x \bullet s_y,$$

where $\bullet$ on the right hand side refers to the natural reflection space structure on the set $\mathrm{Inv}(\mathrm{Bij}(M))$ of involutions in the group $\mathrm{Bij}(M)$ of permutations of $M$ (Example 1.2(a)).

### 1.2. Powers and geodesics in reflection spaces.

DEFINITION 1.4. (Quadratic representation and powers) Let $(M, \bullet)$ be a reflection space and $e \in M$ be a base point. The map

$$P = P_e \colon M \to \mathrm{Bij}(M), \quad P_e(m) := s_m s_e$$

is called the *quadratic representation of $M$ with respect to $e$*. For $x \in M$, we define the *powers with respect to $e$* by $x^0 := e$, $x^1 := x$ and

$$x^{n+2} := P(x)x^n = x \bullet (e \bullet x) \quad \text{for} \quad n \geq 0, \quad \text{and} \quad x^n := x^{-n} \quad \text{for} \quad n \in -\mathbb{N}.$$

An easy induction then shows that

$$x^n \bullet x^m = x^{2n-m} \quad \text{for} \quad n, m \in \mathbb{Z}. \tag{1.5}$$

We also note that, if $f \colon (M, \bullet) \to (M', \bullet)$ is a morphism of reflection spaces and $e' = f(e)$, then

$$f(x^n) = f(x)^n \quad \text{for} \quad x \in M, n \in \mathbb{Z}. \tag{1.6}$$

Note that (1.5) means that the map $(\mathbb{Z}, \bullet) \to (M, \bullet), n \mapsto x^n$ is a morphism of reflection spaces if $\mathbb{Z}$ carries the canonical reflection space structure (Example 1.2(a)).

DEFINITION 1.5. If $(M, \bullet)$ is a (topological) reflection space, then we call a (continuous) morphism $\gamma \colon (\mathbb{R}, \bullet) \to (M, \bullet)$ of reflection spaces a *geodesic*.

THEOREM 1.6. (One-parameter group theorem for geodesics in reflection spaces) *Let $G$ be a topological group. Then the geodesics $\gamma \colon \mathbb{R} \to G$ with $\gamma(0) = g$ are the curves of the form*

$$\gamma(t) = \eta(t)g,$$

*where $\eta \in \mathrm{Hom}(\mathbb{R}, G)$ is a continuous one-parameter group. The range of $\gamma$ is contained in* $\mathrm{Inv}(G)$ *if and only if*

$$g \in \mathrm{Inv}(G) \quad \textit{and} \quad g\eta(t)g^{-1} = \eta(-t) \quad \textit{for} \quad t \in \mathbb{R}.$$

PROOF. Since right multiplication with $g^{-1}$ is an automorphism of the reflection space $(G, \bullet)$, we may w.l.o.g. assume that $g = e$ and show that in this case the geodesics are the continuous one-parameter groups.

Clearly, every one-parameter group $\gamma \colon \mathbb{R} \to G$ is also a morphism of reflection spaces, hence a geodesic. Suppose, conversely, that $\gamma$ is a geodesic with $\gamma(0) = e$. From (1.6) and the relation $t^n = nt$ in the pointed reflection space $(\mathbb{R}, \bullet, 0)$, it follows that

$$\gamma(nt) = \gamma(t)^n \quad \text{for} \quad t \in \mathbb{R}, n \in \mathbb{Z}.$$

It follows in particular, that the restriction of $\gamma$ to any cyclic subgroup $\mathbb{Z}t \subseteq \mathbb{R}$ is a group homomorphism. Applying this to $t = \frac{1}{n}$, $n \in \mathbb{N}$, we see that $\gamma|_{\mathbb{Q}} \colon \mathbb{Q} \to G$ is a group homomorphism. Now the continuity of $\gamma$ implies that $\gamma$ is a homomorphism.

The second assertion is trivial. □

For the case of injective geodesics in topological reflection spaces, the preceding result is contained in [**FHHK17**, Lemma 2.20]. Later, but independently, D. Oeh obtained Theorem 1.6 in his MSc thesis [**Oe17**, Cor. 2.28].

**1.3. Dilation spaces.** Although the reflection space structure is a key bridge between manifold geometry and transformation groups ([**Lo67**]), there are natural situations where one has additional structures encoded by a family of actions $\mu_x \colon \Sigma \to \mathrm{Diff}(M)$ of a given Lie group $\Sigma$ on $M$ such that $x$ is fixed under $\mu_x$ and the family $(\mu_x)_{x\in M}$ satisfies a certain compatibility condition similar to (S3). This leads to the notion of a $\Sigma$-space introduced by O. Loos in ([**Lo72**]). Here we shall need only the special case $\Sigma = \mathbb{R}^\times$, so that we shall speak of dilation spaces. Restricting to the subgroup $\{\pm 1\} \subseteq \mathbb{R}^\times$, we then obtain a reflection space, so that dilation spaces are reflection spaces with additional point symmetries encoded in $\mathbb{R}^\times$-actions parametrized by the points of $M$.

DEFINITION 1.7. Let $M$ be a set and suppose we are given a map

$$\mu\colon M \times \mathbb{R}^\times \times M \to M, \quad (x, r, y) \mapsto x \bullet_r y := r_x(y) := \mu_r(x, y) := \mu(x, r, y)$$

with the following properties:

(D1) $r_x(x) = x$ for every $x \in M$ and $r \in \mathbb{R}^\times$.

(D2) $r_x \circ s_x = (rs)_x$ for $x \in M$, $r, s \in \mathbb{R}^\times$.

(D3) $r_x(y \bullet_s z) = r_x(y) \bullet_s r_x(z)$ for $x, y \in M$, $r, s \in \mathbb{R}^\times$, i.e., $r_x \in \mathrm{Aut}(M, \mu)$.

Then we call $(M, \mu)$ a *dilation space.*

REMARK 1.8. (a) If $(M, \mu)$ is a dilation space, then $(M, \mu_{-1})$ is a reflection space.

(b) In [**Be00**, Def. VI.2.1] the notion of a *ruled space* is defined as a smooth dilation space $(M, \mu)$ with the additional property that, for every $r \in \mathbb{R}^\times$, the tangent map $T_x(r_x)$ is diagonalizable with eigenvalues $r$ and $r^{-1}$. This ensures that $(M, \mu_{-1})$ is a symmetric space. We refer to [**Be00**, Thm. VI.2.2] for a characterization of the ruled spaces among symmetric spaces.

EXAMPLE 1.9. (a) For every group $G$, the space $M := \mathrm{Hom}(\mathbb{R}^\times, G)$ is a dilation space with

$$(\gamma \bullet_r \eta)(s) := \gamma(r)\eta(s)\gamma(r)^{-1} \quad \text{for} \quad \gamma, \eta \in M, r, s \in \mathbb{R}^\times. \tag{1.7}$$

Here (D1/2) are clear. For (D3) we calculate

$$\begin{aligned}\big(\gamma \bullet_r (\eta \bullet_s \zeta)\big)(t) &= \gamma(r)\eta(s)\zeta(t)\eta(s)^{-1}\gamma(r)^{-1}\\ &= \big(\gamma(r)\eta(s)\gamma(r)^{-1}\big)\big(\gamma(r)\zeta(t)\gamma(r)^{-1}\big)\big(\gamma(r)\eta(s)^{-1}\gamma(r)^{-1}\big)\\ &= [(\gamma \bullet_r \eta) \bullet_s (\gamma \bullet_r \zeta)](t).\end{aligned}$$

For $r = -1$, (1.7) specializes to the reflection space structure given by

$$(\gamma \bullet \eta)(s) := \gamma(-1)\eta(s)\gamma(-1)^{-1}.$$

(b) For every $\gamma \in \mathrm{Hom}(\mathbb{R}^\times, G)$, the conjugacy orbit $G.\gamma = \{\gamma^g \colon g \in G\}$ with $\gamma^g(t) := g\gamma(t)g^{-1}$ is a dilation subspace. This follows directly from

$$\begin{aligned}(\gamma^{g_1} \bullet_r \gamma^{g_2})(s) &= g_1\gamma(r)g_1^{-1}g_2\gamma(s)g_2^{-1}g_1\gamma(r)^{-1}g_1^{-1}\\ &= (g_1\gamma(r)g_1^{-1}g_2)\gamma(s)(g_1\gamma(r)g_1^{-1}g_2)^{-1}.\end{aligned}$$

Note that $G.\gamma = G_1.\gamma$ follows from the fact that $\gamma(-1) \in G_\gamma$.

(c) If $G$ is a Lie group with Lie algebra $\mathfrak{g}$, we represent any smooth $\gamma \in \mathrm{Hom}(\mathbb{R}^\times, G)$ by the pair $(\gamma'(0), \gamma(-1)) \in \mathfrak{g} \times \mathrm{Inv}(G)$. The corresponding set of pairs is

$$\mathcal{G} := \{(x, \sigma) \in \mathfrak{g} \times \mathrm{Inv}(G) \colon \mathrm{Ad}_\sigma x = x\}$$

and the reflection structure takes on this set the form

$$(x,\sigma) \bullet (y,\eta) = (\mathrm{Ad}_\sigma y, \sigma\eta\sigma) = (\mathrm{Ad}_\sigma y, \sigma \bullet \eta).$$

The additional dilation space structure from (a) is given by

$$(x,\sigma) \bullet_{e^t} (y,\eta) = (e^{t\,\mathrm{ad}\,x}y, \exp(tx)\eta\exp(-tx)).$$

For $(x,\sigma) \in \mathcal{G}$, the $G$-orbit is the set

$$\mathcal{O}_{(x,\sigma)} := \{(\mathrm{Ad}_g x, g\sigma g^{-1}) \colon g \in G\} \tag{1.8}$$

which has an obvious fiber bundle structure $G \times_{G^\sigma} \mathrm{Ad}_{G^\sigma} x \to G/G^\sigma$ over the symmetric space $G/G^\sigma$.

In many interesting cases (see f.i. [**NÓ17**, §4.3]), we actually have $G_x \subseteq G^\sigma$. This is in particular the case if $\mathrm{Ad}_\sigma = e^{\pi i\,\mathrm{ad}\,x}$ and $G = G_0\{e,\sigma\}$ ([**NÓ17**]). Then

$$\mathcal{O}_{(x,\sigma)} \cong \mathrm{Ad}_G x \tag{1.9}$$

simply is an adjoint orbit in $\mathfrak{g}$.

On homogeneous spaces, dilation space structures can sometimes be constructed along the lines of Example 1.2(b) if one considers subgroups with a central one-parameter group:

PROPOSITION 1.10. *Let $G$ be a group and $\alpha\colon \mathbb{R}^\times \to \mathrm{Aut}(G)$ be a homomorphism. Then*

$$g \bullet_r h := g\alpha_r(g^{-1}h)$$

*defines on $G$ the structure of a dilation space for which left translations $\lambda_g(x) = gx$ act by automorphisms. If $H \subseteq G$ is a subgroup fixed pointwise by $\alpha$, then the space $G/H$ of left $H$-cosets is a dilation space with respect to*

$$gH \bullet_r uH := g\alpha_r(g^{-1}u)H \quad \text{for} \quad g,u \in G.$$

PROOF. Here (D1) is trivial and (D2) follows from $r_g = \lambda_g\alpha_r\lambda_g^{-1}$ for $g \in G$. For (D3) we observe that

$$g \bullet_r (h \bullet_s u) = g\alpha_r\big(g^{-1}h\alpha_s(h^{-1}u)\big) = g\alpha_r\big(g^{-1}h)\alpha_{rs}(h^{-1}u)$$

equals

$$\begin{aligned}(g \bullet_r h) \bullet_s (g \bullet_r u) &= g\alpha_r(g^{-1}h)\alpha_s\big(\alpha_r(g^{-1}h)^{-1}g^{-1}g\alpha_r(g^{-1}u)\big)\\ &= g\alpha_r(g^{-1}h)\alpha_s\big(\alpha_r(h^{-1}u)\big) = g\alpha_r(g^{-1}h)\alpha_{rs}(h^{-1}u).\end{aligned}$$

The second assertion follows immediately from the first one because the binary operations $\bullet_r$ are obviously well-defined on $G/H$ and (D1-3) for $G/H$ follows immediately from the corresponding relations for $G$. □

By specialization we immediately obtain:

EXAMPLE 1.11. (a) Every real affine space $A$ is a dilation space with respect to

$$a \bullet_r b = a + r(b-a) = (1-r)a + rb \quad \text{for} \quad a,b \in A, r \in \mathbb{R}^\times.$$

(b) Let $V$ be a vector space and $\alpha\colon \mathbb{R}^\times \to \mathrm{GL}(V)$ be a group homomorphism. Then

$$a \bullet_r b = a + \alpha_r(b-a)$$

defines on $V$ the structure of a dilation space.

If $(M, \mu)$ is a dilation space and $e \in M$ is a base point, then $\alpha_e(r)(x) := e \bullet_r x$ defines a homomorphism $\alpha_e \colon \mathbb{R}^\times \to \mathrm{Aut}(M, \bullet_{-1})_e$ whose range is central. Conversely, we have:

COROLLARY 1.12. *Let $(M, \bullet)$ be a reflection space and $G \subseteq \mathrm{Aut}(M, \bullet)$ be a subgroup acting transitively on $M$. We fix a base point $e \in M$ and a homomorphism $\beta \colon \mathbb{R}^\times_+ \to G_e$ with central range. Then $\bullet_{-1} := \bullet$ together with*

$$g.e \bullet_r y := g\beta(r)g^{-1}.y \quad \text{for} \quad g \in G, r \in \mathbb{R}^\times_+, y \in M \tag{1.10}$$

*defines on the reflection space $(M, \bullet)$ the structure of a dilation space on which $G$ acts by automorphisms.*

PROOF. Let $H := G_e$, so that we may identify $M$ with $G/H$. Since the elements $\beta(r)$, $r \in \mathbb{R}^\times$, fix $e$, they commute with the reflection $s_e$. Therefore $\alpha_r(g) := \beta(r)g\beta(r)^{-1}$ for $r \in \mathbb{R}^\times_+$ and $\alpha_{-1}(g) := s_e g s_e$ define a homomorphism $\alpha \colon \mathbb{R}^\times \to \mathrm{Aut}(G)$ and $H = G_e$ is fixed pointwise by each $\alpha_r$. We now obtain with Proposition 1.10 on $G/H$ the structure of a dilation space by

$$gH \bullet_r uH := g\alpha_r(g^{-1}u)H \quad \text{for} \quad g, u \in G.$$

For $r = -1$, this means that

$$gH \bullet_{-1} uH = gs_{eH}g^{-1}us_{eH}H = gs_{eH}g^{-1}uH = g(eH \bullet g^{-1}uH) = gH \bullet uH$$

recovers the given reflection space structure. For $r > 0$ we find

$$gH \bullet_r uH = g\beta(r)g^{-1}u\beta(r)^{-1}H = g\beta(r)g^{-1}uH = (g\beta(r)g^{-1}).uH,$$

and this coincides with (1.10). □

**1.4. Geodesics in dilation spaces.** Since dilation spaces $(M, \mu)$ are reflection spaces with additional structure, geodesics in $(M, \bullet)$ are not always compatible with the dilation structure. The following lemma characterizes those which are.

LEMMA 1.13. *If $(M, \mu)$ is a dilation space and $\lambda \in \mathbb{R}^\times$, then the following are equivalent for a geodesic $\gamma \colon \mathbb{R} \to (M, \bullet)$:*

(a) *$\gamma$ is a morphism of dilation spaces, where the dilation structure on $(\mathbb{R}, \bullet)$ is given by $t \bullet_r s := (1 - r^\lambda)t + r^\lambda s = t + r^\lambda(s - t)$ for $r > 0$.*
(b) *$\gamma((1 - r^\lambda)t + r^\lambda s) = \gamma(t) \bullet_r \gamma(s)$ for $t, s \in \mathbb{R}, r \in \mathbb{R}^\times_+$.*
(c) *$\gamma(r^\lambda s) = \gamma(0) \bullet_r \gamma(s)$ for $s \in \mathbb{R}, r \in \mathbb{R}^\times$.*

PROOF. The equivalence of (a) and (b) is by definition. Further, (c) follows from (b) by specializing to $t = 0$. If (c) is satisfied, then we obtain

$$\begin{aligned}\gamma(t \bullet_r s) &= \gamma(t + r^\lambda(s - t)) = \gamma\big(\tfrac{t}{2} \bullet r^\lambda(t - s)\big)\\ &= \gamma\big(\tfrac{t}{2}\big) \bullet \gamma(r^\lambda(t - s)) \overset{(c)}{=} \gamma\big(\tfrac{t}{2}\big) \bullet_{-1} \big(\gamma(0) \bullet_r \gamma(t - s)\big)\\ &= \big(\gamma\big(\tfrac{t}{2}\big) \bullet_{-1} \gamma(0)\big) \bullet_r \big(\gamma\big(\tfrac{t}{2}\big) \bullet_{-1} \gamma(t - s)\big)\\ &= \gamma\big(\tfrac{t}{2} \bullet 0\big) \bullet_r \gamma\big(\tfrac{t}{2} \bullet (t - s)\big) = \gamma(t) \bullet_r \gamma(t - (t - s)) = \gamma(t) \bullet_r \gamma(s).\end{aligned}$$ □

REMARK 1.14. The main difference between Examples 1.11(a) and (b) is that, for $0 \neq v \in V$, the geodesic $\gamma(t) = tv$ is a morphism of dilation spaces for the $\lambda$-dilation structure on $\mathbb{R}$ if and only if

$$r^\lambda tv = \gamma(r^\lambda t) = 0 \bullet_r tv = \alpha(r)tv \quad \text{for} \quad r \in \mathbb{R}^\times_+, t \in \mathbb{R}.$$

This is equivalent to $\alpha(r)v = r^\lambda v$ for all $r \in \mathbb{R}^\times_+$. Therefore the geodesics which are morphisms of dilation spaces are generated by the elements of the common eigenspace

$$V_\lambda := \bigcap_{r\in\mathbb{R}^\times_+} \ker(\alpha(r) - r^\lambda \mathbf{1}).$$

## 2. The space of standard subspaces

We now apply the general discussion of reflection and dilation spaces to the space $\mathrm{Stand}(\mathcal{H})$ of standard subspaces and its relatives, the space $\mathrm{Mod}(\mathcal{H})$ of pairs of modular objects $(\Delta, J)$ and the space $\mathrm{Hom}_{\mathrm{gr}}(\mathbb{R}^\times, \mathrm{AU}(\mathcal{H}))$ of continuous antiunitary representations of $\mathbb{R}^\times$.

### 2.1. $\mathrm{Stand}(\mathcal{H})$ as a dilation space.

DEFINITION 2.1. A closed real subspace $V \subseteq \mathcal{H}$ is called a *standard subspace* if $V \cap iV = \{0\}$ and $V + iV$ is dense in $\mathcal{H}$. We write $\mathrm{Stand}(\mathcal{H})$ for the set of standard subspaces of $\mathcal{H}$.

For every standard subspace $V \subseteq \mathcal{H}$, we obtain an antilinear unbounded operator

$$S\colon \mathcal{D}(S) := V + iV \to \mathcal{H}, \qquad S(v + iw) := v - iw$$

with $V = \mathrm{Fix}(S) = \ker(S - \mathbf{1})$. The operator $S$ is closed, so that $\Delta_V := S^*S$ is a positive selfadjoint operator. We thus obtain the polar decomposition

$$S = J_V \Delta_V^{1/2},$$

where $J_V$ is an antilinear involution and the modular relation $J_V\Delta_V J_V = \Delta_V^{-1}$ is satisfied (cf. [**Lo08**, Prop. 3.3], [**NÓ16**]).

We write $\mathrm{Mod}(\mathcal{H})$ for the set of pairs $(\Delta, J)$, where $J$ is a *conjugation* (an antilinear bijective isometry) and $\Delta$ is a positive selfadjoint operator with $J\Delta J = \Delta^{-1}$. Then the map

$$\Phi\colon \mathrm{Mod}(\mathcal{H}) \to \mathrm{Stand}(\mathcal{H}), \quad \Phi(\Delta, J) = \mathrm{Fix}(J\Delta^{1/2}) \tag{2.1}$$

is a bijection ([**Lo08**, Prop. 3.2]).

To see more geometric structure on $\mathrm{Stand}(\mathcal{H})$, we have to connect its elements to homomorphisms $\mathbb{R}^\times \to \mathrm{AU}(\mathcal{H})$. This is best done in the context of graded groups and their antiunitary representations.

DEFINITION 2.2. (a) A *graded group* is a pair $(G, \varepsilon_G)$ consisting of a group $G$ and a surjective homomorphism $\varepsilon_G\colon G \to \{\pm 1\}$. We write $G_1 := \ker \varepsilon_G$ and $G_{-1} = G \setminus G_1$, so that

$$G = G_1 \dot\cup G_{-1} \quad \text{and} \quad G_j G_k = G_{jk} \quad \text{for} \quad j, k \in \{\pm 1\}.$$

Often graded groups are specified as pairs $(G, G_1)$, where $G_1$ is a subgroup of index 2, so that we obtain a grading by $\varepsilon_G(g) := 1$ for $g \in G_1$ and $\varepsilon_G(g) := -1$ for $g \in G \setminus G_1$.

If $G$ is a Lie group and $\varepsilon_G$ is continuous, i.e., $G_1$ is an open subgroup, then $(G, \varepsilon_G)$ is called a *graded Lie group*.

If $G$ is a topological group with two connected components, then we obtain a canonical grading for which $G_1$ is the identity component. Concrete examples are

$\mathbb{R}^\times, \mathrm{GL}_n(\mathbb{R}), \mathrm{O}_n(\mathbb{R})$ and the group $\mathrm{AU}(\mathcal{H})$ of unitary and antiunitary operators on a complex Hilbert space $\mathcal{H}$, endowed with the strong operator topology.

(b) A *morphism of graded groups* $\varphi\colon (G,\varepsilon_G) \to (H,\varepsilon_H)$ is a group homomorphism $\varphi\colon G \to H$ with $\varepsilon_H \circ \varphi = \varepsilon_G$, i.e., $\varphi(G_j) \subseteq H_j$ for $j = 1, -1$. We write $\mathrm{Hom}_{\mathrm{gr}}(G,H)$ for the set of (continuous) graded homomorphism between the (topological) graded groups $(G,\varepsilon_G)$ and $(H,\varepsilon_H)$.

(c) For a complex Hilbert space $\mathcal{H}$, the group $\mathrm{AU}(\mathcal{H})$ carries a natural grading defined by $\varepsilon(U) = -1$ if $U$ is antiunitary and $\varepsilon(U) = 1$ if $U$ is unitary. For a topological graded group $(G,\varepsilon_G)$, an *antiunitary representation* $(U,\mathcal{H})$ is a continuous homomorphism $U\colon G \to \mathrm{AU}(\mathcal{H})$ of graded groups, where $\mathrm{AU}(\mathcal{H})$ carries the strong operator topology.

It is easy to see that

$$\Psi\colon \mathrm{Mod}(\mathcal{H}) \to \mathrm{Hom}_{\mathrm{gr}}(\mathbb{R}^\times, \mathrm{AU}(\mathcal{H})), \quad \Psi(\Delta,J)(e^t) := \Delta^{-it/2\pi}, \quad \Psi(\Delta,J)(-1) := J$$

defines a bijection ([**NÓ17**, Lemma 2.22]). Combining this with $\Phi$ from (2.1), we obtain a bijection

$$\mathcal{V} := \Phi \circ \Psi^{-1}\colon \mathrm{Hom}_{\mathrm{gr}}(\mathbb{R}^\times, \mathrm{AU}(\mathcal{H})) \to \mathrm{Stand}(\mathcal{H}). \tag{2.2}$$

THEOREM 2.3. *We obtain the structure of a dilation space*

- *on* $\mathrm{Hom}_{\mathrm{gr}}(\mathbb{R}^\times, \mathrm{AU}(\mathcal{H}))$ *by*

$$(\gamma \bullet_r \eta)(t) := \gamma(r)\eta(t)\gamma(r)^{-1},$$

- *on* $\mathrm{Mod}(\mathcal{H})$ *by*

$$(\Delta_1, J_1) \bullet_r (\Delta_2, J_2) = \begin{cases} (J_1\Delta_2^{-1}J_1, J_1 \bullet J_2) & \text{for } r = -1 \\ (\Delta_1^{-it/2\pi}\Delta_2\Delta_1^{it/2\pi}, \Delta_1^{-it/2\pi}J_2\Delta_1^{it/2\pi}) & \text{for } r = e^t, \end{cases}$$

- *and on* $\mathrm{Stand}(\mathcal{H})$ *by*

$$V_1 \bullet_r V_2 = \begin{cases} J_{V_1}J_{V_2}V_2 & \text{for} \quad r = -1 \\ \Delta_{V_1}^{-it/2\pi}V_2 & \text{for} \quad r = e^t. \end{cases}$$

*The bijections* $\Phi\colon \mathrm{Mod}(\mathcal{H}) \to \mathrm{Stand}(\mathcal{H})$ *and* $\Psi\colon \mathrm{Mod}(\mathcal{H}) \to \mathrm{Hom}_{\mathrm{gr}}(\mathbb{R}^\times, \mathrm{AU}(\mathcal{H}))$ *are isomorphisms of dilation spaces.*

PROOF. First we observe that, for each $r \in \mathbb{R}^\times$ we obtain by $\bullet_r$ a binary operations on the spaces $\mathrm{Hom}_{\mathrm{gr}}(\mathbb{R}^\times, \mathrm{AU}(\mathcal{H}))$, $\mathrm{Mod}(\mathcal{H})$ and $\mathrm{Stand}(\mathcal{H})$, respectively. In particular, $\mathrm{Hom}_{\mathrm{gr}}(\mathbb{R}^\times, \mathrm{AU}(\mathcal{H}))$ is a dilation subspace of $\mathrm{Hom}(\mathbb{R}^\times, \mathrm{AU}(\mathcal{H}))$ (Example 1.9), hence a dilation space. It therefore remains to show that $\Phi$ and $\Psi$ are compatible with all binary operations $\bullet_r$ and this implies in particular that (D1-3) are satisfied on $\mathrm{Mod}(\mathcal{H})$ and $\mathrm{Stand}(\mathcal{H})$.

First we consider $\Psi$. Let $\gamma_j = \Psi(\Delta_j, J_j)$ for $j = 1,2$, $r \in \mathbb{R}^\times$, and $\gamma := \Psi((\Delta_1,J_1) \bullet_r (\Delta_2,J_2))$. For $r = -1$ we then have

$$\gamma(-1) = J_1J_2J_1 = \gamma_1(-1)\gamma_2(-1)\gamma_1(-1) = (\gamma_1 \bullet_{-1} \gamma_2)(-1),$$

and, for $t \in \mathbb{R}$, we have

$$\gamma(e^t) = (J_1\Delta_2^{-1}J_1)^{-it/2\pi} = J_1\Delta_2^{-it/2\pi}J_1 = \gamma_1(-1)\gamma_2(e^t)\gamma_1(-1) = (\gamma_1 \bullet \gamma_2)(e^t).$$

For $r = e^t$, $t \in \mathbb{R}$, the pair $(\Delta_1,J_1) \bullet_r (\Delta_2,J_2)$ is obtained from $(\Delta_2,J_2)$ by conjugating with $\Delta_1^{-it/2\pi} = \gamma_1(e^t)$ and this immediately implies that $\gamma = \gamma_1 \bullet_r \gamma_2$. We conclude that $\Psi$ is an isomorphism of dilation spaces.

Now we turn to $\Phi$. From

$$\begin{aligned}(2.3)\quad \mathrm{Fix}((J_1J_2J_1)(J_1\Delta_2^{-1}J_1)^{1/2}) &= \mathrm{Fix}(J_1J_2\Delta_2^{-1/2}J_1)\\ &= J_1\,\mathrm{Fix}(J_2\Delta_2^{-1/2}) = J_1V_2' = J_1J_2V_2\end{aligned}$$

it follows that $\Phi$ is compatible with $\bullet_{-1}$. For $r = e^t$, $t \in \mathbb{R}$, the pair $(\Delta_1, J_1) \bullet_r (\Delta_2, J_2)$ is obtained from $(\Delta_2, J_2)$ by conjugating with $\Delta_1^{-it/2\pi}$, so that

$$\Phi\big((\Delta_1, J_1) \bullet_r (\Delta_2, J_2)\big) = \Delta_1^{-it/2\pi} V_2.$$

This completes the proof. □

Any antiunitary representation of a graded group $G$ leads to a map $\mathcal{V}_U\colon \mathrm{Hom}_{\mathrm{gr}}(\mathbb{R}^\times, G) \to \mathrm{Stand}(\mathcal{H})$, an observation due to Brunetti, Guido and Longo ([**BGL02**, Thm. 2.5]; see also [**NÓ17**, Prop. 5.6]). The naturality of this map immediately shows that it is a morphism of dilation spaces:

COROLLARY 2.4. (The BGL (Brunetti–Guido–Longo) construction) *Let $(U, \mathcal{H})$ be an antiunitary representation of the graded topological group $(G, \varepsilon_G)$. Then*

$$\mathcal{V}_U := \mathcal{V} \circ U_*\colon \mathrm{Hom}_{\mathrm{gr}}(\mathbb{R}^\times, G) \to \mathrm{Stand}(\mathcal{H}), \quad \gamma \mapsto \mathcal{V}(U \circ \gamma)$$

*is a morphism of dilation spaces.*

PROOF. Since $\Phi$ and $\Psi$ are isomorphisms of dilation spaces, it suffices to observe that

$$U_*\colon \mathrm{Hom}_{\mathrm{gr}}(\mathbb{R}^\times, G) \to \mathrm{Hom}_{\mathrm{gr}}(\mathbb{R}^\times, \mathrm{AU}(\mathcal{H})), \quad \gamma \mapsto U \circ \gamma$$

is a morphism of dilation spaces. But this is a trivial consequence of the fact that $U$ is a morphism of graded topological groups. □

In addition to the dilation structure, the space $\mathrm{Stand}(\mathcal{H})$ carries a natural involution $\theta$:

REMARK 2.5. (The canonical involution)
(a) On $\mathrm{Hom}_{\mathrm{gr}}(\mathbb{R}, \mathrm{AU}(\mathcal{H}))$ the involution $\theta(\gamma) = \gamma^\vee$, $\gamma^\vee(t) := \gamma(t^{-1})$, defines an isomorphism of reflection spaces which is compatible with the dilation space structure in the sense that

$$(\gamma \bullet_r \eta)^\vee = \gamma \bullet_r \eta^\vee \quad \text{for} \quad \gamma, \eta \in \mathrm{Hom}_{\mathrm{gr}}(\mathbb{R}, \mathrm{AU}(\mathcal{H}))$$

because

$$(\gamma \bullet_r \eta)^\vee(s) = \gamma(r)\eta(s^{-1})\gamma(r)^{-1} = (\gamma \bullet_r \eta^\vee)(s).$$

The corresponding automorphism is given on $\mathrm{Stand}(\mathcal{H})$ by $\theta(V) := V'$ and on $\mathrm{Mod}(\mathcal{H})$ by $\theta(\Delta, J) = (\Delta^{-1}, J)$.

The fixed points of $\theta$ correspond to

- graded homomorphisms $\gamma\colon \mathbb{R}^\times \to \mathrm{AU}(\mathcal{H})$ with $\mathbb{R}_+^\times \subseteq \ker\gamma$.
- elements $V \in \mathrm{Stand}(\mathcal{H})$ which are Lagrangian subspaces of the symplectic vector space $(\mathcal{H}, \omega)$, where $\omega(v, w) = \mathrm{Im}\langle v, w\rangle$. As the symplectic orthogonal space $V' := V^{\perp_\omega}$ coincides with $iV^{\perp_{\mathbb{R}}}$, where $\perp_{\mathbb{R}}$ denotes the orthogonal space with respect to the real scalar product $\mathrm{Re}\langle v, w\rangle$, the Lagrangian condition $V = V'$ is equivalent to $V = iV^{\perp_{\mathbb{R}}}$, which is equivalent to $V \oplus iV = \mathcal{H}$ being an orthogonal direct sum.
- pairs $(\Delta, J)$ of modular objects with $\Delta = \mathbf{1}$.

(b) The canonical embedding of the symmetric space $\mathrm{Conj}(\mathcal{H})$: Note that

$$\zeta\colon \mathrm{Conj}(\mathcal{H}) \to \mathrm{Hom}_{\mathrm{gr}}(\mathbb{R}^\times, \mathrm{AU}(\mathcal{H})), \quad \zeta(J)(-1) := J, \quad \zeta(J)(e^t) = \mathbf{1} \quad \text{for} \quad t \in \mathbb{R}$$

defines a morphism of reflection spaces whose range is the set $\mathrm{Hom}_{\mathrm{gr}}(\mathbb{R}^\times, \mathrm{AU}(\mathcal{H}))^\theta$ of $\theta$-fixed points. We likewise obtain morphisms of reflection spaces

$$\mathrm{Conj}(\mathcal{H}) \to \mathrm{Mod}(\mathcal{H}), \quad J \mapsto (\mathbf{1}, J), \qquad \mathrm{Conj}(\mathcal{H}) \to \mathrm{Stand}(\mathcal{H}), J \mapsto \mathrm{Fix}(J) = \mathcal{H}^J.$$

REMARK 2.6. For $\mathcal{H} = \mathbb{C}^n$, the space $\mathrm{Stand}(\mathbb{C}^n) \cong \mathrm{GL}_n(\mathbb{C})/\mathrm{GL}_n(\mathbb{R})$ carries a natural symmetric space structure corresponding to $\tau(g) = \overline{g}$ on $\mathrm{GL}_n(\mathbb{C})$ and given by

$$g\,\mathrm{GL}_n(\mathbb{R}) \sharp h\,\mathrm{GL}_n(\mathbb{R}) := g\tau(g^{-1}h)\,\mathrm{GL}_n(\mathbb{R}), \qquad \text{resp.,} \qquad g\mathbb{R}^n \sharp h\mathbb{R}^n = g\overline{g}^{-1}\overline{h}\mathbb{R}^n.$$

This reflection structure is different from the one defined above by $s_{V_1}V_2 = V_1 \bullet V_2 = J_1J_2V_2$. In fact, if $J_1 = J_2$, then $J_1J_2V_2 = V_2$ shows that $V_2$ is a fixed point of $s_{V_1}$ and thus $V_1$ is not isolated in $\mathrm{Fix}(s_{V_1})$, whereas this is the case for the reflections defined by $\sharp$ (cf. Example 1.2(b)).

**2.2. Loos normal form of** $\mathrm{Stand}(\mathcal{H})$**.** The unitary group $\mathrm{U}(\mathcal{H})$ acts on the reflection space $(\mathrm{Stand}(\mathcal{H}), \bullet)$ by automorphisms and the morphism of reflection spaces

$$q\colon \mathrm{Stand}(\mathcal{H}) \to \mathrm{Conj}(\mathcal{H}), \quad q(V) := J_V$$

is equivariant with respect to the conjugation action on $\mathrm{Conj}(\mathcal{H})$ which is transitive.

For the involutive automorphism $\tau(g) := JgJ$ of $G := \mathrm{U}(\mathcal{H})$ we have $G^\tau \cong \mathrm{O}(\mathcal{H}^J)$. For $g_1, g_2 \in G$ and $V_1, V_2 \in \mathrm{Stand}_J(\mathcal{H})$, we have

$$g_1V_1 \bullet g_2V_2 = J_{g_1V_1}J_{g_2V_2}g_2V_2 = (g_1Jg_1^{-1})(g_2Jg_2^{-1})g_2V_2 = g_1\tau(g_1^{-1}g_2)V_2.$$

PROPOSITION 2.7. *Let $J \in \mathrm{Conj}(\mathcal{H})$ and write*

$$\mathrm{Stand}_J(\mathcal{H}) := \{V \in \mathrm{Stand}(\mathcal{H})\colon J_V = J\} = q^{-1}(J)$$

*for the $q$-fiber of $J$. Then the map*

$$\mathcal{P}\colon \mathrm{U}(\mathcal{H}) \times \mathrm{Stand}_J(\mathcal{H}) \to \mathrm{Stand}(\mathcal{H}), \quad (g, V) \mapsto gV$$

*is surjective and factors to a bijection*

$$\overline{\mathcal{P}}\colon \mathrm{U}(\mathcal{H}) \times_{\mathrm{O}(\mathcal{H}^J)} \mathrm{Stand}_J(\mathcal{H}) \to \mathrm{Stand}(\mathcal{H}), \quad [g, V] \mapsto gV.$$

*It is an $\mathrm{U}(\mathcal{H})$-equivariant isomorphism of reflection spaces if the space on the left carries the reflection space structure given by*

$$[g_1, V_1] \bullet [g_2, V_2] = [g_1\tau(g_1^{-1}g_2), V_2].$$

*In particular,* $\mathrm{Stand}_J(\mathcal{H})$ *is a trivial reflection subspace of* $\mathrm{Stand}(\mathcal{H})$ *on which the product is given by $V_1 \bullet V_2 = V_2$.*

We thus obtain a "normal form" of the reflection space $\mathrm{Stand}(\mathcal{H})$ similar to the one in Example 1.2(f).

PROOF. The surjectivity of $\mathcal{P}$ follows from the transitivity of the action of $\mathrm{U}(\mathcal{H})$ on $\mathrm{Conj}(\mathcal{H})$, which in turn follows from the existence of an orthonormal basis of $\mathcal{H}$ fixed pointwise by $J$. The second assertion follows from the fact that $\mathrm{O}(\mathcal{H}^J)$ is the stabilizer of $J \in \mathrm{Conj}(\mathcal{H})$. □

The subset $\mathrm{Stand}_J(\mathcal{H})$ corresponds to the set of all positive selfadjoint operators $\Delta$ with $J\Delta J = \Delta^{-1}$. For $A := i\log\Delta$, this means that $JAJ = A$, so that $A$ corresponds to a skew-adjoint operator on the real Hilbert space $\mathcal{H}^J$, hence, by the real version of Stone's Theorem to the infinitesimal generator of a continuous one-parameter group in $\mathrm{O}(\mathcal{H}^J)$. We thus obtain a bijection $\mathrm{Stand}_J(\mathcal{H}) \to \mathrm{Hom}(\mathbb{R}, \mathrm{O}(\mathcal{H}^J))$.

REMARK 2.8. The dilations on $\mathrm{Stand}(\mathcal{H})$ corresponding to $V$ are implemented by $J_V$ and the unitary operators $(\Delta_V^{it})_{t\in\mathbb{R}}$. Since $\mathrm{AU}(\mathcal{H})$ acts on $\mathrm{Conj}(\mathcal{H})$, these one-parameter groups act naturally on $\mathrm{Conj}(\mathcal{H})$ but they do not give rise to a dilation space structure because they do commute with the stabilizer $\mathrm{O}(\mathcal{H}^J) \cong \mathrm{U}(\mathcal{H})_J$ of $J$. In addition, the dilation groups depend on the pair $(\Delta, J)$ and not only on $J$.

**2.3. Geodesics in** $\mathrm{Stand}(\mathcal{H})$**.** In a reflection space, we have a canonical notion of a geodesics. Although we do not specify a topology on $\mathrm{Stand}(\mathcal{H})$, the space $\mathrm{Conj}(\mathcal{H}) \subseteq \mathrm{AU}(\mathcal{H})$ carries the strong operator topology, and this immediately provides a natural continuity requirement for geodesics in $\mathrm{Stand}(\mathcal{H})$.

PROPOSITION 2.9. (Geodesics in $\mathrm{Stand}(\mathcal{H})$) *Let* $\gamma\colon \mathbb{R} \to \mathrm{Stand}(\mathcal{H})$ *be a geodesic with* $\gamma(0) = V$ *such that the corresponding geodesic* $(J_{\gamma(t)})_{t\in\mathbb{R}}$ *in* $\mathrm{Conj}(\mathcal{H})$ *is strongly continuous. Then there exists a strongly continuous unitary one-parameter group* $(U_t)_{t\in\mathbb{R}}$ *satisfying*

$$J_V U_t J_V = U_{-t} \quad \textit{for} \quad t \in \mathbb{R} \tag{2.4}$$

*such that*

$$\gamma(t) = U_{t/2}V \quad \textit{for} \quad t \in \mathbb{R}.$$

*The* $U_t$ *are uniquely determined by the relation*

$$J_{\gamma(t)} = U_{t/2} J U_{t/2}^{-1} = U_t J, \qquad \textit{resp.} \quad U_t = J_{\gamma(t)} J. \tag{2.5}$$

PROOF. Consider the base point $e := V \in \mathrm{Stand}(\mathcal{H})$ and the morphism

$$q\colon \mathrm{Stand}(\mathcal{H}) \to \mathrm{Conj}(\mathcal{H}), \quad W \mapsto J_W$$

of reflection spaces. By assumption, $\overline{\gamma} := q \circ \gamma\colon \mathbb{R} \to \mathrm{Conj}(\mathcal{H})$ is a strongly continuous geodesic, hence of the form $\overline{\gamma}(t) = U_t J_V$ for some continuous unitary one-parameter group $(U_t)_{t\in\mathbb{R}}$ (Theorem 1.6). Since the range of $\overline{\gamma}$ consists of involutions in $\mathrm{AU}(\mathcal{H})$, (2.4) follows. Further,

$$\gamma(t) = \gamma(t/2) \bullet \gamma(0) = J_{\gamma(t/2)} J_V V = \overline{\gamma}(t/2) J_V V = U_{t/2} V.$$

The relation (2.5) follows immediately from (2.4). □

DEFINITION 2.10. We call a geodesic $\gamma\colon \mathbb{R} \to \mathrm{Stand}(\mathcal{H})$ with $\gamma(0) = V$ *dilation invariant* if it is invariant under the corresponding one-parameter group $(\Delta_V^{it})_{t\in\mathbb{R}}$ of modular automorphisms, i.e., there exists an $\alpha \in \mathbb{R}$, such that

$$\Delta_V^{-is/2\pi}\gamma(t) = \gamma(e^{\alpha s} t) \quad \text{for} \quad s, t \in \mathbb{R}.$$

PROPOSITION 2.11. *Consider the non-constant geodesic* $\gamma(t) = U_{t/2}V$ *through* $V$ *and the unitary one-parameter group* $W_s := \Delta_V^{-is/2\pi}$ *implementing the dilations in* $V$*. If* $\gamma$ *is dilation invariant, then there exists an* $\alpha \in \mathbb{R}$ *with*

$$W_s U_t W_{-s} = U_{e^{\alpha s} t} \quad \textit{for} \quad t, s \in \mathbb{R}.$$

*Then*

$$U_{(0,-1)} := J_V \quad \text{and} \quad U_{(b,e^s)} := U_b W_s \tag{2.6}$$

*defines an antiunitary representation of the group* $G_\alpha := \mathbb{R} \rtimes_\zeta \mathbb{R}^\times$ *with* $\zeta(-1)x = -x$ *and* $\zeta(e^t)x = e^{\alpha t}x$*. Conversely, for every antiunitary representation* $(U, \mathcal{H})$ *of* $G_\alpha$*, the restriction to* $\{0\} \times \mathbb{R}^\times$ *specifies a standard subspace* $V$ *and* $\gamma(t) := U_{(t/2,1)}V$ *is a dilation invariant geodesic with* $\gamma(0) = V$*.*

Note that, for $\alpha \neq 0$, the group $G_\alpha$ is isomorphic to $\mathrm{Aff}(\mathbb{R})$ and otherwise to $\mathbb{R}^2 \rtimes_\sigma \{\pm 1\}$ with $\sigma(-1)(x, y) = (-x, y)$.

PROOF. The dilation invariance of $\gamma$ implies the existence of an $\alpha \in \mathbb{R}$ with $W_s\gamma(t) = \gamma(e^{\alpha s}t)$ for $s, t \in \mathbb{R}$. For the corresponding unitary one-parameter group $U$, this leads to

$$W_s U_t W_{-s} = W_s J_{\gamma(t)} J_V W_{-s} = J_{\gamma(e^{\alpha s}t)} J_V = U_{e^{\alpha s}t} \quad \text{for} \quad s, t \in \mathbb{R}.$$

Therefore (2.6) defines an antiunitary representation of $G_\alpha$. The converse is clear. □

## 3. The order on Stand($\mathcal{H}$)

As a set of subsets of $\mathcal{H}$, the space Stand($\mathcal{H}$) carries a natural order structure, defined by set inclusion. We shall see below that non-trivial inclusions $V_1 \subset V_2$ arise only if both modular operators $\Delta_{V_1}$ and $\Delta_{V_2}$ are unbounded. Therefore inclusions of standard subspaces appear only if $\mathcal{H}$ is infinite dimensional. Here a natural question is to understand when a non-constant geodesic $\gamma \colon \mathbb{R} \to \mathrm{Stand}(\mathcal{H})$ is monotone with respect to the natural order on $\mathbb{R}$. In general this seems to be hard to characterize, but for dilation invariant geodesics, Proposition 2.11 can be combined with the Borchers–Wiesbrock Theorem ([**NÓ17**, Thms. 3.13, 3.15]) which provides a complete answer in terms of the positive/negative energy condition on the corresponding antiunitary representation of $\mathrm{Aff}(\mathbb{R})$.

### 3.1. Monotone dilation invariant geodesics.

LEMMA 3.1. *If* $V_1 \subset V_2$ *is a proper inclusion of standard subspaces, then both operators* $\Delta_{V_j}$*,* $j = 1, 2$*, are unbounded.*

PROOF. Suppose that $\Delta_{V_1}$ is bounded. Then $S_1 := J_{V_1}\Delta_{V_1}^{1/2}$ is also bounded and thus $\mathcal{H} = \mathcal{D}(S_1) = V_1 + iV_1$. As $V_1 \subseteq V_2$ and $V_2 \cap iV_2 = \{0\}$, the inclusion cannot be strict.

If $\Delta_{V_2} = \Delta_{V_2'}^{-1}$ is bounded, this argument shows that the inclusion $V_2' \subseteq V_1'$ cannot be strict, hence $V_2' = V_1'$ and thus $V_2 = V_2'' = V_1'' = V_1$. □

From [**Lo08**, Prop. 3.10] we recall:

LEMMA 3.2. *If* $V_1 \subseteq V_2$ *for two standard subspaces and* $V_1$ *is invariant under the modular automorphisms* $(\Delta_{V_2}^{it})_{t\in\mathbb{R}}$*, then* $V_1 = V_2$*.*

THEOREM 3.3. *Let* $\gamma(t) = U_{t/2}V$ *be a non-constant dilation invariant geodesic with* $U_t = e^{itH}$ *and* $\alpha = 1$*. Then* $\gamma$ *is decreasing if and only if* $H \geq 0$*.*

PROOF. By Proposition 2.11 we obtain an antiunitary representation of $\mathrm{Aff}(\mathbb{R})$, and now the assertion follows from [**NÓ17**, Thm. 3.13] or [**Lo08**, Thm. 3.17]. □

REMARK 3.4. Comparing Proposition 2.11 with [**NÓ17**, Thm. 3.22], it follows that two different standard subspaces $V_0, V_1 \in \mathrm{Stand}(\mathcal{H})$ lie on a dilation invariant geodesic with $\alpha = 1$ if and only if they have a $+$-modular intersection (see [**NÓ17**, §3.5] for details).

PROBLEM 3.5. Find a characterization of the monotone geodesics in $\mathrm{Stand}(\mathcal{H})$. By Lemma 3.1 it is necessary that $\Delta_V$ is unbounded. For dilation invariant geodesics with $\alpha = 1$, Theorem 3.3 provides a characterization in terms of the positive/negative spectrum condition on $U$. In this case the representation theory of $\mathrm{Aff}(\mathbb{R})$ even implies that, apart from the subspace of fixed points of the translation group, the operator $\Delta_V$ must be equivalent to the multiplication operator $(Mf)(x) = xf(x)$ on some space $L^2(\mathbb{R}^\times, \mathcal{K})$, where $\mathcal{K}$ is a Hilbert space counting multiplicity ([**NÓ17**, §2.4.1], [**Lo08**, Thm. 2,8]). We also note that the fixed point space $\mathcal{H}_0 := \ker(\Delta_V - \mathbf{1})$ leads to an orthogonal decomposition $\mathcal{H} = \mathcal{H}_0 \oplus \mathcal{H}_1$ and $V = V_0 \oplus V_1$ with $\Delta_V = \mathbf{1} \oplus \Delta_{V_1}$ such that $\Delta_{V_1}$ has no non-zero fixed vectors.

Since it seems quite difficult to address this problem directly, it is natural to consider subspaces of $\mathrm{Stand}(\mathcal{H})$ which are more accessible. Such subspaces can be obtained from an antiunitary representation $(U, \mathcal{H})$ of a graded Lie group $(G, \varepsilon_G)$ and a fixed $\gamma \in \mathrm{Hom}_{\mathrm{gr}}(\mathbb{R}^\times, G)$ from the image $\mathcal{O}_V := G_1.V$ of the conjugacy class $G_1.\gamma$ under the $G_1$-equivariant morphism $\mathcal{V}_U \colon \mathrm{Hom}_{\mathrm{gr}}(\mathbb{R}^\times, G) \to \mathrm{Stand}(\mathcal{H})$ of dilation spaces, where $V := \mathcal{V}_U(\gamma)$ (Corollary 2.4). Then

$$S_V := \{g \in G_1 \colon U_g V \subseteq V\}$$

is a closed subsemigroup of $G_1$ with $G_{1,V} = S_V \cap S_V^{-1}$ and $S_V$ determines an order structure on $G_1/G_{1,V}$ by $gG_{1,V} \leq g'G_{1,V}$ if $g \in g'S_V$ (see [**HN93**, §4] for background material on semigroups and ordered homogeneous spaces) for which the inclusion $G_1/G_{1,V} \hookrightarrow \mathrm{Stand}(\mathcal{H})$ is an order embedding. Note that $\mathcal{O}_V$ is a $G_1$-equivariant quotient of $G_1/G_{1,\gamma} \cong G_1.\gamma \subseteq \mathrm{Hom}_{\mathrm{gr}}(\mathbb{R}^\times, G)$. In the following subsection we explain how these spaces and their order structure can be obtained quite explicitly for an important class of examples including the important case where $\gamma$ is a Lorentz boost associated to a wedge domain in Minkowski space.

### 3.2. Conformal groups of euclidean Jordan algebras.

DEFINITION 3.6. A finite dimensional real vector space $E$ endowed with a symmetric bilinear map $E \times E \to E, (a,b) \mapsto a \cdot b$ is said to be a *Jordan algebra* if

$$x \cdot (x^2 \cdot y) = x^2 \cdot (x \cdot y) \quad \text{for} \quad x, y \in E.$$

If $L(x)y = xy$ denotes the left multiplication, then $E$ is called *euclidean* if the trace form $(x, y) \mapsto \mathrm{tr}(L(xy))$ is positive definite.

Every finite dimensional euclidean Jordan algebra is a direct sum of simple ones and simple euclidean Jordan algebras permit a nice classification ([**FK94**, §V.3]). They are of the following types:

- $\mathrm{Herm}_n(\mathbb{K})$, $\mathbb{K} = \mathbb{R}, \mathbb{C}, \mathbb{H}$, with $A \cdot B := \frac{AB+BA}{2}$.
- $\mathrm{Herm}_3(\mathbb{O})$, where $\mathbb{O}$ denotes the 8-dimensional alternative division algebra of octonions.
- $\Lambda_n := \mathbb{R} \times \mathbb{R}^{n-1}$ with $(t,v)(t',v') = (tt' + \langle v, v' \rangle, tv' + t'v)$. Then the trace form is a Lorentz form, so that we can think of $\Lambda_n$ as the $n$-dimensional Minkowski space, where the first component corresponds to the time coordinate.

Let $E$ be a euclidean Jordan algebra. Then $C_+ := \{v^2 \colon v \in E\}$ is a pointed closed convex cone in $E$ whose interior is denoted $E_+$. The Jordan inversion $j_E(x) = x^{-1}$ acts by a rational map on $E$. The *causal group* $G_1 := \mathrm{Cau}(E)$ is the group of birational maps on $E$ generated by the linear automorphism group $\mathrm{Aut}(E_+)$ of the open cone $E_+$, the map $-j_E$ and the group of translations. It is an index two subgroup of the *conformal group* $G := \mathrm{Conf}(E)$ generated by the structure group $H := \mathrm{Aut}(E_+) \cup -\mathrm{Aut}(E_+)$ of $E$ ([**FK94**, Prop. VIII.2.8]), $j_E$ and the translations. For any $g \in G$ and $x \in E$ in which $g(x)$ is defined, the differential $\mathrm{d}g(x)$ is contained in $H$. This specifies a group grading $\varepsilon_G \colon G \to \{\pm 1\}$ for which $\ker \varepsilon_G = G_1 = \mathrm{Cau}(E)$ (see [**Be96**, Thm. 2.3.1] and [**Be00**] for more details on causal groups). It also follows that an element of $G$ defines a linear map if and only if it belongs to the structure group $H$.

The conformal completion $E_c$ of $E$ is a compact smooth manifold containing $E$ as an open dense submanifold on which $G$ acts transitively. By analytic extension, it can be identified with the Shilov boundary of the corresponding tube domain $E + iE_+ \subseteq E_{\mathbb{C}}$ ([**FK94**, Thm. X.5.6], [**Be96**, Thms. 2.3.1, 2.4.1]). The Lie algebra $\mathfrak{g}$ of $G$ has a natural 3-grading

$$\mathfrak{g} = \mathfrak{g}_1 \oplus \mathfrak{g}_0 \oplus \mathfrak{g}_{-1},$$

where $\mathfrak{g}_1 \cong E$ corresponds to the space of constant vector fields on $E$ (generating translations), $\mathfrak{g}_0 = \mathfrak{h}$ is the Lie algebra of $H$ (the structure algebra of $E$) which corresponds to linear vector field, and $\mathfrak{g}_{-1}$ corresponds to certain quadratic vector fields which are conjugate under the inversion $j_E$ to constant ones ([**FK94**, Prop. X.5.9]).

We have a canonical homomorphism

$$\gamma \colon \mathbb{R}^\times \to H \subseteq G, \quad r \mapsto r\,\mathrm{id}_E \tag{3.1}$$

which is graded because $\gamma(-1) = -\mathrm{id}_E$ maps $E_+$ to $-E_+$. Note that $h = \gamma'(0) \in \mathfrak{g}_0$ satisfies

$$\mathfrak{g}_j = \{x \in \mathfrak{g} \colon [h, x] = jx\} \quad \text{for} \quad j = -1, 0, 1 \tag{3.2}$$

and

$$\mathrm{Ad}_{\gamma(r)} x_j = r^j x_j \quad \text{for} \quad r \in \mathbb{R}^\times, x_j \in \mathfrak{g}_j. \tag{3.3}$$

For the involution $\tau(g) := \gamma(-1)g\gamma(-1)$, resp., $\tau(g)(x) = -g(-x)$ for $x, g(-x) \in E \subseteq E_c$, we then have $G_\gamma = H \subseteq G^\tau$ and since by (3.3) both groups have the same Lie algebra $\mathfrak{g}_0$, the homogeneous space $M := G/H$ is symmetric (Example 1.2(b)).

To determine the homogeneous space $G_1/G_{1,\gamma} \cong G_1.\gamma = G.\gamma \in \mathrm{Hom}_{\mathrm{gr}}(\mathbb{R}^\times, G)$, we first determine the stabilizer group $G_\gamma$ and derive some information on related subgroups.

LEMMA 3.7. *The following assertions hold:*

(i) $H = G_\gamma = \{g \in G \colon \mathrm{Ad}_g h = h\}$ *and* $G_{1,\gamma} = H_1 = \mathrm{Aut}(E_+)$.

(ii) $-j_E$ *is contained in the identity component* $G_0$ *of* $G_1$.

(iii) $G^\tau = H \rtimes \{\mathbf{1}, j_E\} = H_1 \rtimes \{\pm\mathbf{1}, \pm j_E\}$ *and* $G_1^\tau = H_1 \rtimes \{\mathbf{1}, -j_E\}$.

PROOF. (i) As $H \subseteq G_\gamma \subseteq G_h := \{g \in G \colon \mathrm{Ad}_g h = h\}$ is obvious, we have to show that every element $g \in G_h$ acts by a linear map on $E \subseteq E_c$. Then the assertion follows from $\mathrm{d}g(x) \in H$ for every $g \in G$ and $x \in E$.

For every $v \in E \subseteq E_c$ we have $\lim_{t\to\infty} \exp(-th).v = 0$, and this property determines the point 0 as the unique attracting fixed point of the flow defined by

$t \mapsto \exp(-th)$ on $E_c$. We conclude that $G_h$ fixes 0. Likewise $\infty := j_E(0) \in E_c$ is the unique attracting fixed point of the flow defined by $t \mapsto \exp(th)$, and so $G_h$ fixes $\infty$ as well. This implies that $G_h$ acts on $E$ by affine maps fixing 0, hence by linear maps ([**Be96**, Thm. 2.1.4]).

(ii) Let $e \in E$ be the unit element of the Jordan algebra $E$. Then $-j_E(z) := -z^{-1}$ is the point reflection in the base point $ie$ of the hermitian symmetric space $T_{E_+} = E + iE_+$ with the holomorphic automorphism group $G_1 \cong \mathrm{Aut}(T_{E_+})$ ([**FK94**, Thm. X.5.6]). Let $K := G_{1,ie}$ denote the stabilizer group of $ie$ in $G_1$. Then $K$ is maximally compact in $G_1$ and its Lie algebra $\mathfrak{k}$ contains a central element $Z$ with $\exp(Z) = s$. This follows easily from the realization of $T_{E_+}$ as the unit ball $\mathcal{D} \subseteq E_{\mathbb{C}}$ of the spectral norm by the Cayley transform $p\colon T_{E_+} \to \mathcal{D}, p(z) := (z - ie)(z + ie)^{-1}$ which maps $ie$ to 0 ([**FK94**, p. 190]). Now the connected circle group $\mathbb{T}$ acts on $\mathcal{D}$ by scalar multiplications and the assertion follows.

(iii) Since the Lie algebra of $G^\tau$ is $\mathfrak{g}^\tau = \mathfrak{g}_0$, the group $\mathrm{Ad}_{G^\tau}$ leaves $\mathfrak{z}(\mathfrak{g}_0) = \mathbb{R}h$ invariant. If $g \in G^\tau$ and $\mathrm{Ad}_g h = \lambda h$, then $\mathrm{Spec}(\mathrm{ad}\, h) = \{-1, 0, 1\}$ implies $\lambda \in \{\pm 1\}$. If $\lambda = 1$, then $g \in G_h = H$, and if $\lambda = -1$, then we likewise obtain $g^{-1} j_E \in H$. □

Let $\theta := \mathrm{Ad}_{-j_E} \in \mathrm{Aut}(\mathfrak{g})$ be the involution induced by the map $-j_E \in G_1 = \mathrm{Cau}(E)$ which is a Cartan involution of $\mathfrak{g}$. It satisfies $\theta(h) = -h$ for the element $h = \gamma'(0)$ defining the grading and thus $\theta(\mathfrak{g}_j) = \mathfrak{g}_{-j}$ for $j = -1, 0, 1$. Therefore

$$C := C_+ + \theta(C_+) \subseteq \mathfrak{q} := \mathfrak{g}_1 \oplus \mathfrak{g}_{-1}$$

is a proper $\mathrm{Ad}(H_1)$-invariant closed convex cone and $\mathrm{Ad}_h C = -C$ for $h \in H \setminus H_1$. By Lawson's Theorem ([**HN93**, Thm. 7.34, Cor. 7.35], [**HÓ96**]) $S_C := H_1 \exp(C)$ is a closed subsemigroup of $G_1$ which defines on $G_1/H_1$ a natural order structure by $gG_{1,V} \leq g'G_{1,V}$ if $g \in g'S_V$ which is invariant under the action of $G_1$ and reversed by elements $g \in G \setminus G_1$.

THEOREM 3.8. (Koufany) *We have the equalities of semigroups*

$$S_C = S_{E_+} := \{g \in G_1 \colon gE_+ \subseteq E_+\} = \exp(C_+)\,\mathrm{Aut}(E_+)\exp(\theta(C_+))$$

*and in particular* $S_C \cap S_C^{-1} = G_{E_+} = H_1 = \mathrm{Aut}(E_+)$.

PROOF. By [**Ko95**, Thm. 4.9], we have in the identity component $G_0$ of $G$ the equality

$$S_{E_+} \cap G_0 = \exp(C_+)(\mathrm{Aut}(E_+) \cap G_0)\exp(\theta(C_+)) = (\mathrm{Aut}(E_+) \cap G_0)\exp(C) = S_C \cap G_0.$$

The definition of $G_1 = \mathrm{Cau}(E)$ and Lemma 3.7(ii) imply that

$$G_1 = G_0\,\mathrm{Aut}(E_+)\{\mathrm{id}, -j_E\} = G_0\,\mathrm{Aut}(E_+). \tag{3.4}$$

This implies that

$$S_{E_+} = (S_{E_+} \cap G_0)\,\mathrm{Aut}(E_+) = \exp(C_+)\,\mathrm{Aut}(E_+)\exp(\theta(C_+))$$

and likewise, by Lemma 3.7(i),

$$S_C = H_1 \exp(C) = \mathrm{Aut}(E_+)(\mathrm{Aut}(E_+) \cap G_0)\exp(C) = \mathrm{Aut}(E_+)(S_{E_+} \cap G_0) = S_{E_+}.$$

□

In view of the fact that, in Quantum Field Theory standard subspaces are associated to domains in space-time, it is interesting to observe that the ordered space $(G_1/H_1, \leq)$ can be realized as a set of subsets of $E_c \cong G_1/P^-$, where $P^- = H_1 \exp(\mathfrak{g}_{-1})$ is the stabilizer of $0 \in E \subseteq E_c$ in $G_1$ ([**Be96**, Thm. 2.1.4(ii)]).

COROLLARY 3.9. *The map*

$$\Xi\colon G_1/H_1 \hookrightarrow 2^{E_c}, \quad g_1H_1 \mapsto g_1E_+$$

*is an order embedding.*

PROOF. Because of the $G_1$-equivariance of $\Xi$, this follows from $S_C = S_{E_+}$, which implies that $g_1E_+ \subseteq g_2E_+$ is equivalent to $g_2^{-1}g_1 \in S_C$, i.e., to $g_1H_1 \leq g_2H_1$ in $G_1/H_1$. □

REMARK 3.10. For the case where $E = \mathbb{R}^{1,d-1}$ is $d$-dimensional Minkowski space, the preceding results lead to the set $\mathcal{W} = G.E_+$ of *conformal wedge domains* in the conformal completion $E_c$. It contains in particular the standard right wedge

$$W_R = \{(x_0, x_1, \ldots, x_{d-1})\colon x_1 > |x_0|\} \subseteq E$$

and all its images under the Poincaré group (cf. [**NÓ17**, Exs. 5.15]).

For $E = \mathbb{R}$, we obtain in particular the set of open intervals in $E_c \cong \mathbb{S}^1$.

Finally, we connect the ordered symmetric space $G_1/H_1$ to $\mathrm{Stand}(\mathcal{H})$ by using antiunitary positive energy representations.

DEFINITION 3.11. We call an antiunitary representation $(U, \mathcal{H})$ of $(G, \varepsilon_G)$ a *positive energy representation* if there exists a non-zero $x \in C_+ \subseteq \mathfrak{g}_1$ for which the selfadjoint operator $-i\mathrm{d}U(x)$ has non-negative spectrum.

REMARK 3.12. (a) For every antiunitary representation of $G$, the set

$$W_U := \{x \in \mathfrak{g}\colon\ -i\mathrm{d}U(x) \geq 0\}$$

is a closed convex invariant cone in $\mathfrak{g}$ which is invariant under the adjoint action of $G_1$ and any $g \in G \setminus G_1$ satisfies $\mathrm{Ad}_g W_U = -W_U$.

(b) Since the Lie algebra $\mathfrak{g}$ is simple, it contains a pair $W_{\min} \subseteq W_{\max}$ of non-zero closed convex invariant cones and any other proper invariant convex cone $W$ satisfies

$$W_{\min} \subseteq W \subseteq W_{\max} \quad \text{or} \quad W_{\min} \subseteq -W \subseteq W_{\max}$$

([**HN93**, Thm. 7.25]). As $W_{\min} \cap \mathfrak{g}_1 = W_{\max} \cap \mathfrak{g}_1 \in \{\pm C_+\}$ by [**HNO94**, Prop. II.7, Thm. II.10, Prop. III.7], for an antiunitary representation $(U, \mathcal{H})$ of $G$ with $W_U \neq \{0\}$ either $U$ or its dual $U^*$ satisfies the positive energy condition.

For a concrete classification of antiunitary positive energy representation we refer to [**NÓ17b**]. By [**NÓ17**, Thm. 2.11], this classification can be reduced to the unitary highest weight representations of $G_1$, resp., its identity component, which have been determined by Enright, Howe and Wallach. We refer to the monograph [**Ne00**] for a systematic exposition of this theory.

THEOREM 3.13. *Let $(U, \mathcal{H})$ be an antiunitary positive energy representation of $G$ for which $\mathrm{d}U$ is non-zero, so that $W_U$ is a non-zero proper invariant cone. Let $V := \mathcal{V}_U(\gamma) \in \mathrm{Stand}(\mathcal{H})$ be the standard subspace corresponding to $\gamma$ under the BGL map* (Corollary 2.4). *Then*

$$S_C = S_V := \{g \in G_1\colon U_gV \subseteq V\}$$

*and this implies that the BGL-map* $\mathcal{V}_U: G_1.\gamma \cong G_1/H_1 \to \mathcal{O}_V = U_{G_1}.V \subseteq \mathrm{Stand}(\mathcal{H})$ *defines an isomorphism of ordered dilation spaces.*

PROOF. By assumption $W_U$ is a proper closed convex invariant cone in $\mathfrak{g}$ and in Remark 3.12 we have seen that $W_U \cap \mathfrak{g}_1 \in \{\pm C_+\}$, so that the positive energy condition leads to $W_U \cap \mathfrak{g}_1 = C_+$. For $x \in C_+$ we have $[h, x] = x$ and $-i\mathrm{d}U(x) \geq 0$, so that $\mathbb{R}x + \mathbb{R}h$ is a 2-dimensional Lie algebra isomorphic to $\mathfrak{aff}(\mathbb{R})$. Therefore Theorem 3.3 implies $\exp(\mathbb{R}_+x) \subseteq S_V$ and we even see that

$$\{x \in \mathfrak{g}_1 \colon \exp(\mathbb{R}_+x) \subseteq S_V\} = W_U \cap \mathfrak{g}_1 = C_+. \tag{3.5}$$

As $\theta = \mathrm{Ad}_{-j_E} \in \mathrm{Ad}_{G_1}$ (Lemma 3.7), it leaves $W_U$ invariant. We conclude with Koufany's Theorem 3.8 that $S_{E_+} = S_C = \exp(C_+)H_1\exp(\theta(C_+)) \subseteq S_V$. Finally, we use the maximality of the subsemigroup $S_{E_+} \subseteq G_1$ (Theorem A.1) to see that $S_C = S_V$. □

## 4. Open problems

PROBLEM 4.1. Let $(G, \varepsilon_G)$ be a graded Lie group with two connected components, $\gamma: \mathbb{R}^\times \to G$ be a graded smooth homomorphism and $(U, \mathcal{H})$ be an antiunitary representation of $G$. Then the $G_1$-invariant cone $W_U \subseteq \mathfrak{g}$ can be analyzed with the well-developed theory of invariant cones in Lie algebras (see [**HN93**, §7.2] and also [**Ne00**]).

- Let $V := \mathcal{V}_U(\gamma)$. Is it possible to determine when the order structure on the subset $U_{G_1}V = \mathcal{V}_U(G_1.\gamma) \subseteq \mathrm{Stand}(\mathcal{H})$ is non-trivial? Theorems 3.3 and 3.13 deal with very special cases.
- Is it possible to determine the corresponding order, which is given by the subsemigroup $S_V \subseteq G_1$, intrinsically in terms of $\gamma$? Here the difficulty is that $G_1.\gamma \cong G_1/G_{1,\gamma}$ carries no obvious order structure.

  In this context results on maximality of subsemigroups of Lie groups constitute important tools. In addition to [**HN95**] that we shall use to prove Theorem A.1 below, results on maximal subsemigroups have been developed by J. D. Lawson in[**La94**] and by A. Neumann and G. Ólafsson in [**NÓ00**], where Lawson's approach is applied in the context of symmetric spaces.

PROBLEM 4.2. In several papers Wiesbrock develops a quite general program how to generate Quantum Field Theories, resp., von Neumann algebras of local observables from finitely many modular automorphism groups ([**Wi93**, **Wi93b**, **Wi97**, **Wi98**]). This contains in particular criteria for three modular groups corresponding to three standard subspaces $(V_j)_{j=1,2,3}$ to generate groups isomorphic to the Poincaré group in dimension 2 or to $\mathrm{PSL}_2(\mathbb{R})$ ([**NÓ17**, Thm. 3.19]). On the level of von Neumann algebras there are also criteria for finitely many modular groups to define representations of $\mathrm{SO}_{1,3}(\mathbb{R})^\uparrow$ or the connected Poincaré group $P(4)^\uparrow_+$ ([**KW01**]).

It would be interesting to see how these criteria can be expressed in terms of the geometry of finite dimensional totally geodesic dilation subspaces of $\mathrm{Stand}(\mathcal{H})$.

## Appendix A. Maximality of the compression semigroup of the cone

In this appendix we prove the maximality of the semigroup $S_{E_+}$ in the causal group $\mathrm{Cau}(E)$ of a simple euclidean Jordan algebra $E$ which is used in the proof of Theorem 3.13.

THEOREM A.1. *If $E$ is a simple euclidean Jordan algebra and $E_+ \subseteq E$ the open positive cone, then the subsemigroup $S_{E_+}$ of $G_1 = \mathrm{Cau}(E)$ is maximal, i.e., any subsemigroup of $G_1$ properly containing $S_{E_+}$ coincides with $G_1$.*

PROOF.

STEP 1 (Reduction to connected groups). First we recall from (3.4) that $G_1 = G_0 \operatorname{Aut}(E_+)$. As $\operatorname{Aut}(E_+) \subseteq S_{E_+} \cap S_{E_+}^{-1}$, it therefore suffices to show that $S_{E_+}^0 := S_{E_+} \cap G_0$ is a maximal subsemigroup of the identity component $G_0$.

STEP 2. We want to derive the assertion from [**HN95**, Thm. V.4]. In [**HN95**] one considers a connected semisimple Lie group $G$, a parabolic subgroup $P$ and an involutive automorphism $\tau$ of $G$. In loc. cit. it is assumed that the symmetric Lie algebra $(\mathfrak{g}, \tau)$ is irreducible (there are no non-trivial $\tau$-invariant ideals) and that, for a $\tau$-invariant Cartan decomposition $\mathfrak{g} = \mathfrak{k} \oplus \mathfrak{p}$ and the $\tau$-eigenspace decomposition $\mathfrak{g} = \mathfrak{h} \oplus \mathfrak{q}$, the center of the Lie algebra

$$\mathfrak{h}^a := (\mathfrak{h} \cap \mathfrak{k}) \oplus (\mathfrak{q} \cap \mathfrak{p}) \quad \text{satisfies} \quad \mathfrak{z}(\mathfrak{h}^a) \cap \mathfrak{q} \cap \mathfrak{p} \neq \{0\}. \tag{A.1}$$

Then the conclusion of [**HN95**, Thm. V.4] is that, if

- $G^\tau P$ is open in $G$,
- the subsemigroup $S(G^\tau, P) := \{g \in G \colon gG^\tau P \subseteq G^\tau P\}$ has non-empty interior, and
- $G = \langle \exp_{G_{\mathbb{C}}} \mathfrak{g}_{\mathfrak{g}} \rangle$ in the simply connected complex group $G_{\mathbb{C}}$ with Lie algebra $\mathfrak{g}_{\mathbb{C}}$,

then $S(G^\tau, P)$ is maximal in $G$.

We next explain how the assumption that $G_{\mathbb{C}}$ is simply connected can be weakened. Suppose that $G$ injects into its universal complexification $G_{\mathbb{C}}$ (which is always the case if it has a faithful finite dimensional representation). Let $q_{\mathbb{C}} \colon \widetilde{G}_{\mathbb{C}} \to G_{\mathbb{C}}$ denote the simply connected covering group and, as $G$ is connected, the integral subgroup $G^\sharp := \langle \exp_{\widetilde{G}_{\mathbb{C}}} \mathfrak{g} \rangle \subseteq \widetilde{G}_{\mathbb{C}}$ satisfies $q_{\mathbb{C}}(G^\sharp) = G$ and $\ker q_{\mathbb{C}}$ is a finite central subgroup of $\widetilde{G}_{\mathbb{C}}$. Consider the covering map $q := q_{\mathbb{C}}|_{G^\sharp} \colon G^\sharp \to G$. Then $P^\sharp := q^{-1}(P)$ is a parabolic subgroup of $G^\sharp$ and $G/P \cong G^\sharp/P^\sharp$. Let $\tau$ also denote the involution of $G^\sharp$ obtained by first extending $\tau$ from $G$ to a holomorphic involution of $G_{\mathbb{C}}$, then lifting it to $\widetilde{G}_{\mathbb{C}}$ and then restricting to $G^\sharp$. Now $H' := q((G^\sharp)^\tau) \subseteq G^\tau$ is an open subgroup satisfying $q((G^\sharp)^\tau P^\sharp) = H'P \subseteq G^\tau P$. As $\ker(q) \subseteq P^\sharp$ and $q$ is surjective, we even obtain

$$(G^\sharp)^\tau P^\sharp = q^{-1}(H'P). \tag{A.2}$$

By [**HN95**, Thm. V.4], $S_1 := S((G^\sharp)^\tau, P^\sharp)$ is a maximal subsemigroup of $G^\sharp \subseteq \widetilde{G}_{\mathbb{C}}$. As $\ker q \subseteq P^\sharp$, we have $S_1 = q^{-1}(S_2)$ for $S_2 := q(S_1)$. Further (A.2) shows that

$$S_2 = \{g \in G \colon gH'P \subseteq H'P\}.$$

Now the maximality of $S_1$ in $G^\sharp$ immediately implies the maximality of $S_2$ in $G$.

STEP 3 (Application to causal groups of Jordan algebras). First we verify the regularity condition (A.1). Here the Lie algebra $\mathfrak{g}$ is simple, which implies in particular that $(\mathfrak{g}, \tau)$ is irreducible.

A natural Cartan involution of $\mathfrak{g}$ is given by $\theta := \mathrm{Ad}_{-j_E}$ which satisfies $\theta(h) = -h$, hence $\theta(\mathfrak{g}_j) = \mathfrak{g}_{-j}$ for $j \in \{-1, 0, 1\}$. Then $\mathfrak{h} = \mathfrak{g}^\tau = \mathfrak{g}_0$ inherits the Cartan decomposition

$$\mathfrak{h} = \mathfrak{str}(E) = \mathfrak{aut}(E) \oplus L(E), \quad \text{where} \quad L(x)y = xy \quad \text{for} \quad x, y \in E,$$

where $\mathfrak{aut}(E)$ is the Lie algebra of the automorphism group $\mathrm{Aut}(E)$ of the Jordan algebra $E$, which coincides with the stabilizer group $H_e$ of the Jordan identity $e$ in $H$. This shows that

$$\mathfrak{h}^a = \mathfrak{aut}(E) \oplus \{x - \theta(x) \colon x \in \mathfrak{g}_1\}.$$

Then the centralizer of $\mathfrak{aut}(E)$ in $E \cong \mathfrak{g}_1$ is $\mathbb{R}e$, and therefore

$$\{x - \theta(x) \colon x \in \mathfrak{g}_1\}^{\mathfrak{aut}(E)} = \mathbb{R}(e - \theta(e)).$$

To verify (A.1), it remains to show that the element $e - \theta(e)$ is central in $\mathfrak{h}^a$, i.e., that it commutes with with all elements of the form $u - \theta(u)$, $u \in \mathfrak{g}_1$. As $\mathfrak{g}_{\pm 1}$ are abelian subalgebras of $\mathfrak{g}$, we have

$$[e - \theta(e), u - \theta(u)] = -[e, \theta(u)] - [\theta(e), u] = -[e, \theta(u)] - \theta([e, \theta(u)]).$$

So it suffices to show that $[e, \theta(u)] \in \mathfrak{h}^{-\theta}$. The vector field on $E$ corresponding to $\theta(u)$ is given by

$$X(z) = P(z, z)u \quad \text{with} \quad P(x, y) = L(x)L(y) + L(y)L(x) - L(xy)$$

and therefore $[e, \theta(u)]$ corresponds to the linear vector field

$$E \to E, \quad z \mapsto \mathrm{d}X(z)e = 2P(z, e)u = 2L(z)u = 2L(u)z.$$

Since this is given by a Jordan multiplication, it belongs to $\mathfrak{h}^{-\theta}$ ([**FK94**, Prop. X.5.8]). This proves that $(\mathfrak{g}, \tau)$ satisfies the regularity condition (A.1).

STEP 4 (The maximality of $S_{E_+}$). Let $G_0$ denote the identity component of $G$. Then the stabilizer $P := G_{0,e}$ of the Jordan identity $e \in E_+ \subseteq E \subseteq E_c$ is a parabolic subgroup and $G_0/P \cong E_c$ is a flag manifold of $G_0$.

As above, let $\tau(g) := \gamma(-1)g\gamma(-1)$, resp., $\tau(g)(x) = -g(-x)$, as a birational map on $E$, and observe that

$$(G_0)^\tau = (H \cap G_0)\{\mathbf{1}, -j_E\} = (\mathrm{Aut}(E_+) \cap G_0)\{\mathbf{1}, -j_E\} \tag{A.3}$$

by Lemma 3.7.

The subgroup $H' \subseteq (G_0)^\tau$ from Step 2 consists of elements which are images of elements in the simply connected connected complex group $\widetilde{G}_{\mathbb{C}}$ fixed under the involution $\tau$. The group $\widetilde{G}_{\mathbb{C}}$ acts by birational maps on the complex Jordan algebra $E_{\mathbb{C}}$ and since $\widetilde{G}_{\mathbb{C}}$ is simply connected, the subgroup $G_{\mathbb{C}}^\tau$ of $\tau$-fixed points in $G_{\mathbb{C}}$ is connected ([**Lo69**, Thm. IV.3.4]). Therefore elements of $G_{\mathbb{C}}^\tau$ act on $E_{\mathbb{C}}$ by elements of the complex group $\exp_{G_{\mathbb{C}}}(\mathfrak{g}_{0,\mathbb{C}})$, hence by linear maps. This shows that $H'$ acts on $E$ by linear maps and thus $H' \subseteq \mathrm{Aut}(E_+)$ follows from (A.3). Further, $H'$ contains $(G^\tau)_0 = \mathrm{Aut}(E_+)_0$ and thus $H'.e = E_+$, as a subset of $E_c$. This means that $H'P/P$ corresponds to $E_+$, and therefore the maximality of $S_{E_+} = S_2$ follows from Step 2.

□

## Acknowledgment

We are most grateful to Wolfgang Bertram for illuminating discussions on the subject matter of this note and for pointing out several crucial references, such as [**Lo67**]. We also thank Jan Möllers for suggestions to improve earlier drafts of the manuscript.

## References

[Ar99] Huzihiro Araki, *Mathematical theory of quantum fields*, International Series of Monographs on Physics, vol. 101, Oxford University Press, New York, 1999. Translated from the 1993 Japanese original by Ursula Carow-Watamura. MR1799198

[BW92] Hellmut Baumgärtel and Manfred Wollenberg, *Causal nets of operator algebras*, Mathematische Lehrbücher und Monographien, II. Abteilung: Mathematische Monographien [Mathematical Textbooks and Monographs, Part II: Mathematical Monographs], vol. 80, Akademie-Verlag, Berlin, 1992. Mathematical aspects of algebraic quantum field theory. MR1339763

[Be96] Wolfgang Bertram, *Un théorème de Liouville pour les algèbres de Jordan* (French, with English and French summaries), Bull. Soc. Math. France **124** (1996), no. 2, 299–327. MR1414541

[Be00] Wolfgang Bertram, *The geometry of Jordan and Lie structures*, Lecture Notes in Mathematics, vol. 1754, Springer-Verlag, Berlin, 2000. MR1809879

[Bo92] H.-J. Borchers, *The CPT-theorem in two-dimensional theories of local observables*, Comm. Math. Phys. **143** (1992), no. 2, 315–332. MR1145798

[BGL02] R. Brunetti, D. Guido, and R. Longo, *Modular localization and Wigner particles*, Rev. Math. Phys. **14** (2002), no. 7-8, 759–785. Dedicated to Professor Huzihiro Araki on the occasion of his 70th birthday. MR1932665

[EN15] Mohamed Elhamdadi and Sam Nelson, *Quandles—an introduction to the algebra of knots*, Student Mathematical Library, vol. 74, American Mathematical Society, Providence, RI, 2015. MR3379534

[FK94] Jacques Faraut and Adam Korányi, *Analysis on symmetric cones*, Oxford Mathematical Monographs, The Clarendon Press, Oxford University Press, New York, 1994. Oxford Science Publications. MR1446489

[FHHK17] Freyn, W., T. Hartnick, M. Horn and R. Köhl, *Kac–Moody symmetric spaces*, Preprint, arXiv:math.GR:1830310 11 Mar 2017

[Ha96] Rudolf Haag, *Local quantum physics*, 2nd ed., Texts and Monographs in Physics, Springer-Verlag, Berlin, 1996. Fields, particles, algebras. MR1405610

[HN93] Joachim Hilgert and Karl-Hermann Neeb, *Lie semigroups and their applications*, Lecture Notes in Mathematics, vol. 1552, Springer-Verlag, Berlin, 1993. MR1317811

[HN95] Joachim Hilgert and Karl-Hermann Neeb, *Maximality of compression semigroups*, Semigroup Forum **50** (1995), no. 2, 205–222. MR1315512

[HNO94] Joachim Hilgert, Karl-Hermann Neeb, and Bent Ørsted, *The geometry of nilpotent coadjoint orbits of convex type in Hermitian Lie algebras*, J. Lie Theory **4** (1994), no. 2, 185–235. MR1337191

[HÓ96] Joachim Hilgert and Gestur Ólafsson, *Causal symmetric spaces*, Perspectives in Mathematics, vol. 18, Academic Press, Inc., San Diego, CA, 1997. Geometry and harmonic analysis. MR1407033

[KW01] Ralf Kähler and Hans-Werner Wiesbrock, *Modular theory and the reconstruction of four-dimensional quantum field theories*, J. Math. Phys. **42** (2001), no. 1, 74–86. MR1808767

[Ko95] Khalid Koufany, *Semi-groupe de Lie associé à un cône symétrique* (French, with English and French summaries), Ann. Inst. Fourier (Grenoble) **45** (1995), no. 1, 1–29. MR1324123

[La94] J. D. Lawson, *Maximal Ol′shanskiĭ semigroups*, J. Lie Theory **4** (1994), no. 1, 17–29. MR1326949

[Lo67] Ottmar Loos, *Reflexion spaces and homogeneous symmetric spaces*, Bull. Amer. Math. Soc. **73** (1967), 250–253. MR0217742

[Lo67b] Ottmar Loos, *Spiegelungsräume und homogene symmetrische Räume* (German), Math. Z. **99** (1967), 141–170. MR0212742

[Lo69] Ottmar Loos, *Symmetric spaces. I: General theory*, W. A. Benjamin, Inc., New York-Amsterdam, 1969. MR0239005

[Lo72] Ottmar Loos, *An intrinsic characterization of fibre bundles associated with homogeneous spaces defined by Lie group automorphisms*, Abh. Math. Sem. Univ. Hamburg **37** (1972), 160–179. MR0307124

[Lo08] Roberto Longo, *Real Hilbert subspaces, modular theory,* $\mathrm{SL}(2,\mathbb{R})$ *and CFT*, Von Neumann algebras in Sibiu, Theta Ser. Adv. Math., vol. 10, Theta, Bucharest, 2008, pp. 33–91. MR2512325

[Ne00] Karl-Hermann Neeb, *Holomorphy and convexity in Lie theory*, De Gruyter Expositions in Mathematics, vol. 28, Walter de Gruyter & Co., Berlin, 2000. MR1740617

[NÓ16] Neeb, K.-H., G. Ólafsson, *KMS conditions, standard subspaces and reflection positivity on the circle group*, arXiv:math-ph:1611.00080; submitted

[NÓ17] —, *Antiunitary representations and modular theory*, to appear in "50th Seminar Sophus Lie", Eds. K. Grabowska, J. Grabowski, A. Fialowski and K.-H. Neeb, Banach Center Publications, arXiv:math-RT:1704.01336

[NÓ17b] —, *Antiunitary representations of hermitian groups*, in preparation

[Nel16] Sam Nelson, *What is ... a quandle?*, Notices Amer. Math. Soc. **63** (2016), no. 4, 378–380. MR3444659

[NÓ00] Andreas Neumann and Gestur Ólafsson, *Minimal and maximal semigroups related to causal symmetric spaces*, Semigroup Forum **61** (2000), no. 1, 57–85. MR1839215

[Oe17] Oeh, D., *Antiunitary representations of symmetric spaces*, Master Thesis, University Erlangen–Nürnberg, 2017

[Wi93] Hans-Werner Wiesbrock, *Half-sided modular inclusions of von-Neumann-algebras*, Comm. Math. Phys. **157** (1993), no. 1, 83–92. MR1244859

[Wi93b] Hans-Werner Wiesbrock, *Conformal quantum field theory and half-sided modular inclusions of von Neumann algebras*, Comm. Math. Phys. **158** (1993), no. 3, 537–543. MR1255426

[Wi97] Hans-Werner Wiesbrock, *Symmetries and modular intersections of von Neumann algebras*, Lett. Math. Phys. **39** (1997), no. 2, 203–212. MR1437753

[Wi98] Hans-Werner Wiesbrock, *Modular intersections of von Neumann algebras in quantum field theory*, Comm. Math. Phys. **193** (1998), no. 2, 269–285. MR1618159

Department Mathematik, Universität Erlangen-Nürnberg, Cauerstrasse 11, 91058 Erlangen, Germany

*Email address*: neeb@math.fau.de

Contemporary Mathematics
Volume **714**, 2018
https://doi.org/10.1090/conm/714/14328

# Pure spinors and a construction of the $E_*$-Lie algebras

Marcus J. Slupinski and Robert J. Stanton

*This paper is dedicated to our friend and collaborator Gestur Ólafsson.*

ABSTRACT. Let $(V, g)$ be a $2n$-dimensional hyperbolic space and $C(V, g)$ its Clifford algebra. $C(V, g)$ has a $\mathbb{Z}$-grading, $C^k$, and an algebra isomorphism $C(V, g) \cong End(S)$, $S$ the space of spinors. É. Cartan defined operators $L_k : End(S) \to C^k$ which are involved in the definition of pure spinors. We shall give a more refined study of the operator $L_2$, in fact, obtain explicit formulae for it in terms of spinor inner products and combinatorics, as well as the matrix of it in a basis of pure spinors. Using this information we give a construction of the exceptional Lie algebras $\mathfrak{e}_6, \mathfrak{e}_7, \mathfrak{e}_8$ completely within the theory of Clifford algebras and spinors.

## 1. Introduction

Constructions of exceptional Lie algebras over quite general fields have been given by many people and from various perspectives. While the list is too long to give, error free, we must mention Freudenthal and Tits. The perspective of this paper is that of spinors. When the base field $k$ is $\mathbb{R}$, the classification by É. Cartan of irreducible Riemannian symmetric spaces of the noncompact type already provides an example, the case of $\mathfrak{e}_8$, for which the pair $(\mathfrak{so}(16, \mathbb{R}), S_\pm)$ occurs. More recently, J. F. Adams [**Ad**] gave a construction of compact exceptional Lie groups using compact spin groups and the relationship of some to Jordan algebras. Moroianu-Semmelman [**MoSe**] gave a construction of exceptional Lie algebras of compact type by refining Kostant's [**K**] invariant 4-tensor characterization of certain holonomy representations and coupled with the compact spinor material from [**Ad**]. Our point of view is to present natural properties of Clifford algebras and their spinors for a hyperbolic space over a very general field $k$, and then to derive the existence of the $E_*$-series using these properties and the combinatorics of pure spinors, thus a construction intrinsic to spinor algebra.

The paper is essentially self-contained and written with Lie theorists in mind, such as a master like Gestur, hence includes some standard material on spinors

2010 *Mathematics Subject Classification.* Primary 15A66, 17B05.
*Key words and phrases.* Spinors, exceptional Lie algebras.
The first author acknowledges the recurring support from the Math. Research Inst., OSU, of our collaboration.
The second author is grateful to IRMA, Université de Strasbourg, for its hospitable and stimulating environment.

known to experts. We begin with a Clifford algebra $C(V,g)$ and its basic isomorphism with $End(S)$, $S$ the spinors. Then we relate properties of $C(V,g)$ and $S$, including fundamental material about the spinor norm. The key tool in the paper is the operator introduced by Cartan, $L_2 : S \times S \to C_2$. After we give a new description of $L_2$, we obtain an explicit formula for $L_2$ in terms of Clifford elements. The formula is intrinsic to Clifford theory as the operator is completely specified by various spinor norms and combinatorics. Then using the basis of pure spinors we compute the matrix of $L_2$ and express its entries in terms of spinor norms and combinatorics. This treatment is completely general for a hyperbolic space $(V,g)$ and a field $k$ of characteristic not 2 or 3.

In the last section we specialise the formula for the matrix of $L_2$ to three specific dimensions and show that various entries of the matrix vanish for combinatorial reasons yielding a Jacobi identity for the various Lie algebras in the $E_*$-series.

There are several potential future directions. The choice of a hyperbolic $g$ was made to avoid field extensions of $k$ - indeed there are metrics of other signature that could be considered. Also, the combinatorics that arise in the computations mirror properties of the Weyl group quotient that parametrises Schubert cells in the flag variety of projectivised pure spinors. We did not consider whether other topological properties of the cells are responsible for the various combinatorial identities. Finally, since the spinor algebra is a universal linear construction, we expect the spinor algebra constructions, in particular $L_2$, to transfer to vector bundles.

## 2. Background on spinors

Let $V$ be a $2n$-dimensional vector space over a field $k$ of characteristic not 2 or 3. We shall assume that $V$ has a nondegenerate symmetric bilinear form $g$ of Witt index $n$ (i.e. a hyperbolic form). The hyperbolic case allows us to give a rather complete presentation of the results without any base extension of $k$. This was highlighted by Chevalley and today seems even more relevant. A good reference for much of the basic material of this section is [**Ch**].

**2.1. Clifford algebra.** Let $C = C(V,g)$ be the Clifford algebra of $V$ with respect to $g$. Then $C$ has the usual $\mathbb{Z}_2$ grading $C = C_+ \dot{\oplus} C_-$ inherited from the tensor algebra of $V$. As $g$ is hyperbolic, $C$ is isomorphic to the algebra of $2^n \times 2^n$ matrices over $k$. We can choose a $2^n$-dimensional $k$-vector space $S$, up to equivalence, called the space of spinors, and obtain an algebra isomorphism

$$C \cong End(S).$$

Hence $C$ has a natural trace that we denote $Tr : C(V,g) \to k$.

The vector space $V$ is naturally included in $C_-$ so, from now on, we consider $V$ as a subset of $C$. By the universal property of $\Lambda^*(V)$, the exterior algebra of $V$, one can extend the inclusion of $V$ into $C_-$ to an $O(V,g)$- equivariant linear (but not algebra) isomorphism $Q : \Lambda^*(V) \to C$ by defining

$$Q(v_1 \wedge \cdots \wedge v_k) = \frac{1}{k!} \sum_{\sigma} (-1)^{\sigma} v_{\sigma(1)} \cdots v_{\sigma(k)},$$

at least if $k$ is of characteristic 0. If $v_1, \ldots, v_k$ are orthogonal this formula implies that

$$Q(v_1 \wedge \cdots \wedge v_k) = v_1 \cdots v_k$$

and one uses this property to characterize $Q$ when $k$ is of positive characteristic (see [**Ch**]).

We set $C^k = Q(\Lambda^k(V))$. Hence if $\{v_1, \dots, v_{2n}\}$ is a basis of $V$ and $i_1 < i_2 < \dots < i_k$ then the collection $\{Q(v_{i_1} \wedge \dots \wedge v_{i_k})\}$ is a basis of $C^k$. The collection of subspaces $C^k$ then give $C$ the structure of a $\mathbb{Z}$ graded vector space.

$C$ is also a filtered algebra where $D^k$ is generated by products of at most $k$ elements of $V$. We have then an isomorphism of the associated $\mathbb{Z}$ graded space determined by the filtration onto the $\mathbb{Z}$ graded $Q(\Lambda^*(V))$.

The following commutator relations are well known:

$$[C^1, C^1] \subseteq C^2, \quad [C^2, C^m] \subseteq C^m, \quad [C^2, C^{2n}] = 0.$$

Consequently, $C^2$, $C^1 \oplus C^2$ and $C^2 \oplus C^{2n}$ are Lie algebras.

The composition with $Q$ of any $\mathfrak{o}(V,g)$-equivariant isomorphism

$$\mathfrak{o}(V,g) \cong \Lambda^2(V)$$

defines a Lie algebra isomorphism $\mathfrak{o}(V,g) \cong C^2$.

Similarly one shows easily that the orthogonal Lie algebra of a vector space of dimension $2n+1$ of maximal Witt index is isomorphic to $C^1 \oplus C^2$.

The canonical anti-automorphism of order 2 of $C(V,g)$, namely the one extending $v \mapsto v$ for $v \in V$, is inherited from the tensor algebra. It will be denoted $x \mapsto x^T$. Using the canonical anti-automorphism $T$ and the trace $Tr$ one can give $C$ a norm, namely $\|c\|^2 = Tr(c^T c)$.

PROPOSITION 2.1. *Let $g_\Lambda$ be the natural extension of $g$ to $\Lambda^*(V)$. Then*

$$2^n g_\Lambda(\alpha, \beta) = Tr(Q(\alpha)^T Q(\beta)),$$

*i.e. $Q$ is a multiple of an isometry.*

REMARK 2.2. *By an orthonormal basis of $V$ we mean a basis $\{e_1, \dots, e_{2n}\}$ which satisfies*

$$g(e_i, e_j) = \pm\delta_{ij}.$$

*Orthonormal bases exist because $(V,g)$ is isometric to an orthogonal sum of hyperbolic planes. Later we will use ordered orthonormal bases.*

For the $\mathbb{Z}$ grading $C = \oplus C^k$ and with respect to an orthonormal basis there is a formula for the projection $\pi_k : C \to C^k$:

$$\pi_k(c) = \frac{1}{2^n} \sum_{i_1 < \dots < i_k} g(e_{i_1}, e_{i_1}) \dots g(e_{i_k}, e_{i_k}) Tr(e_{i_k} \dots e_{i_1} c) e_{i_1} \dots e_{i_k},$$

as follows easily from the fact that $\{Q(v_{i_1} \wedge \dots \wedge v_{i_k})\}$ is a basis of $C^k$ and $Q(e_{i_1} \wedge \dots \wedge e_{i_k}) = e_{i_1} \cdots e_{i_k}$.

There is a natural $\mathbb{Z}_2$ grading of $S$, $S = S_1 \oplus S_2$, into a direct sum of two $2^{n-1}$ dimensional subspaces compatible with the $\mathbb{Z}_2$ graded action of $C$, i.e.,

$$C_+ \cdot S_1 \subseteq S_1, \quad C_+ \cdot S_2 \subseteq S_2, \quad C_- \cdot S_1 \subseteq S_2 \text{ and } C_- \cdot S_2 \subseteq S_1.$$

Elements of $S_1$ or $S_2$ are called half-spinors. An element of $C$ that implements such a grading of $S$, i.e. which is the identity on one half-spinor space and minus the identity on the other, will be called a grading element. They are usually denoted by $\varepsilon$ and are elements of $C^{2n}$. Note that if $\{e_1, \dots, e_{2n}\}$ is an orthonormal basis of $V$ then

$$\varepsilon = e_1 \dots e_{2n}$$

satisfies $\varepsilon^2 = 1$ and is a grading element. The graded decomposition of $S$ corresponding to this particular $\varepsilon$ will be denoted $S = S_+ \oplus S_-$.

PROPOSITION 2.3. *If $\varepsilon$ is a grading element, then for all $1 \le k \le 2n$ and all $c \in C$,*

$$\pi_{2n-k}(\varepsilon c) = \varepsilon \pi_k(c).$$

PROOF. This is essentially III.4.3 in [**Ch**]. □

**2.2. Spinor norms.** Recall that $V$ is of dimension $2n$ and that $g$ is hyperbolic.

DEFINITION 2.4. *A spinor norm is a bilinear map $B : S \times S \to k$ such that*

$$B(v \cdot \phi, \psi) = B(\phi, v \cdot \psi) \quad \forall v \in V, \forall \phi, \psi \in S.$$

*Hence, given a non-zero spinor norm $B$, the canonical anti-automorphism of $C$, $x \mapsto x^T$, is, via the isomorphism $C \cong End(S)$, the transpose relative to $B$.*

PROPOSITION 2.5. *(É. Cartan) The space of spinor norms is one dimensional.*

The symmetry type of a spinor norm, symmetric or alternating, as well as the type of its restriction to the half-spinor spaces is summarized in the next result.

PROPOSITION 2.6. *Let $B$ be a spinor norm.*
*(i) If $n \equiv 0 \mod 4$ then $B$ is symmetric and even, i.e., $B(S_1, S_2) = 0$.*
*(ii) If $n \equiv 1 \mod 4$ then $B$ is symmetric and odd, i.e., $B(S_1, S_1) = B(S_2, S_2) = 0$.*
*(iii) If $n \equiv 2 \mod 4$ then $B$ is antisymmetric and even.*
*(iv) If $n \equiv 3 \mod 4$ then $B$ is antisymmetric and odd.*

We shall fix a non-degenerate spinor norm, $B$. It is natural to have the relationship of the $\mathbb{Z}$ grading of $C$ to $B$. If $\phi, \psi \in S$, $1 \le k \le 2n$ and $v_1, \ldots, v_k \in V$ are orthogonal, it is clear that

$$B(v_1 \cdots v_k \cdot \phi, \psi) = (-1)^{\frac{1}{2}k(k-1)} B(\phi, v_1 \cdots v_k \cdot \psi).$$

Hence we have

COROLLARY 2.7. *If $k \equiv 2$ or $3$ (mod 4), then*

$$B(c \cdot \phi, \psi) + B(\phi, c \cdot \psi) = 0 \quad \forall c \in C^k,$$

*i.e., a spinor norm is invariant under the action of $C^k$ iff $k \equiv 2$ or $k \equiv 3$ (mod 4).*

Since $B$ can have either symmetry type, symmetric or alternating, we denote by $\mathfrak{aut}(S, B)$ the endomorphisms of $S$ that leave invariant the spinor norm.

COROLLARY 2.8. *Identifying $C(V, g)$ with $End(S)$, we have*

$$\bigoplus_{k \equiv 2 \text{ or } 3 \,(\mathrm{mod}\, 4)} C^k = \mathfrak{aut}(S, B).$$

PROOF. By Corollary 2.7 the LHS is included in the RHS. The result follows from a dimension count for the corresponding symmetry type of $B$:

$$\sum_{k \equiv 2 \text{ or } 3 \,(\mathrm{mod}\, 4)} \binom{2n}{k} = 2^{n-1}(2^n - 1) = \dim \mathfrak{so}(S, B) \quad (n \equiv 0, 1 \mod 4)$$

and

$$\sum_{k\equiv 2 \text{ or } 3 \pmod 4} \binom{2n}{k} = 2^{n-1}(2^n+1) = \dim \mathfrak{sp}(S,B) \quad (n \equiv 2,3 \mod 4).$$

□

COROLLARY 2.9. *If $\varepsilon$ is any grading element and we set*

$$\mathfrak{aut}_{\pm}(S,B) = \{c \in \mathfrak{aut}(S,B) : c\varepsilon = \pm\varepsilon c\}$$

*then*

$$\bigoplus_{k\equiv 2 \,(\mathrm{mod}\, 4)} C^k = \mathfrak{aut}_+(S,B), \qquad \bigoplus_{k\equiv 3 \,(\mathrm{mod}\, 4)} C^k = \mathfrak{aut}_-(S,B).$$

PROOF. This is immediate from the previous Corollary. □

From Prop. 2.6 we see that the spinor norm $B$ is even if and only if $n$ is even. In this case $B$ restricts to nondegenerate forms $B_1$ and $B_2$ on the half-spinor spaces $S_1$ and $S_2$ respectively. Thus for $n$ even and grading element $\varepsilon$, the Lie algebra $\mathfrak{aut}_+(S,B)$ is a direct product

$$\mathfrak{aut}_+(S,B) \cong \mathfrak{aut}(S_1,B_1) \oplus \mathfrak{aut}(S_2,B_2).$$

To realise this decomposition of $\mathfrak{aut}_+(S,B)$ in the Clifford algebra we use the grading element.

COROLLARY 2.10. *Let $n$ be even. Let $\varepsilon \in C(V,g)$ be a grading element and $S = S_1 \oplus S_2$ the associated grading (the $\pm 1$ eigenspaces of $\varepsilon$). Set $\varepsilon_{\pm} = \frac{(1\pm\varepsilon)}{2}$.*

*(i) Then*

$$\bigoplus_{k\equiv 2 \,(\mathrm{mod}\, 4)} C^k = \left( \bigoplus_{k\equiv 2 \,(\mathrm{mod}\, 4), k\le n} C^k\varepsilon_+ \right) \oplus \left( \bigoplus_{k\equiv 2 \,(\mathrm{mod}\, 4), k\le n} C^k\varepsilon_- \right).$$

*(ii) With respect to the decomposition $S = S_1 \oplus S_2$, the two summands of (i) are:*

$$\bigoplus_{k\equiv 2 \,(\mathrm{mod}\, 4),\, k\le n} C^k\varepsilon_+ = \left( \begin{array}{c|c} \mathfrak{aut}(S_1,B_1) & 0 \\ \hline 0 & 0 \end{array} \right)$$

*and*

$$\bigoplus_{k\equiv 2 \,(\mathrm{mod}\, 4),\, k\le n} C^k\varepsilon_- = \left( \begin{array}{c|c} 0 & 0 \\ \hline 0 & \mathfrak{aut}(S_2,B_2) \end{array} \right)$$

PROOF. Part (i) basically follows from the fact that

$$1 = \frac{1}{2}(1+\varepsilon) + \frac{1}{2}(1-\varepsilon)$$

decomposes the identity of $C$ as a sum of two orthogonal idempotents. Part (ii) is straightforward. □

REMARK 2.11. *If $k < n$, both $C^k\varepsilon_+$ and $C^k\varepsilon_-$ are isomorphic to $\Lambda^k(V)$ as absolutely irreducible $\mathfrak{so}(V,g)$-representations. If $k = n$, É. Cartan showed that $C^n\varepsilon_+$ and $C^n\varepsilon_-$ are absolutely irreducible, non-isomorphic representations of the same dimension. The proposition therefore gives an explicit reduction of $\mathfrak{aut}(S_1,B_1)$ and $\mathfrak{aut}(S_2,B_2)$ into their $\mathfrak{so}(V,g)$-irreducible components.*

### 2.3. Tensor product $S \otimes S$.

As usual, a choice of $B$ on $S$ gives a $C^2$-equivariant isomorphism $\tau : S \otimes S \to End(S)$:

$$\tau(\phi \otimes \psi)(\xi) = B(\phi, \xi)\psi.$$

Since

$$B(\tau(\phi \otimes \psi)(\xi), \eta) = (-1)^{\frac{1}{2}n(n-1)} B(\xi, \tau(\psi \otimes \phi)(\eta)),$$

it follows that

$$\begin{cases} \tau(\phi \otimes \psi - \psi \otimes \phi) \in \mathfrak{aut}(S, B) & \text{if} \quad n \equiv 0, 1 \mod 4, \\ \tau(\phi \otimes \psi + \psi \otimes \phi) \in \mathfrak{aut}(S, B) & \text{if} \quad n \equiv 2, 3 \mod 4. \end{cases}$$

Identifying $End(S)$ with the Clifford algebra $C$ and using the preceding Remark one can reduce symmetric and antisymmetric spinors as $\mathfrak{so}(V, g)$-representations.

PROPOSITION 2.12.
*(i) If $n \equiv 0, 1 \mod 4$ then $\tau$ induces $\mathfrak{so}(V, g)$-equivariant isomorphisms:*

$$\Lambda^2(S) \cong \bigoplus_{k \equiv 2 \text{ or } 3 \,(\mathrm{mod}\, 4)} C^k, \qquad S^2(S) \cong \bigoplus_{k \equiv 0 \text{ or } 1 \,(\mathrm{mod}\, 4)} C^k.$$

*(ii) If $n \equiv 2, 3 \mod 4$ then $\tau$ induces $\mathfrak{so}(V, g)$-equivariant isomorphisms:*

$$\Lambda^2(S) \cong \bigoplus_{k \equiv 0 \text{ or } 1 \,(\mathrm{mod}\, 4)} C^k, \qquad S^2(S) \cong \bigoplus_{k \equiv 2 \text{ or } 3 \,(\mathrm{mod}\, 4)} C^k.$$

## 3. Cartan's operator $L_2$

The operators $L_2$ and $L_{2n}$ to be defined in this section are among the operators $L_k$ that appear in Cartan and Chevalley, where mostly they are used to characterise pure spinors. The interesting properties of $L_2$ to be described herein appear to be new.

Composing $\tau : S \otimes S \to End(S)$ with the projection $\pi_2 : C \to C^2$ we can define a $C^2$-equivariant map $L_2 : S \times S \to C^2$ :

$$L_2(\phi, \psi) = \frac{1}{2^n} \sum_{i<j} g(e_i, e_i) g(e_j, e_j) B(\phi, e_j e_i \cdot \psi) e_i e_j, \tag{1}$$

where $\{e_1, \ldots, e_{2n}\}$ is any orthonormal basis of $V$.

PROPOSITION 3.1.
*(i) If $n \equiv 0 \mod 4$ then $L_2$ is antisymmetric and even, i.e.* $L_2(S_1, S_2) = 0$.
*(ii) If $n \equiv 1 \mod 4$ then $L_2$ is antisymmetric and odd, i.e.*

$$L_2(S_1, S_1) = L_2(S_2, S_2) = 0.$$

*(iii) If $n \equiv 2 \mod 4$ then $L_2$ is symmetric and even.*
*(iv) If $n \equiv 3 \mod 4$ then $L_2$ is symmetric and odd.*

$L_2$ has an interesting formulation in terms of orbit maps.

DEFINITION 3.2. *For $\phi \in S$ we define $\phi : V \to S$ and $\phi^* : S \to V$ by*

$$\phi(v) = v \cdot \phi, \qquad \phi^*(\psi) = \sum_i g(e_i, e_i) B(\phi, e_i \cdot \psi) e_i \tag{2}$$

*where $\{e_1, \ldots, e_{2n}\}$ is any orthonormal basis of $V$.*

The maps $\phi$ and $\phi^*$ are adjoints for the respective norms, i.e.,

$$B(\phi(v), \psi) = g(v, \phi^*(\psi)). \tag{3}$$

Thus we have

$$\begin{cases} g(\phi^* \circ \psi(v_1), v_2) = B(\phi(v_2), \psi(v_1)), \\ g(\psi^* \circ \phi(v_2), v_1) = B(\psi(v_1), \phi(v_2)), \end{cases}$$

hence

$$g(\phi^* \circ \psi(v_1), v_2) = (-1)^{\frac{1}{2}n(n-1)} g(\psi^* \circ \phi(v_2), v_1).$$

Similarly, equation (3) implies

$$B(\psi \circ \phi^*(\xi), \eta) = g(\phi^*(\xi), \psi^*(\eta)) = B(\phi \circ \psi^*(\eta), \xi),$$

hence

$$B(\psi \circ \phi^*(\xi), \eta) = (-1)^{\frac{1}{2}n(n-1)} B(\xi, \phi \circ \psi^*(\eta)).$$

The next Proposition follows immediately from the above.

PROPOSITION 3.3. *Let $\phi, \psi \in S$ be spinors. Then*

$$(-1)^{\frac{1}{2}n(n-1)}\psi^* \circ \phi - \phi^* \circ \psi, \qquad (-1)^{\frac{1}{2}n(n-1)}\psi^* \circ \phi + \phi^* \circ \psi$$

*are respectively antisymmetric and symmetric endomorphisms of $(V, g)$, while*

$$(-1)^{\frac{1}{2}n(n-1)}\psi \circ \phi^* - \phi \circ \psi^*, \qquad (-1)^{\frac{1}{2}n(n-1)}\psi \circ \phi^* + \phi \circ \psi^*$$

*are respectively antisymmetric and symmetric endomorphisms of $(S, B)$.*

Consequently the first expression of Proposition 3.3 defines a bilinear form on $S$ with values in $\mathfrak{so}(V, g)$, and even/odd depending on $n$ mod 4. By Proposition 3.1, the bilinear form $L_2$ also takes values in $\mathfrak{so}(V, g)$ and can be seen to have the same parity properties. Analogously, the third expression also defines a bilinear form on $S$ with the same parity properties as $L_2$ but taking values in $\mathfrak{aut}(S, B)$. Taken together, they suggest a type of curvature operator. The exact relationship between them will be described next.

In fact it will be convenient to 'renormalise' $L_2$ as follows:

DEFINITION 3.4. *Define $\tilde{L}_2 : S \times S \to C^2$ by*

$$\tilde{L}_2 = 2^{n-1} L_2$$

PROPOSITION 3.5.
*(i) For all $\phi, \psi \in S$,*

$$2\tilde{L}_2(\phi, \psi) = (-1)^{\frac{1}{2}n(n-1)}\psi^* \circ \phi - \phi^* \circ \psi$$

*in the sense that for all $v \in V$,*

$$2[\tilde{L}_2(\phi, \psi), v] = (-1)^{\frac{1}{2}n(n-1)}\psi^* \circ \phi(v) - \phi^* \circ \psi(v). \tag{4}$$

*(ii) For all $\phi, \psi \in S$ and all $v \in V$,*

$$2B(\phi, \psi)v = (-1)^{\frac{1}{2}n(n-1)}\psi^* \circ \phi(v) + \phi^* \circ \psi(v). \tag{5}$$

PROOF. Let $\{e_1, \cdots, e_{2n}\}$ be an orthonormal basis of $V$. For any $e_k$ in this basis, by equation (1) we have

$$[\tilde{L}_2(\phi, \psi), e_k] = \frac{1}{2} \sum_{i<j} g(e_i, e_i) g(e_j, e_j) B(\phi, e_j e_i \cdot \psi)[e_i e_j, e_k],$$

and using

$$[e_ie_j, e_k] = \begin{cases} 0 & \text{if } i \neq k, j \neq k, \\ -2g(e_k, e_k)e_j & \text{if } i = k, \\ 2g(e_k, e_k)e_i & \text{if } j = k, \end{cases}$$

this simplifies to

$$[\tilde{L}_2(\phi, \psi), e_k] = \sum_{i \neq k} g(e_i, e_i)g(e_k, e_k)^2 B(\phi, e_ke_i \cdot \psi)e_i,$$

which, since $g(e_k, e_k)^2 = 1$, reduces to

$$[\tilde{L}_2(\phi, \psi), e_k] = \sum_{i \neq k} g(e_i, e_i)B(\phi, e_ke_i \cdot \psi)e_i. \tag{6}$$

To calculate the RHS of equation (4) acting on $e_k$ we have

$$\begin{aligned}(-1)^{\frac{1}{2}n(n-1)}\psi^* \circ \phi(e_k) &= (-1)^{\frac{1}{2}n(n-1)} \sum_i g(e_i, e_i)B(\psi, e_ie_k \cdot \phi)e_i \\ &= \sum_i g(e_i, e_i)B(e_ie_k \cdot \phi, \psi)e_i \\ &= \sum_i g(e_i, e_i)B(\phi, e_ke_i \cdot \psi)e_i, \qquad (7)\end{aligned}$$

and

$$\phi^* \circ \psi(e_k) = \sum_i g(e_i, e_i)B(\phi, e_ie_k \cdot \psi)e_i. \tag{8}$$

Since $e_k$ is an arbitrary basis element, both parts of the proposition follow from equations (6), (7) and (8).

□

In the same way there is a $C^2$-equivariant map $L_{2n} : S \times S \to C^{2n}$ obtained by composing $\tau : S \otimes S \to \mathrm{End}(S) = C$ and $\pi_{2n} : C \to C^{2n}$. Explicitly,

$$L_{2n}(\phi, \psi) = \frac{1}{2^n} g(e_1, e_1) \ldots g(e_{2n}, e_{2n})B(\phi, e_{2n} \ldots e_1 \cdot \psi)e_1 \ldots e_{2n}, \tag{9}$$

where $\{e_1, \ldots, e_{2n}\}$ is any orthonormal basis of $V$. The symmetry properties of $L_{2n}$ follow readily from the preceding.

PROPOSITION 3.6.
*(i) If $n \equiv 0 \mod 4$ then $L_{2n}$ is symmetric and even.*
*(ii) If $n \equiv 1 \mod 4$ then $L_{2n}$ is antisymmetric and odd.*
*(iii) If $n \equiv 2 \mod 4$ then $L_{2n}$ is symmetric and even.*
*(iv) If $n \equiv 3 \mod 4$ then $L_{2n}$ is antisymmetric and odd.*

### 3.1. Graded spinor norms.

Spinor norms are invariant under $C^2$ but not under $C^1$, as follows from Corollary 2.7. In order to get something invariant under the action of the Lie algebra $C^1 \oplus C^2$ one can use a grading element.

PROPOSITION 3.7. *Let $B$ be a spinor norm and $\varepsilon \in C$ be a grading element. Define the associated graded spinor norm $B_\varepsilon : S \times S \to k$ by*

$$B_\varepsilon(\phi, \psi) = B(\varepsilon \cdot \phi, \psi) \, \forall \phi, \psi \in S.$$

*Then*

$$B_\varepsilon(v\cdot\phi,\psi) = -B_\varepsilon(\phi, v\cdot\psi)\ \forall v\in V,\ \forall\phi,\psi\in S,$$

*and graded spinor norms are characterised by this property.*

If $\phi,\psi\in S$, $1\le k\le 2n$ and $v_1,\dots,v_k\in V$ are orthogonal, it is clear that

$$B_\varepsilon(v_1\dots v_k\cdot\phi,\psi) = (-1)^{\frac{1}{2}k(k+1)}B_\varepsilon(\phi, v_1\dots v_k\cdot\psi)$$

and hence we have

COROLLARY 3.8. *A graded spinor norm is invariant for the action of the Lie algebra $C^1\oplus C^2$.*

PROPOSITION 3.9. *Let $B_\varepsilon$ be a graded spinor norm.*
*(i) If $n\equiv 0 \mod 4$ then $B_\varepsilon$ is symmetric and even.*
*(ii) If $n\equiv 1 \mod 4$ then $B_\varepsilon$ is antisymmetric and odd.*
*(iii) If $n\equiv 2 \mod 4$ then $B_\varepsilon$ is antisymmetric and even.*
*(iv) If $n\equiv 3 \mod 4$ then $B_\varepsilon$ is symmetric and odd.*

COROLLARY 3.10. *If $\phi,\psi\in S$, $1\le k\le 2n$ and $v_1,\dots,v_k\in V$ are orthogonal, then*

$$B_\varepsilon(\phi, v_1\dots v_k\cdot\psi) = (-1)^{\frac{1}{2}n(n+1)+\frac{1}{2}k(k+1)}B_\varepsilon(\psi, v_1\dots v_k\cdot\phi).$$

Using $B_\varepsilon$ we can define a $C^1\oplus C^2$-equivariant map $\tau_\varepsilon : S\times S\to \mathrm{End}(S)$ by

$$\tau_\varepsilon(\phi,\psi)(\xi) = B_\varepsilon(\phi,\xi)\psi.$$

One can now repeat the preceding but using the graded versions.

Using $B_\varepsilon$ we can define a $C^1\oplus C^2$-equivariant map $L_\varepsilon : S\times S\to C^1\oplus C^2$ by composing $\tau_\varepsilon : S\otimes S\to \mathrm{End}(S) = C$ and $\pi_1\oplus\pi_2 : C\to C^1\oplus C^2$. (To avoid excessive notation we suppress the subscripts $1,2$.) Explicitly,

$$L_\varepsilon(\phi,\psi) = \frac{1}{2^n}\left(\sum_i g(e_i,e_i)B_\varepsilon(\phi, e_i\cdot\psi)e_i + \sum_{i<j} g(e_i,e_i)g(e_j,e_j)B_\varepsilon(\phi, e_je_i\cdot\psi)e_ie_j\right),$$

where $\{e_1,\dots,e_{2n}\}$ is any orthonormal basis of $V$.

PROPOSITION 3.11.
*(i) If $n\equiv 0 \mod 4$ then $L_\varepsilon$ is antisymmetric, $\pi_2\circ L_\varepsilon$ is even and $\pi_1\circ L_\varepsilon$ is odd.*
*(ii) If $n\equiv 1 \mod 4$ then $L_\varepsilon$ is symmetric, $\pi_2\circ L_\varepsilon$ is odd and $\pi_1\circ L_\varepsilon$ is even.*
*(iii) If $n\equiv 2 \mod 4$ then $L_\varepsilon$ is symmetric, $\pi_2\circ L_\varepsilon$ is even and $\pi_1\circ L_\varepsilon$ is odd.*
*(iv) If $n\equiv 3 \mod 4$ then $L_\varepsilon$ is antisymmetric, $\pi_2\circ L_\varepsilon$ is odd and $\pi_1\circ L_\varepsilon$ is even.*

## 4. Maximal isotropic subspaces and polarisations

A maximal isotropic subspace of $V$ is an $n$-dimensional subspace $\mathcal{I}$ of $V$ such that the restriction of $g$ to $\mathcal{I}$ vanishes. By Witt's theorem the group $O(V,g)$ acts transitively on the collection of maximal isotropic subspaces. The stabiliser of $\mathcal{I}$, $S(\mathcal{I})$, is a maximal parabolic subgroup of $O(V,g)$ (see e.g. [**Wo**]). The natural map from $S(\mathcal{I})$ to $GL(\mathcal{I})$ is surjective, giving rise to the exact sequence of groups

$$1\to A\to S(\mathcal{I})\to GL(\mathcal{I})\to 1.$$

A description of $A$ can be obtained as follows.

If $a \in A$, then $(a - Id_V)(V) \subseteq \mathcal{I}$, hence there is a unique $\alpha : V \to V$ such that

$$(i) \quad a = Id_V + \alpha \tag{10}$$

$$(ii) \quad \alpha|_{\mathcal{I}} = 0,\ \alpha(V) \subseteq \mathcal{I} \text{ and } g(\alpha(v), w) + g(v, \alpha(w)) = 0. \tag{11}$$

This shows that $A$ is abelian and $a \mapsto \alpha$ identifies it with a subgroup of the vector space $Hom(V/\mathcal{I}, \mathcal{I})$.

Since $\mathcal{I}$ is maximal isotropic the metric defines a duality pairing $V/\mathcal{I} \otimes \mathcal{I} \to k$, and hence there is an $S(\mathcal{I})$-equivariant isomorphism

$$Hom(V/\mathcal{I}, \mathcal{I}) \cong \mathcal{I} \otimes \mathcal{I}.$$

One checks that the composition of maps $A \hookrightarrow Hom(V/\mathcal{I}, \mathcal{I}) \cong \mathcal{I} \otimes \mathcal{I}$ is an $S(\mathcal{I})$-equivariant isomorphism of $A$ with $\Lambda^2(\mathcal{I})$, the space of antisymmetric two tensors on $\mathcal{I}$. Notice that $A$ acts trivially on $A$, $\mathcal{I}$, and $V/\mathcal{I}$ so that this factors to a $S(\mathcal{I})/A \cong GL(\mathcal{I})$-equivariant isomorphism

$$A \cong \Lambda^2(\mathcal{I}), \tag{12}$$

as a module for $GL(\mathcal{I})$. Finally, under this isomorphism the cone of decomposable elements in $\Lambda^2(\mathcal{I})$ is the image of the set $T$ of elements $a$ of $A$ with the property that there exist $v, w \in \mathcal{I}$ such that

$$a(x) = x + g(x, v)w - g(x, w)v \quad \forall x \in V.$$

Alternatively, $T$ is the set of elements of $A$ which are the identity on some $2n - 2$ dimensional subspace of $V$ containing $\mathcal{I}$. The subspace determines the group element essentially uniquely and then restriction of the above isomorphism to $T$ corresponds to the Plücker embedding of its orthogonal in $\Lambda^2(\mathcal{I})$.

We can realise $S(\mathcal{I})$ as a group of affine transformations of an affine space of which $A$ is the group of translations. Consider the exact sequence of vector spaces:

$$0 \to \mathcal{I} \to V \xrightarrow{p} V/\mathcal{I} \to 0,$$

and let

$$\begin{aligned} \mathcal{A} &= \{s \in Hom(V/\mathcal{I}, V) : p \circ s = Id_{V/\mathcal{I}} \text{ and } \mathrm{Im}\, s \text{ is maximal isotropic}\} \\ &= \{s \in Hom(V/\mathcal{I}, V) : g(s \circ p(v), i) = g(v, i)\, \forall v \in V, i \in \mathcal{I} \\ &\qquad\qquad \text{and } \mathrm{Im}\, s \text{ is maximal isotropic}\} \end{aligned}$$

be the space of isotropic splittings. (One can also identify $\mathcal{A}$ with the space of maximal isotropic complements of $\mathcal{I}$ by $s \in \mathcal{A} \mapsto Im\, s$).

This is not a linear subspace of $Hom(V/\mathcal{I}, V)$ but it is stable under the natural action of $S(\mathcal{I})$:

$$f \mapsto g \circ f \circ g^{-1}$$

and then the group $A$ acts on $\mathcal{A}$ as follows: if $s \in \mathcal{A}$ and if we write $a \in A$ as $a = Id_V + \alpha$, then $a$ maps $s$ to $s' = s + \alpha$. To check that $s' \in \mathcal{A}$ first note that $p \circ s' = p \circ s = Id_{V/\mathcal{I}}$ and so $s'$ is a splitting. Further,

$$g(s' \circ p(v), s' \circ p(w)) = g(s \circ p(v), \alpha \circ p(w)) + g(\alpha \circ p(v), s \circ p(w)) \tag{13}$$

$$= g(v, \alpha \circ p(w)) + g(\alpha \circ p(v), w) \tag{14}$$

$$= g(v, \alpha(w)) + g(\alpha(v), w) = 0, \tag{15}$$

and hence $s' \in \mathcal{A}$. It is clear that $s \mapsto s'$ defines a free group action of $A$ on $\mathcal{A}$.

To see that $A$ acts transitively, take $s, s' \in \mathcal{A}$. Then $\alpha = s' - s \in Hom(V/\mathcal{I}, \mathcal{I})$ and

$$0 = g(s' \circ p(v), s' \circ p(w)) = g(s \circ p(v), \alpha \circ p(w)) + g(\alpha \circ p(v), s \circ p(w)) \tag{16}$$
$$= g(v, \alpha \circ p(w)) + g(\alpha \circ p(v), w), \tag{17}$$

and the difference $s' - s$ of the two isotropic splittings is in $A$.

With respect to this affine structure the group $S(\mathcal{I})$ acts by affine transformations on $\mathcal{A}$. The tangent space at any point of $\mathcal{A}$ is canonically isomorphic to $A$ and hence carries a cone structure induced by $T \subset A$ and this is clearly preserved by the affine action of $S(\mathcal{I})$. From the theory of 3-graded Lie algebras one can see that $S(\mathcal{I})$ is exactly the group of affine transformations of $\mathcal{A}$ which preserve this cone structure.

REMARK 4.1. *Of course for vector spaces $V$ defined over $\mathbb{R}$ or $\mathbb{C}$ some of the above is standard (e.g.* **[Wo]***).*

Let $\mathcal{I}$ be a fixed maximal isotropic subspace of $V$. An $\mathcal{I}$-polarisation of $(V, g)$ is a decomposition $V = \mathcal{I} \oplus \mathcal{E}$ such that $\mathcal{E}$ is a maximal isotropic subspace of $V$, i.e. an element of $\mathcal{A}$ above. Choose a basis $i_1, \ldots, i_n$ of $\mathcal{I}$ and let $e_1, \ldots, e_n$ be the basis of $\mathcal{E}$ satisfying

$$g(e_a, i_b) = \frac{1}{2}\delta_{ab}.$$

The Clifford algebra relations $xy + yx = 2g(x, y)Id$ imply that for all $1 \le a, b \le n$,

$$i_a i_b + i_b i_a = 0$$
$$e_a e_b + e_b e_a = 0$$
$$i_a e_b + e_b i_a = \delta_{ab}. \tag{18}$$

In terms of this basis the grading operator has the following expression:

PROPOSITION 4.2.

$$\varepsilon = (i_1 - e_1)(i_1 + e_1) \ldots (i_n - e_n)(i_n + e_n), \text{ in particular } \varepsilon \in C^{2n}.$$

For a maximal isotropic subspace $\mathcal{I}$, an $\mathcal{I}$-polarisation $V = \mathcal{I} \oplus \mathcal{E}$, and bases $\{i_a\}, \{e_a\}$ of $\mathcal{I}$ and $\mathcal{E}$ as above, since $g(i_a \pm e_a, i_b \pm e_b) = \pm\delta_{ab}$ and $g(i_a + e_a, i_b - e_b) = 0$, the set $\{i_a \pm e_a : 1 \le a \le n\}$ is an orthonormal basis of $V$ as used in §1. For $1 \le a \le n$, set

$$E_a = e_a + i_a, E_{\bar{a}} = e_a - i_a.$$

Then $\{E_1, E_{\bar{1}}, \ldots, E_n, E_{\bar{n}}\}$ is an orthonormal basis with $g(E_a, E_a) = 1$ and $g(E_{\bar{a}}, E_{\bar{a}}) = -1$. We order this basis by

$$1 < \bar{1} < 2 < \bar{2} < \ldots n < \bar{n}.$$

The use of $e_i$ in two different ways in a basis, as in §1 and here in §5, hopefully does not lead to confusion.

## 5. Pure spinors

É. Cartan found a beautiful relationship between the maximal isotropic subspaces of $V$ and a distinguished subset of spinors in $S$. More precisely, he showed that to each maximal isotropic subspace $\mathcal{I}$ there is a unique (up to scalar multiplication) nonzero element $\mathbf{v}_{\mathcal{I}} \in S$ such that

$$i_a \cdot \mathbf{v}_{\mathcal{I}} = 0 \quad \forall 1 \le a \le n.$$

In the language of É. Cartan $\mathbf{v}_\mathcal{I}$ is called a pure spinor, and in the language of physics a vacuum.

Take an $\mathcal{I}$-polarisation of $(V, g)$, so that $V = \mathcal{I} \oplus \mathcal{E}$, and with bases as in §4. Then the grading operator, $\varepsilon \in C$, defined there determines the spaces of half-spinors

$$\varepsilon \cdot \psi_\pm = \pm\psi_\pm \quad \forall \psi_\pm \in S_\pm.$$

Repeated use of (18) shows that $\mathbf{v}_\mathcal{I}$ is in the half-spinor space $S_+$ associated to $\varepsilon$.

A basis for $S$ is obtained by applying succesive "creation operators" $e_a$ to the "vacuum" $\mathbf{v}_\mathcal{I}$, so that $S_+$ is then the space of "even particle states", $S_-$ the space of "odd particle states":

$$S_+ =< \mathbf{v}_\mathcal{I}, e_{i_1}e_{i_2}\dots e_{i_k} \cdot \mathbf{v}_\mathcal{I} : 1 \le i_1 < i_2 < \dots i_k \le n, \text{ k is even} >$$
$$S_- =< e_{i_1}e_{i_2}\dots e_{i_k} \cdot \mathbf{v}_\mathcal{I} : 1 \le i_1 < i_2 < \dots i_k \le n, \text{ k is odd} > .$$

(19)

Notice that $e_{i_1}e_{i_2}\dots e_{i_k} \cdot \mathbf{v}_\mathcal{I}$ is a pure spinor with

$$\mathrm{Ann}_V(e_{i_1}e_{i_2}\dots e_{i_k} \cdot \mathbf{v}_\mathcal{I}) =< e_p, i_q : p \in \{i_1, \cdots, i_k\},\ q \notin \{i_1, \cdots, i_k\} >,$$

a maximal isotropic subspace.

REMARK 5.1. *The basis of pure spinors is an effective computational tool in spinor algebra. Essentially, its use converts computations to combinatorial statements about the parameters for pure spinors.*

An easy example of this is the matrix of the spinor norm $B$. It is helpful to compare this to Proposition 2.6.

PROPOSITION 5.2. *If $K = \{k_1, \dots, k_p\}$ and $J = \{j_1, \dots, j_q\}$ are ordered subsets of $\{1, 2, \dots, n\}$ we set $e_K = e_{k_1}\dots e_{k_p}$, $e_J = e_{j_1}\dots e_{j_q}$ and $e_\emptyset = 1$. Then*

(20)

(a) $B(e_K \cdot \mathbf{v}_\mathcal{I}, e_J \cdot \mathbf{v}_\mathcal{I}) \neq 0 \Rightarrow K \cap J = \emptyset$ *and* $K \cup J = \{1, 2, \dots, n\}$ *i.e.* $J = K^c$,

(21)

(b) $B(\mathbf{v}_\mathcal{I}, e_1e_2e_3\dots e_n \cdot \mathbf{v}_\mathcal{I}) \neq 0.$

For another easy example, take an $\mathcal{I}$-polarisation $V = \mathcal{I} \oplus \mathcal{E}$ and define the following element of the Lie algebra $C^2$:

$$H = \frac{1}{2}\sum_1^n (e_a i_a - i_a e_a). \tag{22}$$

This is independent of the choice of bases $\{i_a\}, \{e_a\}$ above. It is an element of $C^2$ since the set $\{e_a e_b, i_a i_b : 1 \le a < b \le n\} \cup \{e_a i_b - i_b e_a : 1 \le a, b \le n\}$ is a basis of $C^2$.

PROPOSITION 5.3. *For $1 \le a \le n$ and $1 \le i_1 < \dots < i_k \le n$,*

$$(a)\quad [H, e_a] = e_a; \quad (b)\quad [H, i_a] = -i_a;$$
$$(c)\quad He_{i_1}e_{i_2}\dots e_{i_k} \cdot \mathbf{v}_\mathcal{I} = (k - \frac{n}{2})e_{i_1}e_{i_2}\dots e_{i_k} \cdot \mathbf{v}_\mathcal{I}.$$

$H$ is a useful substitute for what is called the number operator in physics, $N = \sum_{a=1}^{a=n} e_a i_a$, which is not in $C^2$.

## 6. Computations of $\tilde{L}_2$

### 6.1. The operator $\tilde{L}_2$.

Recall the definition of the operator $L_2$:

$$L_2(\phi,\psi) = \frac{1}{2^n}\sum_{i<j} g(e_i,e_i)g(e_j,e_j)B(\phi, e_je_i\cdot\psi)e_ie_j.$$

Using the basis $\{E_a, E_{\bar{b}}\}$ from §4 we obtain an alternative expression for $\tilde{L}_2 = 2^{n-1}L_2$ which is more convenient for the computation of the matrix of $\tilde{L}_2$.

PROPOSITION 6.1. *In terms of the basis of* $C^2 : \{e_ae_b, i_ai_b : 1\le a<b\le n\}\cup\{e_ai_b - i_be_a : 1\le a,b\le n\}$ *we have*

$$\begin{aligned}\tilde{L}_2(\psi_1,\psi_2) &= \sum_{a\neq b} B(e_ae_b\cdot\psi_1,\psi_2)i_ai_b\\ &+\sum_{a\neq b} B(i_ai_b\cdot\psi_1,\psi_2)e_ae_b + \sum_{a\neq b} B(e_ai_b\cdot\psi_1,\psi_2)(i_ae_b - e_bi_a)\\ &+\frac{1}{2}\sum_a B((e_ai_a - i_ae_a)\cdot\psi_1,\psi_2)(i_ae_a - e_ai_a).\end{aligned}\tag{23}$$

PROOF. In a basis $\{E_a, E_{\bar{b}}\}$

$$\tilde{L}_2(\psi_1,\psi_2) = \frac{1}{2}\sum_{i<j} g(E_i,E_i)g(E_j,E_j)B(\psi_1, E_jE_i\cdot\psi_2)E_iE_j$$

and using the ordered basis this sum can be split into two subsums:

$$\frac{1}{2}\sum_i g(E_i,E_i)g(E_{\bar{i}},E_{\bar{i}})B(\psi_1, E_{\bar{i}}E_i\cdot\psi_2)E_iE_{\bar{i}}\tag{24}$$

and

$$\begin{aligned}\frac{1}{2}\sum_{i<j}\Big(&g(E_i,E_i)g(E_j,E_j)B(\psi_1,E_jE_i\cdot\psi_2)E_iE_j\\ &+g(E_i,E_i)g(E_{\bar{j}},E_{\bar{j}})B(\psi_1,E_{\bar{j}}E_i\cdot\psi_2)E_iE_{\bar{j}}\\ &+g(E_{\bar{i}},E_{\bar{i}})g(E_j,E_j)B(\psi_1,E_jE_{\bar{i}}\cdot\psi_2)E_{\bar{i}}E_j\\ &+g(E_{\bar{i}},E_{\bar{i}})g(E_{\bar{j}},E_{\bar{j}})B(\psi_1,E_{\bar{j}}E_{\bar{i}}\cdot\psi_2)E_{\bar{i}}E_{\bar{j}}\Big).\end{aligned}\tag{25}$$

Since

$$E_aE_{\bar{a}} = -e_ai_a + i_ae_a,\quad E_{\bar{a}}E_a = -i_ae_a + e_ai_a$$

the sum (24) reduces to

$$\begin{aligned}&-\frac{1}{2}\sum_a B(\psi_1,(e_ai_a - i_ae_a)\cdot\psi_2)(i_ae_a - e_ai_a)\\ &=\frac{1}{2}\sum_a B((e_ai_a - i_ae_a)\cdot\psi_1,\psi_2)(i_ae_a - e_ai_a)\end{aligned}\tag{26}$$

and this is the last term in (23).

To simplify (25) we first observe that

$$\begin{aligned}E_aE_b &= e_{ab} + e_ai_b + i_ae_b + i_ai_b, & E_aE_{\bar{b}} &= e_ae_b - e_ai_b + i_ae_b - i_ai_b,\\ E_{\bar{a}}E_b &= e_{ab} + e_ai_b - i_ae_b - i_ai_b, & E_{\bar{a}}E_{\bar{b}} &= e_ae_b - e_ai_b - i_ae_b + i_ai_b,\end{aligned}\tag{27}$$

Hence for fixed $a < b$, the coefficient of $e_{ab}$ in (25) is

$$\frac{1}{2}\left(B(\psi_1, E_b E_a \cdot \psi_2) - B(\psi_1, E_{\bar{b}} E_a \cdot \psi_2) - B(\psi_1, E_b E_{\bar{a}} \cdot \psi_2) + B(\psi_1, E_{\bar{b}} E_{\bar{a}} \cdot \psi_2)\right)$$

which can be written

$$\frac{1}{2} B(\psi_1, (E_b - E_{\bar{b}})(E_a - E_{\bar{a}}) \cdot \psi_2),$$

that is

$$2B(\psi_1, i_b i_a \cdot \psi_2) = 2B(i_a i_b \cdot \psi_1, \psi_2).$$

Summing over all $a < b$ we get

$$2 \sum_{1 \le a < b \le n} B(i_a i_b \cdot \psi_1, \psi_2) e_{ab}$$

which is the second term in (23). Similarly, looking at the coefficients of $i_a i_b$ and $i_a e_b - e_b i_a$ in (25) we get the first and third terms of (23). □

### 6.2. The matrix of $\tilde{L}_2$.

Next we use the basis of pure spinors to simplify the expression. Let $\mathbf{v}_{\mathcal{I}}$ be a pure spinor defined by the maximal isotropic subspace $\mathcal{I}$. Recall from §5 the basis of pure spinors and take $\psi_1 = e_I \cdot \mathbf{v}_{\mathcal{I}}$ and $\psi_2 = e_J \cdot \mathbf{v}_{\mathcal{I}}$. Most terms in the formula for $\tilde{L}_2(e_I \cdot \mathbf{v}_{\mathcal{I}}, e_J \cdot \mathbf{v}_{\mathcal{I}})$ vanish by Proposition 5.2:

(a) $B(e_a e_b e_I \cdot \mathbf{v}_{\mathcal{I}}, e_J \cdot \mathbf{v}_{\mathcal{I}}) \neq 0 \quad \Leftrightarrow \quad I \cap J = \emptyset$ and $I^c \cap J^c = \{a, b\}$.

(b) $B(i_a i_b e_I \cdot \mathbf{v}_{\mathcal{I}}, e_J \cdot \mathbf{v}_{\mathcal{I}}) \neq 0 \quad \Leftrightarrow \quad I \cap J = \{a, b\}$ and $I^c \cap J^c = \emptyset$.

(c) $B(e_a i_b e_I \cdot \mathbf{v}_{\mathcal{I}}, e_J \cdot \mathbf{v}_{\mathcal{I}}) \neq 0$ and $a \neq b \quad \Leftrightarrow \quad I \cap J = \{b\}$ and $I^c \cap J^c = \{a\}$.

(d) $B((e_a i_a - i_a e_a) e_I \cdot \mathbf{v}_{\mathcal{I}}, e_J \cdot \mathbf{v}_{\mathcal{I}}) \neq 0 \quad \Leftrightarrow \quad I \cap J = \emptyset$ and $I^c \cap J^c = \emptyset$.

REMARK 6.2. *Note that* $(I \cap J) \cup (I^c \cap J^c) = (I \Delta J)^c$ *where* $I \Delta J$ *denotes the symmetric difference of the sets* $I$ *and* $J$.

From this we can calculate the matrix of $\tilde{L}_2(e_I \cdot \mathbf{v}_{\mathcal{I}}, e_J \cdot \mathbf{v}_{\mathcal{I}})$ in the basis of particle states $\{e_K \cdot \mathbf{v}_{\mathcal{I}}\}$.

PROPOSITION 6.3. $\tilde{L}_2(e_I \cdot \mathbf{v}_{\mathcal{I}}, e_J \cdot \mathbf{v}_{\mathcal{I}}) \not\equiv 0$ *iff* $I$ *and* $J$ *satisfy one of (a), (b), (c), (d) above. In those cases in terms of the basis* $\{e_K \cdot \mathbf{v}_{\mathcal{I}}\}$ *we have*

(a) *If* $I \cap J = \emptyset$ *and* $I^c \cap J^c = \{a, b\}$ *then*

$$\tilde{L}_2(e_I \cdot \mathbf{v}_{\mathcal{I}}, e_J \cdot \mathbf{v}_{\mathcal{I}}) e_K \cdot \mathbf{v}_{\mathcal{I}} = 2B(e_a e_b e_I \cdot \mathbf{v}_{\mathcal{I}}, e_J \cdot \mathbf{v}_{\mathcal{I}}) i_a i_b e_K \cdot \mathbf{v}_{\mathcal{I}}.$$

(b) *If* $I \cap J = \{a, b\}$ *and* $I^c \cap J^c = \emptyset$ *then*

$$\tilde{L}_2(e_I \cdot \mathbf{v}_{\mathcal{I}}, e_J \cdot \mathbf{v}_{\mathcal{I}}) e_K \cdot \mathbf{v}_{\mathcal{I}} = 2B(i_a i_b e_I \cdot \mathbf{v}_{\mathcal{I}}, e_J \cdot \mathbf{v}_{\mathcal{I}}) e_a e_b e_K \cdot \mathbf{v}_{\mathcal{I}}.$$

(c) *If* $I \cap J = \{b\}$ *and* $I^c \cap J^c = \{a\}$ *then*

$$\tilde{L}_2(e_I \cdot \mathbf{v}_{\mathcal{I}}, e_J \cdot \mathbf{v}_{\mathcal{I}}) e_K \cdot \mathbf{v}_{\mathcal{I}} = 2B(e_a i_b e_I \cdot \mathbf{v}_{\mathcal{I}}, e_J \cdot \mathbf{v}_{\mathcal{I}}) i_a e_b e_K \cdot \mathbf{v}_{\mathcal{I}}.$$

(d) *If* $I \cap J = \emptyset$ *and* $I^c \cap J^c = \emptyset$ *then*

$$\tilde{L}_2(e_I \cdot \mathbf{v}_{\mathcal{I}}, e_J \cdot \mathbf{v}_{\mathcal{I}}) e_K \cdot \mathbf{v}_{\mathcal{I}} = \frac{1}{2} B(e_I \cdot \mathbf{v}_{\mathcal{I}}, e_J \cdot \mathbf{v}_{\mathcal{I}})(n - 2|I \cap K| - 2|I^c \cap K^c|) e_K \cdot \mathbf{v}_{\mathcal{I}}.$$

REMARK 6.4. *Given two pure spinors $\psi, \psi'$ such that $\tilde{L}_2(\psi, \psi') \not\equiv 0$, the intersection properties of the associated maximal isotropic subspaces* $\mathrm{Ann}_V(\psi), \mathrm{Ann}_V(\psi')$ *determine $\tilde{L}_2(\psi, \psi')$ up to a constant:*

- *if $B(\psi, \psi') = 0$ then dim*$(\mathrm{Ann}_V(\psi) \cap \mathrm{Ann}_V(\psi')) = 2$ *and $\tilde{L}_2(\psi, \psi')$ is proportional to $Q(\omega)$ for any nonzero $\omega \in \Lambda^2(\mathrm{Ann}_V(\psi) \cap \mathrm{Ann}_V(\psi'))$ (see* **[EC]***).*
- *if $B(\psi, \psi') \neq 0$ then dim*$(\mathrm{Ann}_V(\psi) \cap \mathrm{Ann}_V(\psi')) = 0$ *and $\tilde{L}_2(\psi, \psi')$ is proportional to the operator $H$ associated to the polarisation $V = \mathrm{Ann}_V(\psi) \oplus \mathrm{Ann}_V(\psi')$ (see* (22)*).*

*This is a weaker but 'geometric' version of Proposition* 6.3*. For example, if $\psi = e_I \cdot \mathbf{v}_{\mathcal{I}}$ and $\psi' = e_J \cdot \mathbf{v}_{\mathcal{I}}$ are as in Proposition* 6.3 *(a), we have*

$$B(\psi, \psi') = 0, \quad \mathrm{Ann}_V(\psi) \cap \mathrm{Ann}_V(\psi') = < i_a, i_b >$$

*and this result implies that $\tilde{L}_2(\psi, \psi')$ is proportional to $Q(i_a \wedge i_b) = i_a i_b$, whereas more importantly Proposition* 6.3 *also gives the constant of proportionality.*

There is still some simplification possible in the parameters $I, J, K$. Looking at the above more closely we see that $\tilde{L}_2(e_I \cdot \mathbf{v}_{\mathcal{I}}, e_J \cdot \mathbf{v}_{\mathcal{I}}) e_K \cdot \mathbf{v}_{\mathcal{I}}$ is 'symmetric' in $I, J, K$ in the following sense.

COROLLARY 6.5.
*(i) If either $I \cap J \cap K \neq \emptyset$ or $I^c \cap J^c \cap K^c \neq \emptyset$ then*

$$\tilde{L}_2(e_I \cdot \mathbf{v}_{\mathcal{I}}, e_J \cdot \mathbf{v}_{\mathcal{I}}) e_K \cdot \mathbf{v}_{\mathcal{I}} = 0.$$

*(ii) If $I \cap J \cap K = \emptyset$ and $I^c \cap J^c \cap K^c = \emptyset$ then*

$$\tilde{L}_2(e_I \cdot \mathbf{v}_{\mathcal{I}}, e_J \cdot \mathbf{v}_{\mathcal{I}}) e_K \cdot \mathbf{v}_{\mathcal{I}} \text{ is proportional to } e_{(I \cap J) \cup (J \cap K) \cup (K \cap I)} \cdot \mathbf{v}_{\mathcal{I}}.$$

Now three subsets $I, J, K$ of $\{1, 2, \ldots, n\}$ satisfying the conditions

$$I \cap J \cap K = I^c \cap J^c \cap K^c = \emptyset \tag{28}$$

define a partition of $\{1, 2, \ldots, n\}$ into six disjoint subsets:

$$\{1, 2, \ldots, n\} = (I \cap J) \cup (J \cap K) \cup (K \cap I) \cup (I^c \cap J^c) \cup (J^c \cap K^c) \cup (K^c \cap I^c)$$

and in terms of this partition

$$\begin{aligned} I &= (I \cap J) \cup (K \cap I) \cup (J^c \cap K^c) \\ J &= (J \cap K) \cup (I \cap J) \cup (K^c \cap I^c) \\ K &= (K \cap I) \cup (J \cap K) \cup (I^c \cap J^c). \end{aligned} \tag{29}$$

The simplest example of three subsets satisfying the conditions (28) is given by three pairwise disjoint subsets of $\{1, 2, \ldots, n\}$ whose union is $\{1, 2, \ldots, n\}$. In fact this is the general case.

PROPOSITION 6.6. *Let $I, J, K$ be three oriented subsets of $\{1, 2, \ldots, n\}$. Suppose that $I \cap J \cap K = \emptyset$ and $I^c \cap J^c \cap K^c = \emptyset$. Then there is a polarisation $V = \mathcal{I}' \oplus \mathcal{E}'$ and oriented subsets $I', J', K'$ of $\{1, 2, \ldots, n\}$ such that*
*(i) $e_I \cdot \mathbf{v}_{\mathcal{I}} = e'_{I'} \cdot \mathbf{v}_{\mathcal{I}'}$, $\quad e_J \cdot \mathbf{v}_{\mathcal{I}} = e'_{J'} \cdot \mathbf{v}_{\mathcal{I}'}$ and $\quad e_K \cdot \mathbf{v}_{\mathcal{I}} = e'_{K'} \cdot \mathbf{v}_{\mathcal{I}'}$.*
*(ii) $I' \cap J' = \emptyset$, $K' = I'^c \cap J'^c$, and $I' \cup J' \cup K' = \{1, 2, \ldots, n\}$.*
*(iii) If $I, J, K$ are of the same parity then $I', J', K'$ are of the same parity, i.e $S_+$ or $S_-$.*

PROOF. Set

$$\begin{aligned}\mathcal{I}' &= \text{Vect} < e_a, i_b : a \in (I^c \cap J^c) \cup (J^c \cap K^c) \cup (K^c \cap I^c),\\ &\qquad b \in (I \cap J) \cup (J \cap K) \cup (K \cap I) >,\\ \mathcal{E}' &= \text{Vect} < e_a, i_b : a \in (I \cap J) \cup (J \cap K) \cup (K \cap I),\\ &\qquad b \in (I^c \cap J^c) \cup (J^c \cap K^c) \cup (K^c \cap I^c) > .\end{aligned}$$

Then it is clear that $V = \mathcal{I}' \oplus \mathcal{E}'$ is a polarisation, and that

$$\mathbf{v}_{\mathcal{I}'} := e_{(I\cap J)\cup(J\cap K)\cup(K\cap I)} \cdot \mathbf{v}_{\mathcal{I}}$$

is a pure spinor defined by $\mathcal{I}'$. It is equally clear that if

$$\begin{aligned}I' &= (J^c \cap K^c) \cup (J \cap K),\\ J' &= (K^c \cap I^c) \cup (K \cap I),\\ K' &= (I^c \cap J^c) \cup (I \cap J),\end{aligned}$$

then $I' \cap J' = J' \cap K' = K' \cap I' = \emptyset$ and $I' \cup J' \cup K' = \{1, 2, \dots, n\}$.

Define

$$e'_a = \begin{cases} e_a & \text{if } a \in (I^c \cap J^c) \cup (J^c \cap K^c) \cup (K^c \cap I^c),\\ i_a & \text{if } a \in (I \cap J) \cup (J \cap K) \cup (K \cap I)\end{cases}$$

and

$$i'_a = \begin{cases} e_a & \text{if } a \in (I \cap J) \cup (J \cap K) \cup (K \cap I),\\ i_a & \text{if } a \in (I^c \cap J^c) \cup (J^c \cap K^c) \cup (K^c \cap I^c).\end{cases}$$

Then $\{e'_a, i'_a : 1 \le a \le n\}$ satisfy the Clifford relations

$$\begin{aligned}i'_a i'_b + i'_b i'_a &= 0\\ e'_a e'_b + e'_b e'_a &= 0\\ i'_a e'_b + e'_b i'_a &= \delta_{ab}\end{aligned} \tag{30}$$

and up to signs,

$$e'_{I'} \cdot \mathbf{v}_{\mathcal{I}'} = e_{J^c\cap K^c} i_{J\cap K} e_{(I\cap J)\cup(J\cap K)\cup(K\cap I)} \cdot \mathbf{v}_{\mathcal{I}} = e_{J^c\cap K^c} e_{(I\cap J)\cup(K\cap I)} \cdot \mathbf{v}_{\mathcal{I}}.$$

Since $(J^c \cap K^c) \cup (I \cap J) \cup (K \cap I) = I$ this means (up to signs)

$$e'_{I'} \cdot \mathbf{v}_{\mathcal{I}'} = e_I \cdot \mathbf{v}_{\mathcal{I}}.$$

Similarly $e'_{J'} \cdot \mathbf{v}_{\mathcal{I}'} = e_J \cdot \mathbf{v}_{\mathcal{I}}$ and $e'_{K'} \cdot \mathbf{v}_{\mathcal{I}'} = e_K \cdot \mathbf{v}_{\mathcal{I}}$. This proves (i). Parts (ii) and (iii) follow immediately. □

PROPOSITION 6.7. *Let $I, J, K$ be three oriented subsets of $\{1, 2, \dots, n\}$ that are pairwise disjoint, $I \cap J = J \cap K = K \cap I = \emptyset$, and $I^c \cap J^c \cap K^c = \emptyset$. Then*

*(i)* $|I| + |J| + |K| = n$.

*(ii) If $|K|$ is not equal to 0 or 2, then $\tilde{L}_2(e_I \cdot \mathbf{v}_{\mathcal{I}}, e_J \cdot \mathbf{v}_{\mathcal{I}}) e_K \cdot \mathbf{v}_{\mathcal{I}} = 0$.*

*(iii) If $|K| = 0$ then*

$$\tilde{L}_2(e_I \cdot \mathbf{v}_{\mathcal{I}}, e_J \cdot \mathbf{v}_{\mathcal{I}}) e_K \cdot \mathbf{v}_{\mathcal{I}} = B(e_I \cdot \mathbf{v}_{\mathcal{I}}, e_J \cdot \mathbf{v}_{\mathcal{I}})(|I| - \frac{n}{2}) \mathbf{v}_{\mathcal{I}}.$$

*(iv) If $|K| = 2$ and $K = \{\overrightarrow{ba}\}$ then*

$$\tilde{L}_2(e_I \cdot \mathbf{v}_{\mathcal{I}}, e_J \cdot \mathbf{v}_{\mathcal{I}}) e_K \cdot \mathbf{v}_{\mathcal{I}} = 2B(e_{ab} e_I \cdot \mathbf{v}_{\mathcal{I}}, e_J \cdot \mathbf{v}_{\mathcal{I}}) \mathbf{v}_{\mathcal{I}}.$$

PROOF. Part (i) is clear since as we observed before $I, J, K$ define a partition of $\{1, \cdots, n\}$. Parts (ii), (iii) and (iv) follow from Proposition 6.3 since $I \cap J = \emptyset$ and $K = I^c \cap J^c$. □

In §8 we will need an expression for $L_{2n}$. Recall from (9)

$$L_{2n}(\psi_1, \psi_2) = \frac{1}{2^n} g(e_1, e_1) \dots g(e_{2n}, e_{2n}) B(\psi_1, e_{2n} \dots e_1 \cdot \psi_2) e_1 \dots e_{2n}.$$

A simplification of this in terms of pure spinors is rather straightforward.

PROPOSITION 6.8.

$$L_{2n}(\psi_1, \psi_2) = \frac{1}{2^n} B(\psi_1, \varepsilon \cdot \psi_2) \varepsilon.$$

**6.3. Résumé.**
Let

- $(V, g)$ be a $2n$-dimensional vector space with a hyperbolic metric $g$;
- $S$ be a space of spinors (i.e., we identify $C(V, g)$ with $\mathrm{End}(S)$ for some $2^n$-dimensional vector space $S$;
- $B : S \times S \to k$ be a spinor norm (Cartan form).
- $V = \mathcal{I} \oplus \mathcal{E}$ be a polarisation of $V$;
- $\mathbf{v}_{\mathcal{I}}$ be a pure spinor associated to $\mathcal{I}$ (i.e., $v \cdot \mathbf{v}_{\mathcal{I}} = 0$ for all $v \in \mathcal{I}$);
- $i_1, \dots, i_n$ and $e_1, \dots, e_n$ be bases of $\mathcal{I}$ and $\mathcal{E}$ respectively such that

$$\begin{aligned} i_a i_b + i_b i_a &= 0 \\ e_a e_b + e_b e_a &= 0 \\ i_a e_b + e_b i_a &= \delta_{ab}; \end{aligned} \tag{31}$$

- $\tilde{L}_2 : S \times S \to C^2(V, g)$ be the normalised projection operator:

$$\begin{aligned} \tilde{L}_2(\psi_1, \psi_2) = &\sum_{a \neq b} B(e_a e_b \psi_1, \psi_2) i_a i_b \\ &+ \sum_{a \neq b} B(i_a i_b \psi_1, \psi_2) e_a e_b + \sum_{a \neq b} B(e_a i_b \psi_1, \psi_2)(i_a e_b - e_b i_a) \\ &+ \frac{1}{2} \sum_a B((e_a i_a - i_a e_a) \cdot \psi_1, \psi_2)(i_a e_a - e_a i_a). \end{aligned} \tag{32}$$

Then for all oriented subsets $I, J, K$ of $\{1, \dots, n\}$ we have shown that:

- $\tilde{L}_2(e_I \cdot \mathbf{v}_{\mathcal{I}}, e_J \cdot \mathbf{v}_{\mathcal{I}}) e_K \cdot \mathbf{v}_{\mathcal{I}} = 0$ unless $I \cap J \cap K = \emptyset$ and $I^c \cap J^c \cap K^c = \emptyset$.
- If $I \cap J \cap K = \emptyset$ and $I^c \cap J^c \cap K^c = \emptyset$ there is a polarisation $V = \mathcal{I}' \oplus \mathcal{E}'$ and oriented subsets $I', J', K'$ of $\{1, \cdots, n\}$ such that

$$\begin{aligned} &(i) \quad e_I \cdot \mathbf{v}_{\mathcal{I}} = e_{I'} \cdot \mathbf{v}_{\mathcal{I}'}, \quad e_J \cdot \mathbf{v}_{\mathcal{I}} = e_{J'} \cdot \mathbf{v}_{\mathcal{I}'}, \quad e_K \cdot \mathbf{v}_{\mathcal{I}} = e_{K'} \cdot \mathbf{v}_{\mathcal{I}'}. \\ &(ii) \quad I' \cap J' = \emptyset \text{ and } K' = I'^c \cap J'^c. \end{aligned} \tag{33}$$

PROPOSITION 6.9.
*Let $I, J, K$ be three oriented subsets of $\{1, 2, \dots, n\}$ satisfying $I \cap J = \emptyset$ and $K = I^c \cap J^c$. Then*
*(i) $|I| + |J| + |K| = n$.*
*(ii) If $|K|$ is not equal to $0$ or $2$, then $\tilde{L}_2(e_I \cdot \mathbf{v}_{\mathcal{I}}, e_J \cdot \mathbf{v}_{\mathcal{I}}) e_K \cdot \mathbf{v}_{\mathcal{I}} = 0$.*

*(iii) If* $|K| = 0$ *then*

$$\tilde{L}_2(e_I \cdot \mathbf{v}_{\mathcal{I}}, e_J \cdot \mathbf{v}_{\mathcal{I}})e_K \cdot \mathbf{v}_{\mathcal{I}} = B(e_I \cdot \mathbf{v}_{\mathcal{I}}, e_J \cdot \mathbf{v}_{\mathcal{I}})(|I| - \frac{n}{2})\mathbf{v}_{\mathcal{I}}.$$

*(iv) If* $|K| = 2$ *and* $K = \{\overrightarrow{ba}\}$ *then*

$$\tilde{L}_2(e_I \cdot \mathbf{v}_{\mathcal{I}}, e_J \cdot \mathbf{v}_{\mathcal{I}})e_K \cdot \mathbf{v}_{\mathcal{I}} = 2B(e_{ab}e_I \cdot \mathbf{v}_{\mathcal{I}}, e_J \cdot \mathbf{v}_{\mathcal{I}})\mathbf{v}_{\mathcal{I}}.$$

- In all cases $\tilde{L}_2(e_I \cdot \mathbf{v}_{\mathcal{I}}, e_J \cdot \mathbf{v}_{\mathcal{I}})e_K \cdot \mathbf{v}_{\mathcal{I}}$ is proportional to $e_{(I\cap J)(J\cap K)(K\cap I)} \cdot \mathbf{v}_{\mathcal{I}}$.

## 7. Potential Lie algebra structures

As Cartan's operator $\tilde{L}_2$ (or $\tilde{L}_2 + L_{2n}$) maps from $S_i \times S_i$ (or $S \times S$) to $C^2$ (or $C^2 \oplus C^{2n}$) it provides a natural candidate for a type of "curvature"operator on $S_i$ (or $S$). Cartan's calculation of curvature operators for symmetric spaces then motivates possible Lie triple system structures.

If $n \equiv 0 \bmod 4$, we can now define ($i = 1, 2$) a unique antisymmetric map

$$(C^2 \oplus S_i) \times (C^2 \oplus S_i) \to C^2 \oplus S_i$$

such that

(34) $$[A, B] = AB - BA \quad \text{if } A, B \in C^2;$$

(35) $$[A, \psi] = A \cdot \psi \quad \text{if } A \in C^2,\ \psi \in S_i;$$

(36) $$[\phi, \psi] = \tilde{L}_2(\phi, \psi) \quad \text{if } \phi, \psi \in S_i.$$

Similarly, If $n \equiv 1 \bmod 4$, we can define a unique antisymmetric map

$$(C^2 \oplus C^{2n} \oplus S) \times (C^2 \oplus C^{2n} \oplus S) \to C^2 \oplus C^{2n} \oplus S$$

such that

(37) $$[A, B] = AB - BA \quad \text{if } A, B \in C^2 \oplus C^{2n}$$

(38) $$[A, \psi] = A \cdot \psi \quad \text{if } A \in C^2 \oplus C^{2n},\ \psi \in S$$

(39) $$[\phi, \psi] = \tilde{L}_2(\phi, \psi) + L_{2n}(\phi, \psi) \quad \text{if } \phi, \psi \in S.$$

The question is: do these brackets define Lie algebra structures on $C^2 \oplus S_i$ $(n = 0 \bmod 4)$ and $C^2 \oplus C^{2n} \oplus S$ $(n = 1 \bmod 4)$ respectively? Since $C^2, C^2 \oplus C^{2n}$ are Lie algebras , since $S_i, S$ are representations and since $\tilde{L}_2 : S_i \times S_i \to C^2$ and $\tilde{L}_2 + L_{2n} : S \times S \to C^2 \oplus C^{2n}$ are equivariant maps, this will be the case if and only if the following Jacobi identities are satisfied:

$$\tilde{L}_2(\psi_1, \psi_2) \cdot \psi_3 + \tilde{L}_2(\psi_2, \psi_3) \cdot \psi_1 + \tilde{L}_2(\psi_3, \psi_1) \cdot \psi_2 = 0 \quad (n = 0 \bmod 4)$$

and respectively if for some $a, b \in k^*$

$$(a\tilde{L}_2 + bL_{2n})(\psi_1, \psi_2) \cdot \psi_3 + (a\tilde{L}_2 + bL_{2n})(\psi_2, \psi_3) \cdot \psi_1 + (a\tilde{L}_2 + bL_{2n})(\psi_3, \psi_1) \cdot \psi_2 = 0 \quad (n = 1 \bmod 4).$$

Note that if $n = 8$, $C^2 \oplus S_i$ is of dimension 248 and that if $n = 5$, $C^2 \oplus C^{2n} \oplus S$ is of dimension 78.

## 8. Spinor constructions of exceptional Lie algebras

### 8.1. Construction of split $e_8$.

Let $(V, g)$ be a sixteen-dimensional vector space with a nondegenerate hyperbolic symmetric bilinear form $g$. Choose a 256-dimensional space of spinors $S = S_1 \oplus S_2$ and an isomorphism $C(V, g) \cong End(S)$. Since $n = 8 = 0 \bmod 4$, $B : S \times S \to k$ is even symmetric and $\tilde{L}_2 : S \times S \to C^2(V, g)$ is even antisymmetric (Proposition 2.6 and Proposition 3.1).

On the 248-dimensional vector space

$$E = C^2(V, g) \oplus S_1$$

following the procedure above we define $[\ ,\ ] : E \otimes E \to E$ to be the unique antisymmetric bilinear map such that:

$$[A, B] = AB - BA \quad \text{if } A, B \in C^2; \tag{40}$$

$$[A, \psi] = A \cdot \psi \quad \text{if } A \in C^2,\ \psi \in S_1; \tag{41}$$

$$[\phi, \psi] = \tilde{L}_2(\phi, \psi) \quad \text{if } \phi, \psi \in S_1. \tag{42}$$

where $\tilde{L}_2$ is given by (23). The bracket $[\ ,\ ]$ defines a Lie algebra structure on $E$ iff

$$\tilde{L}_2(\psi_1, \psi_2) \cdot \psi_3 + \tilde{L}_2(\psi_2, \psi_3) \cdot \psi_1 + \tilde{L}_2(\psi_3, \psi_1) \cdot \psi_2 = 0 \quad \forall \psi_1, \psi_2, \psi_3 \in S_1. \tag{43}$$

Choose a polarisation $V = \mathcal{I} \oplus \mathcal{E}$ of $V$ such that a pure spinor $\mathbf{v}_\mathcal{I}$ corresponding to $\mathcal{I}$ is in $S_1$ and bases $i_1, \cdots, i_n$ and $e_1, \cdots, e_n$ of respectively $\mathcal{I}$ and $\mathcal{E}$ such that

$$\begin{aligned} i_a i_b + i_b i_a &= 0 \\ e_a e_b + e_b e_a &= 0 \\ i_a e_b + e_b i_a &= \delta_{ab}. \end{aligned} \tag{44}$$

Then

$$\{e_I \cdot \mathbf{v}_\mathcal{I} : \quad |I| \text{ is even or } \emptyset\}$$

is a basis of $S_1$ and to prove the Jacobi identity (43) it is sufficient to prove that

$$\tilde{L}_2(e_I \cdot \mathbf{v}_\mathcal{I}, e_J \cdot \mathbf{v}_\mathcal{I}) \cdot e_K \cdot \mathbf{v}_\mathcal{I} + \tilde{L}_2(e_J \cdot \mathbf{v}_\mathcal{I}, e_K \cdot \mathbf{v}_\mathcal{I}) \cdot e_I \cdot \mathbf{v}_\mathcal{I} + \tilde{L}_2(e_K \cdot \mathbf{v}_\mathcal{I}, e_I \cdot \mathbf{v}_\mathcal{I}) \cdot e_J \cdot \mathbf{v}_\mathcal{I} = 0 \tag{45}$$

for all even subsets $I, J, K$ of $\{1, 2, 3, 4, 5, 6, 7, 8\}$.

By the résumé above, all terms in this equation vanish unless $I \cap J \cap K = \emptyset$ and $I^c \cap J^c \cap K^c = \emptyset$. So in fact we need only prove (45) for even subsets of $\{1, 2, 3, 4, 5, 6, 7, 8\}$ satisfying these two conditions. But then, again by the résumé, by changing the polarisation if necessary, we can always assume that

$$I \cap J = \emptyset, \quad K = I^c \cap J^c$$

and then all terms in (45) vanish unless one of the sets $I, J, K$ has 0 or 2 elements. So in the end, to prove that the bracket $[\ ,\ ]$ defines a Lie algebra structure on $E$ it remains to prove only that

$$\tilde{L}_2(e_I \cdot \mathbf{v}_\mathcal{I}, e_J \cdot \mathbf{v}_\mathcal{I}) \cdot e_K \cdot \mathbf{v}_\mathcal{I} + \tilde{L}_2(e_J \cdot \mathbf{v}_\mathcal{I}, e_K \cdot \mathbf{v}_\mathcal{I}) \cdot e_I \cdot \mathbf{v}_\mathcal{I} + \tilde{L}_2(e_K \cdot \mathbf{v}_\mathcal{I}, e_I \cdot \mathbf{v}_\mathcal{I}) \cdot e_J \cdot \mathbf{v}_\mathcal{I} = 0 \tag{46}$$

for those oriented subsets $I, J, K$ of $\{1,2,3,4,5,6,7,8\}$ such that

- $|I|, |J|$ and $|K|$ are even;
- $|I| + |J| + |K| = 8$;
- $I \cap J = J \cap K = K \cap I = \emptyset$;
- One of $|I|, |J|$ or $|K|$ is equal to 0 or 2.

Up to permutations of $I, J$ and $K$ the only possibilities are

(i) $|I| = 0$, $|J| = 0$ and $|K| = 8$;
(ii) $|I| = 0$, $|J| = 2$ and $|K| = 6$;
(iii) $|I| = 0$, $|J| = 4$ and $|K| = 4$;
(iv) $|I| = 2$, $|J| = 2$ and $|K| = 4$.

Since $|K| \neq 0$ and $|K| \neq 2$ in all four cases we have

$$\tilde{L}_2(e_I \cdot \mathbf{v}_{\mathcal{I}}, e_J \cdot \mathbf{v}_{\mathcal{I}}) \cdot e_K \cdot \mathbf{v}_{\mathcal{I}} = 0$$

and hence proving (46) reduces to proving that

$$\tilde{L}_2(e_J \cdot \mathbf{v}_{\mathcal{I}}, e_K \cdot \mathbf{v}_{\mathcal{I}}) \cdot e_I \cdot \mathbf{v}_{\mathcal{I}} + \tilde{L}_2(e_K \cdot \mathbf{v}_{\mathcal{I}}, e_I \cdot \mathbf{v}_{\mathcal{I}}) \cdot e_J \cdot \mathbf{v}_{\mathcal{I}} = 0 \tag{47}$$

for all subsets $I, J, K$ of $\{1,2,3,4,5,6,7,8\}$ satisfying one of (i), (ii), (iii) or (iv).

CASE **(i)**. *We have $I = J = \emptyset$ and $K = \{1,2,3,4,5,6,7,8\}$. By Proposition 6.9 (iii) this means that*

$$\tilde{L}_2(e_J \cdot \mathbf{v}_{\mathcal{I}}, e_K \cdot \mathbf{v}_{\mathcal{I}}) \cdot e_I \cdot \mathbf{v}_{\mathcal{I}} = -4B(\mathbf{v}_{\mathcal{I}}, e_K \cdot \mathbf{v}_{\mathcal{I}})\mathbf{v}_{\mathcal{I}}$$

*and*

$$\tilde{L}_2(e_K \cdot \mathbf{v}_{\mathcal{I}}, \mathbf{v}_{\mathcal{I}}) \cdot e_J \cdot \mathbf{v}_{\mathcal{I}} = 4B(e_K \cdot \mathbf{v}_{\mathcal{I}}, \mathbf{v}_{\mathcal{I}})\mathbf{v}_{\mathcal{I}}.$$

*Hence*

$$\tilde{L}_2(e_J \cdot \mathbf{v}_{\mathcal{I}}, e_K \cdot \mathbf{v}_{\mathcal{I}}) \cdot e_I \cdot \mathbf{v}_{\mathcal{I}} + \tilde{L}_2(e_K \cdot \mathbf{v}_{\mathcal{I}}, e_I \cdot \mathbf{v}_{\mathcal{I}}) \cdot e_J \cdot \mathbf{v}_{\mathcal{I}} = 0$$

*since $B$ is symmetric.*

CASE **(ii)**. *We have $I = \emptyset$ and without loss of generality we can suppose $J = \{1,2\}$ and $K = \{3,4,5,6,7,8\}$. By Proposition 6.9(iii) this means*

$$\tilde{L}_2(e_J \cdot \mathbf{v}_{\mathcal{I}}, e_K \cdot \mathbf{v}_{\mathcal{I}}) \cdot e_I \cdot \mathbf{v}_{\mathcal{I}} = -2B(e_{12}\mathbf{v}_{\mathcal{I}}, e_{345678} \cdot \mathbf{v}_{\mathcal{I}})\mathbf{v}_{\mathcal{I}}$$

*and by Proposition 6.9(iv) that*

$$\tilde{L}_2(e_K \cdot \mathbf{v}_{\mathcal{I}}, e_I \cdot \mathbf{v}_{\mathcal{I}}) \cdot e_J \cdot \mathbf{v}_{\mathcal{I}} = 2B(e_{21} \cdot e_{345678} \cdot \mathbf{v}_{\mathcal{I}}, \mathbf{v}_{\mathcal{I}})\mathbf{v}_{\mathcal{I}}.$$

*Hence from $B(e_{12} \cdot \mathbf{v}_{\mathcal{I}}, e_{345678} \cdot \mathbf{v}_{\mathcal{I}}) = B(e_2 \cdot \mathbf{v}_{\mathcal{I}}, e_1 e_{345678} \cdot \mathbf{v}_{\mathcal{I}}) = B(\mathbf{v}_{\mathcal{I}}, e_2 e_1 e_{345678} \cdot \mathbf{v}_{\mathcal{I}})$ and the fact that $B$ is symmetric it follows that*

$$\tilde{L}_2(e_J \cdot \mathbf{v}_{\mathcal{I}}, e_K \cdot \mathbf{v}_{\mathcal{I}}) \cdot e_I \cdot \mathbf{v}_{\mathcal{I}} + \tilde{L}_2(e_K \cdot \mathbf{v}_{\mathcal{I}}, e_I \cdot \mathbf{v}_{\mathcal{I}}) \cdot e_J \cdot \mathbf{v}_{\mathcal{I}} = 0$$

CASE **(iii)**. *We have $I = \emptyset$ and without loss of generality we can suppose $J = \{1,2,3,4\}$ and $K = \{5,6,7,8\}$. By Proposition 6.9 (iii) this means ($4 = \frac{8}{2}!$) that*

$$\tilde{L}_2(e_J \cdot \mathbf{v}_{\mathcal{I}}, e_K \cdot \mathbf{v}_{\mathcal{I}}) \cdot e_I \cdot \mathbf{v}_{\mathcal{I}} = 0$$

*and by Proposition 6.9 (ii) that*

$$\tilde{L}_2(e_K \cdot \mathbf{v}_{\mathcal{I}}, \mathbf{v}_{\mathcal{I}}) \cdot e_J \cdot \mathbf{v}_{\mathcal{I}} = 0.$$

*It follows immediately that*

$$\tilde{L}_2(e_J \cdot \mathbf{v}_{\mathcal{I}}, e_K \cdot \mathbf{v}_{\mathcal{I}}) \cdot e_I \cdot \mathbf{v}_{\mathcal{I}} + \tilde{L}_2(e_K \cdot \mathbf{v}_{\mathcal{I}}, e_I \cdot \mathbf{v}_{\mathcal{I}}) \cdot e_J \cdot \mathbf{v}_{\mathcal{I}} = 0$$

CASE **(iv)**. *We can suppose without loss of generality that* $I = \{1,2\}$, $J = \{3,4\}$ *and* $K = \{5,6,7,8\}$. *By Proposition* 6.9*(iv) this means that*

$$\tilde{L}_2(e_J \cdot \mathbf{v}_{\mathcal{I}}, e_K \cdot \mathbf{v}_{\mathcal{I}}) \cdot e_I \cdot \mathbf{v}_{\mathcal{I}} = -2B(e_{12}e_{34} \cdot \mathbf{v}_{\mathcal{I}}, e_{5678} \cdot \mathbf{v}_{\mathcal{I}})\mathbf{v}_{\mathcal{I}}$$

*and that*

$$\tilde{L}_2(e_K \cdot \mathbf{v}_{\mathcal{I}}, e_I \cdot \mathbf{v}_{\mathcal{I}}) \cdot e_J \cdot \mathbf{v}_{\mathcal{I}} = 2B(e_{43}e_{5678} \cdot \mathbf{v}_{\mathcal{I}}, e_{12} \cdot \mathbf{v}_{\mathcal{I}})\mathbf{v}_{\mathcal{I}}.$$

*Hence from* $B(e_{12}e_{34} \cdot \mathbf{v}_{\mathcal{I}}, e_{5678} \cdot \mathbf{v}_{\mathcal{I}}) = B(\mathbf{v}_{\mathcal{I}}, e_4e_3e_2e_1e_{5678} \cdot \mathbf{v}_{\mathcal{I}})$ *and the fact that* $B$ *is symmetric it follows that*

$$\tilde{L}_2(e_J \cdot \mathbf{v}_{\mathcal{I}}, e_K \cdot \mathbf{v}_{\mathcal{I}}) \cdot e_I \cdot \mathbf{v}_{\mathcal{I}} + \tilde{L}_2(e_K \cdot \mathbf{v}_{\mathcal{I}}, e_I \cdot \mathbf{v}_{\mathcal{I}}) \cdot e_J \cdot \mathbf{v}_{\mathcal{I}} = 0.$$

PROPOSITION 8.1. *The Lie algebra constructed above is simple.*

PROOF. The Lie bracket we have just defined on the 248-dimensional vector space

$$E = C^2(V,g) \oplus S_1$$

has the following properties:

(i) $C^2(V,g)$ is a simple Lie subalgebra;
(ii) the bracket of $C^2(V,g)$ with $S_1$ defines a nontrivial irreducible representation of $C^2(V,g)$ on $S_1$;
(iii) $\dim(C^2(V,g)) > \dim(S_1)$;
(iv) $[S_1, S_1] = C^2(V,g)$ .

We denote by $\pi_{C^2} : E \to C^2(V,g)$ and $\pi_{S_1} : E \to S_1$ the projections defined by the direct sum decomposition $E = C^2(V,g) \oplus S_1$. Clearly these are $C^2(V,g)$-equivariant maps whose respective kernels are $S_1$ and $C^2(V,g)$.

If $\mathcal{I}$ is a nonzero ideal in $E$ then $\pi_{C^2}(\mathcal{I})$ cannot be equal to $\{0\}$ - if so then $\mathcal{I} \subseteq S_1$ which would imply $\mathcal{I} = S_1$ (cf. (ii)) and this is impossible since $S_1$ is not an ideal (cf. (iv)). From the equivariance of $\pi_{C^2}$ and the irreducibility of $C^2(V,g)$ it then follows that

$$\pi_{C^2}(\mathcal{I}) = C^2(V,g) \tag{48}$$

and similarly, since $C^2(V,g)$ is not an ideal (cf. (ii)), we have

$$\pi_{S_1}(\mathcal{I}) = S_1. \tag{49}$$

The rank theorem for $\pi_{C^2} : \mathcal{I} \to C^2(V,g)$ and (48) imply

$$\dim(\mathcal{I}) \geq \dim(C^2(V,g))$$

and by (iii) this means

$$\dim(\mathcal{I} \cap C^2(V,g)) > 0$$

which by (i) implies

$$\mathcal{I} \cap C^2(V,g) = C^2(V,g). \tag{50}$$

It now follows from (49), (50) and the rank theorem for $\pi_{S_1} : \mathcal{I} \to S_1$ that

$$\dim(\mathcal{I}) = \dim(\mathcal{I} \cap C^2(V, g)) + \dim(\pi_{S_1}(\mathcal{I})) = \dim(C^2(V, g)) + \dim(S_1) = \dim(E)$$

and hence

$$\mathcal{I} = E.$$

□

**8.2. Construction of split $e_7$.**

Let $(V, g)$ be a twelve-dimensional vector space with a nondegenerate hyperbolic symmetric bilinear form $g$. Choose a 64-dimensional space of spinors $S = S_1 \oplus S_2$ and an isomorphism $C(V, g) \cong End(S)$. Since $n = 6 = 2 \bmod 4$, $B : S \times S \to k$ is even antisymmetric and $\tilde{L}_2 : S \times S \to C^2(V, g)$ is even symmetric (Proposition 2.6 and Proposition 3.1). Hence $(S_1, B, C^2(V, g), L)$ is a symplectic representation of the Lie algebra $C^2(V, g)$ possessing a natural equivariant symmetric bilinear form $\tilde{L}_2$ with values in $C^2(V, g)$.

Following [**Fa**] or [**GSSR**] we can define a Lie bracket on the 133-dimensional space

$$E = C^2(V, g) \oplus sl(2, k) \oplus S_1 \otimes k^2$$

if $\tilde{L}_2$ (or a multiple of $\tilde{L}_2$) satisfies the equation

$$\tilde{L}_2(\psi_1, \psi_2) \cdot \psi_3 - \tilde{L}_2(\psi_1, \psi_3) \cdot \psi_2 = -B(\psi_1, \psi_2)\psi_3 + B(\psi_1, \psi_3)\psi_2 + 2B(\psi_2, \psi_3)\psi_1 \quad \forall \psi_1, \psi_2, \psi_3 \in S_1. \tag{51}$$

Choose a polarisation $V = \mathcal{I} \oplus \mathcal{E}$ of $V$ such that a pure spinor $\mathbf{v}_{\mathcal{I}}$ corresponding to $\mathcal{I}$ is in $S_1$ and bases $i_1, \cdots, i_n$ and $e_1, \cdots, e_n$ of respectively $\mathcal{I}$ and $\mathcal{E}$ such that

$$\begin{aligned} i_a i_b + i_b i_a &= 0 \\ e_a e_b + e_b e_a &= 0 \\ i_a e_b + e_b i_a &= \delta_{ab}. \end{aligned} \tag{52}$$

Then

$$S_1 = \mathrm{Vect}\{e_I \cdot \mathbf{v}_{\mathcal{I}} : \quad |I| \text{ is even}\}$$

and to prove the identity (51) it is sufficient to prove that

$$\begin{aligned} &\tilde{L}_2(e_I \cdot \mathbf{v}_{\mathcal{I}}, e_J \cdot \mathbf{v}_{\mathcal{I}}) \cdot e_K \cdot \mathbf{v}_{\mathcal{I}} - \tilde{L}_2 e_I \cdot \mathbf{v}_{\mathcal{I}}, e_K \cdot \mathbf{v}_{\mathcal{I}}) \cdot e_J \cdot \mathbf{v}_{\mathcal{I}} \\ &= -B(e_I \cdot \mathbf{v}_{\mathcal{I}}, e_J \cdot \mathbf{v}_{\mathcal{I}}) e_K \cdot \mathbf{v}_{\mathcal{I}} + B(e_I \cdot \mathbf{v}_{\mathcal{I}}, e_K \cdot \mathbf{v}_{\mathcal{I}}) e_J \cdot \mathbf{v}_{\mathcal{I}} \\ &\quad + 2B(e_J \cdot \mathbf{v}_{\mathcal{I}}, e_K \cdot \mathbf{v}_{\mathcal{I}}) e_I \cdot \mathbf{v}_{\mathcal{I}} \end{aligned} \tag{53}$$

for all even oriented subsets $I, J, K$ of $\{1, 2, 3, 4, 5, 6\}$.

By the résumé above, the two terms on the LHS of this equation vanish unless $I \cap J \cap K = \emptyset$ and $I^c \cap J^c \cap K^c = \emptyset$ and by Proposition 5.2 the three terms on the RHS vanish if $I \cap J \cap K \neq \emptyset$. Hence we need only prove (53) for even subsets of $\{1, 2, 3, 4, 5, 6\}$ satisfying $I \cap J \cap K = \emptyset$ and $I^c \cap J^c \cap K^c = \emptyset$ . But then, again by the résumé, by changing the polarisation if necessary, we can always assume that

$$I \cap J = \emptyset, \quad K = I^c \cap J^c$$

and then all terms on the LHS of (53) vanish unless one of the sets $I, J, K$ has 0 or 2 elements. So in fact to prove (53) for all even oriented subsets $I, J, K$ of

$\{1,2,3,4,5,6\}$ it is sufficient to prove that

$$\begin{aligned}
&\tilde{L}_2(e_I\cdot\mathbf{v}_\mathcal{I}, e_J\cdot\mathbf{v}_\mathcal{I})\cdot e_K\cdot\mathbf{v}_\mathcal{I} - \tilde{L}_2(e_I\cdot\mathbf{v}_\mathcal{I}, e_K\cdot\mathbf{v}_\mathcal{I})\cdot e_J\cdot\mathbf{v}_\mathcal{I}\\
&= -B(e_I\cdot\mathbf{v}_\mathcal{I}, e_J\cdot\mathbf{v}_\mathcal{I})e_K\cdot\mathbf{v}_\mathcal{I} + B(e_I\cdot\mathbf{v}_\mathcal{I}, e_K\cdot\mathbf{v}_\mathcal{I})e_J\cdot\mathbf{v}_\mathcal{I}\\
&\qquad + 2B(e_J\cdot\mathbf{v}_\mathcal{I}, e_K\cdot\mathbf{v}_\mathcal{I})e_I\cdot\mathbf{v}_\mathcal{I}
\end{aligned}\tag{54}$$

for all oriented subsets $I, J, K$ of $\{1,2,3,4,5,6\}$ such that

- $|I|, |J|$ and $|K|$ are even;
- $|I|+|J|+|K| = 6$;
- $I\cap J = J\cap K = K\cap I = \emptyset$;
- One of $|I|, |J|$ or $|K|$ is equal to 0 or 2.

Up to permutations of $I, J, K$ the only possibilities are

(i) $|I| = 0$, $|J| = 0$ and $|K| = 6$;
(ii) $|I| = 0$, $|J| = 2$ and $|K| = 4$;
(iii) $|I| = 2$, $|J| = 2$ and $|K| = 2$.

CASE **(i)**. *We have* $I = J = \emptyset$ *and* $K = \{1,2,3,4,5,6\}$. *By Proposition* 6.9*(ii) this means that*

$$\tilde{L}_2(e_I\cdot\mathbf{v}_\mathcal{I}, e_J\cdot\mathbf{v}_\mathcal{I})\cdot e_K\cdot\mathbf{v}_\mathcal{I} = 0$$

*and by Proposition* 6.9*(iii) that*

$$\tilde{L}_2(e_I\cdot\mathbf{v}_\mathcal{I}, e_K\cdot\mathbf{v}_\mathcal{I})\cdot e_J\cdot\mathbf{v}_\mathcal{I} = -3B(\mathbf{v}_\mathcal{I}, e_{123456}\cdot\mathbf{v}_\mathcal{I})\mathbf{v}_\mathcal{I}.$$

*Hence the LHS of* (53) *is*

$$3B(\mathbf{v}_\mathcal{I}, e_{123456}\cdot\mathbf{v}_\mathcal{I})\mathbf{v}_\mathcal{I}$$

*which is equal to the RHS of* (53) *since by Proposition* 5.2,

$$-B(e_I\cdot\mathbf{v}_\mathcal{I}, e_J\cdot\mathbf{v}_\mathcal{I})e_K\cdot\mathbf{v}_\mathcal{I} + B(e_I\cdot\mathbf{v}_\mathcal{I}, e_K\cdot\mathbf{v}_\mathcal{I})e_J\cdot\mathbf{v}_\mathcal{I} + 2B(e_J\cdot\mathbf{v}_\mathcal{I}, e_K\cdot\mathbf{v}_\mathcal{I})e_I\cdot\mathbf{v}_\mathcal{I}$$

*reduces to*

$$0 + B(\mathbf{v}_\mathcal{I}, e_{123456}\cdot\mathbf{v}_\mathcal{I})\mathbf{v}_\mathcal{I} + 2B(\mathbf{v}_\mathcal{I}, e_{123456}\cdot\mathbf{v}_\mathcal{I})e_I\cdot\mathbf{v}_\mathcal{I} = 3B(\mathbf{v}_\mathcal{I}, e_{123456}\cdot\mathbf{v}_\mathcal{I})\mathbf{v}_\mathcal{I}.$$

CASE **(ii)**. *We have* $I = \emptyset$ *and without loss of generality we can suppose* $J = \{1,2\}$ *and* $K = \{3,4,5,6\}$. *By Proposition* 6.9*(ii) this means that*

$$\tilde{L}_2(e_I\cdot\mathbf{v}_\mathcal{I}, e_J\cdot\mathbf{v}_\mathcal{I})\cdot e_K\cdot\mathbf{v}_\mathcal{I} = 0$$

*and by Proposition* 6.9*(iv) that*

$$\tilde{L}_2(e_I\cdot\mathbf{v}_\mathcal{I}, e_K\cdot\mathbf{v}_\mathcal{I})\cdot e_J\cdot\mathbf{v}_\mathcal{I} = 2B(e_{21}\mathbf{v}_\mathcal{I}, e_{3456}\cdot\mathbf{v}_\mathcal{I})\mathbf{v}_\mathcal{I}.$$

*Hence the LHS of* (53) *is*

$$-2B(e_{21}\mathbf{v}_\mathcal{I}, e_{3456}\cdot\mathbf{v}_\mathcal{I})\mathbf{v}_\mathcal{I}$$

*which is equal to the RHS of* (53) *since by Proposition* 5.2,

$$-B(e_I\cdot\mathbf{v}_\mathcal{I}, e_J\cdot\mathbf{v}_\mathcal{I})e_K\cdot\mathbf{v}_\mathcal{I} + B(e_I\cdot\mathbf{v}_\mathcal{I}, e_K\cdot\mathbf{v}_\mathcal{I})e_J\cdot\mathbf{v}_\mathcal{I} + 2B(e_J\cdot\mathbf{v}_\mathcal{I}, e_K\cdot\mathbf{v}_\mathcal{I})e_I\cdot\mathbf{v}_\mathcal{I}$$

*reduces to*

$$0 + 0 + 2B(e_{12}.\mathbf{v}_\mathcal{I}, e_{3456}\cdot\mathbf{v}_\mathcal{I})\mathbf{v}_\mathcal{I} = -2B(e_{21}\mathbf{v}_\mathcal{I}, e_{3456}\cdot\mathbf{v}_\mathcal{I})\mathbf{v}_\mathcal{I}.$$

CASE **(iii)**. *Without loss of generality we can suppose* $I = \{1,2\}$, $J = \{3,4\}$ *and* $K = \{5,6\}$. *By Proposition* 6.9*(iv) this means that*

$$\tilde{L}_2(e_I \cdot \mathbf{v}_\mathcal{I}, e_J \cdot \mathbf{v}_\mathcal{I}) \cdot e_K \cdot \mathbf{v}_\mathcal{I} = 2B(e_{65}e_{12} \cdot \mathbf{v}_\mathcal{I}, e_{34} \cdot \mathbf{v}_\mathcal{I})\mathbf{v}_\mathcal{I}$$

*and that*

$$\tilde{L}_2(e_I \cdot \mathbf{v}_\mathcal{I}, e_K \cdot \mathbf{v}_\mathcal{I}) \cdot e_J \cdot \mathbf{v}_\mathcal{I} = 2B(e_{43}e_{12} \cdot \mathbf{v}_\mathcal{I}, e_{56} \cdot \mathbf{v}_\mathcal{I})\mathbf{v}_\mathcal{I}.$$

*Hence the LHS of* (53) *is*

$$2B(e_{65}e_{12} \cdot \mathbf{v}_\mathcal{I}, e_{34} \cdot \mathbf{v}_\mathcal{I})\mathbf{v}_\mathcal{I} - 2B(e_{43}e_{12} \cdot \mathbf{v}_\mathcal{I}, e_{56} \cdot \mathbf{v}_\mathcal{I})\mathbf{v}_\mathcal{I} = 0.$$

*The RHS of* (53) *also vanishes since each term of*

$$-B(e_I \cdot \mathbf{v}_\mathcal{I}, e_J \cdot \mathbf{v}_\mathcal{I})e_K \cdot \mathbf{v}_\mathcal{I} + B(e_I \cdot \mathbf{v}_\mathcal{I}, e_K \cdot \mathbf{v}_\mathcal{I})e_J \cdot \mathbf{v}_\mathcal{I} + 2B(e_J \cdot \mathbf{v}_\mathcal{I}, e_K \cdot \mathbf{v}_\mathcal{I})e_I \cdot \mathbf{v}_\mathcal{I}$$

*vanishes by Proposition* 5.2.

PROPOSITION 8.2. *The Lie algebra constructed above is simple.*

PROOF. The Lie bracket we have just defined on the 133-dimensional space

$$E = C^2(V,g) \oplus sl(2,k) \oplus S_1 \otimes k^2$$

has the properties:

(i) $C^2(V,g)$ and $sl(2,k)$ are commuting simple Lie subalgebras;
(ii) the bracket of $C^2(V,g) \oplus sl(2,k)$ with $S_1 \otimes k^2$ defines a faithful, irreducible representation of $C^2(V,g) \oplus sl(2,k)$ on $S_1 \otimes k^2$;
(iii) $[S_1 \otimes k^2, S_1 \otimes k^2] = C^2(V,g) \oplus sl(2,k)$.

Pick any standard semisimple $h$ in $sl(2,k)$. Then $\mathrm{ad}(h) : E \to E$ is diagonalisable with eigenvalues $\{0, \pm 1, \pm 2\}$ and:

- $C^2(V,g) \oplus sl(2,k) = E_+$ is the sum of the eigenspaces corresponding to even eigenvalues;
- $S_1 \otimes k^2 = E_-$ is the sum of the eigenspaces corresponding to odd eigenvalues.

Let $\mathcal{I}$ be a nonzero ideal in $E$. Then $[h, \mathcal{I}] \subseteq \mathcal{I}$ and hence

$$\mathcal{I} = \mathcal{I} \cap E_+ \oplus \mathcal{I} \cap E_-.$$

If $\mathcal{I} \cap E_- = \{0\}$ then $\mathcal{I} = \mathcal{I} \cap E_+ \neq \{0\}$ which is impossible since no nontrivial ideal of $E$ can be contained in $E_+$ ($E_+$ acts faithfully on $E_-$ by (ii)). Hence $\mathcal{I} \cap E_- \neq \{0\}$.

If $\mathcal{I} \cap E_- \neq \{0\}$ then in fact $\mathcal{I} \cap E_- = E_-$ since $\mathcal{I} \cap E_-$ is stable under $E_+$ and $E_-$ is an irreducible representation of $E_+$ (cf (ii)). However if $\mathcal{I}$ contains $E_-$ it contains $E_+$ by (iii) and hence $\mathcal{I} = E$.

□

**8.3. Construction of split $e_6$.**

Let $(V,g)$ be a ten-dimensional vector space with a nondegenerate hyperbolic symmetric bilinear form $g$. Choose a 32-dimensional space of spinors $S = S_1 \oplus S_2$ and an isomorphism $C(V,g) \cong End(S)$. Since $n = 5 = 1 \pmod 4$, $B : S \times S \to k$ is odd symmetric (Proposition 2.6), $\tilde{L}_2 : S \times S \to C^2(V,g)$ is odd antisymmetric (Proposition 3.1) and $L_{10} : S \times S \to C^{10}(V,g)$ is odd antisymmetric (Proposition 3.6).

Consider the 78-dimensional vector space

$$E = C^{10}(V,g) \oplus C^2(V,g) \oplus S.$$

Following the procedure in §7 define $[\ ,\ ] : E \otimes E \to E$ to be the unique antisymmetric bilinear map such that:

$$[A, B] = AB - BA \qquad \text{if } A, B \in C^2 \oplus C^{10} \tag{55}$$

$$[A, \psi] = A \cdot \psi \qquad \text{if } A \in C^2 \oplus C^{10},\ \psi \in S \tag{56}$$

$$[\psi_1, \psi_2] = 2\tilde{L}_2(\psi_1, \psi_2) + 96L_{10}(\psi_1, \psi_2) \qquad \text{if } \psi_1, \psi_2 \in S. \tag{57}$$

Since $C^2, C^2 \oplus C^{10}$ are Lie algebras , since $S$ is a representation and since $\tilde{L}_2 : S \times S \to C^2$ and $2\tilde{L}_2 + 96L_{10} : S \times S \to C^2 \oplus C^{10}$ are equivariant maps, this map defines a Lie bracket on $E$ if and only if the following Jacobi identies are satisfied:

$$\begin{aligned}(2\tilde{L}_2 + 96L_{10})(\psi_1, \psi_2) \cdot \psi_3 + (2\tilde{L}_2 + 96L_{10})(\psi_2, \psi_3) \cdot \psi_1 \\ +(2\tilde{L}_2 + 96L_{10})(\psi_3, \psi_1) \cdot \psi_2 = 0 \\ \forall \psi_1, \psi_2, \psi_3 \in S.\end{aligned} \tag{58}$$

Recall that if $\varepsilon \in C^{10}(V, g)$ is a grading operator then

$$L_{10}(\psi_1, \psi_2) = \frac{1}{2^5} B(\psi_1, \varepsilon \cdot \psi_2)\varepsilon$$

so that this is equivalent to

$$\begin{aligned}2\tilde{L}_2(\psi_1, \psi_2) \cdot \psi_3 + 2\tilde{L}_2(\psi_2, \psi_3) \cdot \psi_1 + 2\tilde{L}_2(\psi_3, \psi_1) \cdot \psi_2 = \\ -3B(\psi_1, \varepsilon \cdot \psi_2)\varepsilon \cdot \psi_3 - 3B(\psi_2, \varepsilon \cdot \psi_3)\varepsilon \cdot \psi_1 - 3B(\psi_3, \varepsilon \cdot \psi_1)\varepsilon \cdot \psi_2 \\ \forall \psi_1, \psi_2, \psi_3 \in S.\end{aligned} \tag{59}$$

A purist would say that in this example we should use the graded spinor norm described in §3.1. However, the calculations in §6 were all done with the usual spinor norm. For the sake of ease of verifying the result we shall not use the graded norm but just the usual spinor norm.

Choose a polarisation $V = \mathcal{I} \oplus \mathcal{E}$ of $V$ such that a pure spinor $\mathbf{v}_{\mathcal{I}}$ corresponding to $\mathcal{I}$ is in $S_1$ and bases $i_1, \cdots, i_n$ and $e_1, \cdots, e_n$ of respectively $\mathcal{I}$ and $\mathcal{E}$ such that

$$\begin{aligned} i_a i_b + i_b i_a &= 0 \\ e_a e_b + e_b e_a &= 0 \\ i_a e_b + e_b i_a &= \delta_{ab}. \end{aligned} \tag{60}$$

Then

$$S_1 = \mathrm{Vect}\{e_I \cdot \mathbf{v}_{\mathcal{I}} : \ |I| \text{ is even}\}, \quad S_2 = \mathrm{Vect}\{e_I \cdot \mathbf{v}_{\mathcal{I}} : \ |I| \text{ is odd}\}$$

and to prove the identity (59) it is sufficient to prove that

$$\begin{aligned} 2\tilde{L}_2(e_I \cdot \mathbf{v}_{\mathcal{I}}, e_J \cdot \mathbf{v}_{\mathcal{I}}) \cdot e_K \cdot \mathbf{v}_{\mathcal{I}} + 2\tilde{L}_2(e_J \cdot \mathbf{v}_{\mathcal{I}}, e_K \cdot \mathbf{v}_{\mathcal{I}}).e_I \cdot \mathbf{v}_{\mathcal{I}} \\ + 2\tilde{L}_2(e_K \cdot \mathbf{v}_{\mathcal{I}}, e_I \cdot \mathbf{v}_{\mathcal{I}}) \cdot e_J \cdot \mathbf{v}_{\mathcal{I}} \\ = -3B(e_I \cdot \mathbf{v}_{\mathcal{I}}, \varepsilon \cdot e_J \cdot \mathbf{v}_{\mathcal{I}})\varepsilon \cdot e_K \cdot \mathbf{v}_{\mathcal{I}} - 3B(e_J \cdot \mathbf{v}_{\mathcal{I}}, \varepsilon \cdot e_K \cdot \mathbf{v}_{\mathcal{I}})\varepsilon \cdot e_I \cdot \mathbf{v}_{\mathcal{I}} \\ - 3B(e_K \cdot \mathbf{v}_{\mathcal{I}}, \varepsilon \cdot e_I \cdot \mathbf{v}_{\mathcal{I}})\varepsilon \cdot e_J \cdot \mathbf{v}_{\mathcal{I}} \end{aligned} \tag{61}$$

for all oriented subsets $I, J, K$ of $\{1, 2, 3, 4, 5\}$.

If $|I|, |J|, |K|$ are of the same parity, all terms in this expression vanish ($\tilde{L}_2$ and $B$ are odd) and the identity is true. If $|I|, |J|, |K|$ are not of the same parity two of them must be even and one odd, and without loss of generality we can suppose

$|I|, |J|$ are even and $|K|$ is odd. Since $\tilde{L}_2$ and $B$ are odd the identity (61) then reduces to

$$2\tilde{L}_2(e_J \cdot \mathbf{v}_\mathcal{I}, e_K \cdot \mathbf{v}_\mathcal{I})\cdot e_I \cdot \mathbf{v}_\mathcal{I} + 2\tilde{L}_2(e_K \cdot \mathbf{v}_\mathcal{I}, e_I \cdot \mathbf{v}_\mathcal{I}) \cdot e_J \cdot \mathbf{v}_\mathcal{I} =$$
$$3B(e_J \cdot \mathbf{v}_\mathcal{I}, \varepsilon \cdot e_K \cdot \mathbf{v}_\mathcal{I})\varepsilon \cdot e_I \cdot \mathbf{v}_\mathcal{I} - 3B(e_K \cdot \mathbf{v}_\mathcal{I}, \varepsilon \cdot e_I \cdot \mathbf{v}_\mathcal{I})\varepsilon \cdot e_J \cdot \mathbf{v}_\mathcal{I}. \tag{62}$$

By the résumé above, the two terms on the LHS of this equation vanish unless $I \cap J \cap K = \emptyset$ and $I^c \cap J^c \cap K^c = \emptyset$ and by Proposition 5.2 the three terms on the RHS vanish if $I \cap J \cap K \neq \emptyset$.

If $I \cap J \cap K = \emptyset$ and $I^c \cap J^c \cap K^c = \emptyset$ then by changing the polarisation if necessary, we can always assume that

$$I \cap J = \emptyset, \quad K = I^c \cap J^c,$$

and then all terms on the LHS of (62) vanish unless one of the sets $I, J, K$ has 0 or 2 elements. Up to permutations of $I, J, K$ the only possibilities are

(i) $|I| = 0$, $|J| = 0$ and $|K| = 5$;
(ii) $|I| = 0$, $|J| = 2$ and $|K| = 3$;
(iii) $|I| = 0$, $|J| = 4$ and $|K| = 1$;
(iv) $|I| = 2$, $|J| = 2$ and $|K| = 1$.

CASE **(i)**. *We have $I = J = \emptyset$ and $K = \{1, 2, 3, 4, 5\}$. By Proposition* 6.9*(iii) this means that*

$$2\tilde{L}_2(e_J \cdot \mathbf{v}_\mathcal{I}, e_K \cdot \mathbf{v}_\mathcal{I}) \cdot e_I \cdot \mathbf{v}_\mathcal{I} = 2B(\mathbf{v}_\mathcal{I}, e_{12345} \cdot \mathbf{v}_\mathcal{I})(0 - \frac{5}{2})\mathbf{v}_\mathcal{I}$$

*and that*

$$2\tilde{L}_2(e_K \cdot \mathbf{v}_\mathcal{I}, e_I \cdot \mathbf{v}_\mathcal{I}) \cdot e_J \cdot \mathbf{v}_\mathcal{I} = 2B(e_{12345} \cdot \mathbf{v}_\mathcal{I}, \mathbf{v}_\mathcal{I})(5 - \frac{5}{2})\mathbf{v}_\mathcal{I}$$

*Hence the LHS of* (62) *vanishes since $B$ is symmetric as does the RHS for the same reason.*

CASE **(ii)**. *We have $I = \emptyset$ and without loss of generality we can suppose $J = \{1, 2\}$ and $K = \{3, 4, 5\}$. By Proposition* 6.9*(iii) this means that*

$$2\tilde{L}_2(e_J \cdot \mathbf{v}_\mathcal{I}, e_K \cdot \mathbf{v}_\mathcal{I}) \cdot e_I \cdot \mathbf{v}_\mathcal{I} = 2B(e_{12}\mathbf{v}_\mathcal{I}, e_{345} \cdot \mathbf{v}_\mathcal{I})(2 - \frac{5}{2})\mathbf{v}_\mathcal{I},$$

*and by Proposition* 6.9*(iv) that*

$$2\tilde{L}_2(e_K \cdot \mathbf{v}_\mathcal{I}, e_I \cdot \mathbf{v}_\mathcal{I}) \cdot e_J \cdot \mathbf{v}_\mathcal{I} = 4B(e_{21} \cdot e_{345} \cdot \mathbf{v}_\mathcal{I}, \mathbf{v}_\mathcal{I})\mathbf{v}_\mathcal{I}.$$

*Hence the LHS of* (62) *reduces to*

$$3B(e_{21} \cdot e_{345} \cdot \mathbf{v}_\mathcal{I}, \mathbf{v}_\mathcal{I})\mathbf{v}_\mathcal{I}.$$

*The RHS of* (62) *is*

$$3B(e_J \cdot \mathbf{v}_\mathcal{I}, \varepsilon \cdot e_K \cdot \mathbf{v}_\mathcal{I})\varepsilon \cdot e_I \cdot \mathbf{v}_\mathcal{I} - 3B(e_K \cdot \mathbf{v}_\mathcal{I}, \varepsilon \cdot e_I \cdot \mathbf{v}_\mathcal{I})\varepsilon \cdot e_J \cdot \mathbf{v}_\mathcal{I}$$

*which by Proposition* 5.2 *also reduces to*

$$3B(e_{12} \cdot \mathbf{v}_\mathcal{I}, e_{345} \cdot \mathbf{v}_\mathcal{I})\mathbf{v}_\mathcal{I} - 0 = 3B(e_{21} \cdot e_{345} \cdot \mathbf{v}_\mathcal{I}, \mathbf{v}_\mathcal{I})\mathbf{v}_\mathcal{I}.$$

CASE **(iii)**. *We have $I = \emptyset$ and without loss of generality we can suppose $J = \{1, 2, 3, 4\}$ and $K = \{5\}$. By Proposition* 6.9*(iii) this means that*

$$2\tilde{L}_2(e_J \cdot \mathbf{v}_\mathcal{I}, e_K \cdot \mathbf{v}_\mathcal{I}) \cdot e_I \cdot \mathbf{v}_\mathcal{I} = 2B(e_{1234} \cdot \mathbf{v}_\mathcal{I}, e_5 \cdot \mathbf{v}_\mathcal{I})(4 - \frac{5}{2})\mathbf{v}_\mathcal{I}$$

*and by Proposition* 6.9*(ii) that*

$$2\tilde{L}_2(e_K \cdot \mathbf{v}_\mathcal{I}, e_I \cdot \mathbf{v}_\mathcal{I}) \cdot e_J \cdot \mathbf{v}_\mathcal{I} = 0$$

*Hence the LHS of* (62) *reduces to*

$$3B(e_{12345} \cdot \mathbf{v}_\mathcal{I}, \mathbf{v}_\mathcal{I})\mathbf{v}_\mathcal{I}.$$

*The RHS of* (62) *is*

$$3B(e_J \cdot \mathbf{v}_\mathcal{I}, \varepsilon \cdot e_K \cdot \mathbf{v}_\mathcal{I})\varepsilon \cdot e_I \cdot \mathbf{v}_\mathcal{I} - 3B(e_K \cdot \mathbf{v}_\mathcal{I}, \varepsilon \cdot e_I \cdot \mathbf{v}_\mathcal{I})\varepsilon \cdot e_J \cdot \mathbf{v}_\mathcal{I}$$

*which by Proposition* 5.2 *also reduces to*

$$3B(e_{1234}\mathbf{v}_\mathcal{I}, e_5 \cdot \mathbf{v}_\mathcal{I})\mathbf{v}_\mathcal{I} - 0 = 3B(e_{12345} \cdot \mathbf{v}_\mathcal{I}, \mathbf{v}_\mathcal{I})\mathbf{v}_\mathcal{I}.$$

CASE **(iv)**. *We can suppose without loss of generality that* $I = \{1,2\}$, $J = \{3,4\}$ *and* $K = \{5\}$. *By Proposition* 6.9*(iv) this means*

$$2\tilde{L}_2(e_J \cdot \mathbf{v}_\mathcal{I}, e_K \cdot \mathbf{v}_\mathcal{I}) \cdot e_I \cdot \mathbf{v}_\mathcal{I} = 4B(e_{21} \cdot e_{34} \cdot \mathbf{v}_\mathcal{I}, e_5 \cdot \mathbf{v}_\mathcal{I})\mathbf{v}_\mathcal{I}$$

*and*

$$2\tilde{L}_2(e_K \cdot \mathbf{v}_\mathcal{I}, e_I \cdot \mathbf{v}_\mathcal{I}) \cdot e_J \cdot \mathbf{v}_\mathcal{I} = 4B(e_{43} \cdot e_5 \cdot \mathbf{v}_\mathcal{I}, e_{12} \cdot \mathbf{v}_\mathcal{I})\mathbf{v}_\mathcal{I}.$$

*Since*

$$B(e_{21} \cdot e_{34} \cdot \mathbf{v}_\mathcal{I}, e_5 \cdot \mathbf{v}_\mathcal{I})\mathbf{v}_\mathcal{I} = B(e_{34} \cdot e_{21} \cdot \mathbf{v}_\mathcal{I}, e_5 \cdot \mathbf{v}_\mathcal{I})\mathbf{v}_\mathcal{I} = B(e_{21} \cdot \mathbf{v}_\mathcal{I}, e_{43} \cdot e_5 \cdot \mathbf{v}_\mathcal{I})\mathbf{v}_\mathcal{I}$$

*and* $B$ *is symmetric, the LHS of* (62) *vanishes as do all terms on the RHS by Proposition* 5.2.

PROPOSITION 8.3. *The Lie algebra constructed above is simple.*

PROOF. The Lie bracket we have just defined on the 78-dimensional space

$$E = C^2(V, g) \oplus C^{10}(V, g) \oplus S$$

has the properties:

(i) $C^2(V, g)$ is a simple Lie subalgebra and $C^{10}(V, g)$ is its one-dimensional commutant in $E$;
(ii) there exists $\varepsilon \in C^{10}(V, g)$ such that; $\mathrm{ad}(\varepsilon) : E \to E$ is diagonalisable with eigenvalues $\{0, \pm 1\}$ and $S = E_{-1} \oplus E_1$ is the decomposition of $S$ into faithful, non-isomorphic irreducible $C^2(V, g)$-modules.
(iii) $[E_{-1}, E_1] = C^2(V, g) \oplus C^{10}(V, g)$.

Let $\mathcal{I}$ be a nonzero ideal in $E$. Then $[\varepsilon, \mathcal{I}] \subseteq \mathcal{I}$ and hence

$$\mathcal{I} = \mathcal{I} \cap E_0 \oplus \mathcal{I} \cap E_{-1} \oplus \mathcal{I} \cap E_1.$$

If $\mathcal{I} \cap E_{-1} = \mathcal{I} \cap E_1 = \{0\}$ then $\mathcal{I} = \mathcal{I} \cap E_0 \neq \{0\}$ which is impossible since no nontrivial ideal of $E$ can be contained in $E_0$ by (ii) ($E_0$ acts faithfully on $S$ by (ii)). Hence either $\mathcal{I} \cap E_{-1} \neq \{0\}$ or $\mathcal{I} \cap E_1 \neq \{0\}$.

If $\mathcal{I} \cap E_{-1} \neq \{0\}$ then in fact $\mathcal{I} \cap E_{-1} = E_{-1}$ since $\mathcal{I} \cap E_{-1}$ is stable under $E_0$ and $E_{-1}$ is an irreducible representation of $E_0$ (cf (ii)). However if $\mathcal{I}$ contains $E_{-1}$ it contains $E_0$ by (iii) and hence $\mathcal{I} = E$. Similarly, if $\mathcal{I} \cap E_1 \neq \{0\}$ then $\mathcal{I} = E$ and the proposition is proved. □

## References

[Ad] J. F. Adams, *Lectures on exceptional Lie groups*, Chicago Lectures in Mathematics, University of Chicago Press, Chicago, IL, 1996. With a foreword by J. Peter May; Edited by Zafer Mahmud and Mamoru Mimura. MR1428422

[EC] É. Cartan, *Leçons sur la théorie des spineurs* Vol. I, Actualités Sci. Indust. No. 643, Hermann, Paris, (1938); Vol. II, Actualités Sci. Indust. No. 701, (1938).

[Ch] Claude Chevalley, *The algebraic theory of spinors and Clifford algebras*, Springer-Verlag, Berlin, 1997. Collected works. Vol. 2; Edited and with a foreword by Pierre Cartier and Catherine Chevalley; With a postface by J.-P. Bourguignon. MR1636473

[Fa] John R. Faulkner, *A construction of Lie algebras from a class of ternary algebras*, Trans. Amer. Math. Soc. **155** (1971), 397–408. MR0294424

[K] Bertram Kostant, *On invariant skew-tensors*, Proc. Nat. Acad. Sci. U. S. A. **42** (1956), 148–151. MR0077990

[MoSe] Andrei Moroianu and Uwe Semmelmann, *Invariant four-forms and symmetric pairs*, Ann. Global Anal. Geom. **43** (2013), no. 2, 107–121. MR3019158

[Se] George B. Seligman, *Rational methods in Lie algebras*, Marcel Dekker, Inc., New York-Basel, 1976. Lecture Notes in Pure and Applied Mathematics, Vol. 17. MR0427394

[GSSR] Marcus J. Slupinski and Robert J. Stanton, *The geometry of special symplectic representations*, J. Algebra **428** (2015), 149–189. MR3314290

[Wo] Joseph A. Wolf, *Unitary representations of maximal parabolic subgroups of the classical groups*, Mem. Amer. Math. Soc. **8** (1976), no. 180, iii+193. MR0444847

IRMA, Université de Strasbourg, 7 rue René Descartes, F-67084 Strasbourg Cedex France

*Email address*: marcus.slupinski@math.unistra.fr

1937 Beverly Road, Columbus, Ohio 43221

*Email address*: stanton.2@osu.edu

Contemporary Mathematics
Volume **714**, 2018
https://doi.org/10.1090/conm/714/14327

# Representations on partially holomorphic cohomology spaces, revisited

Joseph A. Wolf

Abstract. This is a semi–expository update and rewrite of my 1974 AMS AMS Memoir describing Plancherel formulae and partial Dolbeault cohomology realizations for standard tempered representations for general real reductive Lie groups. Even after so many years, much of that Memoir is up to date, but of course there have been a number of refinements, advances and new developments, most of which have applied to smaller classes of real reductive Lie groups. Here we rewrite that AMS Memoir in in view of these advances and indicate the ties with some of the more recent (or at least less classical) approaches to geometric realization of unitary representations.

## 0. Introduction

In the 1960's I initiated the study of the orbit structure for the action of a general real semisimple Lie group $G$ on a complex flag manifold $X = G_{\mathbb{C}}/Q$ [**50**]. Here $Q$ is a parabolic subgroup of the complexification $G_{\mathbb{C}}$ of $G$. In the early 1970's I associated the various series of standard tempered representations of $G$ to certain $G$–orbits in $X$. I showed how to construct the standard tempered representations as the natural action of $G$ on certain square integrable partially holomorphic cohomology spaces, corresponding to the CR structure of the orbit and to partially holomorphic complex vector bundles over the orbit [**51**]. At the time these geometric realizations required the infinitesimal characters of the standard tempered representations be sufficiently nonsingular [**40**]; but that condition is no longer needed [**41**]. This simplifies the theory and clarifies our exposition.

The advances in the theory include (i) getting rid of the "sufficiently nonsingular" condition, as mentioned above; (ii) a complete description of the irreducible constituents of the standard tempered representations ([**33**], [**31**], [**32**]) and their analytic continuation ([**47**], [**48**]); (iii) a better approach to the Plancherel Formula ([**27**], [**28**]); and (iv) the Atlas project [**1**]. Most of the advances (ii), (iii) and (iv) apply to smaller classes of groups than the ones we consider here, specifically to groups of Harish–Chandra class or even the smaller class of real reductive linear algebraic groups. In this paper we go over the material of the original AMS Memoir, indicating the principal changes in view of those developments. We also fix one error and a few trivial typos.

---

2010 *Mathematics Subject Classification.* Primary 32L25; Secondary 22E46, 32L10.
Research partially supported by the Simons Foundation.

Semisimple representation theory has benefited from increased use of algebraic methods, with more emphasis on Harish–Chandra modules than on representations of the group itself. This has led to better results in many cases, such as the advances (i) and (ii) mentioned above, but has tended to sever the connection with differential geometry. Papers that bridge the gap to some extent include [**26**], [**42**], [**43**], [**53**] and [**54**]. While one must be aware of those results we do not attempt to describe them here, first because of limitations of space, and second because many of them have not yet been extended to general real reductive Lie groups.

**0.1.** In this paper we work with the class of *general real reductive Lie groups.* Those are the Lie groups $G$ whose Lie algebra $\mathfrak{g}$ is reductive and such that (a) if $g \in G$ then $\mathrm{Ad}(g)$ is an inner automorphism of the complexified Lie algebra $\mathfrak{g}_{\mathbb{C}}$ and (b) $G$ has a closed normal abelian subgroup $Z$ such that (i) $Z$ centralizes the identity component $G^0$, (ii) $|G/ZG^0| < \infty$ and (iii) $Z \cap G^0$ is co-compact in the center $Z_{G^0}$ of $G^0$. These conditions are inherited by reductive components of parabolic subgroups, and the class of groups that satisfy them includes both Harish–Chandra's class $\mathcal{H}$ and all connected real semisimple Lie groups. See [**51**, §0.3] for a discussion. We work out geometric realizations for all classes in the unitary dual $\widehat{G}$ except for a set of Plancherel measure zero, and we express the Plancherel formula in terms of them.

The first part is the construction and analysis of standard tempered representations of general real reductive Lie groups. Roughly speaking this is a matter of extending Harish–Chandra's work from groups of class $\mathcal{H}$, in other word from the case where $|G/G^0| < \infty$ and $[G^0, G^0]$ has finite center. We construct a series of representations for each conjugacy class of Cartan subgroups of $G$, describe the characters of those representations, and use them for a Plancherel formula. This material is in §§2 through 5.

The second part is the geometric realization of the representations just constructed. Given a Cartan subgroup $H \subset G$ we construct partially complex homogeneous spaces $Y$ of $G$, partially holomorphic $G$–homogeneous vector bundles $\mathbb{E} \to Y$, and Hilbert spaces $\mathcal{H}_2^{0,q}(Y;\mathbb{E})$ of square integrable partially harmonic $\mathbb{E}$–valued $(0,q)$–forms on $Y$. If $[\pi] \in \widehat{G}$ belongs to the series for the $G$–conjugacy class of $H$ then it is realized by a representation of $G$ by its natural action on one of the $\mathcal{H}_2^{0,q}(Y;\mathbb{E})$. If $G/Z$ is compact and connected this reduces to the Bott–Borel–Weil Theorem. If $H/Z$ is compact this includes the discrete series realizations of Schmid and Narasimhan–Okamoto. This material is in §§6 through 8.

In §1 we illustrate the geometric realization procedure by carrying it out for the "principal series", corresponding to the conjugacy class of minimal parabolic subgroups $B \subset G$. We construct an Iwasawa decomposition $G = KAN$ where $Z \subset K$, and $B = MAN$ where $M$ is the centralizer of $A$ in $K$. We construct certain subgroups $U \subset M$ that contain Cartan subgroups of $M$, partially complex homogeneous spaces $Y = G/UAN$ and $G$–equivariant fibrations $Y \to Y/M = G/B$ whose fibers are maximal complex subvarieties of $Y$. Given a unitary equivalence class $[\mu] \in \widehat{U}$ and a functional $\sigma \in \mathfrak{a}^*$ we construct a $G$–homogeneous bundle $\mathbb{E}_{\mu,\sigma} \to Y$ that is holomorphic over the fibers of $Y \to Y/M$. We extend the Bott–Borel–Weil Theorem to groups that are compact modulo their center and identify the sheaf cohomology representation $\eta_\mu^q$ of $M$ on $H^q(M/U; \mathcal{O}(\mathbb{E}_{\mu,\sigma}|_{M/U}))$. Given a unitary equivalence class $[\eta] \in \widehat{M}$ we find all $[\mu] \in \widehat{U}$ and all $q \geqq 0$ such that

$\eta^q_\mu \in [\eta]$. That done we show that the principal series class

$$[\pi_{\eta,\sigma}] = [\mathrm{Ind}\,_B^G(\eta \otimes e^{i\sigma})], \text{ where } (\eta \otimes e^{i\sigma})(man) = e^{i\sigma}(a)\eta(m)$$

is realized the natural action of $G$ on a certain square integrable partially holomorphic cohomology space $\mathcal{H}_2^{0,q}(Y; \mathbb{E}_{\mu,\sigma})$. The representations of $U$ and $M$ come out of the Peter–Weyl Theorem and are characterized by Cartan's highest weight theory. In the general case we substitute an extension of Harish–Chandra's theory of the discrete series, based on results of Schmid. In §1 the $\eta^q$ are identified by a mild extension of the Bott–Borel–Weil Theorem. In the general case the matter is delicate and we need detailed information on distribution characters of representations induced from parabolic subgroups.

**0.2.** Here is a more detailed description of the contents of this paper and a comparison with the earlier AMS Memoir, except §1, which is described above.

In §2 we record the basic facts on square integrable representations of a locally compact unimodular group $G$ relative to a unitary character $\zeta \in \widehat{Z}$, Here $Z$ is a closed normal abelian subgroup. Denote

$$L_2(G/Z, \zeta) = \{f : G \to \mathbb{C} \mid f(gz) = \zeta(z)^{-1} \text{ and } \int_{G/Z} |f(g)|^2 d(gZ) < \infty\}$$

and $\widehat{G}_\zeta = \{[\pi] \in \widehat{G} \mid \pi(gz) = \zeta(z)\pi(g)\}$. Then $\widehat{G} = \bigcup_{\widehat{Z}} \widehat{G}_\zeta$ and $L_2(G) = \int_{\widehat{Z}} L_2(G/Z, \zeta) d\zeta = \int_{\widehat{G}_\zeta} \mathcal{H}_\pi \widehat{\otimes} \mathcal{H}_\pi^* \, d_\zeta[\pi]$ where $d_\zeta[\pi]$ is the Plancherel measure on $\widehat{G}_\zeta$. We say *$\zeta$-discrete* for the classes in $\widehat{G}_{\zeta-disc}$. The set of all these classes forms the *$\zeta$-discrete series* $\widehat{G}_{\zeta-disc}$ of $G$, and the *relative discrete series* is the union $\widehat{G}_{disc} = \bigcup \widehat{G}_{\zeta-disc}$. If $[\pi] \in \widehat{G}_\zeta$ then $[\pi] \in \widehat{G}_{\zeta-disc}$ if and only if its coefficient functions $\phi_{u,v}(g) = \langle u, \pi(g)v\rangle$ satisfy $|\phi_{u,v}| \in L_2(G/Z)$. When $Z$ is compact, $\widehat{G}_{disc}$ is the usual discrete series of $G$.

Let $U$ be a closed subgroup of $G$ with $Z \subset U$ and $U/Z$ compact. We give a short proof that $\widehat{U} = \widehat{U}_{disc}$ and that every class in $\widehat{U}$ is finite dimensional. Then we write down the relative Plancherel formula for the $L_2(G/Z, \zeta)$ and the absolute Plancherel formula for $L_2(G)$.

In §3 we see how Harish–Chandra's theory of the discrete series for his class $\mathcal{H}$ extends to our class of general real reductive Lie groups. This was originally done in [**51**, §3] by looking at central extensions $1 \to S \to G[\zeta] \to ZG^0/Z \to 1$ where $S$ is the circle group $\{s \in \mathbb{C} \mid |s| = 1\}$ and $G[\zeta] = \{S \times ZG^0\}/\{(\zeta(z)^{-1}, z) \mid z \in Z\}$. That led to a bijection $\widehat{G[\zeta]}_1 \to (\widehat{ZG^0})_\zeta$. We showed that $G[\zeta]$ is a connected reductive Lie group with compact center and verified that Harish–Chandra's discrete series theory applies to such groups. Here we go directly and apply results of R. Herb and the author ([**27**], [**28**]). In particular we see that $\widehat{G}_{disc}$ is nonempty if and only if $G/Z$ has a compact Cartan subgroup, and in that case one has the expected infinitesimal and global character formulae.

In §4 we construct a series of unitary representations of $G$ for every conjugacy class of Cartan subgroups $H \subset G$. For lack of a better term we continue to refer to the series for $\{\mathrm{Ad}(g)H \mid g \in G\}$ as the "$H$–series" of $G$. If $H/Z$ is compact then the $H$–series is the relative discrete series. If $H/Z$ is maximally noncompact it is the principal series. If $G/G^0$ and the center of $[G^0, G^0]$ are finite, in other words if $G$ is of class $\mathcal{H}$, then the various $H$–series are just the standard tempered series constructed by Harish–Chandra and used by him to decompose $L_2(G)$. Our constructions and results first came as straightforward extensions of results of Harish–Chandra, Hirai and Lipsman, but now they follow directly from Herb and the author [**27**].

If $H$ is a Cartan subgroup of $G$ we construct a "Cartan involution" $\theta$ of $G$ that leaves $H$ stable. Thus $\theta^2 = 1$, $\theta(H) = H$, $K = \{g \in G \mid \theta(g) = g\}$ contains $Z$, and $K/Z$ is a maximal compact subgroup of $G/Z$. Further $H = T \times A$ where $T = H \cap K$ and $A = \exp(\mathfrak{a})$, $\mathfrak{a} = \{\xi \in \mathfrak{h} \mid \theta(\xi) = -\xi\}$. Choose a positive $\mathfrak{a}$–root system $\Sigma_{\mathfrak{a}}^+$ on $\mathfrak{g}$, define $\mathfrak{n} = \sum_{\phi \in \Sigma_{\mathfrak{a}}^+} \mathfrak{g}^{-\phi}$, and let $N$ be the analytic subgroup of $G$ for $\mathfrak{n}$. All this defines

$$P = \{g \in G \mid \mathrm{Ad}(g)N = N\}, \text{ cuspidal parabolic subgroup of } G.$$

The centralizer $Z_G(A) = M \times A$ with $\theta(M) = M$, $P = MAN$, $N$ is the unipotent radical of $P$, and $MA$ is the reductive part of $P$. The groups $M$ and $MA$ inherit our working conditions from $G$: they belong to the class of general real reductive Lie groups. Further, $T$ is a Cartan subgroup of $M$, and $T/Z$ is compact, so $\widehat{M}_{disc}$ is not empty. If $[\eta] \in \widehat{M}$ and $\sigma \in \mathfrak{a}^*$ then $(\eta \otimes e^{i\sigma})(ma) = e^{i\sigma}(a)\eta(m)$ defines an irreducible unitary representation of $P$. The *$H$–series* of $G$ consists of the unitary equivalence classes $[\pi_{\eta,\sigma}] = [\mathrm{Ind}_P^G(\eta \otimes e^{i\sigma})]$ where $[\eta] \in \widehat{M}_{disc}$ and $\sigma \in \mathfrak{a}^*$.

We compute central, infinitesimal and distribution characters of the classes $[\mathrm{Ind}_P^G(\eta \otimes e^{i\sigma})]$. In particular we see that the $H$–series classes are finite sums of irreducibles, and that the $H$–series of $G$ depends only on the conjugacy class of $H$, independent of the choice of $\Sigma_{\mathfrak{a}}^+$. We also see that if $H_1$ and $H_2$ are non–conjugate Cartan subgroups of $H$ then the $H_1$–series and the $H_2$–series are disjoint.

In §5 we describe the Plancherel formula for $G$. There is a sharp improvement ([**27**] and [**28**]) on our original argument [**51**, §5]. It gives a precise description of the character formula and shows that the densities $m_{j,\zeta,\nu}$ in the Plancherel formula are restrictions of meromorphic functions. However that proof is rather technical, involving machinery that takes some space to describe, and we do not need the meromorphicity properties of the $m_{j,\zeta,\nu}$. For that reason we simply state the result as proved in [**51**, §5]. That ends the first part of this paper.

§6 starts the second part of this paper, the geometric realization of the representations involved in the Plancherel formula, based on the action of $G$ on complex flag manifolds $X = \overline{G}_{\mathbb{C}}/Q$. Here $\overline{G} = G/Z_G(G^0)$ is the adjoint group, $\overline{G}_{\mathbb{C}}$ is its complexification, and $Q$ is a parabolic subgroup of $\overline{G}_{\mathbb{C}}$. Then $G$ acts naturally on $X$, as in [**50**], through the lift of the action of $\overline{G}$. The main points for us are the concept of holomorphic arc component and measurable orbit. The orbits over which we realize a general $H$–series representation of $G$ are measurable, and the relative discrete series representations of $M$ are constructed over holomorphic arc components of measurable orbits. We end §6 with a classification and structure theory for the orbits $G(x) \subset X$ over which we have geometric realizations of $H$–series representations of $G$.

In §7 we work out the partially holomorphic cohomology realizations of relative discrete series representations of $G$. Suppose that $G$ has a Cartan subgroup $H$ with $H/Z$ compact. Then we have complex flag manifolds $X = \overline{G}_{\mathbb{C}}/Q$ with orbits $Y = G(x) \subset X$ such that $Y = G/U$ and $H \subset U$ with $U/Z$ compact, and all such $G$–orbits on $X$ are open. They are defined by a choice of positive $\mathfrak{h}_{\mathbb{C}}$–root system $\Sigma^+$ and a subset $\Phi$ of the corresponding simple root system, such that $\Phi$ is the simple root system for $\mathfrak{u}_{\mathbb{C}}$. For each $[\mu] \in \widehat{U}$ we have a $G$–homogeneous hermitian holomorphic vector bundle $\mathbb{E} \to Y$. Let $q \geqq 0$. Then we have the Hilbert space $\mathcal{H}_2^{0,q}(Y;\mathbb{E})$ of $\mathbb{E}$–valued square integrable harmonic $(0,q)$–forms on $Y$, and $G$ acts on $\mathcal{H}_2^{0,q}(Y;\mathbb{E})$ by a unitary representation $\pi_\mu^q$. Write $\Theta_{\pi_\mu^q,disc}$ for the sum of the distribution characters of the irreducible summands of $\pi_\mu^q$. Let $\rho$ denote half the sum of the positive roots. Thus $\pi_\mu^q = \pi_{\chi,\lambda+\rho}$ in the notation of §3 where $[\chi] \in \widehat{Z_G(G^0)}_\zeta$ and $\lambda+\rho \in i\mathfrak{h}^*$ integrates to a unitary character on $H^0$ that agrees with $\zeta$ on $Z_{G^0}$. We can (and do) arrange this in such a way that $\lambda$ is $\mathfrak{u}$–dominant, i.e. $\langle\lambda,\phi\rangle \geqq 0$ for every $\phi\in\Phi$. Note $U = Z_G(G^0)U^0$ so $[\mu] = [\chi\otimes\mu^0]$.

$$q(\lambda+\rho) = \#\{\alpha \text{ compact } \mid \langle\lambda+\rho,\alpha\rangle < 0\} + \#\{\alpha \text{ noncompact } \mid \langle\lambda+\rho,\alpha\rangle > 0\}.$$

We prove

(i) $\sum_{q\geqq 0}(-1)^q\Theta_{\pi_\mu^q,disc} = (-1)^{|\Sigma^+|+q(\lambda+\rho)}\Theta_{\chi,\lambda+\chi}$,

(ii) If $q\neq q(\lambda+\rho)$ then $\mathcal{H}_2^{0,q}(Y;\mathbb{E}) = 0$, and (iii) $[\pi_\mu^{q(\lambda+\rho)}] = [\pi_{\chi,\lambda+\rho}] \in \widehat{G}_{disc}$.

This improvement over [**51**, §7] is mostly due to the improvement of [**41**] over [**40**], but it still relies on the Plancherel formula.

In §8 we work out the geometric realization for all standard tempered representations. Fix a Cartan subgroup $H = T\times A$ and a corresponding cuspidal parabolic subgroup $P = MAN \subset G$. The $H$–series classes are realized over measurable orbits $Y = G(x) \subset \overline{G}_{\mathbb{C}}/Q = X$ such that (i) The $G$–normalizer $N_{[x]}$ of the holomorphic arc component $S_{[x]}$ through $x$ is an open subgroup of $P$ and (ii) $T\subset U = \{m\in M \mid m(x) = x\}$ with $U/Z$ compact. (Such pairs always exist and were classified at the end of §6.) Then $G$ has isotropy subgroup $UAN$ at $x$, $N_{[x]} = M^\dagger AN$ where $M^\dagger = Z_M(M^0)M^0$, and $U = Z_M(M^0)U^0$ with $U\cap M^0 = U^0$.

Let $\rho_{\mathfrak{a}} = \frac{1}{2}\sum_{\phi\in\Sigma_{\mathfrak{a}}^+}(\dim\mathfrak{g}^\phi)\phi$ where $\mathfrak{m} = \sum_{\phi\in\Sigma_{\mathfrak{a}}^+}\mathfrak{g}^{-\phi}$. Given $[\mu]\in\widehat{U}$ and $\sigma\in\mathfrak{a}^*$ we have the $G$–homogeneous vector bundle $\mathbb{E}_{\mu,\sigma}\to G/UAN = Y$ associated to $[\mu\otimes e^{\rho_{\mathfrak{a}}+i\sigma}] \in \widehat{UAN}$. $\mathbb{E}_{\mu,\sigma}$ carries a $K$–invariant hermitian metric and it is holomorphic over every holomorphic arc component $S_{[x]}$ of $Y$. That leads to the Hilbert spaces $\mathcal{H}_2^{0,q}(Y;\mathbb{E}_{\mu,\sigma})$ of all square integrable partially harmonic $\mathbb{E}_{\mu,\sigma}$–valued $(0,q)$–forms on $Y$. $G$ acts there by a unitary representation $\pi_{\mu,\sigma}^q$, and we prove $[\pi_{\mu,\sigma}^q] = [\operatorname{Ind}_{N_{[x]}}^G(\eta_\mu^q\otimes e^{i\sigma})]$ where $\eta_\mu^q$ is the representation of $M^\dagger$ on $\mathcal{H}_2^{0,q}(S_{[x]};\mathbb{E}_{\mu,\sigma}|_{S_{[x]}})$.

Decompose $[\mu] = [\chi\otimes\mu^0]$ where $[\chi]\in\widehat{Z_M(M^0)}$. Let $\nu$ be the highest weight of $[\mu^0]$. Then $\chi = e^\nu$ on the center of $M^0$. Our main result, Theorem 8.3.2, is

*(i) The $H$–series constituents of $\pi_{\mu,\sigma}^q$ are just its irreducible subrepresentations, and $\pi_{\mu,\sigma}^q$ has well defined distribution character $\Theta_{\pi_{\mu,\sigma}^q}$.*

*(ii) $\sum_{q\geqq 0}(-1)^q\Theta_{\pi_{\mu,\sigma}^q} = (-1)^{|\Sigma_{\mathfrak{t}}^+|+q_M(\nu+\rho_{\mathfrak{t}})}\Theta_{\pi_{\chi,\nu+\rho_{\mathfrak{t}},\sigma}}$ where $[\pi_{\chi,\nu+\rho_{\mathfrak{t}},\sigma}]$ is the $H$–series class as denoted in §4.*

*(iii) If $q\neq q_M(\nu+\rho_{\mathfrak{t}})$ then $\mathcal{H}_2^{0,q}(Y;\mathbb{E}_{\mu,\sigma}) = 0$*

*(iv) $[\pi_{\mu,\sigma}^{q_M(\nu+\rho_{\mathfrak{t}})}]$ is the $H$–series representation class $[\pi_{\chi,\nu+\rho_{\mathfrak{t}},\sigma}]$.*

That gives explicit geometric realizations for the $H$–series classes of unitary representations of $G$. The improvement over [**51**, §8] comes from the improvement of §7 over [**51**, §7].

## 1. The Principal Series

The first example of geometric realization for general real reductive Lie groups is given by the "principal series", combining the Bott–Borel–Weil Theorem with Mackey's theory of unitary induction.

**1.1.** Let $M$ be a reductive Lie group, i.e. $\mathfrak{m} = \mathfrak{z}_{\mathfrak{m}} \oplus [\mathfrak{m}, \mathfrak{m}]$ where $\mathfrak{z}_{\mathfrak{m}}$ is the center of $\mathfrak{m}$. We assume that $M$ has a closed normal abelian subgroup $Z$ such that (i) $M/Z$ is compact and (ii) if $m \in M$ then $\mathrm{Ad}(m)$ is an inner automorphism on $\mathfrak{g}_{\mathbb{C}}$. From (ii), $ZM^0$ has finite index in $M$ and $Z \cap M^0$ is co-compact in the center $Z_{M^0}$ of $M^0$.

As usual, if $\eta$ is a unitary representation of $M$ then $[\eta]$ is its unitary equivalence class. The set of all such equivalence classes $[\eta]$, with $\eta$ irreducible, is the *unitary dual* $\widehat{M}$. If $E$ is a close central subgroup of $M$ and $\xi \in \widehat{E}$ then $\widehat{M}_\xi := \{[\eta] \in \widehat{M} \mid \eta|_E \text{ is a multiple of } \xi\}$.

PROPOSITION 1.1.1. *Let $Z_M(M^0)$ denote the centralizer of $M^0$ in $M$. Then*

1. $Z_M(M^0) \cap M^0 = Z_{M^0}$ *and* $M = Z_M(M^0)M^0$.
2. *If $[\eta] \in \widehat{M}$ then $\eta$ is finite dimensional.*
3. *If $[\eta] \in \widehat{M}$ then there exist unique $[\xi] \in \widehat{Z_{M^0}}$, $[\chi] \in \widehat{Z_M(M^0)}_\xi$ and $[\eta^0] \in \widehat{(M^0)}_\xi$ such that $[\eta] = [\chi \otimes \eta^0]$.*
4. *Let $\mathfrak{t}$ be a Cartan subalgebra of $\mathfrak{m}$, $\Sigma_{\mathfrak{t}}^+$ a positive root system, $\rho_{\mathfrak{t}}$ half the sum of the positive roots, $T^0 = \exp(\mathfrak{t})$, and*

$$L_{\mathfrak{m}}^+ = \{\nu \in i\mathfrak{t}^* \mid e^{\nu - \rho_{\mathfrak{t}}} \in \widehat{T^0} \text{ is well defined and } \langle \nu, \phi \rangle > 0 \text{ for every } \phi \in \Sigma_{\mathfrak{t}}^+\}.$$

*Then there is a bijection $\nu \mapsto [\eta_\nu^0]$ of $L_{\mathfrak{m}}^+$ onto $\widehat{M^0}$ where $\nu - \rho_{\mathfrak{t}}$ is the highest weight of $\eta_\nu^0$. Further, $[\eta_\nu^0] \in \widehat{(M^0)}_\xi$ where $\xi = e^{\nu-\rho_{\mathfrak{t}}}|_{Z_{M^0}}$.*

PROOF. If $m \in M$ then $\mathrm{Ad}(m)$ is an inner automorphism on $M^0$, so $M = Z_M(M^0)M^0$, in particular $Z_{M^0}$ is central in $M$.

Let $[\eta] \in \widehat{M}$. Then $\eta|_{Z_{M^0}}$ is a multiple of a unitary character $\xi$ on $Z_{M^0}$. As $M^0/Z_{M^0}$ and $Z_M(M^0)/Z_{M^0}$ are compact, $M^0$ and $Z_M(M^0)$ are of type I with every irreducible representation finite dimensional; see §2.4 below. Now $[\eta|_{Z_M(M^0)}]$ is a multiple of a finite dimensional class $[\chi] \in \widehat{Z_M(M^0)}_\xi$ and $[\eta|_{M^0}]$ is a multiple of a finite dimensional class $[\eta^0] \in \widehat{M^0}_\xi$. Assertions (2) and (3) follow, and assertion (4) boils down to the highest weight theory of $\mathfrak{m}$. □

The Bott–Borel–Weil Theorem extends from compact connected Lie groups to give geometric realizations of the classes in $\widehat{M}$.

The homogeneous Kähler manifolds of $M$ are the manifolds $S_\Phi$ given by

$$\begin{aligned} &\Pi_{\mathfrak{t}} : \text{ simple } \mathfrak{t}_{\mathbb{C}}\text{–root system on } \mathfrak{m}_C \text{ for } \Sigma_{\mathfrak{t}}^+, \\ &\Phi : \text{ arbitrary subset of } \Pi_{\mathfrak{t}} \\ &\mathfrak{z}_\Phi = \{x \in \mathfrak{t} \mid \Phi(x) = 0\} \text{ and } Z_\Phi^0 = \exp(\mathfrak{z}_\Phi) \subset T^0 \\ &U_\Phi \text{ is the } M\text{–centralizer of } Z_\Phi^0 \text{ and } S_\Phi = M/U_\Phi \end{aligned} \tag{1.1.2}$$

Note that $\Phi$ is a simple $\mathfrak{t}_{\mathbb{C}}$–root system for $\mathfrak{u}_\Phi$. Using Proposition 1.1.1(1), construct

$$\begin{aligned} &\overline{M} := M/Z_M(M^0) = M^0/Z_{M^0} \text{ compact connected Lie group.} \\ &\overline{T} := T/Z_M(M^0) = T^0/Z_{M^0} \text{ maximal torus in } \overline{M}. \end{aligned} \tag{1.1.3}$$

This leads to the homogeneous Kähler structure on $S_\Phi$. $\overline{M}$ has complexification $\overline{M}_{\mathbb{C}} = \mathrm{Int}(\mathfrak{m}_{\mathbb{C}})$, inner automorphism group of $\mathfrak{m}_{\mathbb{C}}$. The group $\overline{U}_\Phi := U_\Phi / Z_M(M^0)$ is connected, and we define

$$\text{(1.1.4)} \qquad \begin{aligned} &\mathfrak{r}_\Phi := \bar{\mathfrak{u}}_{\Phi,\mathbb{C}} + \sum\nolimits_{\phi \in \Sigma_{\mathfrak{t}}^+} \overline{\mathfrak{m}}^{-\phi} \text{ subalgebra of } \overline{\mathfrak{m}}_{\mathbb{C}}, \\ &R_\Phi \text{ is the complex analytic subgroup of } \overline{M}_{\mathbb{C}} \text{ for } \mathfrak{r}_\Phi . \end{aligned}$$

Then $R_\Phi$ is closed in $\overline{M}_{\mathbb{C}}$, and $M$ acts on $\overline{M}_{\mathbb{C}}/R_\Phi$ by the projection $m \mapsto \overline{m}$ of $M$ onto $\overline{M}$. The orbit $M(1R_\Phi)$ is closed because $\overline{M}$ is compact, and one checks that it has the same real dimension as $\overline{M}_{\mathbb{C}}/R_\Phi$. Thus $M$ is transitive on $\overline{M}_{\mathbb{C}}/R_\Phi$, and $mU_\Phi \mapsto \overline{m}R_\Phi$ is a covering space projection. But $\overline{M}_{\mathbb{C}}/R_\Phi$ is simply connected.

LEMMA 1.1.5. *The map $mU_\Phi \mapsto \overline{m}R_\Phi$ is an $M$–equivariant bijection of $S_\Phi = M/U_\Phi$ onto $\overline{M}_{\mathbb{C}}/R_\Phi$.*

Now the complex presentation $S_\Phi = \overline{M}_{\mathbb{C}}/R_\Phi$ defines an $M$–homogeneous complex structure on $S_\Phi$, and the coboundary of any $\mathfrak{u}_\Phi$–regular element of $\mathfrak{z}_\Phi^*$ is an $M$–homogeneous Kähler metric on $S_\Phi$.

The irreducible $M$–homogeneous holomorphic vector bundles $\mathbb{E}_\mu \to S_\Phi$ are constructed as follows. Let $[\mu] \in \widehat{U}_\Phi$ and let $E_\mu$ denote its representation space. Let $\mathbb{E}_\mu \to S_\Phi$ denote the associated complex vector bundle. The group $M^{(1)} = [M^0, M^0]$ is compact, connected and semisimple, so the projection. $M \to \overline{M}$ restricts to a finite covering $M^{(1)} \to \overline{M}$, and that complexifies to a finite holomorphic covering $p : M_{\mathbb{C}}^{(1)} \to \overline{M}_{\mathbb{C}}$. Denote $R_\Phi^{(1)} = p^{-1}(R_\Phi)$. It is connected. Now $\mu^{(1)}$ has a unique completely reducible holomorphic extension $\mu^{(1)}$ to $R_\Phi^{(1)}$. That defines an $M_{\mathbb{C}}^{(1)}$–homogeneous holomorphic vector bundle $\mathbb{E}_\mu = \mathbb{E}_{\mu_{\mathbb{C}}^{(1)}} \to M_{\mathbb{C}}^{(1)}/R_\Phi^{(1)} = \overline{M}_{\mathbb{C}}/R_\Phi = S_\Phi$. That bundle structure is stable under the action of $M$, proving

LEMMA 1.1.6. *$\mathbb{E}_\mu \to S_\Phi$ is an $M$–homogeneous holomorphic vector bundle.*

The sheaf $\mathcal{O}(\mathbb{E}_\mu) \to S_\Phi$ of germs of holomorphic sections of $\mathbb{E}_\mu \to S_\Phi$ defines

$$\text{(1.1.7)} \qquad \begin{aligned} &H_2^{0,q}(S_\Phi, \mathbb{E}_\mu) : L_2 \text{ harmonic } (0,q)\text{–forms on } S_\Phi \text{ with values in } \mathbb{E}_\Phi, \\ &\eta_\mu^q : \text{ representation of } M \text{ on } H^q(S_\Phi; \mathcal{O}(\mathbb{E}_\mu)) \text{ and on } H_2^{0,q}(S_\Phi, \mathbb{E}_\mu) \end{aligned}$$

Simple connectivity of $S_\Phi$ implies $U_\Phi \cap M_\Phi^0 = U_\Phi^0$. With Proposition 1.1.1,

$$\text{(1.1.8)} \qquad \begin{aligned} &\widehat{U_\Phi} = \bigcup\nolimits_{\xi \in \widehat{Z_{M^0}}} (\widehat{U_\Phi})_\xi \text{ where} \\ &\widehat{(U_\Phi)}_\xi = \{[\chi \otimes \mu^0] \mid [\chi] \in \widehat{Z_M(M^0)}_\xi \text{ and } [\mu^0] \in \widehat{(U_\Phi^0)}_\xi\}. \end{aligned}$$

The Bott–Borel–Weil Theorem [**3**] extends to our setting as follows.

PROPOSITION 1.1.9. *Let $[\mu] = [\chi \otimes \mu^0] \in \widehat{U}_\Phi$ as in (1.1.8) and let $\beta$ be the highest weight of $\mu^0$.*

1. *If $\langle \beta + \rho_{\mathfrak{t}}, \phi \rangle = 0$ for some $\phi \in \Sigma_{\mathfrak{t}}^+$ then $H^q(S_\Phi; \mathcal{O}(\mathbb{E}_\mu)) = 0$ for all integers $q$.*

2. *If $\langle \beta + \rho_{\mathfrak{t}}, \phi \rangle \neq 0$ for every $\phi \in \Sigma_{\mathfrak{t}}^+$ define $q_0$ to be the number of roots $\phi \in \Sigma_{\mathfrak{t}}^+$ for which $\langle \beta + \rho_{\mathfrak{t}}, \phi \rangle < 0$, and let $\nu \in L_{\mathfrak{m}}^+$ that is $W(M^0, T^0)$–conjugate to $\beta + \rho_{\mathfrak{t}}$. Then $[\eta_\mu^{q_0}] = [\chi \otimes \eta_\nu^0]$ and $H^q(S_\Phi; \mathcal{O}(\mathbb{E}_\mu)) = 0$ for every integer $q \neq q_0$.*

PROOF. $[\mu] = [\chi \otimes \mu^0]$ has representation space $E_\mu = E_\chi \otimes E_{\mu^0}$. As $Z_M(M^0)$ acts trivially on $S_\Phi$ the associated bundle $\mathbb{E}_\mu = E_\chi \otimes \mathbb{E}_{\mu^0}$. This reduces the proof to the case where $M$ is connected.

Now that $M$ is connected, $M = Z_M^0 M^{(1)}$ where $M^{(1)} = [M, M]$ is compact, connected and semisimple and where $F := Z_M^0 \cap M^{(1)}$ is finite. Now $U_\Phi = Z_M^0 U_\Phi^{(1)}$ where $U_\Phi^{(1)} = U_\Phi \cap M^{(1)}$ and $Z_M^0 \cap U^{(1)} = F$. Split $[\mu] = [\varepsilon \otimes \mu^{(1)}]$ where $[\varepsilon] \in \widehat{Z_M^0}$ and $[\mu^{(1)}] \in \widehat{U_\Phi^{(1)}}$ give the same character $\varepsilon|_F$ on $F$. As above $[\eta_\mu^q] = [\varepsilon \otimes \eta_{\mu^{(1)}}^q]$ where $\eta_{\mu^{(1)}}^q$ induces the representation of $M^{(1)}$ on $H^q(S_\Phi; \mathcal{O}(\mathbb{E}_\mu))$. We have reduced the proof to the case where $M$ is compact, connected and semisimple, which is [**3**]. □

**1.2.** Let $G$ be a general real reductive Lie group. So $\mathfrak{g} = \mathfrak{c} \oplus [\mathfrak{g}, \mathfrak{g}]$ where $\mathfrak{c}$ is the center and the derived algebra $[\mathfrak{g}, \mathfrak{g}]$ is semisimple. We recall the global conditions:

(i) if $m \in M$ then $\mathrm{Ad}(g)$ is an inner automorphism on $\mathfrak{g}_\mathbb{C}$, and

(1.2.1) (ii) $G$ has a closed normal abelian subgroup $Z$ such that

$Z$ centralizes $G^0$, $|G/ZG^0| < \infty$, and $Z \cap G^0$ is co-compact in $Z_{G^0}$

Fix a Cartan involution $\theta$ of $G$. Its fixed point set $K = G^\theta$ contains $Z_G(G^0)$ and

(1.2.2) $K/Z_G(G^0)$ is a maximal compact subgroup of $G/Z_G(G^0)$.

Let $\mathfrak{a}$ be a maximal abelian subspace of $\{x \in \mathfrak{g} \mid \theta(x) = -x\}$. Any two choices are $\mathrm{Ad}_G(K)$–conjugate, $\mathrm{Ad}(\mathfrak{a})$ is diagonalizable on $\mathfrak{g}$, and $\mathfrak{g}$ is the direct sum the joint eigenspaces $\mathfrak{g}^\phi = \{x \in \mathfrak{g} \mid [a, x] = \phi(a)x \text{ for all } a \in \mathfrak{a}\}$. The *$\mathfrak{a}$–root system* of $\mathfrak{g}$ is $\Sigma_\mathfrak{a} := \{\text{joint eigenvalues } \phi \mid \mathfrak{g}^\phi \neq 0\}$. Choose a positive subsystem $\Sigma_\mathfrak{a}^+$; any two choices are conjugate by an element of the $K$–normalizer of $\mathfrak{a}$. The pair $(\mathfrak{a}, \Sigma_\mathfrak{a}^+)$ specifies

$$N: \text{ analytic subgroup of } G \text{ for } \mathfrak{n} = \textstyle\sum_{\alpha \in \Sigma_\mathfrak{a}^+} \mathfrak{g}^{-\alpha}, \qquad (1.2.3)$$

$$A = \exp(\mathfrak{a}), M = Z_K(A), \text{ and } B = \{g \in G \mid Ad(g)N = N\}.$$

Then $MA = M \times A = Z_G(A)$, $B = MAN$ is a *minimal parabolic* subgroup of $G$ and $G = KAN$ is the *Iwasawa decomposition*. Further, $M$ and $MA$ satisfy (1.2.1).

$G$ has a Cartan subgroup $H = T \times A$ where $T$ is a Cartan subgroup of $M$. The corresponding positive root system $\Sigma^+$ satisfies $\Sigma_\mathfrak{a}^+ = \{\gamma|_\mathfrak{a} \mid \gamma \in \Sigma^+ \text{ and } \gamma|_\mathfrak{a} \neq 0\}$ and $\Sigma_\mathfrak{t}^+ = \{\gamma|_\mathfrak{t} \mid \gamma \in \Sigma^+ \text{ and } \gamma|_\mathfrak{a} = 0\}$.

Let $[\chi \otimes \eta_\nu^0] \in \widehat{M}$ and $\sigma \in \mathfrak{a}^*$. Together they specify

$$[\alpha_{\chi,\nu,\sigma}] \in \widehat{B} \text{ by } \alpha_{\chi,\nu,\sigma}(man) = (\chi \otimes \eta_\nu^0)(m)e^{i\sigma}(a). \qquad (1.2.4)$$

The corresponding *principal series representation* of $G$ is

$$\pi_{\chi,\nu,\sigma} = \mathrm{Ind}_B^G(\alpha_{\chi,\nu,\sigma}), \text{ unitarily induced representation.} \qquad (1.2.5)$$

We construct partially holomorphic cohomology spaces $Y_\Phi$ and use them for geometric constructions of the principal series representations $[\pi_{\chi,\nu,\sigma}]$. Let $\Pi$ be the simple $(\mathfrak{t} + \mathfrak{a})$ root system for $\Sigma^+$ Then $\Phi \subset \Pi_\mathfrak{t} \subset \Pi$. Denote

$$\overline{G} = G/Z_G(G^0), \text{ so } \overline{G}_\mathbb{C} = \mathrm{Int}(\mathfrak{g}_\mathbb{C}) \text{ and } \overline{\mathfrak{g}} = \mathfrak{g}/\mathfrak{c}. \qquad (1.2.6)$$

$\mathfrak{q}_\Phi = (\mathfrak{u}_\Phi/\mathfrak{c}) + \mathfrak{a}_\mathbb{C} + \sum_{\gamma \in \Sigma^+} \mathfrak{g}^{-\gamma}$, is a parabolic subalgebra of $\overline{\mathfrak{g}}_\mathbb{C}$. Let $Q_\Phi$ denote the corresponding parabolic subgroup of $\overline{G}_\mathbb{C}$; it is the $\overline{G}_\mathbb{C}$–normalizer of $N$.

That defines the complex homogeneous projective variety,

$$X_\Phi = \overline{G}_{\mathbb{C}}/Q_\Phi\,, \text{ complex flag manifold.} \tag{1.2.7}$$

As $Q_\Phi$ is its own normalizer in $\overline{G}_{\mathbb{C}}$ we can identify $X_\Phi$ with the set of all $\overline{G}_{\mathbb{C}}$–conjugates of $Q_\Phi$. As discussed in §1.1, $G$ acts on $X_\Phi$ by holomorphic diffeomorphisms, through the homomorphism $G \to \overline{G}$. Denote

$$x_\Phi = 1 \cdot Q_\Phi \in X_\Phi \text{ and } Y_\Phi = G(x_\Phi).$$

Then, as in Lemma 1.1.5,

$$U_\Phi AN = \{g \in G \mid g(x_\Phi) = x_\Phi\}, \text{ so } Y_\Phi = U_\Phi AN. \tag{1.2.8}$$

In particular $Y_\Phi$ contains

$$S_\Phi := M(x_\Phi) = (MAN)(x_\Phi), \text{ compact complex submanifold of } X_\Phi\,.$$

Since the minimal parabolic subgroup $B$ is its own $G$–normalizer one can look on the Lie algebra level to see that $B = \{g \in G \mid gS_\Phi = S_\Phi\}$.

LEMMA 1.2.9. *If $g \in G$ then $gS_\Phi$ is a complex submanifold of $X_\Phi$ contained in $Y_\Phi$. If $S \subset Y_\Phi$ is a connected complex submanifold of $X_\Phi$ then $S$ is contained in one of the $gS_\Phi$, $Y_\Phi \to G/B = K/M$ is a well defined equivariant fibration, and $gS_\Phi$ is the fiber over $gB$.*

PROOF. In the terminology of §6.4 the topological component of $x_\Phi$ in $S_\Phi$ is the holomorphic arc component of $Y_\Phi$ through $x_\Phi$. Let $\tau$ denote complex conjugation. Then $\mathfrak{q}_\Phi + \tau\mathfrak{q}_\Phi = (\mathfrak{m}/\mathfrak{c})_{\mathbb{C}} + \mathfrak{a}_{\mathbb{C}} + \mathfrak{n}_{\mathbb{C}} = (\mathfrak{b}/\mathfrak{c})_{\mathbb{C}}$ subalgebra of $\overline{\mathfrak{g}}_{\mathbb{C}}$. In other words $Y_\Phi$ is integrable in the sense of §6.4. Also, $\mathfrak{q}_\Phi$ has $\tau$–stable Levi component $(\mathfrak{u}_\Phi/\mathfrak{c})_{\mathbb{C}}$. Now the assertion is a special case of some results from §6.4. □

We now construct the partially holomorphic bundles and cohomology spaces for the principal series. Let $[\mu] \in \widehat{U_\Phi}$, $\sigma \in \mathfrak{a}^*$ and $\rho_{\mathfrak{a}} = \frac{1}{2}\sum_{\phi \in \Sigma_{\mathfrak{a}}^+}(\dim \mathfrak{g}^{-\phi})\phi \in \mathfrak{a}^*$. Then we have a representation $\gamma_{\mu,\sigma}$ of $U_\Phi AN$ on $E_\mu$, and the associated $G$–homogeneous vector bundle, given by

$$\gamma_{\mu,\sigma}(uan) = e^{\rho_{\mathfrak{a}} + i\sigma}(a)\mu(u) \text{ and } \mathbb{E}_{\mu,\sigma} \to G/U_\Phi AN = Y_\Phi\,. \tag{1.2.10}$$

Note that $\mathbb{E}_{\mu,\sigma}|_{S_\Phi} = \mathbb{E}_\mu$. If $g \in G$ then $\mathbb{E}_{\mu,\sigma}|_{gS_\Phi} \to gS_\Phi$ is an $\mathrm{Ad}(g)B$–homogeneous holomorphic vector bundle. As $[\mu]$ is unitary and $K$ permutes $\{gS_\Phi\}$ transitively, $\mathbb{E}_{\mu,\sigma}$ carries a $K$–invariant hermitian metric, which we will use without comment.

If $y \in Y_\Phi$, say $y = g(x_\Phi)$, we view the holomorphic tangent space to $gS_\Phi$ at $g(x_\Phi)$ as a subspace $T_{g(x_\Phi)}$ of the complexified tangent space of $Y_\Phi$. These subspaces define a sub-bundle $\mathbb{T} \to Y_\Phi$ of the complexified tangent bundle of $Y_\Phi$. Note that $\mathbb{T} \to Y_\Phi$ is $G$–homogeneous and is holomorphic over every $gS_\Phi$. Then

$$A^{0,q}(Y_\Phi, \mathbb{E}_{\mu,\sigma}) = \{C^\infty \text{ sections of } \mathbb{E}_{\mu,\sigma} \otimes \Lambda^q(\overline{\mathbb{T}}^*) \to Y_\Phi\} \tag{1.2.11}$$

is the space of $C^\infty$ *partially $(0,q)$–forms on $Y_\Phi$ with values in $\mathbb{E}_{\mu,\sigma}$*. From any $K$–invariant hermitian metric on $\mathbb{T}$, thus also on $\mathbb{E}_{\mu,\sigma} \otimes \Lambda^q(\overline{\mathbb{T}}^*)$, we have Kodaira–Hodge operators

$$\Lambda^{0,q}(Y_\Phi, \mathbb{E}_{\mu,\sigma}) \xrightarrow{\#} \Lambda^{n,n-q}(Y_\Phi, \mathbb{E}^*_{\mu,\sigma}) \xrightarrow{\widetilde{\#}} \Lambda^{0,q}(Y_\Phi, \mathbb{E}_{\mu,\sigma}) \tag{1.2.12}$$

where $n = \dim_{\mathbb{C}} S_\Phi$. The $\overline{\partial}$ operator of $X_\Phi$ induces the $\overline{\partial}$ on each $gS_\Phi$. They fit together to define operators $\overline{\partial} : A^{0,q}(Y_\Phi, \mathbb{E}_{\mu,\sigma}) \to A^{0,q+1}(Y_\Phi, \mathbb{E}_{\mu,\sigma})$. Each $A^{0,q}(Y_\Phi, \mathbb{E}_{\mu,\sigma})$ is a pre–Hilbert space with inner product

$$\langle \alpha, \beta \rangle = \int_{K/M} d(kM) \int_{kS_\Phi} \alpha \,\bar{\wedge}\, \# \beta \tag{1.2.13}$$

where $\bar{\wedge}$ signifies exterior product followed by contraction $E_\mu \otimes E_\mu^* \to \mathbb{C}$. Define

$$L_2^{0,q}(Y_\Phi, \mathbb{E}_{\mu,\sigma}) : \text{ Hilbert space completion of } A^{0,q}(Y_\Phi, \mathbb{E}_{\mu,\sigma}). \tag{1.2.14}$$

Then $\overline{\partial}$ has formal adjoint $\overline{\partial}^* = -\widetilde{\#}\overline{\partial}\#$ there, and that leads to the essentially self adjoint partial Kodaira–Hodge operators

$$\square = (\overline{\partial} + \overline{\partial}^*)^2 = \overline{\partial}\overline{\partial}^* + \overline{\partial}^*\overline{\partial} \text{ on } L_2^{0,q}(Y_\Phi, \mathbb{E}_{\mu,\sigma}). \tag{1.2.15}$$

The *partially harmonic* $(0,q)$*-forms on* $Y_\Phi$ *with values in* $\mathbb{E}_{\mu,\sigma}$ are the elements of

$$H^{0,q}(Y_\Phi, \mathbb{E}_{\mu,\sigma}) = \{\omega \in L_2^{0,q}(Y_\Phi, \mathbb{E}_{\mu,\sigma}) \mid \square\,\omega = 0\}. \tag{1.2.16}$$

Note that $H^{0,q}(Y_\Phi, \mathbb{E}_{\mu,\sigma})$ consists of all Borel–measurable sections $\omega$ of $\mathbb{E}_{\mu,\sigma} \otimes \Lambda^q(\overline{\mathbb{T}}^*) \to Y_\Phi$ such that (i) for almost all $k \in K$, $\omega|_{kS_\Phi}$ is harmonic in the ordinary sense and (ii) the $L_2$ norms satisfy $\int_{K/M} ||\omega|^2_{kS_\Phi} d(kM) < \infty$. In particular $H^{0,q}(Y_\Phi, \mathbb{E}_{\mu,\sigma})$ is a closed subspace of $L_2^{0,q}(Y_\Phi, \mathbb{E}_{\mu,\sigma})$ and the natural action of $G$ on $H^{0,q}(Y_\Phi, \mathbb{E}_{\mu,\sigma})$ is a continuous representation $\pi^q_{\mu,\sigma}$.

THEOREM 1.2.17. *The representation* $\pi^q_{\mu,\sigma}$ *of* $G$ *on* $H^{0,q}(Y_\Phi, \mathbb{E}_{\mu,\sigma})$ *is unitary. Let* $[\mu] = [\chi \otimes \mu^0]$ *as in* (1.1.8) *and let* $\beta$ *be the highest weight of* $\mu^0$.

1. *If* $\langle \beta + \rho_{\mathfrak{t}}, \phi \rangle = 0$ *for some* $\phi \in \Sigma_{\mathfrak{t}}^+$ *then* $H^{0,q}(Y_\Phi, \mathbb{E}_{\mu,\sigma}) = 0$ *for all* $q$.
2. *If* $\langle \beta + \rho_{\mathfrak{t}}, \phi \rangle \neq 0$ *for every* $\phi \in \Sigma_{\mathfrak{t}}^+$, *let* $q_0$ *be the number of roots* $\phi \in \Sigma_{\mathfrak{t}}^+$ *for which* $\langle \beta + \rho_{\mathfrak{t}}, \phi \rangle < 0$, *and let* $\nu$ *be the unique element of* $L_{\mathfrak{m}}^+$ *conjugate to* $\beta + \rho_{\mathfrak{t}}$ *by an element of the Weyl group* $W(M^0, T^0)$. *Then* $[\pi^{q_0}_{\mu,\sigma}] = [\pi_{\chi,\nu,\sigma}]$, *principal series class, and* $H^{0,q}(Y_\Phi, \mathbb{E}_{\mu,\sigma}) = 0$ *for every* $q \neq q_0$.
3. *In particular, given a principal series class* $[\pi_{\chi,\nu,\sigma}]$, *we can realize it on* $H^{0,0}(Y_\Phi, \mathbb{E}_{\mu,\sigma})$ *where* $\mu = [\chi \otimes \mu^0]$ *and* $\mu^0$ *has highest weight* $\nu - \rho_{\mathfrak{t}}$;.

PROOF. Let $\tilde{\pi}^q_{\mu,\sigma}$ denote the representation of $G$ on $L_2^{0,q}(Y_\Phi, \mathbb{E}_{\mu,\sigma})$. Factor $\gamma_{\mu,\sigma} = {}'\gamma_{\mu,\sigma} \cdot e^{\rho_{\mathfrak{a}}}$ where ${}'\gamma_{\mu,\sigma}(uan) = e^{i\sigma}(a)\mu(u)$ is unitary and $e^{\rho_{\mathfrak{a}}}$ compensates non–unimodularity of $U_\Phi AN$. Thus $\tilde{\pi}^q_{\mu,\sigma} = \mathrm{Ind}_{U_\Phi AN}^G(\gamma_{\mu,\sigma})$ is unitary, so its subrepresentation $\pi^q_{\mu,\sigma} = \mathrm{Ind}_{U_\Phi AN}^G(\gamma_{\mu,\sigma})$ is unitary.

Write $E^q_\mu$ for $E_\mu \otimes \Lambda^q(\overline{T}^*_{x_\Phi})$, so $L_2^{0,q}(Y_\Phi, \mathbb{E}_{\mu,\sigma})$ is the space of all measurable $f : G \to E^q_\mu$ such that $f(guan) = (\gamma_{\mu,\sigma}(uan)^{-1} \otimes \Lambda^q \mathrm{Ad}(uan)_{\mathfrak{n}}^{-1})f(g)$ and $\int_{K/U_\Phi} ||f(k)||^2 d(kU_\Phi) < \infty$. Note $\tilde{\pi}^q_{\mu,\sigma} = \mathrm{Ind}_B^G(\tilde{\psi})$ where $\tilde{\psi} = \mathrm{Ind}\, U_\Phi AN^B({}'\gamma_{\mu,\sigma})$. The representation space elements annihilated by the partial Laplacian $\square$ correspond to the subspace of the representation space of $\tilde{\psi}$ annihilated by the full Laplacian of $\mathbb{E}_{\mu,\sigma}|_{S_\Phi}$, thus $\pi^q_{\mu,\sigma} = \mathrm{Ind}_B^G(\psi)$ where $\psi$ represents $B$ on $H^{0,q}(S_\Phi, \mathbb{E}_{\mu,\sigma}|_{S_\Phi})$. Denote the representation of $M$ on $H^{0,q}(S_\Phi, \mathbb{E}_\mu)$ by $\eta^q_\mu$. Now $\psi(man) = \eta^q_\mu(m)e^{i\sigma}(a)$. The theorem now follows from Proposition 1.1.9. $\square$

## 2. General Notion of Relative Discrete Series

We recall some basic facts on the relative discrete series of a locally compact group $G$. Left Haar measure is denoted $dg$, and the left regular representation of $G$ on $L_2(G)$ is $\ell(g)f](x) = f(g^{-1}x)$. If $\pi$ is an irreducible unitary representation of $G$ then $\mathcal{H}_\pi$ is its representation space and $[\pi]$ its unitary equivalence class. Its coefficients are the functions $\phi_{u,v}(g) = \langle u, \pi(g)v\rangle$. The set of all equivalence classes of irreducible unitary representations is $\widehat{G}$.

**2.1.** Let $Z$ be a closed normal abelian subgroup of $G$. It has left regular representation $\ell^Z = \int_{\widehat{Z}} \zeta\, d\zeta$. The left regular representation of $G$ decomposes as

$$\ell = \mathrm{Ind}_{\{1\}}^G(1) = \mathrm{Ind}_Z^G(\ell_Z) = \int_{\widehat{Z}} \ell_\zeta d\zeta = \int_{\widehat{Z}} \mathrm{Ind}_Z^G(\zeta) d\zeta \tag{2.1.1}$$

where $\ell_\zeta := \mathrm{Ind}_Z^G(\zeta)$ is the left regular representation of $G$ on the Hilbert space

$$L_2(G/Z,\zeta) = \{f : G \to \mathbb{C} \mid f(gz) = \zeta(z)^{-1}f(g) \text{ and } \int_{G/Z} |f(g)|^2 dg < \infty\}. \tag{2.1.2}$$

Thus $L_2(G) = \int_{\widehat{Z}} L_2(G/Z,\zeta)$.

Given $\zeta \in \widehat{Z}$ let $\widehat{G}_\zeta = \{[\pi] \in \widehat{G} \mid \zeta \text{ is a subrepresentation of } \pi|_Z\}$. Thus $\ell_\zeta$ is a direct integral over $\widehat{G}_\zeta$. We say that $[\pi] \in \widehat{G}$ is *$\zeta$–discrete* if $\pi$ is a subrepresentation of $\ell_\zeta$. The $\zeta$–discrete classes form the *$\zeta$–discrete series* $\widehat{G}_{\zeta-disc}$. The relative (to $Z$) discrete series is $\widehat{G}_{disc} = \bigcup \widehat{G}_{\zeta-disc}$.

**2.2.** Let the closed normal abelian subgroup $Z$ be central in $G$. If $[\pi] \in \widehat{G}$ there is a character $\zeta_\pi \in \widehat{Z}$ such that $\pi|_Z$ is a multiple of $\zeta_\pi$. So $\widehat{G}_{disc} = \bigcup \widehat{G}_{\zeta-disc}$ is disjoint.

If $[\pi] \in \widehat{G}$ and $\zeta \in \widehat{Z}$ the following are equivalent: (i) there exists $0 \neq u \in \mathcal{H}_\pi$ with $\phi_{u,u} \in L_2(G/Z,\zeta)$, (ii) if $u, v \in \mathcal{H}_\pi$ then $\phi_{u,v} \in L_2(G/Z,\zeta)$, (iii) $[\pi] \in \widehat{G}_{\zeta-disc}$. Under these conditions there is a number $\deg(\pi) > 0$, the *formal degree* of $[\pi]$, such that

$$\begin{aligned} &\langle \phi_{u,v}, \phi_{u',v'}\rangle = \tfrac{1}{\deg(\pi)}\langle u, u'\rangle\overline{\langle v, v'\rangle} \text{ for } u, u', v, v' \in \mathcal{H}_\pi \\ &\langle \phi_{u,v}, \phi_{u',v'}\rangle = 0 \text{ for } u, v \in \mathcal{H}_\pi \text{ and } u', v' \in \mathcal{H}_{\pi'} \text{ with } [\pi] \neq [\pi]. \end{aligned} \tag{2.2.1}$$

If $G$ is compact, $Z = \{1\}$ and we normalize $\int_G dg = 1$, then $\widehat{G} = \widehat{G}_{disc}$, $\deg\pi$ is the degree in the usual sense, and (2.2.1) reduces to the Frobenius–Schur Relations.

Define $L_p(G/Z,\zeta)$ in the same way, integration over $G/Z$, for $1 \leqq p < \infty$. Since $Z$ is central $(f * h)(x) = \int_{G/Z} f(g)h(g^{-1}x)dx$ gives a well defined convolution product $L_1(G/Z,\zeta) \times L_p(G/Z,\zeta) \to L_p(G/Z,\zeta)$. The first line of (2.2.1) can be expressed $\phi_{u,v} * \phi_{u',v'} = \frac{1}{\deg(\pi)}\langle u, v'\rangle\phi_{u',v} \in L_2(G/Z,\zeta)$.

These results are due to Godement [**7**] for compact $Z$, to Harish–Chandra [**15**] for semisimple $G$. See Dixmier's exposition [**6**, §14] of Godement or apply Rieffel's results [**38**] to the convolution algebra $L_1(G/Z,\zeta) \cap L_2(G/Z,\zeta)$.

**2.3.** We can relax the condition that $Z$ be central in $G$, requiring only that $G$ have a finite index subgroup $J$ that centralizes $Z$. Then we may assume $Z \subset J$ and $J$ normal in $G$. If $\zeta \in \widehat{Z}$ denote $\ell_\zeta^J = \mathrm{Ind}_Z^J(\zeta)$, so $\ell_\zeta = \mathrm{Ind}_J^G(\ell_\zeta^J)$. Since $|G/J| < \infty$

the restriction $\ell_\zeta|_J = \sum_{xJ\in G/J} \ell^J_{\zeta_x}$, finite sum, where $\zeta_x(z) = \zeta(x^{-1}zx)$. Now

$$\begin{aligned}\widehat{G}_{disc} &= \{[\pi] \in \widehat{G} \mid \pi|_J \text{ has a subrepresentation } [\psi] \in \widehat{J}_{disc}\} \text{ and} \\ \widehat{J}_{disc} &= \{[\psi] \in \widehat{J} \mid \psi \text{ is a subrepresentation of } \pi|_J \text{ for some } [\pi] \in \widehat{G}_{disc}\}\end{aligned} \tag{2.3.1}$$

In our applications, if $[\psi] \in \widehat{J}_{disc}$ then $\mathrm{Ind}_J^G(\psi)|_J$ is a finite sum of mutually inequivalent representations, so $[\psi] \mapsto [\mathrm{Ind}_J^G(\psi)]$ maps $\widehat{J}_{disc}$ onto $\widehat{G}_{disc}$.

**2.4.** Later we will construct the relative discrete series $\widehat{G}_{disc}$ on homogeneous vector bundles $\mathbb{E} \to G/U$ where $Z \subset U$ and $U/Z$ is compact. For that we will need a mild extension of the Peter–Weyl Theorem.

LEMMA 2.4.1. *Let $U$ be a locally compact group and $Z$ a closed central subgroup such that $U/Z$ is compact. Then every class $[\chi] \in \widehat{U}$ is finite dimensional.*

PROOF. Let $S$ be the circle group $\{s \in \mathbb{C} \mid |s| = 1\}$ and $1_S \in \widehat{S}$ the character $1_S(s) = s$. Given $\zeta \in \widehat{Z}$ define $U[\zeta] = \{S \times U\}/\{(\zeta(z)^{-1}, z) \mid z \in Z\}$. If $[\chi] \in \widehat{U}_\zeta$ now $[1_S \otimes \chi] \in \widehat{S \times U}$ factors through $U[\zeta]$. But $U[\zeta]$ is an extension $1 \to S \to U[\zeta] \to U/Z \to 1$ of a compact group by a compact group, so it is compact. Thus $1_S \otimes \chi$ is finite dimensional. □

LEMMA 2.4.2. *If $\zeta \in \widehat{Z}$ then $\widehat{U}_\zeta = \widehat{U}_{\zeta-disc}$.*

PROOF. Coefficients of $[\chi] \in \widehat{U}_\zeta$ are in $L_2(U/Z, \zeta)$ because $U/Z$ is compact. □

LEMMA 2.4.3. *Let $\zeta \in \widehat{Z}$, $[\chi] \in \widehat{U}_\zeta$, $E_\chi$ the representation space of $\chi$, and $L_\chi$ the space of coefficients. Then $L_2(U/Z, \zeta)$ is the orthogonal direct sum $\sum_{\widehat{U}_\zeta} L_\chi = \sum_{\widehat{U}_\zeta} E_\chi \otimes E_\chi^*$ where $U \times U$ acts on $L_\chi = E_\chi \otimes E_\chi^*$ by $(\chi \otimes 1) \boxtimes (1 \otimes \chi^*)$.*

PROOF. Let $q : U \to U[\zeta]$ be the restriction to $U$ of the projection $S \times U \to U[\zeta]$ in the proof of Lemma 2.4.1. The assertions follow from the part concerning $\widehat{U[\zeta]}_{1_S}$ in the standard Peter–Weyl Theorem for $U[\zeta]$ □

Normalize Haar measures by $\int_U du = \int_Z dz \int_{U/Z} d(uZ)$ and $\int_{U/Z} d(uZ) = 1$. $C_c(U)$ denotes the space of continuous compactly supported functions on $U$. If $f \in C_c(U)$ then $\chi(f) : E_\chi \to E_\chi$ by $\chi(f)v = \int_U f(u)\chi(u)v\, du$.

PROPOSITION 2.4.4. *Let $\zeta \in \widehat{Z}$, $[\chi] \in \widehat{U}_\zeta$. If $f \in C_c(U)$ then* trace $\chi(f) = \int_U f(u)$ trace $\chi(u)\, du$ *and* dim $\chi(= \dim E_\chi)$ *is the formal degree* $\deg \chi$. *Further, orthogonal projection $L_2(U/Z, \zeta) \to L_\chi$ is given by right (or left) convolution with* $(\dim \chi)$ trace $\overline{\chi}$.

PROOF. Let $\{v_1, \dots, v_n\}$ be an orthonormal basis of $E_\chi$ and $\phi_{i,j} = \langle v_i, \chi(v_j)\rangle$. Then trace $\chi(f) = \sum \langle \chi(f)v_i, v_i\rangle = \sum \int_U \langle f(u)\chi(u)v_i, v_i\rangle du = \int_U f(u)$ trace $\chi(u)du$. That is the first assertion. For the second, use $q : U \to U[\zeta]$ as in the proof of Lemma 2.4.3 to see $\dim \chi = \dim(1_S \otimes \chi) = \deg(1_S \otimes \chi) = \deg \chi$. For the last, use $\phi_{u,v} * \phi_{u',v'} = \frac{1}{\deg(\pi)} \langle u, v'\rangle \phi_{u',v} \in L_2(G/Z, \zeta)$ (from the first line of (2.2.1)) to calculate $1 = \langle \text{trace } \chi, \text{trace } \chi\rangle = \sum_{i,j} \langle \phi_{i,i}, \phi_{j,j}\rangle = n/\deg(\chi)$. □

Combining (2.1.2), Lemma 2.4.3 and Proposition 2.4.4 we have the Plancherel Formula for $U$. If $f \in C_c(U)$ and $\zeta \in \widehat{Z}$ we denote $f_\zeta(u) = \int_Z f(uz)\zeta(z)dz$.

PROPOSITION 2.4.5. $L_2(U) = \int_{\widehat{Z}} \left(\sum_{\widehat{U}_\zeta} E_\chi \otimes E_\chi^*\right) d\zeta$. *If* $f \in C_c(U)$ *and* $u \in U$ *then* $f(u) = \int_{\widehat{Z}} \left(\sum_{\widehat{U}_\zeta} \text{trace } \chi((r(u)f)_\zeta) \dim \chi\right) d\zeta$ *where* $r(u)f(u') = f(uu')$.

## 3. Relative Discrete Series for Reductive Groups

Harish–Chandra's theory of the discrete series ([**18**], [**19**]) extends from his class $\mathcal{H}$ to our class of reductive Lie groups that contains all connected reductive groups and has certain hereditary properties. We describe that class in §3.1. Certain points of Harish–Chandra's general character theory are recalled in §3.2. In §3.3 we show how to reduce certain questions from connected reductive groups in general to the case of compact center. We use that in §3.4 to extend Harish–Chandra's theory of the discrete series to all connected reductive Lie groups, in particular to all connected semisimple Lie groups. Then in §3.5 we obtain the discrete series for our class of groups described in §3.1.

**3.1.** From now on, $G$ is a reductive Lie group in the sense that its Lie algebra is reductive: $\mathfrak{g} = \mathfrak{c} \oplus [\mathfrak{g}, \mathfrak{g}]$ where $\mathfrak{c}$ is the center and $\mathfrak{g}' = [\mathfrak{g}, \mathfrak{g}]$ is semisimple.

DEFINITION 3.1.1. Suppose that $G$ satisfies the condition

$$\text{if } g \in G \text{ then } \mathrm{Ad}(g) \text{ is an inner automorphism on } \mathfrak{g}_{\mathbb{C}}$$

and that $G$ has a closed normal abelian subgroup $Z$ such that

$$Z \text{ centralizes } G^0,\ G/ZG^0 \text{ is finite, and } ZG^0/(Z \cap G^0) \text{ is compact.}$$

Then we say that $G$ is a *general real reductive Lie group*, i.e., that $G$ of class $\widetilde{\mathcal{H}}$.

If $G/G^0$ is finite then $Z_{G^0}$ satisfies the conditions for $Z$ in the definition. The Harish–Chandra class $\mathcal{H}$ consists of the groups $G \in \widetilde{\mathcal{H}}$ for which $Z$ is finite. In any case, the discrete series relative to $Z$ is independent of choice of group $Z$ that satisfies the conditions of Definition 3.1.1. We use the notation

$$G^\dagger = \{g \in G \mid \mathrm{Ad}(g) \text{ is an inner automorphism on } G^0. \tag{3.1.2}$$

Then $G^\dagger = Z_G(G^0)G^0$, $Z_G(G^0)$ is compact and $G^\dagger/ZG^0$ is finite.

**3.2.** Using the second condition of Definition(3.1.1) we choose a maximal compact subgroup $K/Z$ of $G/Z$. Then $K^0 = K \cap G^0$, $K$ meets every topological component of $G$, and $Z_{G^0} \subset K$. The basis of Harish–Chandra's character theory is

LEMMA 3.2.1. *There is an integer* $n_G \geqq 1$ *such that, if* $[\kappa] \in \widehat{K}$ *and* $[\pi] \in \widehat{G}$, *then the multiplicity* $mult(\kappa, \pi|_K) \leqq n_G \dim(\kappa) < \infty$.

PROOF. This was proved by Harish–Chandra for connected reductive Lie groups [**11**], in particular for $G^0$. If $[\kappa_1] \in \widehat{ZK^0}$ and $[\pi_1] \in \widehat{ZG^0}$ we have $\zeta, \zeta' \in \widehat{Z}, [\kappa_0] \in \widehat{K^0}$ and $[\pi_0] \in \widehat{G^0}$ such that $\kappa_1 = \zeta \otimes \kappa_0$ and $\pi_1 = \zeta' \otimes \pi_o$. Now $mult(\kappa_1, \pi_1|_{ZK^0}) \leqq mult(\kappa_0, \pi_0|_{K^0}) \leqq n_{G^0} \dim(\kappa_0) = n_{G^0} \dim(\kappa)$, so the lemma follows for $ZG^0$. Finally, if $[\kappa] \in \widehat{K}$ and $[\pi] \in \widehat{G}$, we decompose $\kappa|_{ZK^0} = \sum \kappa_i$ and $\pi|_{ZG^0} = \sum \pi_j$ into irreducible constituents, and

$$\begin{aligned} mult(\kappa, \pi|_K) &\leqq \sum_{i,j} mult(\kappa_i, \pi_j|_{ZK^0}) \leqq \sum_{i,j} n_{G^0} \dim(\kappa_i) \\ &= \sum_i n_{G^0} \dim(\kappa) \leqq (n_{G^0}|G/ZG^0|) \dim(\kappa). \end{aligned}$$

□

REMARK 3.2.2. *Using Casselman–Miličić* [**5**] *one can see that* $n_G \leqq |G/ZG^0|$.

The first consequence of Lemma 3.2.1 is that the group $G$ is CCR. In other words, if $[\pi] \in \widehat{G}$ and $f \in L_1(G)$ then $\pi(f) = \int_G f(g)\pi(g)dg$ is a compact operator on $\mathcal{H}_\pi$. In particular $G$ is of type I. The second consequence is that $\pi(f)$ is of trace class for $f \in C_c^\infty(G)$ and that

(3.2.3) $\quad \Theta_\pi : C_c^\infty \to \mathbb{C}$ by $\Theta_\pi(f) = \text{trace } \pi(f)$ is a Schwartz distribution on $G$.

$\Theta_\pi$ is the *global character* or *distribution character* of $\pi$. Classes $[\pi] = [\pi']$ if and only if $\Theta_\pi = \Theta_{\pi'}$.

A differential operator $z$ on $G$ has *transpose* given by $\int_G^t z(f)(g)h(g)dg = \int_G f(g)z(h)(g)dg$. The operator $z$ acts on distributions by $(z\Theta)(f) = \Theta({}^t z(f))$. Given $\Theta$ now $z \mapsto z\Theta$ is linear in $z$. A distribution on $G$ is *invariant* if $\Theta(f) = \Theta(f \cdot \text{Ad}(g))$ for $f \in C_c^\infty(G)$ and $g \in G$.

The universal enveloping algebra $\mathcal{U}(\mathfrak{g})$ is the associative algebra of all left–invariant differential operators on $G$. The center $\mathcal{Z}(\mathfrak{g})$ consists of the bi–invariant operators. That uses (3.1.1). A distribution $\Theta$ on $G$ is an *eigendistribution* if $\dim \mathcal{Z}(\mathfrak{g})(\Theta) \leqq 1$. In that case, using commutativity of $\mathcal{Z}(\mathfrak{g})$, we have an algebra homomorphism $\chi_\Theta : \mathcal{Z}(\mathfrak{g}) \to \mathbb{C}$ defined by $z\Theta = \chi_\Theta(z)\Theta$. If $[\pi] \in \widehat{G}$ its distribution character $\Theta_\pi$ is an invariant eigendistribution on $G$, and the associated homomorphism is

(3.2.4) $\quad \chi_\pi : \mathcal{Z}(\mathfrak{g}) \to \mathbb{C},\ z\Theta_\pi = \chi_\pi(z)\Theta_\pi$, is the *infinitesimal character* of $[\pi]$.

Choose a Cartan subalgebra $\mathfrak{h} \subset \mathfrak{g}$ and let $I(\mathfrak{h}_\mathbb{C})$ denote the algebra of all polynomials on $\mathfrak{h}_\mathbb{C}^*$ invariant under the Weyl group $W(\mathfrak{g}_\mathbb{C}, \mathfrak{h}_\mathbb{C})$. Let $\gamma : \mathcal{Z}(\mathfrak{g}) \to I(\mathfrak{h}_\mathbb{C})$ denote the Harish–Chandra homomorphism [**16**]. If $\lambda \in \mathfrak{h}_\mathbb{C}^*$ then

(3.2.5) $$\chi_\lambda : \mathcal{Z}(\mathfrak{g}) \to \mathbb{C} \text{ by } \chi_\lambda(z) = [\gamma(z)](\lambda)$$

is a homomorphism, every homomorphism $\mathcal{Z}(\mathfrak{g}) \to \mathbb{C}$ is a $\chi_\lambda$, and $\chi_\lambda = \chi_{\lambda'}$ if and only if $\lambda$ and $\lambda'$ are $W(\mathfrak{g}_\mathbb{C}, \mathfrak{h}_\mathbb{C})$–equivalent.

Harish–Chandra [**16**] used the equations $z\Theta_\pi = \chi_\pi(z)\Theta_\pi$ and the description (3.2.5) of $\chi_\pi$ to show that $\Theta_\pi$ is a locally $L_1$ function that is analytic on the *regular set* $G'$. Here $G'$ is the dense open subset of all $g \in G$ for which $\{\xi \in \mathfrak{g} \mid \text{Ad}(g)\xi = \xi\}$ is a Cartan subalgebra. The complement of $G'$ has measure zero in $G$.

The differential equations also show that at most finitely many classes in $\widehat{G}$ can have the same infinitesimal character.

**3.3.** We describe a method for reducing questions of harmonic analysis on $ZG^0$ and $G^0$ to the same questions on connected reductive groups with compact center. With that we extends some of Harish–Chandra's discrete series results [**19**] to $G^0$ in §3.4 and then to $G$ in §3.5. This uses the Mackey central extension $1 \to S \to G[\zeta] \to ZG^0/Z \to 1$ for $\delta\zeta$ as a normalized multiplier on $ZG^0$. Thus $G[\zeta] = \{S \times ZG^0\}/\{(\zeta(z)^{-1}, z) | z \in Z\}$ and $S$ is the circle group $\{s \in \mathbb{C} \mid |s| = 1\}$. $G[\zeta]$ is a connected reductive group with Lie algebra $\mathfrak{s} \oplus (\mathfrak{g}/\mathfrak{z})$ and compact center.

LEMMA 3.3.1. *Let* $p : ZG^0 \to G[\zeta]$ *be the restriction of the projection* $S \times ZG^0 \to G[\zeta]$. *Then* $f \mapsto f \cdot p$ *gives a* $G$*–equivariant isometry of* $L_2(G[\zeta]/S, 1_S)$ *onto* $L_2(ZG^0/Z, \zeta)$.

PROOF. View $f \in L_2(G[\zeta]/S, 1_S)$ as a function on $S \times ZG^0$. If $g \in ZG^0$ and $z \in Z$ then $(f \cdot p)(gz) = f(1, gz) = f(\zeta(z), g) = \zeta(z)^{-1} f(1, g) = \zeta(z)^{-1}[f \cdot p](g)$, and

$\int_{ZG^0/Z} |(f\cdot p)(g)|^2 d(gZ) = \int_{(S\times ZG^0)/(S\times Z)} |f(s,g)|^2 d(sS\times gZ) = \int_{G[\zeta]/S} |f(\overline{g})|^2 d(\overline{g}S)$. Thus $f \mapsto f\cdot p$ is an isometric injection of $L_2(G[\zeta]/S, 1_S)$ into $L_2(ZG^0/Z,\zeta)$. It is surjective because every $f' \in L_2(ZG^0/Z,\zeta)$ has form $f\cdot p$ with $f$ defined on $S\times ZG^0$ by $f(s,g) = s^{-1}f'(g)$. □

THEOREM 3.3.2. *$\varepsilon[\psi] = [\psi\cdot f]$ defines a bijection $\varepsilon = \varepsilon_\zeta : \widehat{G[\zeta]}_{1_S} \to (\widehat{ZG^0})_\zeta$. It maps $\widehat{G[\zeta]}_{1_S-disc}$ onto $\widehat{ZG^0}_{\zeta-disc}$ and carries Plancherel measure of $\widehat{G[\zeta]}_{1_S}$ to Plancherel measure of $(\widehat{ZG^0})_\zeta$. Distribution characters satisfy $\Theta_{\varepsilon[\psi]} = \Theta_{[\psi]}\cdot p$.*

PROOF. View $[\psi] \in \widehat{G[\zeta]}_{1_S}$ as a representation of $S\times ZG^0$. If $z\in Z$ and $g\in ZG^0$ then $\psi\cdot p)(gz) = \psi(1,gz) = \psi(\zeta(z),g) = \zeta(z)\psi(1,g) = \zeta(z)[\psi\cdot p](g)$. Thus the central character of $\psi\cdot p$ restricts to $\zeta$ on $Z$, and $[\psi\cdot p]\in(\widehat{ZG^0})_\zeta$.

Let $[\psi],[\psi'] \in \widehat{G[\zeta]}_{1_S}$ and $b : \mathcal{H}_\psi \to \mathcal{H}_{\psi'}$ an isometry. If $\psi' = b\psi b^{-1}$ the above calculation shows $(\psi' p) = b(\psi p)b^{-1}$. If $b$ intertwines $\psi p$ with $\psi' p$ the same calculation shows that it intertwines $\psi|_{ZG^0}$ with $\psi'|_{ZG^0}$. Then it intertwines $\psi$ and $\psi'$ because $\psi'(s,g) = s\psi'(1,g) = sb\psi(1,g)b^{-1} = b\psi(s,g)b^{-1}$. Now $\varepsilon_\zeta : \widehat{G[\zeta]}_{1_S} \to (\widehat{ZG^0})_\zeta$ is a well defined bijection.

We reduce the proof of Theorem 3.3.2 to the case where $Z\subset G^0$, i.e. where $ZG^0$ is connected. Let $\zeta^0 = \zeta|_{ZG^0}$. Then $G^0 \hookrightarrow G$ induces a commutative diagram

$$\begin{array}{ccccccccc} 1 & \longrightarrow & S & \longrightarrow & G^0[\zeta^0] & \longrightarrow & G^0/(Z\cap G^0) & \longrightarrow & 1 \\ & & \downarrow & & \downarrow{\scriptstyle a} & & \downarrow & & \\ 1 & \longrightarrow & S & \longrightarrow & G[\zeta] & \longrightarrow & ZG^0/Z & \longrightarrow & 1 \end{array}$$

where $a$ is an isomorphism because $S\to S$ and $G^0/(Z\cap G^0)\to ZG^0/Z$ are. That results in a bijection $\varepsilon$ that is the composition $\widehat{G[\zeta]}_{1_S} \xrightarrow{a^*} \widehat{G^0[\zeta^0]}_{1_S} \xrightarrow{\varepsilon^0} (\widehat{G^0})_{\zeta^0} \xrightarrow{b} (\widehat{ZG^0})_\zeta$ where $a^*$ is induced by $a$ and $b[\pi^0] = [\zeta\otimes\pi^0]$. Plancherel measure and relative discrete series are transported by $b$. If $[\pi^0]\in(\widehat{G^0})_{\zeta^0}$, $z\in Z$ and $g\in G^0$ then $\Theta_{b[\pi^0]}(zg) = \zeta(z)\Theta_{\pi^0}(g)$. Thus if $\varepsilon : \widehat{G^0[\zeta^0]}_{1_S} \to (\widehat{G^0})_{\zeta^0}$ has the properties asserted in Theorem 3.3.2 those properties pass over to $\varepsilon : \widehat{G[\zeta]}_{1_S} \to (\widehat{ZG^0})_\zeta$.

Now we assume $Z\subset G^0$, and we further reduce the proof to the case where $G^0$ is simply connected. Let $\tau : \widetilde{G}\to G^0$ be the universal cover, $\widetilde{Z} = \tau^{-1}(Z)$, and $\widetilde{\zeta}$ the corresponding lift of $\zeta$. Then we have $\begin{array}{ccc} \widehat{\widetilde{G}[\widetilde{\zeta}]}_{1_S} & \xrightarrow{\widetilde{\varepsilon}} & \widehat{\widetilde{G}}_{\widetilde{\zeta}} \\ {\scriptstyle i^*}\uparrow & & \uparrow{\scriptstyle j} \\ \widehat{G[\zeta]}_{1_S} & \xrightarrow{\varepsilon} & (\widehat{G^0})_{\zeta_0} \end{array}$ linking $\varepsilon$ to $\widetilde{\varepsilon}$. Everything is preserved by $i^*$, and $j$ transports Plancherel measure and relative discrete series.

We check that $\Theta_{j[\psi]} = \Theta_{[\psi]}\tau$ for $[\psi]\in(\widehat{G^0})_\zeta$. It suffices to test this on functions $f\in C_c^\infty(\widetilde{G})$ in a component $U$ of $\tau^{-1}(U_1)$ where $U_1$ is an open subset of $G^0$ admissible for the covering. Now $f$ is the lift to $U$ of $f_1\in C_c^\infty(U_1)$ and we calculate $\Theta_{j[\psi]}(f) = \text{trace}\int_U f(x)(\psi\cdot\tau)(x)dx = \text{trace}\int_U f_1(\tau x)dx = \text{trace}\int_{U_1} f_1(x_1)\psi(x_1)dx_1 = \Theta_{[\psi]}(f_1)$. That shows $\Theta_{j[\psi]} = \Theta_{[\psi]}\tau$.

We have shown that if $\widetilde{\varepsilon} : \widehat{\widetilde{G}[\widetilde{\zeta}]}_{1_S} \to \widehat{\widetilde{G}}_{\widetilde{\zeta}}$ satisfies Theorem 3.3.2 then the same is true for $\varepsilon : \widehat{G[\zeta]}_{1_S} \to \widehat{G^0}_\zeta$. This reduces the proof of Theorem 3.3.2 to the case where $Z\subset G^0$ and $G^0$ is simply connected. There $G^0 = V\times G_{ss}$ where $V$ is a vector group and $G_{ss}$ is semisimple. We can enlarge $Z$ to $VZ$ and assume $Z = V\times D$

where $D$ is a subgroup of finite index in the center $Z_{ss}$ of $G_{ss}$. That done, $\zeta = \nu \boxtimes \delta$ accordingly, and $G_{ss} \hookrightarrow G^0$ gives an isomorphism $a : G_{ss}[\delta] \cong G^0[\zeta]$. This results in a commutative diagram $\begin{array}{ccc} \widehat{G_{ss}[\delta]}_{1_S} & \xrightarrow{\varepsilon_{ss}} & (\widehat{G_{ss}})_\delta \\ {}_{a^*}\uparrow & & \downarrow_b \\ \widehat{G^0[\zeta]}_{1_S} & \xrightarrow{\varepsilon} & (\widehat{G^0})_{\zeta_0} \end{array}$ where $b[\pi_{ss}] = [\nu \otimes \pi_{ss}]$. As in the reduction to $Z \subset G^0$, if Theorem 3.3.2 holds for $G_{ss}$ it holds for $G$.

We are finally reduced to the case where $Z \subset G^0$ and $G^0$ is semisimple. Then $Z$ is discrete, so $S \times G^0 \to G^0[\zeta]$ is a Lie group covering. The method of reduction to simply connected $G^0$ completes the proof of Theorem 3.3.2. □

**3.4.** We extend Harish–Chandra's description [**19**] of the discrete series of a connected semisimple Lie group. In fact his analysis extends without change to connected reductive Lie groups with compact center, as in the case of $G^0$ when $Z \cap G^0$ is compact. We formulate the results. Recall that $G^0$ is an arbitrary connected reductive Lie group and that $Z \cap G^0$ is co-compact in $Z_{G^0}$.

THEOREM 3.4.1. *$G^0$ has a relative discrete series representation if and only if $G^0/(Z \cap G^0)$ has a compact Cartan subgroup.*

The compact Cartan subgroups of $G^0/(Z \cap G^0)$ have form $H^0/(Z \cap G^0)$. The unitary characters on $H^0$ are in bijective correspondence with

$$L = \{\lambda \in i\mathfrak{h}^* \mid e^\lambda \text{ is well defined on } H^0. \tag{3.4.2}$$

Choose a positive root system $\Sigma^+$ and use the standard notation $\rho = \frac{1}{2}\sum_{\phi \in \Sigma^+} \phi$, $\varpi(\lambda) = \prod_{\phi \in \Sigma^+} \langle \phi, \lambda \rangle$ and $\Delta = \prod_{\phi \in \Sigma^+}(e^{\phi/2} - e^{-\phi/2})$ where $\langle \cdot, \cdot \rangle$ comes from the Killing form. Then $\Sigma^+ \subset L$ so $2\rho \in L$. Passing to a 2–sheeted covering group of $G^0/(Z \cap G^0)$, if necessary, we may assume $\rho \in L$ and $e^\rho(Z \cap G^0) = 1$. $\Delta$ is an analytic function on $H^0$, well defined on $H^0/(Z \cap G^0)$. The *regular set* is $L' = \{\lambda \in L \mid \varpi(\lambda) \neq 0\}$. Note $\rho \in L'$. If $\lambda \in L'$ we define

$$q(\lambda) = |\{\text{compact } \phi \in \Sigma^+ \mid \langle \phi, \lambda \rangle < 0\}| + |\{\text{ noncompact } \phi \in \Sigma^+ \mid \langle \phi, \lambda \rangle > 0\}|. \tag{3.4.3}$$

Thus $(-1)^{q(\lambda)} = (-1)^q \operatorname{sign} \varpi(\lambda)$.

THEOREM 3.4.4. *If $\lambda \in L'$ there is a unique class $[\pi_\lambda] \in \widehat{G^0}_{disc}$ whose distribution character satisfies $\Theta_{\pi_\lambda}|_{H^0 \cap G'} = (-1)^{q(\lambda)} \frac{1}{\Delta} \sum_{w \in W_{G^0}} \det(w) e^{w\lambda}$. Every class in $\widehat{G^0}_{disc}$ is one of these $[\pi_\lambda]$. Classes $[\pi_\lambda] = [\pi_{\lambda'}]$ precisely when $\lambda' \in W_{G^0}(\lambda)$. With a certain normalization of Haar measure on $G^0$, $[\pi_\lambda]$ has formal degree $|\varpi(\lambda)|$.*

COROLLARY 3.4.5. *$[\pi_\lambda] \in (\widehat{G^0})_{disc}$ has dual $[\pi_\lambda^*] = [\pi_{-\lambda}]$, central character $e^{\lambda - \rho}|_{Z_{G^0}}$ and infinitesimal character $\chi_\lambda$ as in (3.2.5); $\chi_\lambda(Casimir) = ||\lambda||^2 - ||\rho||^2$.*

When $Z_{G^0}$ is compact, Theorems 3.4.1 and 3.4.4 reduce to Harish–Chandra's celebrated results [**19**, Theorems 13 and 16]. We describe the reduction.

Let $[\pi] \in (\widehat{G^0})_\zeta$. By Theorem 3.3.2 there exists $[\psi] \in \widehat{G^0[\zeta]}_{1_S}$ such that $\varepsilon_\zeta[\psi] = [\pi]$. In particular [**19**, Theorem 13] $G^0[\zeta]$ has a compact Cartan subgroup, and it must have form $H^0[\zeta]$ where $H^0$ is a Cartan subgroup of $G^0$. Since $H^0[\zeta]$ is compact, so is $H^0/(Z \cap G^0)$. That proves the "only if" part of Theorem 3.4.1.

Conversely, let $H^0/(Z \cap G^0)$ be a compact Cartan subgroup of $G^0/(Z \cap G^0)$ and $\zeta \in \widehat{Z \cap G^0}$, so $H^0[\zeta]$ is a compact Cartan subgroup of $G^0[\zeta]$. Denote
$L[\zeta] = \{\nu \in i\mathfrak{h}[\zeta]^* \mid e^\nu \text{ is well defined on } H^0[\zeta]\}$,

$L[\zeta]'_{1_S} = \{\nu \in L[\zeta] \mid \varpi(\nu) \neq 0 \text{ and } e^\nu|_S = 1_S\}$, and
$L'_\zeta = \{\lambda \in L \mid \varpi(\lambda) \neq 0 \text{ and } e^\lambda|_{Z\cap G^0} = \zeta\}$.
Since $G^0[\zeta]$ is a connected reductive Lie group with compact center, and $e^\rho(S) = 1$, [**19**, Theorem 16] gives a map $\omega_1 : L[\zeta]'_{1_S} \to \widehat{G^0[\zeta]}_{1-disc}$, by $\nu \mapsto [\psi_\nu]$, that satisfies the assertions of Theorem 3.4.4. We construct the corresponding $\omega_\zeta$ : $L'_\zeta \to \widehat{G^0}_{\zeta-disc}$ by $\omega_\zeta \cdot \delta = \varepsilon \cdot \omega_1$ where $\varepsilon$ is the bijection of Theorem 3.3.2 and $\delta : L[\zeta]'_{1_S} \to L'_\zeta$ is defined as follows.

Let $\nu \in L[\zeta]'_{1_S}$. The distribution character of $\varepsilon\omega_1(\nu) = \omega_\zeta\delta(\nu)$ must have $(H^0 \cap G')$–restriction $(-1)^{q(\delta\nu)}\frac{1}{\Delta}\sum \det(w)e^{w\delta\nu} = (-1)^{q(\nu)}\frac{1}{\Delta}\sum \det(w)e^{w\nu}\cdot p$. For that, define $\delta$ by $e^{\delta\nu} = e^\nu \cdot p$, i.e. $\delta\nu = p^*\nu$ under $p : \mathfrak{g} \to \mathfrak{g}[\zeta]$. Since $p$ restricts to an isomorphism of derived algebras, $\delta$ bijects $L[\zeta]'_{1_S}$ to $L'_\zeta$ equivariantly for $W$. Our assertions now go over from $\omega_1$ to $\omega_\zeta$ .

That completes the derivation of Theorems 3.4.1 and 3.4.4 from [**19**].

**3.5.** We extend the description of the relative discrete series from connected reductive groups to the class of real reductive Lie groups specified in §3.1.

LEMMA 3.5.1. *$ZZ_{G^0}$ has finite index in $Z_G(G^0)$. Every class $[\chi] \in \widehat{Z_G(G^0)}$ has dimension* $\dim\chi \leqq |Z_G(G^0)/ZZ_{G^0}|$.

PROOF. The second condition of Definition 3.1.1 shows that every $[\chi] \in \widehat{Z_G(G^0)}$ is a summand of $\operatorname{Ind}_{\widetilde{Z}}^{Z_G(G^0)}(\beta)$ for some $\beta \in \widehat{\widetilde{Z}}$. □

PROPOSITION 3.5.2. *$\widehat{G^\dagger}$ is the disjoint union of the sets*

$$(\widehat{G^\dagger})_\xi = \{[\chi\otimes\pi] \mid [\chi] \in \widehat{Z_G(G^0)}_\xi \text{ and } [\pi] \in (\widehat{G^0})_\xi \text{ where } \xi \in \widehat{Z_{G^0}}\}. \quad (3.5.3)$$

*Here $[\chi\otimes\pi]$ has the same infinitesimal character $\chi_\pi$ and $[\pi]$ and has distribution character $\Theta_{\chi\otimes\pi}(zg) = (\text{trace } \chi(z))\Theta_\pi(g)$ for $z \in Z_G(G^0)$ and $g \in G^0$. Further, $[\chi\otimes\pi] \in (\widehat{G^\dagger})_{disc}$ if and only if $[\pi] \in (\widehat{G^0})_{disc}$.*

PROOF. $\widehat{G^\dagger}$ is the disjoint union of the $(\widehat{G^\dagger})_\xi$ , $\xi \in \widehat{Z_{G^0}}$, because $Z_{G^0}$ is central in $G^\dagger$. Now fix $\xi \in \widehat{Z_{G^0}}$, $[\chi] \in \widehat{Z_G(G^0)}_\xi$ and $[\pi] \in (\widehat{G^0})_\xi$. Note $[\chi\otimes\pi] \in (\widehat{G^\dagger})_\xi$ , and $(\chi\otimes\pi)|_{G^0} = m\pi$ where $m = \dim\chi < \infty$. Thus $\chi\otimes\pi$ has infinitesimal character $\chi_\pi$ , and is discrete relative to $Z$ exactly when $\pi$ is discrete relative to $Z\cap G^0$ .

To prove the formula for $\Theta_{\chi\otimes\pi}$ we need only consider test functions $f \in C_c^\infty(G^\dagger)$ supported in a single coset $z_0G^0$ , and there we compute
$\text{trace }(\chi\otimes\pi)(f) = \text{trace } \int_{G^\dagger} f(zg)\chi(z)\otimes\pi(g)dg = \text{trace } \int_{G^0} f(z_0g)\chi(z_0)\otimes\pi(g)dg$
$= \text{trace }\big(\chi(z_0)\otimes\int_{G^0} f(z_0g)\pi(g)dg\big) = (\text{trace } \chi(z_0))\big(\text{trace } \int_{G^0} f(z_0g)\pi(g)dg\big)$
$= (\text{trace } \chi(z_0))\int_{G^0} f(z_0g)\Theta_\pi(g)dg = \int_{G^0} f(z_0g)(\text{trace } \chi(z_0))\Theta_\pi(g)dg$
$= \int_{G^\dagger} f(zg)(\text{trace } \chi(z))\Theta_\pi(g)d(zg)$, so $\Theta_{\chi\otimes\pi}(zg) = (\text{trace } \chi(z))\Theta_\pi(g)$, as asserted.

Finally let $[\gamma] \in (\widehat{G^\dagger})_\xi$. Since $Z_G(G^0)$ acts trivially on $\widehat{G^0}$ and $G^0$ is of type I, now $\gamma|_{G^0} = mn$ where $[\chi] \in \widehat{Z_G(G^0)}_\xi$. Thus $[\gamma] = [\chi\otimes\pi]$ because $[\chi\otimes\pi]$ is a subrepresentation. That proves (3.5.3). Proposition 3.5.2 is proved. □

Proposition 3.5.2 gives the relative discrete series of $G^\dagger$ in terms of those of $Z_G(G^0)$ and $G^0$. The following lets us go on to $G$.

PROPOSITION 3.5.4. *Let* $[\gamma] = [\chi \otimes \pi] \in \widehat{G^\dagger}$ *and define* $\psi = \operatorname{Ind}_{G^\dagger}^G(\psi)$. *Then* (1) $[\psi]$ *has the same infinitesimal character* $\chi_\pi$ *as* $[\pi]$ *and* (2) $[\pi]$ *has distribution character that is a locally* $L_1$ *function supported in* $G^\dagger$ *and given there by*

$$\Theta_\psi(zg) = \sum\nolimits_{xG^\dagger \in G/G^\dagger} (\text{trace } \chi(x^{-1}zx))\Theta_\pi(x^{-1}gx) \tag{3.5.5}$$

*for* $z \in Z_G(G^0)$ *and* $g \in G^0$. *In particular* $\Theta_\psi$ *is analytic on the regular set* $G'$ *and satisfies* $\Theta_\psi|_{G^\dagger} = \Theta|_{\psi|_{G^\dagger}}$. *Further* (3) *if* $[\pi] \in (\widehat{G^0})_{disc}$ *then* $[\psi] \in \widehat{G}_{disc}$, *and every class in* $\widehat{G}_{disc}$ *is obtained this way.*

PROOF. We follow an argument [**49**, Lemma 4.3.3] of Frobenius for (1) and (2). As $G^\dagger$ is normal and has finite index in $G$, $\Theta_\psi$ exists and is supported in $G^\dagger$, where $\Theta_\psi|_{G^\dagger} = \Theta|_{\psi|_{G^\dagger}}$. Note $\psi|_{G^\dagger} = \sum_{G/G^\dagger} \gamma \cdot \operatorname{Ad}(x^{-1})$. If $z \in Z_G(G^0)$ and $g \in G^0$ now $\Theta_\psi(zg) = \sum \Theta_{\gamma \cdot \operatorname{Ad}(x^{-1})}(zg) = \sum \Theta_\gamma(z^{-1}zx \cdot x^{-1}gx)$. Assertion (2) follows from Proposition 3.5.2.

If $x \in G$ then $\operatorname{Ad}(x)$ is an inner automorphism on $\mathfrak{g}_\mathbb{C}$, hence trivial on $\mathcal{Z}(\mathfrak{g})$, so all the $\gamma \cdot \operatorname{Ad}(x^{-1})$ are the same on $\mathcal{Z}(\mathfrak{g})$. Now $\psi$ has infinitesimal character $\chi_\psi = \chi_\gamma = \chi_\pi$.

Every class in $\widehat{G}_{disc}$ is a subrepresentation of an $[\operatorname{Ind}_{\widehat{G^\dagger}}^G(\gamma)]$, $[\gamma] \in (\widehat{G^\dagger})_{disc}$, because $|G/G^\dagger| < \infty$. If $[\gamma] = [\chi \otimes \pi]$ as in Proposition 3.5.2 then $[\gamma] \in (\widehat{G^\dagger})_{disc}$ is equivalent to $[\pi] \in (\widehat{G^0})_{disc}$. To prove (3) now we need only check that $\psi = \operatorname{Ind}_{G^\dagger}^G(\gamma)$ is irreducible whenever $[\pi] \in (\widehat{G^0})_{disc}$.

Choose a Cartan subgroup $H^0 \subset G^0$ with $H^0/(Z \cap G^0)$ compact. The corresponding Cartan subgroup of $G$ is the centralizer $H$ of $\mathfrak{h}$. Hypothesis (3.1.1) says that the Weyl group $W_G$ is a subgroup of the complex Weyl group $W(\mathfrak{g}_\mathbb{C}, \mathfrak{h}_\mathbb{C})$. As any two compact Cartan subgroups of $G^0/(Z \cap G^0)$ are conjugate we have a system $\{x_1, \dots, x_r\}$ of representatives of $G$ modulo $G^\dagger$ such that each $\operatorname{Ad}(x_i)\mathfrak{h} = \mathfrak{h}$. Now $W_G = \bigcup (x_j H) W_{G^0} \subset W(\mathfrak{g}_\mathbb{C}, \mathfrak{h}_\mathbb{C})$. Let $[\gamma] = [\chi \otimes \pi]$ with $[\pi] \in (\widehat{G^0})_{disc}$. Express $[\pi] = [\pi_\lambda]$ with $\lambda \in L'$. Then $[\pi \cdot \operatorname{Ad}(x_j^{-1})] = [\pi_{\lambda_j}]$ where $\lambda_j = \operatorname{Ad}(x_j^{-1})^*(\lambda)$. Since $\lambda \in L'$ the $\lambda_j$ are distinct modulo the action of $W_{G^0}$. Theorem 3.4.4 now says that the $\pi \cdot \operatorname{Ad}(x_j^{-1})$ are mutually inequivalent. It follows that $\psi$ is irreducible. □

We formulate the extensions of Theorems 3.4.1 and 3.4.4 from $G^0$ to $G$.

THEOREM 3.5.6. $G$ *has a relative discrete series representation if and only if* $G/Z$ *has a compact Cartan subgroup.*

Let $H/Z$ be a compact Cartan subgroup of $G/Z$. Retain the notation of §3.4 for $(\widehat{G^0})_{disc}$ and the notation $G = \bigcup x_j G^\dagger$ where the $x_j$ normalize $H^0$. Write $w_j$ for the element of $W_G$ represented by $x_j$.

THEOREM 3.5.7. *Let* $\lambda \in L'$ *and* $[\chi] \in \widehat{Z_G(G^0)}_\xi$ *where* $\xi = e^{\lambda - \rho}|_{Z_{G^0}}$. *Let* $[\pi] \in (\widehat{G^0})_{disc}$ *as in* Theorem 3.4.4. *Then* $[\pi_{\chi,\lambda}] := [\operatorname{Ind}_{G^\dagger}^G(\chi \otimes \pi_\lambda)]$ *is the unique class in* $\widehat{G}_{disc}$ *whose distribution character satisfies*

$$\Theta_{\pi_{\chi,\lambda}}(zh) = \sum_{1 \leqq j \leqq r} (-1)^{q(w_j\lambda)} \text{trace } \chi(x_j^{-1}zx_j) \cdot \sum_{w \in W_{G^0}} \det(ww_j) e^{ww_j\lambda}(h) \tag{3.5.8}$$

*for* $z \in Z_G(G^0)$ *and* $h \in H^0 \cap G'$. *Every class in* $\widehat{G}_{disc}$ *is one of these* $[\pi_{\chi,\lambda}]$. *Classes* $[\pi_{\chi,\lambda}] = [\pi_{\chi',\lambda'}]$ *precisely when* $([\chi'], \lambda') \in W_G([\chi], \lambda)$. *For appropriate* *normalizations of Haar measures the formal degree* $\deg(\pi_{\chi,\lambda}) = r \cdot \dim(\chi) \cdot |\varpi(\lambda)|$. *Finally,* $[\pi_{\chi,\lambda}]$ *has dual* $[\pi_{\chi^*,-\lambda}]$ *and has infinitesimal character* $\chi_\lambda$ *as in* (3.2.5), *so in particular* $\chi_{[\pi_{\chi,\lambda}]}(Casimir) = ||\lambda||^2 - ||\rho||$.

PROOF. If $\widehat{G}_{disc}$ is nonempty then Propositions 3.5.2 and 3.5.4 show that $(\widehat{G^0})_{disc}$ is not empty, so $G^0/(Z \cap G^0)$ has a compact Cartan subgroup by Theorem 3.4.1. As $G^0/(Z \cap G^0)$ has finite index in $G/Z$ the latter also has a compact Cartan subgroup. If $H/Z$ is a compact Cartan subgroup of $G/Z$ then Theorem 3.5.7 follows directly from Theorem 3.4.4 and Propositions 3.5.2 and 3.5.4. □

## 4. Tempered Series Representations of Reductive Lie Groups

$G$ is a reductive Lie group of the class described in §3.1. In §3 we used the conjugacy class of Cartan subgroups $H$ of $G$, with $H/Z$ compact, to construct the relative discrete series $\widehat{G}_{disc}$. Here we construct a series of unitary representations for every conjugacy class of Cartan subgroups of $G$.

In §4.1 and 4.2 we work out the relation between Cartan involutions $\theta$ of $G$, Cartan subgroups $H$ of $G$, and cuspidal parabolic subgroups $P = MAN$ of $G$. Here $H = T \times A$, $T/Z$ compact and $A$ split/$\mathbb{R}$, and $Z_G(A) = M \times A$ where $M$ is in the class of §3.2 and $T$ is a Cartan subgroup of $M$. Then $H \mapsto P$ gives a bijection from the set of all conjugacy classes of Cartan subgroups of $G$ to the set of all "association classes" of cuspidal parabolic subgroups of $G$.

In §4.3 we describe these representations, calculating infinitesimal and distribution characters. They form the "$H$–series" $\widehat{G}_H$. Then in §4.5 we examine the correspondence $\widehat{M}_{disc} \to \widehat{G}_H$. Finally in §4.5 we look at questions of irreducibility.

**4.1.** $G$ is a real reductive Lie group as in §3.1, $\mathfrak{h}$ is a Cartan subalgebra of $\mathfrak{g}$, and $H = Z_G(\mathfrak{h})$ is the corresponding Cartan subgroup of $G$. If $G^0$ is a linear group, or if $H/Z$ is compact, then $H \cap G^0$ is commutative. In general one only knows that $H^0 = \exp(\mathfrak{h})$ is commutative. We collect some information.

LEMMA 4.1.1. *If* $K/Z$ *is a maximal compact subgroup of* $G/Z$ *then there is a unique involutive automorphism* $\theta$ *of* $G$ *with fixed point set* $K$. *These automorphisms* $\theta$ *are the "Cartan involutions" of* $G$, *and any two are* $\mathrm{Ad}(G^0)$*–conjugate. Every Cartan subgroup of* $G$ *is stable under a Cartan involution.*

LEMMA 4.1.2. *If* $K/Z$ *is a maximal compact subgroup of* $G/Z$ *then* $K^0 = K \cap G^0$, $K$ *meets every component of* $G$, *and* $K = \{g \in G \mid \mathrm{Ad}(g)K^0 = K^0\}$.

These lemmas are standard when $Z = \{1\}$ and $G$ is either linear or semisimple.

PROOF. $Z_{G^0} \subset K$ and $(K \cap G^0)/(Z_{G^0})^0$ is connected, is its own $G^0/(Z_{G^0})^0$–normalizer, and is unique up to conjugacy. The same follows for $K \cap G^0$ in $G^0$. Let $E = \{g \in G \mid \mathrm{Ad}(g)K^0 = K^0\}$; now $E \cap G^0 = K^0 = K \cap G^0$. If $g \in G$ now some $g' \in G^0$ send $\mathrm{Ad}(g)K^0$ to $K^0$, so $E$ meets $gG^0$. Now $K \subset E \subset K$ and Lemma 4.1.2 follows. For Lemma 4.1.1 each simple ideal $\mathfrak{g}_i \subset \mathfrak{g}$ has a unique involution $\theta_i$ with fixed point set $\mathfrak{g}_i \cap \mathfrak{k}$, and we define $\theta$ as their sum with the identity map on the center of $\mathfrak{g}$. Then $\theta$ extends uniquely to the universal cover of $G^0$, and there its fixed point set $\exp(\mathfrak{k})$ contains the center, so $\theta$ extends uniquely to $G^0$ with fixed point set $K^0$. Now $\theta$ extends uniquely to $G = KG^0$ with fixed point set $K$, using Lemma 4.1.2. As any two choices of $K/Z$ are $\mathrm{Ad}(G^0)$–conjugate, that completes the proof of the first statement of Lemma 4.1.1. For the second just note that any two choices of $\mathfrak{k}$ are $\mathrm{Ad}(G^0)$–conjugate. □

Now fix the data: a Cartan subgroup $H$ of $G$, a Cartan involution $\theta$ of $G$ with $\theta(H) = H$, and $K = G^\theta$ fixed point set of $\theta$. We decompose

$$(4.1.3)\qquad \begin{aligned} &\mathfrak{h} = \mathfrak{t} + \mathfrak{a} \text{ into } (\pm 1)\text{–eigenspaces of } \theta|_{\mathfrak{h}} \\ &H = T \times A \text{ where } T = H \cap K \text{ and } A = \exp(\mathfrak{a}). \end{aligned}$$

The $\mathfrak{a}$–*root spaces* of $\mathfrak{g}$ are the $\mathfrak{g}^\phi = \{\xi \in \mathfrak{g} \mid [\alpha, \xi] = \phi(\alpha)\xi \text{ for } \alpha \in \mathfrak{a}\}$ with $0 \neq \phi \in \mathfrak{a}^*$ and $\mathfrak{g}^\phi \neq 0$. The $\mathfrak{a}$–roots are these functionals $\phi$, and $\Sigma_\mathfrak{a}$ denotes the set of all $\mathfrak{a}$–roots. The corresponding $\mathfrak{a}$–root decomposition is $\mathfrak{g} = \mathfrak{z}_\mathfrak{g}(\mathfrak{a}) + \sum_{\phi \in \Sigma_\mathfrak{a}} \mathfrak{g}^\phi$ where $\mathfrak{z}_\mathfrak{g}(\mathfrak{a})$ is the centralizer of $\mathfrak{a}$ in $\mathfrak{g}$. Then it is not too difficult to see that the centralizer $Z_G(A)$ has a unique splitting $Z_G(A) = M \times A$ with $\theta(M) = M$. In particular $\mathfrak{g} = \mathfrak{m} + \mathfrak{a} + \sum_{\phi \in \Sigma_\mathfrak{a}} \mathfrak{g}^\phi$ with $\theta(\mathfrak{m}) = \mathfrak{m}$. The hereditary properties of §3.1 pass down from $G$ to $M$ as follows.

PROPOSITION 4.1.4. *$M$ inherits the conditions of* §3.1 *from $G$: every* $\mathrm{Ad}(m)$ *is inner on $\mathfrak{m}_\mathbb{C}$, $Z$ centralizes $M^0$, $|M/ZM^0| < \infty$, and $Z_{M^0}/(Z \cap M^0)$ is compact. Further, $T/Z$ is a compact Cartan subgroup of $M/Z$.*

The proof of Proposition 4.1.4 requires some information on the $\mathfrak{a}$–root system.

Every $\phi \in \Sigma_\mathfrak{a}$ defines $\phi^\perp := \{\alpha \in \mathfrak{a} \mid \phi(\alpha) = 0\}$. The complement $\mathfrak{a} \setminus \bigcup_{\Sigma_\mathfrak{a}} \phi^\perp$ is a finite union of convex open cones, its topological components, the *Weyl chambers*. A Weyl chamber $\mathfrak{d} \subset \mathfrak{a}$ defines a *positive root system* $\Sigma_\mathfrak{a}^+ = \{\phi \in \Sigma_\mathfrak{a} \mid \phi(\mathfrak{d}) \subset \mathbb{R}^+\}$.

LEMMA 4.1.5. *If $\Sigma_\mathfrak{a}^+$ is a positive $\mathfrak{a}$–root system on $\mathfrak{g}$ and $\Sigma_\mathfrak{t}^+$ is a positive $\mathfrak{t}_\mathbb{C}$–root system on $\mathfrak{m}_\mathbb{C}$ then there is a unique positive $\mathfrak{h}_\mathbb{C}$–root system $\Sigma^+$ on $\mathfrak{g}_\mathbb{C}$ such that $\Sigma_\mathfrak{a}^+ = \{\gamma|_\mathfrak{a} \mid \gamma \in \Sigma^+ \text{ and } \gamma|_\mathfrak{a} \neq 0\}$ and $\Sigma_\mathfrak{t}^+ = \{\gamma|_\mathfrak{t} \mid \gamma \in \Sigma^+ \text{ and } \gamma|_\mathfrak{a} = 0\}$.*

PROOF. Choose ordered bases $\beta_\mathfrak{a}$ of $\mathfrak{a}^*$ and $\beta_\mathfrak{t}$ of $i\mathfrak{t}^*$ whose associated lexicographic orders give $\Sigma_\mathfrak{a}^+$ and $\Sigma_\mathfrak{t}^+$. Then the ordered basis $\{\beta_\mathfrak{a}, \beta_\mathfrak{t}\}$ of $\mathfrak{a}^* + i\mathfrak{t}^*$ gives a lexicographic order whose associated positive $\mathfrak{h}_\mathbb{C}$–root system $\Sigma^+$ has the required properties. Uniqueness of $\Sigma^+$ is similarly straightforward. □

PROOF OF PROPOSITION 4.1.4. $H \subset M \times A$ is a Cartan subgroup of $G$, hence also of $M \times A$, so $T$ is a Cartan subgroup of $M$. $T/Z$ is compact because $K/Z$ is compact.

Let $m \in M$ with $\mathrm{Ad}(m)$ outer on $\mathfrak{m}_\mathbb{C}$. We may move $m$ within $mM^0$ and assume $\mathrm{Ad}(m)\mathfrak{t} = \mathfrak{t}$. As the Weyl group of $(M^0, T^0)$ is simply transitive on the Weyl chambers in $i\mathfrak{t}$, $\mathrm{Ad}(m)$ preserves and acts nontrivially on some $\Sigma_\mathfrak{t}^+$. Choose $\Sigma_\mathfrak{a}^+$; now $\mathrm{Ad}(m)$ preserves and acts nontrivially on the corresponding positive $\mathfrak{h}_\mathbb{C}$–root system of $\mathfrak{g}_\mathbb{C}$, in other words $\mathrm{Ad}(m)$ is outer on $\mathfrak{g}_\mathbb{C}$. That contradicts (3.1.1). Thus $\mathrm{Ad}(m)$ is inner on $\mathfrak{m}_\mathbb{C}$. In particular $M^\dagger := \{m \in M \mid \mathrm{Ad}(m) \text{ inner on } M^0\}$ has finite index in $M$ and $M^\dagger = TM^0$. Now $T/Z$ is a compact subgroup of $M/Z$ such that $M/T = (M/Z)/(T/Z)$ has only finitely any components. Thus $|M/ZM^0| < \infty$. The center of $M^0/(Z \cap M^0)$ is compact because it is a closed subgroup in the torus $T^0/(Z \cap M^0)$, so that center is $Z_{M^0}/(Z \cap M^0)$. □

**4.2.** We apply the considerations of §4.1 to study cuspidal parabolic subgroups.

Retain the splittings (4.1.3) and $Z_G(A) = M \times A$. A positive $\mathfrak{a}$–root system $\Sigma_{\mathfrak{a}}^+$ on $\mathfrak{g}$ defines

$$\mathfrak{n} = \sum_{\phi \in \Sigma_{\mathfrak{a}}^+} \mathfrak{g}^{-\phi} \text{ nilpotent subalgebra of } \mathfrak{g}\,, N = \exp(\mathfrak{n}), P = \{g \in G \mid \mathrm{Ad}(g)N = N\}.$$

LEMMA 4.2.1. *$P$ is a real parabolic subgroup of $G$. It has unipotent radical $P^{unip} = N$ and Levi (reductive) part $P^{red} = M \times A$. Thus $(m, a, n) \mapsto man$ is a real analytic diffeomorphism of $MAN$ onto $P$.*

PROOF. We follow the idea of the proof for the case where $G = G^\dagger$ and $G$ is linear, which follows from the complex case. Let $\pi : G \to \overline{G} = G/ZZ_{G^0}$. That is a real linear algebraic group, so $\pi(P)$ is a parabolic subgroup of $\overline{G}$, normalizer of $\pi(N)$, and $\pi(N) = \pi(P)^{unip}$.

Note $ZZ_{G^0} \subset MA \subset P$. Thus (i) we can choose $\pi(P)^{red}$ to contain $\pi(MA)$ and (ii) $P = P^{red} \cdot N$ semidirect where $P^{red} = \pi^{-1}(\pi(P)^{red})$. Now $MA \subset P^{red}$ and, by dimension, $M^0 A = (P^{red})^0$. $A$ is normal in $P^{red}$ by uniqueness from (4.1.3).

Let $V = \{w \in W(\mathfrak{g}_{\mathbb{C}}, \mathfrak{h}_{\mathbb{C}}) \mid w(\mathfrak{a}) = \mathfrak{a}\,, w|_{\mathfrak{a}} \neq 1\}$. Choose $x \in \mathfrak{a} \setminus \bigcup_{v \in V} \{y \in \mathfrak{a} \mid v(y) = y\}$. Let $g \in P^{red}$. Then $x, \mathrm{Ad}(g)x$ are conjugate by an inner automorphism of $\mathfrak{g}_{\mathbb{C}}$, thus [**39**, Theorem 2.1] conjugate by an inner automorphism on $\mathfrak{g}$. Now $\mathrm{Ad}(g')x = x$ for some $g' \in gG^0$, and $\mathrm{Ad}(g')M = M$. We may assume $\mathrm{Ad}(g')H = H$. Now $g' \in MA$. Thus $P^{red} = MA(P^{red} \cap G^0)$. As $P^{red} \cap G^0 \in Z_G(A)$ now $P^{red} = MA$ as asserted. □

We say that two parabolics in $G$ are *associated* if their reductive parts are $G$–conjugate. Thus the association class of $P = MAN$ is independent of $N$, i.e. is independent of $\Sigma_{\mathfrak{a}}^+$. We say that a parabolic $Q \subset G$ is *cuspidal* if $[(Q^{red})^0, (Q^{red})^0]$ has a Cartan subgroup $E$ such that $E/(E \cap Z_{G^0})$ is compact.

PROPOSITION 4.2.2. *Let $Q$ be a parabolic subgroup of $G$. Then the following are equivalent.*

(i) *$Q$ is a cuspidal parabolic subgroup of $G$.*

(ii) *There exist a Cartan subgroup $H = T \times A$ of $G$ and a positive $\mathfrak{a}$–root system $\Sigma_{\mathfrak{a}}^+$ such that $Q$ is the group $P = MAN$ of* Lemma 4.2.1.

(iii) *$Q^{red}$ has a relative discrete series representation.*

(iv) *$(Q^{red})^0$ has a relative discrete series representation.*

*In particular, the construction $H \mapsto P = MAN$ induces a bijection from the set of all conjugacy classes of Cartan subgroups of $G$ onto the set of all association classes of cuspidal parabolic subgroups of $G$.*

PROOF. Let $\pi : G \to \overline{G} = G/ZZ_{G^0}$ as before. Note $ZZ_{G^0} \subset Q^{red}$. Thus (i), (ii), (iii) each holds for $Q$ exactly when it holds for $\pi(Q)$. Also, (iii) and (iv) are equivalent because $Q^{red}/ZZ_{G^0}$ has only finitely many components. Thus we need only prove the equivalence of (i), (ii) and (iv) when $G$ is a connected centerless semisimple group.

Decompose $Q^{red} = M_Q \times A_Q$, stable under $\theta$, where $A_Q$ is the $\mathbb{R}$–split component of the center of $Q^{red}$. Thus $Q$ is cuspidal if and only if $M_Q$ has a compact Cartan subgroup $T_Q$. That is the case just when $G$ has a Cartan subgroup $H = T_Q \times A_Q$ from which $Q$ is constructed as in Lemma 4.2.1. Thus (i) and (ii) are equivalent. Apply Theorem 3.4.1 to $(Q^{red})^0$. Then (ii) implies (iv) by Proposition 4.1.4 and (iv) implies (i) directly. Now (i), (ii), (iii) and (iv) are equivalent, and the bijection statement follows. □

Two Cartan subalgebras of $\mathfrak{g}$ are conjugate by an inner automorphism of $\mathfrak{g}$ precisely when they are conjugate by an inner automorphism of $\mathfrak{g}_{\mathbb{C}}$ [**39**, Corollary 2.4]. By Proposition 4.2.2 the same holds for association classes of cuspidal parabolic subgoups of $G$. Thus we could use $G^0$–conjugacy, $G^0$–association, or both, in the bijection of Proposition 4.2.2.

**4.3.** We define a series of unitary representations of $G$ for each conjugacy class of Cartan subgroups. Then we work out some generalities on the character theory for that series. The precise character theory is in §4.4.

Retain the notation of §§4.1 and 4.2, including $H = T \times A$, $Z_G(A) = M \times A$ and $P = MAN$. The general unitary equivalence class in $\widehat{P^{red}} = \widehat{M \times A}$ has form $[\eta \otimes e^{i\sigma}]$ where $[\eta] \in \widehat{M}$ and $\sigma \in \mathfrak{a}^*$. That extends to a class $[\eta \otimes e^{i\sigma}] \in \widehat{P}$ that annihilates $N$: $(\eta \otimes e^{i\sigma})(man) = e^{i\sigma}(a)\eta(m)$. Then we have the unitarily induced representation

$$\pi_{\eta,\sigma} = \mathrm{Ind}_P^G(\eta \otimes e^{i\sigma}) \tag{4.3.1}$$

of $G$. The *$H$–series* of $G$ is $\{[\pi_{\eta,\sigma}] \mid [\eta] \in \widehat{M}_{disc} \text{ and } \sigma \in \mathfrak{a}^*\}$. If $H/Z$ is compact then $M = G$ and the $H$–series is just the relative discrete series $\widehat{G}_{disc}$. If $H/ZZ_{G^0}$ is maximally $\mathbb{R}$–split, i.e. if $P$ is a minimal parabolic subgroups of $G$, then the $H$–series is the principal series. We refer to any $H$–series as a *nondegenerate series.* Later we will see that the $H$–series depends only on the conjugacy class of $H$.

Given $\Sigma_{\mathfrak{a}}^+$ define $\rho_{\mathfrak{a}} = \frac{1}{2}\sum_{\phi\in\Sigma_{\mathfrak{a}}^+}(\dim \mathfrak{g}^\phi)\phi$. Then $\mathfrak{a}$ acts (under the adjoint representation of $\mathfrak{g}$) on $\mathfrak{n}$ and on $\mathfrak{p}$ with trace $-2\rho_{\mathfrak{a}}$. Thus $P = MAN$ has modular function $\delta_P$, $\int_P f(xy^{-1})dx = \delta_P(y)\int_P f(x)dx$, given by

$$\delta_P(man) = e^{2\rho_{\mathfrak{a}}}(a) \text{ for } m \in M,\ a \in A,\ n \in N. \tag{4.3.2}$$

Let $[\eta] \in \widehat{M}$ with representation space $E_\eta$, and let $\sigma \in \mathfrak{a}^*$. Then we have the Hilbert space bundle $\mathbb{E}_{\eta,\sigma} \to G/P = K/(K \cap M)$ associated to the non–unitary representation $\eta \otimes e^{\rho_{\mathfrak{a}}+i\sigma}$ of $P$. Here $G$ acts on the bundle but the hermitian metric is invariant only under $K$. We have the $K$–invariant probability measure $d(kZ)$ on $G/P = K/(K \cap Z)$. Thus we have a well defined space of square integrable sections of $\mathbb{E}_{\eta,\sigma} \to G/P$ given by

$$\begin{aligned} L_2(G/P;\mathbb{E}_{\eta,\sigma}) = \ & \text{all Borel–measurable } f : G \to E_\eta \text{ such that} \\ & f(gp) = (\eta \otimes e^{\rho_{\mathfrak{a}}+i\sigma})(p)^{-1}f(g) \text{ and } \int_{K/Z} ||f(k)||^2 d(kZ) < \infty. \end{aligned} \tag{4.3.3}$$

It is a Hilbert space with inner product $\langle f, f'\rangle = \int_{K/Z}\langle f(k), f'(k)\rangle d(kZ)$, and $G$ acts unitarily on it by the representation $(\pi_{\eta,\sigma}(g)(f))(g') = f(g^{-1}g')$ of (4.3.1).

**4.4.** We now describe the distribution character $\Theta_{\pi_{\eta,\sigma}}$ of $\pi_{\eta,\sigma}$ in terms of the character $\Psi_\eta$ of $\eta$. This is based on a minor variation $C_c^\infty(G) \to C_c^\infty(MA)$ of the Harish–Chandra transform.

Let $J$ be a Cartan subgroup of $G$. Then these are equivalent: (i) $J \subset MA$, (ii) $J$ is a Cartan subgroup of $MA$, and (iii) $J = J_M \times A$ where $J_M = J \cap M$ is a Cartan subgroup of $M$. Without loss of generality we may assume $J_M$ stable under the Cartan involution $\theta|_M$ of $M$. Choose a positive $(\mathfrak{j}_M)_{\mathbb{C}}$–root system $\Sigma_{\mathfrak{j}_M}^+$ on $\mathfrak{m}_{\mathbb{C}}$. As in Lemma 4.1.5 there is a unique positive $\mathfrak{j}_{\mathbb{C}}$–root system $\Sigma_{\mathfrak{j}}^+$ on $\mathfrak{g}_{\mathbb{C}}$ such that

$\Sigma_{\mathfrak{a}}^+ = \{\gamma|_{\mathfrak{a}} \mid \gamma \in \Sigma_{\mathfrak{j}}^+ \text{ and } \gamma|_{\mathfrak{a}} \neq 0\}$ and $\Sigma_{\mathfrak{j}_m}^+ = \{\gamma|_{\mathfrak{j}_M} \mid \gamma \in \Sigma_{\mathfrak{j}}^+ \text{ and } \gamma|_{\mathfrak{a}} = 0\}$. Then let $\rho_{\mathfrak{j}} = \frac{1}{2}\sum_{\gamma\in\Sigma_{\mathfrak{j}}^+}\gamma$; so $\rho_{\mathfrak{a}} = \rho_{\mathfrak{j}}|_{\mathfrak{a}}$. Now define

$$(4.4.1) \quad \Delta_{G,J} = \prod\nolimits_{\gamma\in\Sigma_{\mathfrak{a}}^+}(e^{\gamma/2} - e^{-\gamma/2}) \text{ and } \Delta_{M,J_M} = \prod\nolimits_{\phi\in\Sigma_{\mathfrak{j}_M}^+}(e^{\phi/2} - e^{-\phi/2}).$$

LEMMA 4.4.2. *If $\gamma$ is a $\mathfrak{j}_{\mathbb{C}}$ root then $e^\gamma$ is well defined on $J$, unitary on $J\cap K$, and $e^\gamma(Z_G(G^0)) = 1$. If $\phi$ is a $(\mathfrak{j}_M)_{\mathbb{C}}$-root, say $\phi = \gamma|_{\mathfrak{j}_M}$, then $e^\phi = e^\gamma|_{J\cap M}$.*

LEMMA 4.4.3. *We can replace $Z$ by a subgroup of index $\leqq 2$, or replace $G$ by a $\mathbb{Z}_2$ extension, so that the following holds. If $L$ is any Cartan subgroup of $G$ then for any positive $\mathfrak{l}_{\mathbb{C}}$–root system (i) $e^{\rho_{\mathfrak{l}}}$ is well defined on $L$ with $e^{\rho_{\mathfrak{l}}}(Z) = 1$, and (ii) $\Delta_{G,L}$ is a well defined analytic function on $L$. In particular, then, $e^{\rho_{\mathfrak{j}}}$ and $\Delta_{G,J}$ are well defined on $J$, so $e^{\rho_{\mathfrak{j}_M}}$ and $\Delta_{M,J_M}$ are well defined on $J_M$.*

PROOF OF LEMMAS. If $\gamma$ is a $\mathfrak{j}_{\mathbb{C}}$ root then $e^\gamma$ is well defined on the Cartan subgroup $(J/Z_G(G^0))_{\mathbb{C}}$ of the inner automorphism group $\mathrm{Int}(\mathfrak{g}_{\mathbb{C}})$, because that Cartan is connected. Lemma 4.4.2 follows.

With the adjustments of Lemma 4.4.3 we can factor $\mathrm{Ad} : G \to \mathrm{Int}(\mathfrak{g}_{\mathbb{C}})$ as $G \to G/Z \to Q \to \mathrm{Int}(\mathfrak{g}_{\mathbb{C}})$ where $q : Q \to \mathrm{Int}(\mathfrak{g}_{\mathbb{C}})$ is a 1– or 2–sheeted covering with $e^{\rho_{\mathfrak{j}}}$ well defined on $q^{-1}(J/Z_G(G^0))_{\mathbb{C}}$ and $e^{\rho_{\mathfrak{j}}}(Z) = 1$. As any two Cartan subgroups of $\mathrm{Int}(\mathfrak{g}_{\mathbb{C}})$ are conjugate we have $\overline{x} \in \mathrm{Int}(\mathfrak{g}_{\mathbb{C}})$ such that $\mathrm{Ad}(\overline{x})(L/Z_G(G^0))_{\mathbb{C}} = (J/Z_G(G^0))_{\mathbb{C}}$ and $\mathrm{Ad}(\overline{x})^*\rho_{\mathfrak{j}} = \rho_{\mathfrak{l}}$. Now $e^{\rho_{\mathfrak{l}}}$ is well defined on $q^{-1}(L/Z_G(G^0))_{\mathbb{C}}$, thus is well defined on $L$ with $Z$ in its kernel. So also $\Delta_{G,L} = e^{-\rho_{\mathfrak{l}}}\cdot\prod_{\Sigma_{\mathfrak{l}}^+}(e^\gamma - 1)$ is well defined on $L$.

$Q$ was defined so that it has a faithful irreducible holomorphic representation $\psi$ of highest weight $\rho_{\mathfrak{j}}$ relative to $(\mathfrak{j}, \Sigma_{\mathfrak{j}}^+)$. Realize $\psi$ as a subrepresentation of the left multiplication action $\lambda$ of the Clifford algebra on the Lie algebra of $\mathrm{Int}(\mathfrak{g}_{\mathbb{C}})$. The Clifford subalgebra for $M$ is stable under $\lambda(q^{-1}(\mathrm{Ad}_{\mathfrak{g}}(M)))$, and the corresponding representation of $M$ has an irreducible summand of highest weight $\rho_{\mathfrak{j}_M}$. Now $e^{\rho_{\mathfrak{j}_M}}$ and $\Delta_{M,J_M}$ are well defined on $J_M$. □

$G'$ denotes the $G$–regular set in $G$, $M''A$ is the $MA$–regular set in $MA$, and

$$(4.4.4) \quad \begin{array}{l}\mathrm{Car}(G): \text{ the } G\text{–conjugacy classes of Cartan subgroups of } G \text{ and} \\ \mathrm{Car}(MA): \text{ the } MA\text{–conjugacy classes of Cartan subgroups of } MA.\end{array}$$

Then

$$(4.4.5) \quad \begin{array}{l} G' = \bigcup_{L\in\mathrm{Car}(G)} G'_L \text{ where } G'_L = \bigcup_{g\in G}\mathrm{Ad}(g)(L\cap G') \text{ and} \\ M''A = \bigcup_{J\in\mathrm{Car}(MA)} M''_J A \text{ where } M''_J A = \bigcup_{m\in M}(\mathrm{Ad}(m)(J\cap M''))A. \end{array}$$

The following theorem unifies and extends various results of Bruhat [**4**, Ch. III], Harish–Chandra ([**21**, p. 544] and [**22**, §11]), Hirai [**29**, Theorems 1, 2] and Lipsman [**34**, Theorem 9.1]. We assume the adjustment of Lemma 4.4.3. The specialization to $H$–series is in §4.5.

THEOREM 4.4.6. *Let $\zeta \in \widehat{Z}$, $[\eta] \in \widehat{M_\zeta}$ and $\sigma \in \mathfrak{a}^*$. Let $\chi_\nu$ be the infinitesimal character of $[\eta]$ relative to $\mathfrak{t}$ and let $\Psi_\eta$ denote the distribution character of $[\eta]$.*

1. *$[\pi_{\eta,\sigma}]$ has infinitesimal character $\chi_{\nu+i\sigma}$ relative to $\mathfrak{h}$.*

2. *$[\pi_{\eta,\sigma}]$ is a finite sum of classes from $\widehat{G}_\zeta$. In particular $[\pi_{\eta,\sigma}]$ has distribution character $\Theta_{\pi_{\eta,\sigma}}$ that is a locally summable function analytic on the regular set $G'$.*

3. $\Theta_{\pi_{\eta,\sigma}}$ *has support in the closure of* $\bigcup G'_J$ *where* $J$ *runs over a system of representatives of the* $G$*–conjugacy classes of Cartan subgroups of* $MA$.

4. *Let* $J \in \mathrm{Car}(MA)$ *and* $\Xi(J)$ *consist of all* $G$*–conjugates of* $J$ *contained in* $\mathrm{Car}(MA)$. *Enumerate* $\Xi(J) = \{J_1, \dots, J_\ell\}$ *with* $J_i = \mathrm{Ad}(g_i)J$. *If* $h \in J \cap G'$ *denote* $h_i = \mathrm{Ad}(g_i)h$. *Then*

$$\Theta_{\pi_{\eta,\sigma}}(h) = \frac{1}{|\Delta_{G,J}(h)|} \sum_{J_i \in \Xi(J)} \sum_{h' \in N_G(J_i)h_i} \frac{|\Delta_{MA,J_i}(h')|}{|N_{MA}(J_i)h'|} \Psi_\eta(h'_M) e^{i\sigma}(h'_A).$$

{*Note: If* $h \in J^0$ *then the second sum runs over the Weyl group orbit* $W_{G,J_i}(h_i)$. }

COROLLARY 4.4.7. *The class* $[\pi_{\eta,\sigma}]$ *is independent of the choice of parabolic subgroup* $P = MAN$ *associated to the Cartan subgroup* $H = T \times A$ *of* $G$.

The proof of Theorem 4.4.6 is based on the following minor variation of the Harish–Chandra transform $C_c^\infty(G) \to C_c^\infty(MA)$.

PROPOSITION 4.4.8. *Let* $b \in C_c^\infty(G)$ *and define*

$$b_P(ma) = e^{-\rho_{\mathfrak{a}}}(a) \int_{K/Z} \left( \int_N b(kmank^{-1}) dn \right) d(kZ). \tag{4.4.9}$$

*Then* $b_P(C_c^\infty(MA)$, $\pi_{\eta,\sigma}(b)$ *is of trace class, and*

$$\textit{trace}\ \pi_{\eta,\sigma}(b) = \int_{MA} b_P(ma)\Psi_\eta(m) e^{i\sigma}(a)\, dm\, da. \tag{4.4.10}$$

PROOF. Let $K_1$ denote the image of a Borel section of $K \to K/Z$. If $f \in L_2(G/P; \mathbb{E}_{eta,\sigma})$ is continuous it is determined by $f|_{K_1}$. Compute

$$\begin{aligned}(\pi_{\eta,\sigma}(b)f)(k') &= \int_G b(g) f(g^{-1}k') dg = \int_G b(k'g) g(g^{-1} dg \\ &= \int_{K/K\cap M} d(kM) \int_{MAN} b(k'mank^{-1}) f(k(man)^{-1}) e^{-2\rho_{\mathfrak{a}}}(a) dm\, da\, dn \\ &= \int_{K/K\cap M} \left( \int_{MAN} b(k'mank^{-1}) e^{-2\rho_{\mathfrak{a}}}(a) \eta(m) dm\, da\, dn \right) f(k) d(kM).\end{aligned}$$

Define $\Phi_b(k', k) = \int_{MAN} b(k'mank^{-1}) e^{-\rho_{\mathfrak{a}} + i\sigma}(a)\eta(m) dm\, da\, dn : E_\eta \to E_\eta$. Then $\Phi_b(k', km_1) f(km_1) = \Phi_b(k', k) f(k)$ for $m_1 \in M \cap K$. Thus $(\pi_{\eta,\sigma}(b)f)(k') = \int_{K/K\cap M} \Phi_b(k', k) f(k) d(kM)$. As $Z \subset K \cap M$ and $K/Z$ is compact we write this as $(\pi_{\eta,\sigma}(b)f)(k') = \int_{K/Z} \Phi_b(k', k) f(k) d(kZ)$; so

$$\mathrm{trace}\, \pi_{\eta,\sigma}(b) = \int_{K/Z} \mathrm{trace}\, \Phi_b(k, k) d(kZ).$$

Set $\varphi_b(k,m) = \int_{NA} e^{-\rho_{\mathfrak{a}}+i\sigma} b(kmank^{-1})dn\,da$. Then $\varphi_b \in C_c^\infty((K/Z)\times M)$. Noting that we always have absolute convergence we calculate

$$\begin{aligned}\int_{K/Z} \operatorname{trace} \Phi_b(k,k)d(kZ) &= \int_{K/Z} \left(\operatorname{trace} \int_M \Phi_b(k,m)\eta(m)dm\right) d(kz)\\ &= \int_{K/Z} \left(\int_M \Phi_b(k,m)\Psi_\eta(m)dm\right) d(kZ)\\ &= \int_{K/Z} \left(\int_M \left(\int_{NA} e^{-\rho_{\mathfrak{a}}+i\sigma}(a)b(kmank^1)dn\,da\right) \Psi_\eta(m)dm\right) d(kZ)\\ &= \int_M \left(\int_A b_P(ma)e^{i\sigma}(a)\Psi_\eta(m)da\right) dm.\end{aligned}$$

That completes the proof of Proposition 4.4.8 □

Let $\mathcal{Z}(\mathfrak{g})$ and $\mathcal{Z}(\mathfrak{m}+\mathfrak{a})$ denote the respective centers of the enveloping algebras $\mathcal{U}(\mathfrak{g})$ and $\mathcal{U}(\mathfrak{m}+\mathfrak{a})$. Recall the canonical homomorphisms $\gamma_G$ and $\gamma_{MA}$ to Weyl group invariant polynomials. Then [**17**, §12] $\mu_{MA} = \gamma_{MA}^{-1} \cdot \Gamma_G \mathcal{Z}(\mathfrak{g}) \to \mathcal{Z}(\mathfrak{m}+\mathfrak{a})$ has the property that $\mathcal{Z}(\mathfrak{m}+\mathfrak{a})$ has finite rank over its subalgebra $\mu_{MA}(\mathcal{Z}(\mathfrak{g}))$.

Proposition 4.4.8 says $\operatorname{trace} \pi_{\eta,\sigma}(b) = \operatorname{trace}(\eta\otimes e^{i\sigma})(b_P)$ for $b \in C_c^\infty(G)$. Now Harish–Chandra's [**17**, Lemma 52] says $\operatorname{trace} \pi_{\eta,\sigma}(zb) = \operatorname{trace}(\eta\otimes e^{i\sigma})(\mu_{MA}(z)b_P)$ for $z \in \mathcal{Z}(\mathfrak{g})$. From that, the infinitesimal character $\chi_{\pi_{\eta,\sigma}}(z) = \chi_{\eta\otimes e^{i\sigma}}(\mu_{MA}(z)) = \chi_{\nu+i\sigma}(\mu_{MA}(z))$ ($\chi$ for $((\mathfrak{m}+\mathfrak{a})_{\mathbb{C}}, \mathfrak{h}_{\mathbb{C}})) = \gamma_{MA}(\mu_{MA}(z))(\nu+i\sigma) = \gamma_G(z)(\nu+i\sigma) = \chi_{\nu+i\sigma}(z)$ (for $(\mathfrak{g}_{\mathbb{C}}, \mathfrak{h}_{\mathbb{C}})$). That proves the first assertion of Theorem 4.4.6.

To see that $[\pi_{\eta,\sigma}]$ is a direct integral over $\widehat{G}_\zeta$ we set $G^1 = ZG^0$ and $M^1 = M\cap G^1$. Then $[\eta] \in \widehat{M_\zeta}$ gives $[\eta^1] \in \widehat{M^1}_\zeta$ such that $[\eta]$ is a subrepresentation of $[\operatorname{Ind}_{M^1}^1(\eta^1)]$. Thus $\pi_{\eta,\sigma}$ is a subrepresentation of $\operatorname{Ind}_P^G((\operatorname{Ind}_{M^1}^M(\eta^1))\otimes e^{i\sigma}) = \operatorname{Ind}_{G^1}^G(\operatorname{Ind}_{M^1AN}^{G^1}(\eta^1\otimes e^{i\sigma}))$. Here $[\eta^1] \in \widetilde{M^1}_\zeta$, so $\operatorname{Ind}_{M^1AN}^{G^1}(\eta^1\otimes e^{i\sigma})|_Z$ is a multiple of $\zeta$. Then $[\pi_{\eta,\sigma}]$ is a direct integral over $\widehat{G}_\zeta$.

We now prove that there is an integer $n > 0$ such that

$$\pi_{\eta,\sigma}|_K = \sum\nolimits_{\widehat{K_\zeta}} m_\kappa \kappa \text{ where } 0 \leqq m_\kappa \leqq n\cdot\dim\kappa < \infty. \tag{4.4.11}$$

It suffices to prove this for the finite index subgroup $ZG^0$, so we may assume $Z$ central in $G$. Then (4.3.3) and the discussion just above give us a $K$–equivariant injective isometry $r_K : L_2(G/P;\mathbb{E}_{\eta,\sigma}) \to L_2(K/Z;\zeta)$ by $r_K(f)\ f|_K$. As $L_2(K/Z;\zeta) = \sum_{\widehat{K}_\zeta} V_\kappa\otimes V_\kappa^*$ the multiplicity of $V_\kappa$ here is $\dim((V_\kappa^*\otimes E_\eta)^{M\cap K})$. But $\eta|_{M\cap K} = \sum_{(\widehat{M\cap K})_\zeta} m_i\mu_i$ where $0 \leqq m_i \leqq_M \dim\mu_i < \infty$. If $\kappa \in \widehat{K}_\zeta$ then $\dim\kappa < \infty$ so $\kappa_{M\cap K}$ is a finite sum $\sum_{\widehat{M\cap K}_\zeta} m_{\kappa,i}\mu_i$. Now

$$\dim((V_\kappa^*\otimes E_\eta)^{M\cap K}) = \sum m_{\kappa,i}\mu_i \leqq n_M \sum m_{\kappa,i}\dim\mu_i = n_M\dim\kappa < \infty,$$

proving (4.4.11). Note from the proof that $n \leqq n_M|G/ZG^0|$.

We show that $[\pi_{\eta,\sigma}]$ is a finite sum from $\widehat{G}_\zeta$, following Harish–Chandra. The discussion of (3.2.1) shows that we need only consider the case where $G$ is connected. Then $Z$ is central so $[\pi_{\eta,\sigma}]$ has central character $\zeta$ and infinitesimal character $\chi_{\nu+i\sigma}$. By (4.4.11) $\pi_{\eta,\sigma}$ has no nontrivial subrepresentation of infinite multiplicity. Thus it is quasi–simple in the sense of Harish–Chandra [**16**, p. 145]. Consequently it has distribution character $\Theta_{\pi_{\eta,\sigma}}$ that is a locally summable function analytic on the regular set $G'$ [**16**, Theorem 6], and $\pi_{\eta,\sigma} = \sum \pi_j$ discrete direct sum of irreducibles

[**12**, Lemma 2]. Each $[\pi_j] \in \widehat{G}_\zeta$ and each $\chi_{\pi_j} = \chi_{\nu+i\sigma}$. Further, the differential equations $z\Theta = \chi_{\nu+i\sigma}(z)\Theta$ ($z \in \mathcal{Z}(\mathfrak{g})$) constrain the $\Theta_{\pi_j}$ to a finite dimensional space of functions on $G$. Inequivalent classes in $\widehat{G}_\zeta$ have linearly independent distribution characters, so $\pi_{\eta,\sigma} = \sum \pi_j$ involves only finitely many classes from $\widehat{G}_\zeta$. Since the multiplicities $m(\pi_j, \pi_{\eta,\sigma}) < \infty$, $\pi_{\eta,\sigma}$ is a finite sum from $\widehat{G}_\zeta$. That is the second assertion of Theorem 4.4.6.

We now calculate $\Theta_{\pi,\sigma}$ by extending Lipsman's argument [**34**, Theorem 9.1] to our more general situation. Recall (4.4.5) and the definition (4.4.9) of $b_P$.

LEMMA 4.4.12. *Let $L \in \mathrm{Car}(G)$ not conjugate to a Cartan subgroup of $MA$ and $b \in C_c^\infty(G'_L)$. Then $b_P = 0$. On the other hand, if $J \in \mathrm{Car}(MA)$ and $b \in C_c^\infty(G'_J)$ then $b_P \in C_c^\infty(MA \cap G'_J) \subset C_c^\infty((MA)''_J)$.*

PROOF. If $ma \in MA$ and $(\mathrm{Ad}(ma)-1)^{-1}$ is nonsingular on $N$ then [**20**, Lemma 11] gives us $\int_N b(kmank^{-1})dn = |\det(\mathrm{Ad}(ma)^{-1} - 1)_\mathfrak{n}| \int_N b(knman^{-1}k^{-1})dn$. If $b \in C_c^\infty(G'_L)$ where $G'_L$ doesn't meet $MA$ then $b(knman^{-1}k^{-1})$ is identically zero, so $b_P = 0$. If $b \in C_c^\infty(G'_J)$ where $J \in \mathrm{Car}(MA)$ there is a compact set $S \subset G'_J$ such that, if $\int_N b(kmank^{-1})dn \neq 0$ for some $k \in K$ then $ma \in S$. Thus $b_P$ is supported in $S \cap MA \subset G'_J \cap MA \subset (MA)''_J$. □

Let $L \in Car(G) \setminus \mathrm{Car}(MA)$. Let $b \in C_c^\infty(G'_L)$. Combine Proposition 4.4.8 and Lemma 4.4.12 to see $\Theta_{\pi_{\eta,\sigma}}(b) = \int_{MA} b_P(ma)\Psi_\eta(m)e^{i\sigma}(a)dmda = 0$. Thus $\Theta_{\pi_{\eta,\sigma}}|_{G'_L} = 0$. That is the third assertion of Theorem 4.4.6.

Fix $J = J_M \times A \in \mathrm{Car}(MA)$. To compute $\Theta_{\pi_{\eta,\sigma}}|_{G'_J}$ we need a variation on the Weyl Integration Formula. The center $Z_J$ of $J$ is open in $J$ so it inherits Haar measure $dh$. Normalize measure on $G/Z_J$ by $\int_G f(g)dg = \int_{G/Z_J}(\int_{Z_J} f(gh)dh)d(gZ_J)$ and on $MA/Z_J$ by $\int_{MA} F(x)dx = \int_{MA/Z_J}(\int_{Z_J} F(xh)dh)d(xZ_J)$. Extending Harish–Chandra's extension [**19**] of Weyl's argument,

LEMMA 4.4.13. *If $b \in C_c(G'_J)$ and $B \in C_c((MA)''_J)$ then*
$\int_G b(g)dg = \int_{J\cap G'} |N_G(J)(h)|^{-1}(\int_{G/Z_J} b(ghg^{-1})d(gZ_J))|\Delta_{G,J}(h)|^2 dh$ *and*
$\int_{MA} B(x)dx = \int_{J\cap(MA)''} |N_{MA}(J)(h)|^{-1}(\int_{MA/Z_J} b(xhx^{-1})d(xZ_J))|\Delta_{MA,J}(h)|^2 dh.$

PROOF. $(G/Z_J) \times (J \cap G') \to G'_J$, by $(gZ_J, h) \mapsto ghg^{-1}$, is regular, surjective, and $|N_G(J)(h)|$ to one with Jacobian determinant $|\det(\mathrm{Ad}(h)^{-1} - 1)_{\mathfrak{g}/\mathfrak{j}}| = |\prod_{\gamma\in\Sigma_\mathfrak{j}}(e^\gamma - 1)(h)|$ at $(gZ_J, h)$. But $\prod_{\gamma\in\Sigma_\mathfrak{j}}(e^\gamma - 1)(h)$ is the product over $\Sigma_\mathfrak{j}^+$ times the product over $-\Sigma_\mathfrak{j}^+$, which is $(-1)^n\Delta_{G,J}(h)^2$ where $n = |\Sigma_\mathfrak{j}^+|$, so that Jacobian is $|\Delta_{G,J}(h)|^2$. That proves the first equation; the second is similar. □

Given $b \in C_c^\infty(G'_J)$, $\Theta_{\pi_{\eta,\sigma}}(b) = \int_{MA} b_P(ma)\Psi_\eta(m)e^{i\sigma}(a)dmda$ by Proposition 4.4.8. Lemma 4.4.12 ensures convergence. Now, by Lemma 4.4.13,

$$\begin{aligned}\Theta_{\pi_{\eta,\sigma}}(b) = \int_{J\cap G'} & |N_{MA}(J)(h)|^{-1} \\ & \Big(\int_{MA/Z_J} b_P(xhx^{-1}d(xZ_J)\Big)\Psi_\eta(h_M)e^{i\sigma}(h_A)|\Delta_{MA,J}(h)|^2 dh\end{aligned} \tag{4.4.14}$$

where $h = h_A h_M$ along $J = J_M \times A$. As $A$ is central in $MA$,

$$\int_{MA/Z_J} b_P(xhx^{-1})d(xZ_J)$$
$$= e^{-\rho_{\mathfrak{a}}(h_A)} \int_{MA/Z_J} d(xZ_J) \cdot \int_{K/Z} d(kZ) \cdot \int_N b(k.xhx^{-1} \cdot n \cdot k^{-1})dn.$$

Unimodularity of $N$ and [**19**, Lemma 11] say

$$\int_N b(k \cdot maha^{-1}m^{-1} \cdot n \cdot k^{-1})dn = |\det(\mathrm{Ad}(h^{-1})-1)_{\mathfrak{n}}| \int_N b(knma \cdot h \cdot (knma)^{-1}dn.$$

We modify Harish–Chandra's evaluation [**19**, Lemma 12] of $|\det(\mathrm{Ad}(h^{-1})-1)_{\mathfrak{n}}|$ for the case $J = H$. Choose $c \in \mathrm{Int}(\mathfrak{g}_{\mathbb{C}})$ with $c(\mathfrak{g}_{\mathbb{C}}) = \mathfrak{h}_{\mathbb{C}}$, $c(x) = x$ for $x \in \mathfrak{a}$, and $c^*\Sigma^+ = \Sigma_{\mathfrak{j}}^+$, so $c$ also preserves $\mathfrak{m}_{\mathbb{C}}$ and $\mathfrak{n}_{\mathbb{C}}$. Then $\det(\mathrm{Ad}(h^{-1}) - 1)_{\mathfrak{n}}$ $= \det(\mathrm{Ad}(c(h^{-1})) - 1)_{\mathfrak{n}} = \prod_{\gamma \in \Sigma^+ \setminus \Sigma_{\mathfrak{t}}^+} (e^{\gamma}(c(h^{-1})) - 1 = \prod_{\beta \in \Sigma_{\mathfrak{j}}^+ \setminus \Sigma_{\mathfrak{j}_M}^+} (e^{\beta}(h)) - 1)$ $= e^{\rho_{\mathfrak{j}} - \rho_{\mathfrak{j}_M}}(h) \frac{\Delta_{G,J}(h)}{\Delta_{MA,J}(h)} = e^{\rho_{\mathfrak{a}}}(h_A) \frac{\Delta_{G,J}(h)}{\Delta_{MA,J}(h)}$, so $\int_{MA/Z_J} b_P(xhx^{-1})d(xZ_J)$ is equal to

$$\int_{MA/Z_J} d(xZ_J) \cdot \int_{K/Z} d(kZ) \cdot |\Delta_{G,J}(h)/\Delta_{MA,J}(h)| \int_N b(\mathrm{Ad}(knma)h)dn$$
$$= |\tfrac{\Delta_{G,J}(h)}{\Delta_{MA,J}(h)}| \int_{K/K\cap M} \Big(\int_{MNA/Z_J} b(\mathrm{Ad}(kmna)h)dm\, da\, dn\Big) d(kM)$$
$$= |\Delta_{G,J}(h)/\Delta_{MA,J}(h)| \int_{G/Z_J} b(ghg^{-1})d(gZ_J).$$

Substituting this into (4.4.14),

$$\Theta_{\pi_{\eta,\sigma}}(b) = \int_{J\cap G'} |N_{MA}(J)(h)|^{-1} \Big(\int_{G/Z_J} b(ghg^{-1})d(gZ_J)\Big) \Psi_\eta(h_M) e^{i\sigma}(h_A) |\Delta_{G,J}(h)\Delta_{MA,J}(h)|^2 dh \tag{4.4.15}$$

We extend $\Phi_{\eta,\sigma,J}(h) := \frac{1}{|\Delta_{G,J}(h)|} \sum_{h' \in N_G(J)h} \frac{|\Delta_{MA,J}(h')|}{|N_{MA}(J)h'|} \Psi_\eta(h'_M) e^{i\sigma}(h'_A)$ to a class function on $G'_J$ and substitute that into (4.4.15). Thus $\Theta_{\pi_{\eta,\sigma}}(b)$ is

$$\int_{J\cap G'} |N_G(J)(h)|^{-1}|^{-1} Bigl( \int_{G/Z_J} b(ghg^{-1})\Phi_{\eta,\sigma,J}(ghg^{-1})d(gZ_J)\Big) |\Delta_{G,J}(h)|^2 dh.$$

From Lemma 4.4.13 we see that $\Theta_{\pi_{\eta,\sigma}}|_{G'_J}$ is given by $\Phi_{\eta,\sigma,J}$. That proves the character formula and completes the proof of Theorem 4.4.6.

**4.5.** We specialize the results of §4.4 to the $H$–series of $G$, where $[\eta] \in \widehat{M}_{disc}$. The Cartan subgroup $H = T \times A$ and the associated cuspidal parabolic subgroup $P = MAN$ are fixed. The two principal simplifications here are (1) $\Xi(H) = \{H\}$ and (2) the character formulae for $H$–series representations are explicit [**27**].

The choice of $H$ and $P$ specifies $\Sigma_{\mathfrak{a}}^+$ with $\mathfrak{n} = \sum_{\alpha \in \Sigma_{\mathfrak{a}}^+} \mathfrak{g}^{-\alpha}$. Choose $\Sigma_{\mathfrak{t}}^+$ and specify $\Sigma^+$ as in Lemma 4.1.5. We have $\rho$, $\rho_{\mathfrak{a}}$, $\rho_{\mathfrak{t}}$, $\Delta_{G,H}$ and $\Delta_{M,T}$ as in (4.4.1). Make the adjustment of Lemma 4.4.3 if needed, so that $e^\rho \in \widehat{H}$ and $e^{\rho_{\mathfrak{t}}} \in \widehat{T}$ are well defined, and $e^\rho(Z) = e^{\rho_{\mathfrak{t}}}(Z) = 1$. Then $\Delta_{G,H}$ is well defined on $H$ and $\Delta_{M,T}$ is well defined on $M$. Proposition 4.1.4 says that $M$ has relative discrete series as described in §§3.4 and 3.5. It comes out as follows. Let $\varpi_{\mathfrak{t}}(\nu) = \prod_{\phi \in \Sigma_{\mathfrak{t}}^+} \langle \phi, \nu \rangle$ for

$\nu \in \mathfrak{t}_{\mathbb{C}}^*$ and $L_{\mathfrak{t}}'' = \{\nu \in i\mathfrak{t}^* \mid e^\nu \in \widehat{T^0} \text{ and } \varpi_{\mathfrak{t}}(\nu) \neq 0\}$. Every $\nu \in L_{\mathfrak{t}}''$ specifies a class $[\eta_\nu] \in (\widehat{M^0})_{disc}$ whose distribution character satisfies

$$\Psi_{\eta_\nu}|_{T^0 \cap M''} = (-1)^{q_M(\nu)} \tfrac{1}{\Delta_{M,T}} \sum\nolimits_{w \in W(M^0,T^0)} \det(w) e^{w\nu}$$

with $q_M$ defined on $L_{\mathfrak{t}}''$ as in (3.4.3). Every class in $(\widehat{M^0})_{disc}$ is one of these $[\eta_\nu]$. Classes $[\eta_\nu] = [\eta_{\nu'}]$ if and only if $\eta' \in W(M^0, T^0)(\nu)$. Finally, $[\eta_\nu]$ has central character $e^{\nu - \rho_{\mathfrak{t}}}|_{Z_M^0}$ and infinitesimal character $\chi_\nu$ relative to $\mathfrak{t}$.

If $\nu \in L_{\mathfrak{t}}''$ and $[\chi] \in \widehat{Z_M(M^0)}_\xi$ where $\xi = e^{\nu - \rho_{\mathfrak{t}}}|_{Z_M^0}$ then we have

$$\eta_{\chi,\nu} = \operatorname{Ind}_{M^\dagger}^M (\chi \otimes \eta_\nu) \in \widehat{M}_{disc}. \tag{4.5.1}$$

Here recall $M^\dagger := Z_M(M^0)M^0$. Also, $\eta_{\chi,\nu} \in \widehat{M}_{disc}$, and it is the only class there with distribution character given on $Z_M(M^0) \cdot (T^0 \cap M'')$ by

$$\Psi_{\eta_{\chi,\nu}}(zt) =$$
$$= \sum_{1 \leqq j \leqq r} (-1)^{q_M(w_j\nu)} \operatorname{trace} \chi(x_j^{-1} z x_j) \cdot \frac{1}{\Delta_{M,T}} \sum_{w \in W_{M^0,T^0}} \det(ww_j) e^{ww_j\nu}(t) \tag{4.5.2}$$

where there the $w_j = \operatorname{Ad}(x_j T)_{\mathfrak{t}}$ are representatives of $W_{M,T}$ modulo $W_{M^0,T^0}$. Every class in $\widehat{M}_{disc}$ is one of the $[\eta_{\chi,\nu}]$. Classes $[\eta_{\chi,\nu}] = [\eta_{\chi',\nu'}]$ exactly when $([\chi'], \nu') \in W_{M,T}([\chi], \nu)$. Finally, $[\eta_{\chi,\nu}]$ has infinitesimal character $\chi_\nu$ relative to $\mathfrak{t}$. Now we combine this description with Theorem 4.4.6. Recall that the normalizers $N_{MA}(H) = N_M(T) \times A$ and $N_G(H)$ have all orbits finite on $H \cap G'$.

THEOREM 4.5.3. *Let $\nu \in L_{\mathfrak{t}}''$, $\sigma \in \mathfrak{a}^*$ and $[\chi] \in \widehat{Z_M(M^0)}_\xi$ where $\xi = e^{\nu-\rho_{\mathfrak{t}}}|_{Z_M^0}$. Define $\eta_{\chi,\nu}$ and $\Psi_{\eta_{\chi,\nu}}$ by (4.5.1) and (4.5.2). Then $[\pi_{\chi,\nu,\sigma}] := [\operatorname{Ind}_P^G(\eta_{\chi,\nu} \otimes e^{i\sigma})]$ is the unique $H$-series representation class on $G$ whose distribution character satisfies*

$$\Theta_{\pi_{\chi,\nu,\sigma}}(ta) = \frac{|\Delta_{M,T}(t)|}{|\Delta_{G,H}(ta)|} \sum_{w(ta) \in N_G(H)(ta)} |N_M(T)(wt)|^{-1} \Psi_{\eta_{\chi,\nu}}(wt) e^{i\sigma}(wa)$$

*for $t \in T$, $a \in A$ and $ta \in G'$. Every $H$-series class on $G$ is one of the $[\pi_{\chi,\nu,\sigma}]$, and classes $[\pi_{\chi,\nu,\sigma}] = [\pi_{\chi',\nu',\sigma'}]$ if and only if $(\chi', \nu', \sigma') \in W_{G,H}(\chi, \nu, \sigma)$. $[\pi_{\chi',\nu',\sigma'}]$ is a finite sum from $\widehat{G}_\zeta$ where $[\eta_{\chi,\nu}] \in \widehat{M}_\zeta$. The dual $[\pi_{\chi,\nu,\sigma}^*] = [\pi_{\bar{\chi},-\nu,-\sigma}]$. The infinitesimal character is $\chi_{\nu+i\sigma}$ relative to $\mathfrak{h}$, so $[\pi_{\chi,\nu,\sigma}]$ sends the Casimir element of $\mathcal{U}(\mathfrak{g})$ to $||\nu||^2 + ||\sigma||^2 - ||\rho||^2$.*

PROOF. First note that $\Xi(H) = \{H\}$ because any two fundamental Cartan subgroups of $MA$ are $\operatorname{Ad}(M^0)$–conjugate. That eliminates the sum over $\Xi(H)$ expected from Theorem 4.4.6. Now we need only show that $\Theta_{\pi_{\chi,\nu,\sigma}}|_{H \cap G'}$ determines $(\chi, \nu, \sigma)$ modulo $W_{G,H}$. Let $\Theta_{\pi_{\chi,\nu,\sigma}}|_{H \cap G'} = \Theta_{\pi_{\chi',\nu',\sigma'}}|_{H \cap G'}$. By linear independence of characters $e^{i\sigma''}$ on $A$ we may replace $\sigma'$ by any element of $N_G(H)\sigma'$ and assume $\sigma' = \sigma$. Thus, on $H \cap G'$, $\sum |N_M(T)(wt)|^{-1} e^{i\sigma}(wa)(\Psi_{\eta_{\chi,\eta}}(wt) - \Psi_{\eta_{\chi',\eta'}}(wt)) = 0$. Here $|N_M(T)(wt)|$ is locally constant on $T \cap M''$ and the functions $\Psi_{\eta_{\chi'',\eta''}}$ are linearly independent on $T \cap M''$. Thus $\Psi_{\eta_{\chi,\eta}} = \Psi_{\eta_{\chi',\eta'}}$, so $(\chi, \eta) = (\chi', \eta')$, and thus $(\chi', \eta') \in W_{M,T}(\chi, \eta)$. □

COROLLARY 4.5.4. *The $H$-series classes $[\pi_{\chi,\nu,\sigma}]$ are independent of the choice of cuspidal parabolic subgroups $P$ associated to $H$*

The support of $\Theta_{\pi_{\chi,\nu,\sigma}}$ meets the interior of $G'_H$, and by Theorem 4.4.6(3) it determines the conjugacy class of $H$. A stronger result, due to Lipsman [**34**, Theorem 11.1] for connected semisimple groups with finite center, is

THEOREM 4.5.5. *Let $H$ and $'H$ be non-conjugate Cartan subgroups of $G$. Let $[\pi] \in \widehat{G}$ be $H$-series and let $['\pi] \in \widehat{G}$ be $'H$-series. Then the infinitesimal characters $\chi_\pi \neq \chi_{'\pi}$, and $[\pi]$ and $['\pi]$ are disjoint (no composition factors in common).*

PROOF. Take both Cartans are $\theta$-stable, $H = T \times A$ and $'H = 'T \times 'A$. Express $[\pi] = [\pi_{\chi,\nu,\sigma}]$ using $H$ and $['\pi] = [\pi_{'\chi,'\nu,'\sigma}]$ using $'H$. Then $\chi_\pi = \chi_{\nu+i\sigma}$ using $\mathfrak{h}$ and $\chi_{'\pi} = \chi_{'\nu+i'\sigma}$ using $'\mathfrak{h}$.

If $\chi_\pi = \chi_{'\pi}$ there exists $\beta \in \mathrm{Int}(\mathfrak{g}_\mathbb{C})$ such that $\beta(\mathfrak{h}_\mathbb{C}) = '\mathfrak{h}_\mathbb{C}$ and $\beta^*('\nu + '\sigma) = (\nu + i\sigma)$. $\beta^*$ sends real span of roots to real span of roots, so $\beta^*('\nu) = \nu$ and $\beta^*('\sigma) = \sigma$. Further, we may suppose $\beta^*('\Sigma^+) = \Sigma^+$. It follows that $\beta(\mathfrak{h}) = '\mathfrak{h}$. Consequently [**39**, Corollary 2.4] there is an inner automorphism of $\mathfrak{g}$ that sends $\mathfrak{h}$ to $'\mathfrak{h}$, contradicting nonconjugacy of $H$ and $'H$. Thus $\chi_\pi \neq \chi_{'\pi}$. Now $[\pi]$ and $['\pi]$ are disjoint because common factors would have the same infinitesimal character. □

**4.6.** We discuss irreducibility for $H$-series representations. As before fox $H = T \times A$ and $P = MAN$. Let $[\eta] \in \widehat{M}$ have infinitesimal character $\chi_n u$ relative to $\mathfrak{t}_\mathbb{C}$. We say that $[\eta]$ has *real infinitesimal character* if $\langle \phi, \nu \rangle$ is real for every $\phi \in \Sigma_\mathfrak{t}^+$. The classes in $\widehat{M_{disc}}$ have real infinitesimal character.

An element $\sigma \in \mathfrak{a}^*$ is $(\mathfrak{g}, \mathfrak{a})$*-regular* if $\langle \psi, \sigma \rangle \neq 0$ for all $\psi \in \Sigma_\mathfrak{a}$. Choose a minimal parabolic subgroup $P_0 = M_0 A_0 N_0$ of $G$ with $A \subset A_0 = \theta A_0$. The $\mathfrak{a}$-roots are just the nonzero restrictions of the $\mathfrak{h}_\mathbb{C}$-roots, and so they are the nonzero restrictions of the $\mathfrak{a}_0$-roots. If $w \in W(\mathfrak{g}, \mathfrak{a}_0)$ and if $\sigma \in \mathfrak{a}^*$ is $(\mathfrak{g}, \mathfrak{a})$-regular, then $\mathfrak{a}$ is central in the centralizer $\mathfrak{g}^\sigma$ and $w \in W(\mathfrak{g}^\sigma, \mathfrak{a}_0)$, so $w$ is generated by reflections in roots that annihilate $\mathfrak{a}$, forcing $w|_\mathfrak{a}$ to be trivial. In summary,

LEMMA 4.6.1. *If $\sigma \in \mathfrak{a}^*$ then the following conditions are equivalent:* (i) *$\sigma$ is $(\mathfrak{g}, \mathfrak{a})$-regular,* (ii) *If $\phi \in \Sigma^+$ and $\phi|_\mathfrak{a} \neq 0$ then $\langle \phi, \sigma \rangle \neq 0$,* (iii) *If $\psi_0$ is an $\mathfrak{a}_0$-root of $\mathfrak{g}$ and $\psi_0|_\mathfrak{a} \neq 0$ then $\langle \psi_0, \sigma \rangle \neq 0$,* (iv) *If $w \in W(\mathfrak{g}, \mathfrak{a}_0)$ and $w|_\mathfrak{a} \neq 1$ then $w(\sigma) \neq \sigma$.*

The following theorem was proved by Harish–Chandra (unpublished):

THEOREM 4.6.2. *Let $[\eta] \in \widehat{M}$ have real infinitesimal character and let $\sigma \in \mathfrak{a}^*$ be $(\mathfrak{g}, \mathfrak{a})$-regular. Then $[\pi_{\eta,\sigma}] = [\mathrm{Ind}_P^G(\eta \otimes e^{i\sigma})]$ is irreducible.*

COROLLARY 4.6.3. *If $\sigma \in \mathfrak{a}^*$ is $(\mathfrak{g}, \mathfrak{a})$-regular then every $H$-series class $[\pi_{\chi,\nu.\sigma}]$ is irreducible.*

After that, irreducibility were settled by Knapp and Zuckerman ([**31**], [**32**]) for connected reductive real linear algebraic groups (the case where $G$ is connected and is isomorphic to a closed subgroup of some general linear group $GL(n;\mathbb{R})$). In view of Langlands theorem [**33**], that completed the classification of irreducible admissible representations for reductive real linear algebraic groups. For those groups, and more generally for groups of class $\mathcal{H}$, Vogan's treatment of the Kazhdan–Lusztig conjecture and construction and analysis of the KLV polynomials ([**47**], [**48**]) includes a complete analysis of the composition series of any $H$-series representation $[\pi_{\chi,\nu.\sigma}]$. Finally, the *Atlas* software, `http://www.liegroups.org/software/`, allows explicit computation of those composition factors; see [**1**].

## 5. The Plancherel Formula for General Real Reductive Lie Groups

As before, $G$ is a real reductive Lie group that satisfies (1.2.1). The Harish–Chandra class $\mathcal{H}$ consists of all such groups for which $G/G^0$ is finite and the derived group $[G^0, G^0]$ has finite center. We fix a Cartan involution $\theta$ of $G$ and a system $\mathrm{Car}(G) = \{H_1, \ldots, H_\ell\}$ of $\theta$–stable representatives of the conjugacy classes of Cartan subgroups of $G$. Harish–Chandra's announced [**22**, §11] a Plancherel formula for groups of class $\mathcal{H}$: there are unique continuous functions $m_{j,\eta}$ on $\mathfrak{a}_j^*$, meromorphic on $(\mathfrak{a}_j^*)_\mathbb{C}$, invariant under the Weyl group $W(G, H_j)$, such that

$$f(x) = \sum_{1 \leqq j \leqq \ell} \sum_{[\eta] \in (\widehat{M_j})_{disc}} \deg(\eta) \int_{\mathfrak{a}_j^*} \Theta_{\pi_{\eta,\sigma}} r_x(f) m_{j,\eta}(\sigma) d\sigma,$$

absolutely convergent for $x \in G'$ and $f \in C_c^\infty(G)$. This was extended to our class in [**51**] without consideration of meromorphicity. Later Harish–Chandra published a complete treatment for $G$ of class $\mathcal{H}$ and $f$ in the Harish–Chandra Schwartz space $\mathcal{S}(G)$ ([**23**],[**24**],[**25**]). Still later Herb and I extended those results to general real reductive groups, including explicit formulae for the various constants and functions that enter into the Plancherel measure ([**27**], [**28**]).

Here, for lack of space or necessity, I'll only indicate the results from [**51**], because that is all that is needed in §§7 and 8 below.

**5.1.** As above, we have $G$, $\theta$, $K = G^\theta$, $\mathrm{Car}(G) = \{H_1, \ldots, H_\ell\}$, $H_j = T_j \times A_j$, $\Sigma_{\mathfrak{a}_j}^+$ and $P_j = M_j A_j N_j$ with $M_j \times A_j = Z_G(A_j)$. As in §4, $L_j = \{\nu \in i\mathfrak{t}_j^* \mid e^\nu \in \widehat{T_j^0}\}$ and $L_j''$ is its $M_j$–regular set. Fix the $\Sigma_{\mathfrak{t}_j}^+$ and set $\varpi_{\mathfrak{t}_j}(\nu) = \prod_{\phi \in \Sigma_{\mathfrak{t}_j}^+} \langle \phi, \nu \rangle$, so $L_j'' = \{\nu \in L_j \mid \varpi_{\mathfrak{t}_j}(\nu) \neq 0\}$.

If $\zeta \in \widehat{Z}$ then $L_{j,\zeta} = \{\nu \in L_j \mid e^{\nu - \rho_{\mathfrak{t}_j}}|_{Z \cap M_j^0} = \zeta|_{Z \cap M_j^0}\}$ and $L_j'' = L_{j,\zeta} \cap L_j''$. Write $\xi_\nu$ for $e^{\nu - \rho_{\mathfrak{t}_j}}$. Since $ZZ_{M_j^0}$ has finite index in $Z_{M_j}(M_j^0)$ we define finite subsets $S(\nu, \zeta) \subset \widehat{Z_{M_j}(M_j^0)}$ by $S(\nu, \zeta) = \widehat{Z_{M_j}(M_j^0)}_{\zeta \otimes \xi_\nu}$ if $\xi|_{Z \cap M_j^0} = \xi_\nu|_{Z \cap M_j^0}$, $S(\nu, \zeta) = \emptyset$ otherwise. When $\nu \in L_j''$ and $\sigma \in \mathfrak{a}_j^*$ the $H_j$–series classes $[\pi_{\chi,\nu,\sigma}]$ that transform by $\zeta$ are just the ones with $[\chi] \in S(\nu, \zeta)$. Thus we have finite sums

$$\pi_{j,\zeta,\nu+i\sigma} = \sum\nolimits_{S(\nu,\zeta)} (\dim \chi) \pi_{\chi,\nu,\sigma} \text{ and } \Theta_{j,\zeta,\nu+i\sigma} = \sum\nolimits_{S(\nu,\zeta)} (\dim \chi) \Theta_{\pi_{\chi,\nu,\sigma}}.$$

If $\zeta|_{Z \cap M_j^0} \neq \xi_\nu|_{Z \cap M_j^0}$, in other words if $\nu \notin L_{j,\zeta}$, then $\Theta_{j,\zeta,\nu+i\sigma} = 0$. Here is the extension [**51**] of the Harish–Chandra Plancherel Formula ([**21**], [**22**]) to the $\widehat{G}_\zeta$.

THEOREM 5.1.1. *Let $G$ be a general real reductive Lie group* (1.2.1) *and $\zeta \in \widehat{Z}$. Then there are unique Borel–measurable functions $m_{j,\zeta,\nu}$ on $\mathfrak{a}_j^*$, $1 \leqq j \leqq \ell$, defining the* Plancherel measure on $\widehat{G}_\zeta$, *as follows.*

1. *The $m_{j,\zeta,\nu}$ are $W_{G,H_j}$–equivariant: $w^* m_{j,\zeta,\nu}(\sigma) = m_{j,w^*\zeta,w^*\nu}(w^*\sigma)$.*
2. *If $\nu \notin L_{j,\zeta}$ then $m_{j,\zeta,\nu} = 0$.*

3. *Let $f \in L_2(G/Z,\zeta)$ be $C^\infty$ with support compact modulo $Z$. If $x \in G$ define $(r_x f)(g) = f(gx)$. Then*

$$\text{(5.1.2)}\qquad \begin{aligned} &\sum_{1 \leqq j \leqq \ell} \sum_{\nu \in L''_{j,\zeta}} |\varpi_{\mathfrak{t}_j}(\nu)| \int_{\mathfrak{a}_j^*} |\Theta_{j,\zeta,\nu+i\sigma}(r_x f) m_{j,\zeta,\nu}(\sigma)| d\sigma < \infty \text{ and} \\ &f(x) = \sum_{1 \leqq j \leqq \ell} \sum_{\nu \in L''_{j,\zeta}} |\varpi_{\mathfrak{t}_j}(\nu)| \int_{\mathfrak{a}_j^*} \Theta_{j,\zeta,\nu+i\sigma}(r_x f) m_{j,\zeta,\nu}(\sigma) d\sigma. \end{aligned}$$

The following corollary is used for realization of $H$–series representations on spaces of partially harmonic spinors [**52**].

Corollary 5.1.3. *Let $\omega \in \mathcal{U}(\mathfrak{g})$ be the Casimir element. If $c \in \mathbb{R}$ and $\zeta \in \widehat{Z}$ then $\{[\pi] \in \widehat{G}_\zeta \setminus \widehat{G}_{\zeta-disc} \mid \chi_\pi(\omega) = c\}$ has Plancherel measure 0 on $\widehat{G}_\zeta$.*

Corollary 5.1.5 below is needed when we consider spaces of square integrable partially harmonic $(0,q)$–forms in §§7 and 8. It follows from Theorems 4.6.2 and 5.1.1; or one can also derive it from

Lemma 5.1.4. *Let $[\pi]$ be an irreducible constituent of an $H$–series class $[\pi_{\chi,\nu,\sigma}]$ where $\nu + i\sigma \in \mathfrak{h}^*$ is $\mathfrak{g}$–regular. If $G$ has relative discrete series representations, and if $H/Z$ is noncompact, then $\Theta_\pi|_{K \cap G'} = 0$.*

Corollary 5.1.5. *If $G$ has relative discrete series representations and if $\zeta \in \widehat{Z}$ then $\{[\pi] \in \widehat{G}_\zeta \setminus \widehat{G}_{\zeta-disc} \mid \Theta_\pi|_{K\cap G'} \neq 0\}$ has Plancherel measure 0 on $\widehat{G}_\zeta$.*

Corollary 5.1.6. *Fix $\zeta \in \widehat{Z}$. Let $\widehat{G}_{H_j,\zeta}$ denote the set of all $H_j$–series classes $[\pi_{\chi,\nu,\sigma}]$ for $\zeta$ such that $\sigma$ is $(\mathfrak{g},\mathfrak{a}_j)$–regular. Then each $\widehat{G}_{H_j,\zeta} \subset \widehat{G}_\zeta$ and the Plancherel measure on $\widehat{G}_\zeta$ is concentrated on $\bigcup_{1 \leqq j \leqq \ell} \widehat{G}_{H_j,\zeta}$.*

The "absolute" version of Theorem 5.1.1 derives from $f(x) := \int_{\widehat{Z}} f_\zeta(x)\zeta(z)d\zeta$ where $f_\zeta(x) = \int_Z f(xz)\zeta(z)dz$. Given $f \in C_c^\infty(G)$ we apply Theorem 5.1.1 to each $f_\zeta$ and sum over $\widehat{Z}$. The same holds for the corollaries.

## 6. Real Groups and Complex Flags

While $G$ is a general real reductive Lie group (1.2.1) the adjoint representation takes $G$ to a real reductive semisimple Lie group $\overline{G} := G/Z_G(G^0)$. That group has complexification $\overline{G}_\mathbb{C} = \mathrm{Int}(\mathfrak{g}_\mathbb{C})$, the group of inner automorphisms of $\mathfrak{g}_\mathbb{C}$. Notice that $\overline{G}_\mathbb{C}$ is connected. Now $G$ acts on all complex flag manifolds $X = \overline{G}_\mathbb{C}/Q$. Here we recall the part of [**50**] needed for geometric realization of standard induced representations, extending them from $\overline{G}^0$ to $G$ as needed. We discuss holomorphic arc components of $G$–orbits; consider measurable, integrable and flag type orbits; and give a complete analysis of the orbits on which our representations are realized in §§7 and 8.

Notation: $Q$ is used for a (complex) parabolic subgroup of $\overline{G}_\mathbb{C}$ and $P$ is reserved for cuspidal parabolics in $G$. Roots are ordered so that $X = \overline{G}_\mathbb{C}/Q$ has holomorphic tangent space spanned by positive root spaces.

**6.1.** It is standard that the following are equivalent for a closed complex subgroup $Q \subset \overline{G}_{\mathbb{C}}$: (i) $X = \overline{G}_{\mathbb{C}}$ is compact, (ii) $X$ is a compact simply connected Kähler manifold, (iii) $X$ is a $\overline{G}_{\mathbb{C}}$–homogeneous projective algebraic variety, (iv) $X$ is a closed $\overline{G}_{\mathbb{C}}$–orbit in a (finite dimensional) projective representation, and (v) $Q$ contains a Borel subgroup of $\overline{G}_{\mathbb{C}}$. Under these conditions we say that (1) $Q$ is a *parabolic subgroup* of $\overline{G}_{\mathbb{C}}$, (2) $\mathfrak{q}$ is a *parabolic subalgebra* of $\overline{\mathfrak{g}}_{\mathbb{C}}$, and (3) $X = \overline{G}_{\mathbb{C}}/Q$ is a *complex flag manifold* of $\overline{\mathfrak{g}}_{\mathbb{C}}$. Given (i) through (v) $Q$ is the analytic subgroup of $\overline{G}_{\mathbb{C}}$ for $\overline{\mathfrak{q}}_{\mathbb{C}}$, in fact is the $\overline{G}_{\mathbb{C}}$–normalizer of $\overline{\mathfrak{q}}_{\mathbb{C}}$.

Recall the structure. Choose a Cartan subalgebra $\overline{\mathfrak{h}}_{\mathbb{C}}$ of $\overline{\mathfrak{g}}_{\mathbb{C}}$ and a system $\Pi$ of simple $\overline{\mathfrak{h}}_{\mathbb{C}}$–roots on $\overline{\mathfrak{g}}_{\mathbb{C}}$. Any subset $\Phi \subset \Pi$ specifies

- $\Phi^r$: all roots that are linear combinations of elements of $\Phi$;
- $\Phi^u$: all negative roots not contained in $\Phi^r$;
- $\mathfrak{q}^r_\Phi = \overline{\mathfrak{h}}_{\mathbb{C}} + \sum_{\Phi^r} \overline{\mathfrak{g}}^\phi$, $\mathfrak{q}^u_\Phi = \sum_{\Phi^u} \overline{\mathfrak{g}}^\phi$ and $\mathfrak{q}_\Phi = \mathfrak{q}^r_\Phi + \mathfrak{q}^u_\Phi$.

Then $\overline{G}_{\mathbb{C}}$ has analytic subgroups $Q^r_\Phi$ for $\mathfrak{q}^r_\Phi$, $Q^u_\Phi$ for $\mathfrak{q}^u_\Phi$ and $Q_\Phi = Q^u_\Phi \rtimes Q^r_\Phi$ for $\mathfrak{q}_\Phi$. $Q^u_\Phi$ and $\mathfrak{q}^u_\Phi$ are the nilradicals, and $Q^r_\Phi$ and $\mathfrak{q}^r_\Phi$ are the Levi (reductive) components. $\Phi$ is a simple $\overline{\mathfrak{h}}_{\mathbb{C}}$–root system for $\mathfrak{q}^r_\Phi$. $Q_\Phi$ is a parabolic subgroup of $\overline{G}_{\mathbb{C}}$ and every parabolic subgroup of $\overline{G}_{\mathbb{C}}$ is conjugate to exactly one of the $Q_\Phi$. Any parabolic $Q_\Phi$ is its own normalizer in $\overline{G}_{\mathbb{C}}$, so the complex flag manifold $X = \overline{G}_{\mathbb{C}}/Q$ is in one-one correspondence $x \leftrightarrow Q_x$ with the set of $\overline{G}_{\mathbb{C}}$–conjugates of $Q$, by $Q_x = \{\overline{g} \in \overline{G}_{\mathbb{C}} \mid \overline{g}(x) = x\}$. We will make constant use of this identification.

**6.2.** Let $\overline{G}$ be an open subgroup of a real form $\overline{G}_{\mathbb{R}}$ of $\overline{G}_{\mathbb{C}}$, so $\overline{G}^0$ is the real analytic subgroup of $\overline{G}_{\mathbb{C}}$ for $\overline{\mathfrak{g}} = \overline{\mathfrak{g}}_{\mathbb{R}}$. Denote

(6.2.1) $\qquad \tau$ : complex conjugation of $\overline{G}_{\mathbb{C}}$ over $\overline{G}_{\mathbb{R}}$ and of $\overline{\mathfrak{g}}_{\mathbb{C}}$ over $\overline{\mathfrak{g}}_{\mathbb{R}}$.

The isotropy subgroup of $\overline{G}$ at $x \in X$ is $\overline{G} \cap Q_x$. The latter has Lie algebra $\overline{\mathfrak{g}} \cap \mathfrak{q}_x$ which is a real form of $\mathfrak{q}_x \cap \tau\mathfrak{q}_x$. The intersection of any two Borel subgroups contains a Cartan, and using care it follows that we have

(6.2.2) $\qquad$ a Cartan subalgebra $\overline{\mathfrak{h}} \subset \overline{\mathfrak{g}} \cap \mathfrak{q}_x$ of $\overline{\mathfrak{g}}$, a system $\Pi$ of simple $\overline{\mathfrak{h}}_{\mathbb{C}}$–roots of $\overline{\mathfrak{g}}_{\mathbb{C}}$, and a subset $\Phi \subset \Pi$ such that $\mathfrak{q}_x = \mathfrak{q}_\Phi$.

Then we have the key decomposition to understanding $G$–orbits on $X$:

$$\begin{aligned} &\mathfrak{q}_x \cap \tau\mathfrak{q}_x = (\mathfrak{q}_x \cap \tau\mathfrak{q}_x)^r + (\mathfrak{q}_x \cap \tau\mathfrak{q}_x)^u \text{ where } (\mathfrak{q}_x \cap \tau\mathfrak{q}_x)^r = \overline{\mathfrak{h}}_{\mathbb{C}} + \sum_{\Phi^r \cap \tau\Phi^r} \overline{\mathfrak{g}}^\phi \\ &\text{and } (\mathfrak{q}_x \cap \tau\mathfrak{q}_x)^u = \sum_{\Phi^r \cap \tau\Phi^u} \overline{\mathfrak{g}}^\phi + \sum_{\Phi^u \cap \tau\Phi^r} \overline{\mathfrak{g}}^\phi + \sum_{\Phi^u \cap \tau\Phi^u} \overline{\mathfrak{g}}^\phi. \end{aligned} \tag{6.2.3}$$

This shows that $\overline{G}(x)$ has real codimension $|\Phi^u \cap \tau\Phi^u|$ in $X$, in particular that $\overline{G}(x)$ is open in $X$ if and only if $\Phi^u \cap \tau\Phi^u$ is empty, and also that there are only finitely many $\overline{G}$–orbits on $X$. This last shows that $\overline{G}$ has both open and closed orbits.

Recall $\overline{G} = G/Z_G(G^0) = \mathrm{Int}(\mathfrak{g})_{\mathbb{C}}$, so $G$ acts on $X = \overline{G}/Q$ through $G \to \overline{G}$, specifically by $Q_{g(x)} = \mathrm{Ad}(g)Q_x$. Thus $G(x) = \overline{G}(x)$. Now the results on orbits and isotropy of $\overline{G}$ and $\overline{\mathfrak{g}}$ on $X$, apply as well to orbits and isotropy of $G$ and $\mathfrak{g}$.

**6.3.** Let $H$ be a Cartan subgroup of $G$ and $\theta$ a Cartan involution with $\theta(H) = H$. Let $K$ be the fixed point set $G^\theta$. As in (4.1.3) $\mathfrak{h} = \mathfrak{t} + \mathfrak{a}$ and $H = T \times A$ under the action of $\theta$. Thus [**50**, Theorem 4.5] the following conditions are equivalent: (i) $T$ is a Cartan subgroup of $K$, (ii) $\mathfrak{t}$ contains a regular element of $\mathfrak{g}$, and (iii) some simple system $\Pi$ of $\mathfrak{h}_{\mathbb{C}}$–roots satisfies $\tau\Pi = -\Pi$. Then those conditions hold, one says that $\mathfrak{h}$ is a *fundamental Cartan subalgebra* of $\mathfrak{g}$ and $H$ is a *fundamental Cartan subgroup* of $G$. Equivalently, $\mathfrak{h}$ is a *maximally compact Cartan subalgebra* of $\mathfrak{g}$ and $H$ is a *maximally compact Cartan subgroup* of $G$. From (6.2.3),

LEMMA 6.3.1. *$G(x)$ is open in $X$ if and only if there exist a maximally compact Cartan subalgebra $\mathfrak{h} \subset \mathfrak{g}$, a simple $\mathfrak{h}_{\mathbb{C}}$–root system $\Pi$ with $\tau\Pi = -\pi$, and a subset $\Phi \subset \Pi$, such that $\mathfrak{q}_x = \mathfrak{q}_\Phi$.*

An open orbit $G(x)$ carries an invariant Radon measure if and only if the isotropy subgroup at $x$ is reductive, i.e., if and only if the choices in Lemma 6.3.1 can be made so that $\tau\Phi^r = \Phi^r$ and $\tau\Phi^u = -\Phi^u$. Thus [**50**, Theorem 6.3] these conditions are equivalent: (i) $G(x)$ is open in $X$ and has a $G$–invariant positive Radon measure, (ii) $G(x)$ has a $G$–invariant possibly–indefinite Kähler metric, (iii) $\mathfrak{q}_x \cap \tau\mathfrak{q}_x$ is reductive, i.e., $\mathfrak{q}_x \cap \tau\mathfrak{q}_x = \mathfrak{q}_x^r \cap \tau\mathfrak{q}_x^r$ and (iv) $\tau\Phi^r = \Phi^r$ and $\tau\Phi^u = -\Phi^u$. Under those conditions we say that the open orbit $G(x)$ is *measurable*.

A closely related set of equivalent conditions [**50**, Theorem 6.7] is (a) some open $G$–orbit on $X$ is measurable, (b) every open $G$–orbit on $X$ is measurable and (c) if $\mathfrak{q} = \mathfrak{q}_\Phi$ then $\tau\mathfrak{q}$ is conjugate to the opposite parabolic $\mathfrak{q}^r + \mathfrak{q}^{-u}$ where $\mathfrak{q}^{-u} = \sum_{\Phi^u} \mathfrak{g}^{-\phi}$. These conditions are automatic [**50**, Corollary 6.8] if rank $K$ = rank $G$, i.e., if $G$ has relative discrete series representations. In that regard we will need

LEMMA 6.3.2. *Let $U$ be the isotropy subgroup of $G$ at $x \in X$. Suppose that $\mathfrak{q}$ does not contain any nonzero ideal of $\overline{\mathfrak{g}}_{\mathbb{C}}$. Then the following are equivalent.*

1. *$U$ acts on the tangent space to $G(x)$ as a compact group.*
2. *$G(x)$ has a $G$–invariant positive definite hermitian metric.*
3. *$\overline{\mathfrak{g}} \cap \mathfrak{q}_x$ is contained in the fixed point set of a Cartan involution of $\overline{\mathfrak{g}}$*

*Under these conditions, $G(x)$ is open in $X$ and the maximal compact subgroups $\overline{K} \subset \overline{G}$ satisfy* rank $\overline{K}$ = rank $\overline{G}$.

Suppose rank $K$ = rank $G$. Let $G(x)$ be an open orbit, so $\overline{\mathfrak{g}}$ has a Cartan subalgebra $\overline{\mathfrak{h}} \subset \overline{\mathfrak{k}} \cap \mathfrak{q}_x$ where $\overline{H} = H/Z_G(G^0)$ and $\overline{K} = K/Z_G(G^0)$. Let $W_K$, $W_{G_{\mathbb{C}}}$ and $W_{Q_x^r}$ denote Weyl groups relative to $\overline{\mathfrak{h}}$. Then [**50**, Theorem 4.9] the open $G$–orbits on $X$ are enumerated by the double coset space $W_K \backslash W_{G_{\mathbb{C}}} / W_{Q_x^r}$.

**6.4.** We look at the maximal complex analytic pieces of a $G$–orbit on $X$.

Let $V$ be a complex analytic space and $D \subset V$. By *holomorphic arc* in $D$ we mean a holomorphic map $f : \{z \in C \mid |z| < 1\} \to V$ with image in $D$. A *chain of holomorphic arcs* in $D$ is a finite sequence $\{f_1, \dots, f_m\}$ of holomorphic arcs in $D$ such that the image of $f_{k-1}$ meets the image of $f_k$. A *holomorphic arc component* of $D$ is an equivalence class of elements of $D$ under $u \sim v$ if there is a chain $\{f_1, \dots, f_m\}$ of holomorphic arcs in $D$ with $u$ in the image of $f_1$ and $v$ in the image of $f_m$. If $g$ is a holomorphic diffeomorphism of $V$, $g(D) = D$, and $S$ is a holomorphic arc component of $D$ then $g(S)$ is holomorphic arc component of $D$.

Let $L$ be a group of holomorphic diffeomorphisms of $V$ that preserve $D$. If $S$ is a holomorphic arc component of $D$ denote its $L$–normalizer $\{\alpha \in L| \mid \alpha(S) = S\}$ by $N_L(S)$. If $\alpha \in L$ and $\alpha(S)$ meets $S$ then $\alpha(S) = S$. So if $D$ is an $L$–orbit then $S$ is an $N_L(S)$–orbit. It can happen that $D$ is a real submanifold of $V$ but is not a complex submanifold; see [**50**, Example 8.12].

Now we turn to holomorphic arc components of an orbit $G(x) \subset X = \overline{G}_{\mathbb{C}}/Q$. It is a finite union of $G^0$–orbits, which are its topological components. So we have

$$S_{[x]} : \text{ holomorphic arc component of } G(x) \text{ through } x.$$

in the topological component of $x$ in $G(x)$, and its $G$– and $\overline{G}$–normalizers

$$N_{[x]} = \{g \in G \mid gS_{[x]} = S_{[x]}\} \text{ and } \overline{N}_{[x]} = N_{[x]}/Z_G(G^0).$$

The main general fact concerning these groups and their Lie algebras [**50**, Theorems 8.5 and 8.15] is that $\mathfrak{n}_{[x],\mathbb{C}}$ is a $\tau$–stable parabolic subalgebra of $\overline{\mathfrak{g}}_{\mathbb{C}}$, so $N_{[x],\mathbb{C}}$ is a parabolic subgroup of $\overline{G}$, and $\mathfrak{n}_{[x]} = \overline{\mathfrak{g}} \cap \mathfrak{n}_{[x],\mathbb{C}}$ is a parabolic subalgebra of $\overline{\mathfrak{g}}$, so $N_{[x]}$ is a subgroup of finite index in a parabolic subgroup of $\overline{G}$. This ensures $G = KN_{[x]}$ and $\overline{G} = \overline{KN}_{[x]}$. In other words, $K$ and $\overline{K}$ are transitive on the space $G/N_{[x]} = \overline{G}/\overline{N}_{[x]}$ of all holomorphic arc components of $G(x)$.

With $x \in X$ fixed and $\mathfrak{q}_x = \mathfrak{q}_\Phi$ and in (6.2.3) we consider the real linear form $\delta_x = \sum_{\Phi^u \cap \tau\Phi^u} \phi : \overline{\mathfrak{h}} \to \mathbb{R}$. That defines $\mathfrak{q}_{[x]} = \overline{\mathfrak{h}}_{\mathbb{C}} + \sum_{\langle \phi, \delta_x\rangle \geqq 0} \mathfrak{g}^\phi$, a $\tau$–stable parabolic subalgebra of $\mathfrak{g}_{\mathbb{C}}$. Then $(\mathfrak{q}^u_x \cap \tau\mathfrak{q}^u_x) \subset \mathfrak{q}_{[x]} \subset \{\overline{\mathfrak{n}}_{[x],\mathbb{C}} \cap (\mathfrak{q}_x + \tau\mathfrak{q}_x)\}$. Let $\Gamma = \{\phi \in \Delta_{\mathfrak{g}} \mid \langle \phi, \delta_x\rangle < 0, -\phi \notin \Phi^u \cap \tau\Phi^u, \phi + \tau\phi \text{ not a root.}\}$ and define $\mathfrak{m}_{[x]} = \mathfrak{q}_{[x]} + \sum_\Gamma \overline{\mathfrak{g}}^\phi$. Then [**50**, Theorem 8.9] $\mathfrak{m}_{[x]} \subset \{\overline{\mathfrak{n}}_{[x],\mathbb{C}} \cap (\mathfrak{q}_x + \tau\mathfrak{q}_x)\}$ and the following are equivalent: (i) The holomorphic arc components of $G(x)$ are complex submanifolds of $X$, (ii) $\overline{\mathfrak{n}}_{[x],\mathbb{C}} \subset (\mathfrak{q}_x + \tau\mathfrak{q}_x)$, (iii) $\overline{\mathfrak{n}}_{[x],\mathbb{C}} = \mathfrak{m}_{[x]}$, and (iv) $\mathfrak{m}_{[x]}$ is a subalgebra of $\mathfrak{g}_{\mathbb{C}}$. When these hold, we say that the orbit $G(x)$ is *partially complex.*

We will need stronger conditions. An orbit $G(x)$ is *of flag type* if the Zariski closure $\overline{N}_{[x],\mathbb{C}}$ of $S_{[x]}$ is a complex flag manifold, *measurable* if the holomorphic arc components carry positive Radon measures invariant under their normalizers, *integrable* if $(\mathfrak{q}_x + \tau\mathfrak{q}_x)$ is a subalgebra of $\overline{\mathfrak{g}}_{\mathbb{C}}$. Given $\mathfrak{q}_x = \mathfrak{q}_\Phi$ we denote

$$\mathfrak{v}^-_x = \sum_{\Phi^u \cap -\tau\Phi^u} \overline{\mathfrak{g}}^\phi, \ \mathfrak{v}^+_x = \sum_{-\Phi^u \cap -\tau\Phi^u} \overline{\mathfrak{g}}^\phi, \ \mathfrak{v}_x = \mathfrak{v}^-_x + \mathfrak{v}^+_x .$$

Then [**50**, Theorem 9.2] $G(x)$ is measurable if and only if $\overline{\mathfrak{n}}_{[x],\mathbb{C}} = (\mathfrak{q}_x \cap \tau\mathfrak{q}_x) + \mathfrak{v}_x$. It follows that, in $G(x)$ is measurable, then (i) the invariant measure on $S_{[x]}$ comes from an $N_{[x]}$–invariant possibly–indefinite Kähler metric, (ii) $G(x)$ is partially complex and of flag type, and (ii) $G(x)$ is integrable if and only if $\tau\mathfrak{q}^r_x = \mathfrak{q}^r_x$.

On the other hand, if $\tau\mathfrak{q}^r_x = \mathfrak{q}^r_x$ then [**50**, Theorem 9.19] $G(x)$ is measurable $\Leftrightarrow G(x)$ is integrable $\Leftrightarrow G(x)$ is partially complex and of flag type, and under those conditions $\overline{\mathfrak{n}}_{[x],\mathbb{C}} = (\mathfrak{q}_x \cap \tau\mathfrak{q}_x) + \mathfrak{v}_x = \mathfrak{q}_{[x]} = \mathfrak{q}_x + \tau\mathfrak{q}_x$. Open orbits are obviously integrable, partially complex and of flag type. Closed orbits are another matter. There is just one closed $G$–orbit on $X$, every maximal compact subgroup of $\overline{G}$ is transitive on it, and it is connected. There is a problem with [**50**, Theorem 9.12], where it was asserted that the closed orbit always is measurable, hence partially complex (consider [**50**, Example 8.12] applied to $SU(m,m)$). But if $Q$ is a Borel subgroup of $\overline{G}$, then the closed orbit is measurable, hence partially complex and of flag type, and in that case is integrable.

**6.5.** We now describe a class of orbits that plays a key role in the geometric realization of the various nondegenerate series of representations of $G$. Fix a Cartan subgroup $H = T\times A$ of $G$ and an associated cuspidal parabolic $P = MAN$. We need complex flag manifolds $X = \overline{G}_{\mathbb{C}}/Q$ and measurable integrable orbits $Y = G(x) \subset X$ such that the $G$–normalizers of the holomorphic arc components of $Y$ satisfy

(6.5.1) $$N_{[x]} = \{g \in G \mid gS_{[x]} = S_{[x]} \text{ has Lie algebra } \mathfrak{p}\}.$$

As $Y$ is to be measurable $S_{[x]}$ will be an open $M$–orbit in the sub–flag $\overline{M}_{\mathbb{C}}(x)$. So $AN$ will act trivially on $S_{[x]}$ and the isotropy subgroup of $G$ at $x$ will have form

$UAN$ with $T \subset U \subset M$. Finally, we need the condition that

$$U/Z_G(G^0) = \{m \in M \mid m(x) = x\}/Z_G(G^0) \text{ is compact.} \tag{6.5.2}$$

We look at some consequences of (6.5.1) and (6.5.2). Write $\mathfrak{z}_\mathfrak{g}$ for the center of $\mathfrak{g}$, so $\mathfrak{g} \cong \mathfrak{z}_\mathfrak{g} \oplus \overline{\mathfrak{g}}$. Since $Y$ is measurable and integrable, §6.4 would lead to

$$\begin{aligned} &\mathfrak{p}_\mathbb{C} = \mathfrak{z}_{\mathfrak{g},\mathbb{C}} + \mathfrak{q}_x + \tau\mathfrak{q}_x\,,\ \tau\mathfrak{q}_x^r = \mathfrak{q}_x^r\,,\ \mathfrak{n}_\mathbb{C} = \mathfrak{p}_\mathbb{C}^u = \mathfrak{q}_x^u \cap \tau\mathfrak{q}_x^u\,,\\ &(\mathfrak{m}+\mathfrak{a})_\mathbb{C} = \mathfrak{p}_\mathbb{C}^r = (\mathfrak{z}_{\mathfrak{g},\mathbb{C}} + \mathfrak{q}_x^r) + (\mathfrak{q}_x^{-u} \cap \tau\mathfrak{q}_x^u) + (\mathfrak{q}_x^u \cap \tau\mathfrak{q}_x^{-u})\,,\\ &(\mathfrak{u}+\mathfrak{a})_\mathbb{C} = \mathfrak{z}_{\mathfrak{g},\mathbb{C}} + \mathfrak{q}_x^r \text{ and } \mathfrak{m}_\mathbb{C} = \mathfrak{u}_\mathbb{C} + (\mathfrak{q}_x^{-u} \cap \tau\mathfrak{q}_x^u) + (\mathfrak{q}_x^u \cap \tau\mathfrak{q}_x^{-u}). \end{aligned} \tag{6.5.3}$$

Since $S_{[x]}$ a measurable open $M^0$–orbit in the flag $\overline{M}_\mathbb{C}(x)$, (6.5.3) ensures that

(6.5.4) $\mathfrak{r} := \mathfrak{m}_\mathbb{C} + (\mathfrak{z}_{\mathfrak{g},\mathbb{C}} + \mathfrak{q}_x)$ is parabolic in $\mathfrak{m}_\mathbb{C}$ with $\mathfrak{r}^r = \mathfrak{u}_\mathbb{C}$ and $\mathfrak{r}^u = \mathfrak{q}_x^u \cap \tau\mathfrak{q}_x^{-u}$.

The following Proposition shows that (6.5.3) and (6.5.4) give us the parabolics that we need for our geometric realizations.

PROPOSITION 6.5.5. *Let $G$ be a real reductive Lie group in the class $\widetilde{\mathcal{H}}$ of* (1.2.1), *$H = T \times A$, and $P = MAN$ an associated cuspidal parabolic subgroup of $G$. Suppose that* (i) *$\mathfrak{u} \subset \mathfrak{m}$ is the $\mathfrak{m}$–centralizer of a subalgebra of $\mathfrak{t}$ such that $U^0/(U^0 \cap Z_G(G^0))$ is compact,* (ii) *$\mathfrak{r} \subset \mathfrak{m}_\mathbb{C}$ is a parabolic subalgebra with $\mathfrak{r}^r \subset \mathfrak{u}_\mathbb{C}$,* (iii) *$\mathfrak{q}$ is the $\overline{\mathfrak{g}}_\mathbb{C}$–normalizer of $\mathfrak{r}^u + \mathfrak{n}_\mathbb{C}$ and $Q$ is the corresponding analytic subgroup of $\overline{G}_\mathbb{C}$, and* (iv) *$X = \overline{G}_\mathbb{C}/Q$ and $x = 1Q \in X$. Then $Q$ is a parabolic subgroup of $\overline{G}_\mathbb{C}$, $\mathfrak{q}^u = \mathfrak{r}^u + \mathfrak{n}_\mathbb{C}$, and $G(x)$ is a measurable integrable orbit, and $(X, x)$ satisfies* (6.5.1) *and* (6.5.2). *Conversely every pair $(X, x)$ satisfying* (6.5.1) *and* (6.5.2), *$G(x)$ measurable and integrable, $U^0/(U^0 \cap Z_G(G^0))$ compact, is constructed as above.*

PROOF. Denote $\overline{M} = M/Z_G(G^0)$. Let $\overline{R}$ be the analytic subgroup of $\overline{M}_\mathbb{C}$ for $\overline{\mathfrak{r}} := \mathfrak{r}/\mathfrak{z}_{\mathfrak{g},\mathbb{C}}$ and $S = \overline{M}_\mathbb{C}/\overline{R}$. Then $S$ is a complex flag manifold of $\overline{M}_\mathbb{C}$ by (ii). The isotropy subalgebra of $\mathfrak{m}$ at $s = 1\overline{R} \in S$ is $\mathfrak{m} \cap \mathfrak{r}$. It has reductive part $\mathfrak{u}$ by (ii). As $\mathfrak{t} \subset \mathfrak{u}$ by (i), and as $M(s)$ is measurable and open in $S$ by Lemmas 6.3.1 and 6.3.2, we have $\mathfrak{m} \cap \mathfrak{r} = \mathfrak{u}$ and $\mathfrak{r} \cap \tau\mathfrak{r} = \mathfrak{u}_\mathbb{C}$.

Define $\mathfrak{q}$ and $Q$ as in (iii). The contribution to $\mathfrak{q}$ from $\overline{\mathfrak{m}}_\mathbb{C}$ is $\overline{\mathfrak{r}}$, all of $(\mathfrak{a}+\mathfrak{n})_\mathbb{C}$ from $(\mathfrak{a}+\mathfrak{n})_\mathbb{C}$, and 0 from $\mathfrak{n}_\mathbb{C}^-$. So $\mathfrak{q} = (\overline{\mathfrak{u}}_\mathbb{C} + \mathfrak{a}_\mathbb{C}) + (\mathfrak{r}^u + \mathfrak{n}_\mathbb{C})$, thus is parabolic in $\overline{\mathfrak{g}}_\mathbb{C}$. Now $X = \overline{G}_\mathbb{C}/Q$ is a complex flag manifold, $\mathfrak{z}_{\mathfrak{g},\mathbb{C}} + \mathfrak{q}^r = \mathfrak{u}_\mathbb{C} + \mathfrak{a}_\mathbb{C}$, and $\mathfrak{q}^u = \mathfrak{r}^u + \mathfrak{n}_\mathbb{C}$. In particular, $\mathfrak{q} + \tau\mathfrak{q} = \mathfrak{p}_\mathbb{C}/\mathfrak{z}_\mathfrak{g}$ is a subalgebra of $\mathfrak{g}_\mathbb{C}$, so the orbit $G(x) \subset X$ is integrable, and $\tau\mathfrak{q}^r = \mathfrak{r}^r$ so $G(x)$ is measurable with $\overline{\mathfrak{n}}_{[x],\mathbb{C}} = \mathfrak{q} + \tau\mathfrak{q} = \mathfrak{p}_\mathbb{C}/\mathfrak{z}_{\mathfrak{g},\mathbb{C}}$. We conclude $\overline{\mathfrak{n}}_{[x]} = \mathfrak{p}/\mathfrak{z}_\mathfrak{g}$ and so $\mathfrak{n}_{[x]} = \mathfrak{p}$. We have shown that $G(x)$ is a measurable integrable orbit, and $(X, x)$ satisfies (6.5.1) and (6.5.2).

For the converse compare (6.5.1) and (6.5.2) with the construction. □

We enumerate the $(X, x)$ of Proposition 6.5.5. Let $\Pi_\mathfrak{t}$ be a simple $\mathfrak{t}_\mathbb{C}$–roots system on $\mathfrak{m}_\mathbb{C}$ and $\Phi_\mathfrak{t}$ a subset of $\Pi_\mathfrak{t}$. Let $\Pi$ be the simple $\mathfrak{h}_\mathbb{C}$–root system on $\mathfrak{g}_\mathbb{C}$ that contains $\Pi_\mathfrak{t}$ and induces the positive $\mathfrak{a}$–root system used for construction of $P = MAN$. Define $\Phi = \Phi_\mathfrak{t} \cup (\Pi \setminus \Pi_\mathfrak{t})$. The parabolic subalgebras $\mathfrak{q} \subset \overline{\mathfrak{g}}_\mathbb{C}$ of Proposition 6.5.5 are just the $\mathfrak{q}_\Phi$.

COROLLARY 6.5.6. *Given $G(x) \in X$ as in* Proposition 6.5.5, *$M^\dagger$ is the stabilizer $\{m \in M \mid mS_{[x]} = S_{[x]}\}$ of $S_{[x]}$ in $M$. Thus $U \subset M^\dagger$ and $N_{[x]} = M^\dagger AN$.*

PROOF. Let $M^1 = \{m \in M \mid mS_{[x]} = S_{[x]}\}$. Then $M^\dagger \subset M^1$ because $M^\dagger = Z_M(M^0)M^0$ and $Z_M(M^0)$ acts trivially on $S_{[x]}$. The isotropy subgroup $U$ of $M$ at $x$ is in $M^1$ and $M^0$ is transitive on $S_{[x]}$, so $M^1 = UM^0$. Let $u \in U$. All Cartan subalgebras and all Weyl chambers of $\mathfrak{u}$ are $\mathrm{Ad}(U^0)$–conjugate, so we choose a Weyl chamber $\mathfrak{d} \subset i\mathfrak{t}^*$ for $\mathfrak{u}$ and replace $u$ within $uU^0$ so that $\mathrm{Ad}(u)$ preserves $\mathfrak{t}$ and $\mathfrak{d}$. Thus $\mathrm{Ad}(u)$ is an inner automorphism of $\mathfrak{g}_\mathbb{C}$ that is the identity on $\mathfrak{t}$, so $u \in T \subset M^\dagger$. We have shown $M^1 \subset M^\dagger$. As $M^\dagger \subset M^1$ now $M^1 = M^\dagger$. □

COROLLARY 6.5.7. *Given* $G(x) \in X$ *as in* Proposition 6.5.5 *and* $u \in U$, $\mathrm{Ad}(u)$ *is an inner automorphism on* $U^0$.

## 7. Open Orbits and Discrete Series

Let $G$ be a reductive Lie group of our class specified in §3.1. We consider complex flag manifolds $X = \overline{G}_C/Q$ and open orbits $Y = G(x) \subset X$ such that $U = \{g \in G : g(x) = x\}$ is compact modulo $Z$. In §7.1 we see that these pairs $(X, x)$ exist precisely when $G$ has relative discrete series representations, that $U = Z_G(G^0)U^0$ with $U^0 = U \cap G^0$, and that $Y$ has $|G/G^\dagger|$ topological components. If $[\mu] \in \widehat{U}$ we show that the associated $G$–homogeneous hermitian vector bundle $\mathcal{V}_\mu \to Y$ has a unique $G$–homogeneous holomorphic vector bundle structure. That allows us to construct the Hilbert spaces $H_2^{0,q}(Y; \mathcal{V}_\mu)$ of square integrable harmonic $(0, q)$–forms on $Y$ with values in $\mathcal{V}_\mu$, and unitary representations $\pi_\mu^q$ of $G$ on $H_2^{0,q}(\mathcal{V}_\mu)$. The remainder of §7 shows that the $[\pi_\mu^q]$, $q \geqq 0$ and $[\mu] \in \widehat{U}$, are the relative discrete series classes in $\widehat{G}$.

Section 7.2 is the formulation and history of our main result, Theorem 7.2.3. Let $[\mu] \in \widehat{U}$. Then $[\mu] = [\chi \otimes \mu^0]$ where $[\chi] \in \widehat{Z_G(G^0)}$ and $[\mu^0] \in \widehat{U^0}$. Let $\Theta_{\pi_\mu^q}^{disc}$ denote the character of the discrete part of $\pi_\mu^q$. We prove

$$\sum_{q \geqq 0} (-1)^q \Theta_{\pi_\mu^q}^{disc} = (-1)^{n+q(\lambda+\rho)} \Theta_{\pi_{\chi,\lambda+\rho}}$$

where $\lambda$ is the highest weight of $\mu^0$, $n$ is the number of positive roots, and $\rho$ is half the sum of the positive roots. We note that $H_2^{0,q}(\mathcal{V}_\mu) = 0$ for $q \neq q(\lambda + \rho)$, and we show that $[\pi_\mu^{q(\lambda+\rho)}] = [\pi_{\chi,\lambda+\rho}] \in \widehat{G}_{disc}$. Theorem 7.2.3 is proved in §§7.3 through 7.7.

We reduce the proof of Theorem 7.2.3 to the case $G = G^\dagger$ in §7.3, to the case $G = G^0$ in §7.4, and then further to the case where $Q$ is a Borel subgroup of $\overline{G}_C$ in §7.5. In §7.6 we use results of Harish–Chandra and a method of W. Schmid to prove the alternating sum formula for the $\Theta_{\pi_\mu^q}^{disc}$. The vanishing statement comes out of work of Schmid [**41**] cited above. It combines with the alternating sum formula to identify $[\pi_{\chi,\lambda+\rho}] \in \widehat{G}_{disc}$ as the discrete part of $[\pi_\mu^{q(\lambda+\rho)}]$. This trick is due to Narasimhan and Okamoto. Finally we use Corollary 5.7.2 of our Plancherel Theorem to show that $[\pi_\mu^{q(\lambda+\rho)}]$ has no nondiscrete part, so $[\pi_\mu^{q(\lambda+\rho)}] = [\pi_{\chi,\lambda+\rho}]$, completing our proof in §7.7.

**7.1.** $G$ is a Lie group of the class $\widetilde{\mathcal{H}}$ of general real reductive Lie groups defined in §3.1. As explained in §6.2, $\overline{G} = G/Z_G(G^0)$ has complexification $\overline{G}_C = \mathrm{Int}(\mathfrak{g}_C)$, and $G$ acts on the complex flag manifolds of $\overline{G}_C$. To realize the relative discrete series of $G$ we work with

$$\begin{aligned} &X = \overline{G}_C/Q \text{ complex flag manifold of } \overline{G}_C \\ (7.1.1)\qquad &Y = G(x) \subset X \text{ open } G\text{–orbit such that} \\ &\qquad \text{the isotropy subgroup } U \text{ of } G \text{ at } x \text{ is compact modulo } Z. \end{aligned}$$

We collect some immediate consequences of (7.1.1).

LEMMA 7.1.2. *Suppose $(X,x)$ is given as in (7.1.1). Then $U/Z$ contains a compact Cartan subgroup $H/Z$ of $G/Z$, so $G$ has relative discrete series representations. Further, the open orbit $Y = G(x) \subset X$ is measurable and integrable, and $(X,x)$ is the case $P = G$ of (6.7.1). Finally, $U = Z_G(U^0)U^0$, $U \cap G^0 = U^0$, $UG^0 = G^\dagger$, and $G/G^\dagger$ enumerates the topological components of $Y$.*

*Remark.* As a consequence of the second assertion, all possibilities for (7.1.1) are enumerated in the paragraph following Proposition 6.5.5.

PROOF. Isotropy subgroups of $G$ on $X$ all contain Cartan subgroups of $G$ by (6.2.2). Now the first assertion follows from (7.1.1) and Theorem 3.5.6.

$U$ acts on the tangent space at $x$ as $U/Z_G(G^0)$, which is compact by (7.1.1). Thus the orbit $Y = G(x)$ is measurable by Lemma 6.3.2. As open orbits are integrable with $\mathfrak{q}_x + \tau\mathfrak{q}_x = \bar{\mathfrak{g}}_C$. Now we have (6.5.1) and (6.5.2) with $P = M = G = N_{[x]}$.

Let $u \in U$. Corollary 6.5.7 says that $\mathrm{Ad}(u)$ is trivial on some Cartan subalgebra of $\mathfrak{u}$, thus on a Cartan subalgebra of $\mathfrak{g}$. Now $\mathrm{Ad}(u)$ is an inner automorphism of $G^0$, i.e. $u \in G^\dagger$. We have just seen $U \subset G^\dagger = Z_G(G^0)G^0$. On the other hand, open orbits are simply connected, so $U \cap G^0 = U^0$, Thus $U = Z_G(G^0)U^0$ and $UG^0 = G^\dagger$. Since $UG^0$ is the $G$–normalizer of $G^0(x)$, now $G/G^\dagger$ parameterizes the components of $G(x)$. □

The facts about $U$ in Lemma 7.1.2 tell us

$$\widehat{U} = \{[\chi \otimes \mu^0] : [\chi] \in \widehat{Z_G(G^0)} \text{ and } [\mu^0] \in \widehat{U^0}\}; \text{ so}$$

(7.1.3) $\quad$ if $[\mu] \in \widehat{U}$ then its representation space $E_\mu$ has $\dim E_\mu < \infty$ and we have $\mathbb{E}_\mu \to Y$, a $G$–homogeneous hermitian vector bundle.

LEMMA 7.1.4. *There is a unique complex structure on $\mathbb{E}_\mu$ such that $\mathbb{E}_\mu \to Y$ is a $G$–homogeneous holomorphic vector bundle.*

PROOF. The action of $G^0$ on $X$ maps $\mathfrak{g}_\mathbb{C}$ to a Lie algebra of holomorphic vector fields. Define

$$\mathfrak{l} = \{\xi \in \mathfrak{g}_\mathbb{C} \mid \xi_x = 0\},$$

isotropy subalgebra at $x$. The homomorphism $G^0 \to \overline{G}$ induces a homomorphism $\alpha$ of $\mathfrak{g}_C$ onto $\bar{\mathfrak{g}}_C$, and $\mathfrak{l} = \alpha^{-1}(\mathfrak{q}_x)$. Note that $\mathfrak{u}_C = \alpha^{-1}(\mathfrak{q}_x^r)$ reductive subalgebra of $\mathfrak{l}$. Choose a linear algebraic group with Lie algebra $\mathfrak{g}_\mathbb{C}$ and observe that $\alpha$ is a homomorphism of algebraic Lie algebras. Thus $\mathfrak{u}_\mathbb{C}$ is a maximal reductive subalgebra of $\mathfrak{l}$, and there is a nilpotent ideal $\mathfrak{l}^-$ such that $\mathfrak{l} = \mathfrak{u}_\mathbb{C} + \mathfrak{l}^-$ semidirect sum. Observe that $\mathrm{Ad}(u)\mathfrak{l}^- = \mathfrak{l}^-$ for all $u \in U$.

By extension of $\mu$ from U to $\mathfrak{l}$, we mean a (complex linear) representation $\lambda$ of $\mathfrak{l}$ on $V_\mu$ such that

$$\lambda|_\mathfrak{u} = \mu, \text{ i.e., } \lambda(\xi) = \mu(\xi) \text{ for all } \xi \in \mathfrak{u},$$

and

$$\mu(u)\lambda(\xi)\mu(u)^{-1} = \lambda(\mathrm{Ad}(u)\xi) \text{ for all } u \in U \text{ and } \xi \in \mathfrak{l}.$$

Let $\lambda$ be such an extension. Then $\lambda(\mathfrak{l}^-)$ consists of nilpotent linear transformations because $\lambda$ is an algebraic representation of $\mathfrak{l}$, and that implies $\lambda(\mathfrak{l}^-) = 0$ because $\mu$ is irreducible. Thus there is just one extension of $\mu$ from $U$ to $\mathfrak{l}$; it is given by $\lambda(\xi_1 + i\xi_2 + \eta) = \mu(\xi_1) + i\mu(\xi_2)$ where $\xi_1, \xi_2 \in \mathfrak{u}$ and $\eta \in \mathfrak{l}^-$.

Our lemma now follows from the fact [**45**, Theorem 3.6] that the $G$–homogeneous holomorphic vector bundle structures on $\mathbb{E}_\mu \to Y$ are in bijective correspondence with the extensions of $\mu$ from $U$ to $\mathfrak{l}$. □

Using (7.1.1) we fix a $G$–invariant hermitian metric on the complex manifold $Y$. The unitary structure of $E_\mu$ specifies a $G$–invariant hermitian metric on the fibers of $\mathbb{E}_\mu \to Y$. Denote

$$(7.1.5)\qquad \begin{aligned}&A^{p,q}(Y;\mathbb{E}_\mu) = \{C^\infty(p,q)\text{–forms on } Y \text{ with values in } \mathbb{E}_\mu\} \text{ so we have}\\ &\text{Hodge–Kodaira maps } A^{p,q}(Y;\mathbb{E}_\mu) \overset{\#}{\to} A^{n-p,n-q}(Y;\mathbb{E}_\mu^*) \overset{\widetilde{\#}}{\to} A^{p,q}(Y;\mathbb{E}_\mu)\end{aligned}$$

Here $n = \dim_{\mathbb{C}} Y$ and $\mathbb{E}_\mu^* = \mathbb{E}_{\mu^*}$ is the dual bundle. If $\alpha,\ \beta \in A^{p,q}(Y;\mathbb{E}_\mu)$ then $\alpha \wedge \#\beta \in A^{n,n}(Y;\mathbb{E}_\mu \otimes \mathbb{E}_\mu^*)$. The pairing $E_\mu \otimes E_\mu^* \to \mathbb{C}$ sends $\alpha\wedge\#\beta$ to an ordinary $(n,n)$–form on $Y$ that we denote $\alpha\bar{\wedge}\#\beta$. This gives us a pre Hilbert space

$$(7.1.6)\qquad A_2^{p,q}(\mathcal{V}_\mu) = \left\{\alpha \in A^{p,q}(Y;\mathbb{E}_\mu) \,\middle|\, \int_Y \alpha\bar{\wedge}\#\alpha < \infty\right\},\ \langle\alpha,\beta\rangle = \int_Y \alpha\bar{\wedge}\#\beta.$$

The space of *square integrable* $(p,q)$*–forms* on $Y$ with values in $\mathbb{E}_\mu$ is

$$(7.1.7)\qquad L_2^{p,q}(Y;\mathbb{E}_\mu):\ \text{Hilbert space completion of } A_2^{p,q}(Y;\mathbb{E}_\mu).$$

The operator $\overline{\partial} : A^{p,q}(Y;\mathbb{E}_\mu) \to A^{p,q+1}(Y;\mathbb{E}_\mu)$ is densely defined on $L_2^{p,q}(Y;\mathbb{E}_\mu)$ with formal adjoint $\overline{\partial}^* = -\widetilde{\#}\,\overline{\partial}\,\#$. That gives us a second order elliptic operator

$$(7.1.8)\qquad \square = (\overline{\partial} + \overline{\partial}^*)^2 = \overline{\partial}\,\overline{\partial}^* + \overline{\partial}^*\,\overline{\partial} :\ \text{Kodaira–Hodge–Laplacian.}$$

The hermitian metric on $Y$ is complete by homogeneity, so the work of Andreotti and Vesentini [**2**] applies. First, it says that $\square$, with domain consisting of the compactly supported forms in $A^{p,q}(Y;\mathbb{E}_\mu)$, is essentially self adjoint on $L_2^{p,q}(Y;\mathbb{E}_\mu)$. We also write $\square$ for the unique self adjoint extension, which coincides both with the adjoint and the closure. Its kernel

$$(7.1.9)\qquad H_2^{p,q}(Y;\mathbb{E}_\mu) = \{\omega \in L_2^{p,q}(Y;\mathbb{E}_\mu) \mid \square\omega = 0\}$$

consists of the *square integrable harmonic* $(p,q)$*–forms* on $Y$ with values in $\mathbb{E}_\mu$. $H_2^{p,q}(Y;\mathbb{E}_\mu)$ is a closed subspace of $L_2^{p,q}(Y;\mathbb{E}_\mu)$. It is contained in $A_2^{p,q}(Y;\mathbb{E}_\mu)$ by ellipticity of $\square$. Write $c\ell$ for closure. The Andreotti–Vesentini work shows that $\overline{\partial}^*$ has closed range and gives us an orthogonal direct sum

$$(7.1.10)\qquad L_2^{p,q}(Y;\mathbb{E}_\mu) = c\ell\,\overline{\partial}L_2^{p,q-1}(\mathcal{V}_\mu) \oplus \overline{\partial}^*L_2^{p,q+1}(\mathcal{V}_\mu) \oplus H_2^{p,q}(\mathcal{V}_\mu)$$

Here $\overline{\partial}$ has kernel $c\ell\,\overline{\partial}L_2^{p,q-1}(Y;\mathbb{E}_\mu)\oplus H_2^{p,q}(Y;\mathbb{E}_\mu)$ and the kernel of $\overline{\partial}^*$ is the closed subspace $\overline{\partial}^*L_2^{p,q+1}(Y;\mathbb{E}_\mu) \oplus H_2^{p,q}(Y;\mathbb{E}_\mu)$. Thus $H_2^{p,q}(Y;\mathbb{E}_\mu)$ is a square integrable Dolbeault cohomology group.

The metrics and complex structures on $Y$ and $\mathbb{E}_\mu$ are invariant under the action of $G$. Thus $G$ acts on $L_2^{p,q}(Y;\mathbb{E}_\mu)$ by a unitary representation $\widetilde{\pi}_\mu^{p,q}$ that has a subrepresentation

$$(7.1.11)\qquad \pi_\mu^{p,q}:\ \text{unitary representation of } G \text{ on } H_2^{p,q}(\mathbb{E}_\mu).$$

For convenience we also denote $\pi_\mu^q = \pi_\mu^{0,q}$. The program here in §7 is to represent the classes in $\widehat{G}_{disc}$ by the various $\pi_\mu^q$.

**7.2.** Fix a compact Cartan subgroup $H/Z$ of $G/Z$ with $H \subset U$ as in Lemma 7.1.2. Choose a system $\Pi$ of simple $\mathfrak{h}_{\mathbb{C}}$–roots of $\mathfrak{g}_{\mathbb{C}}$ such that $\mathfrak{q}_x = \mathfrak{q}_\Phi$ where $\Phi \subset \Pi$. Let $\Sigma^+$ denote the corresponding positive root system. As usual we pass to a $\mathbb{Z}_2$ extension if necessary so that $e^\rho$ and $\Delta$ are well defined on $H^0$ and define

$$\rho = \frac{1}{2}\sum_{\phi \in \Sigma^+} \phi, \quad \Delta = \prod_{\phi \in \Sigma^+} (e^{\phi/2} - e^{-\phi/2}), \quad \text{and } \widetilde{\omega}(\cdot) = \prod_{\phi\in\Sigma^+} \langle \cdot, \phi\rangle,$$
(7.2.1)
$$L = \{\lambda \in i\mathfrak{h}^* : e^\lambda \text{ is well defined on } H^0\} \text{ and } L' = \{\lambda \in L : \widetilde{\omega}(\lambda) \neq 0\}.$$

Let $\theta$ be the (unique) Cartan involution under which $H$ is stable and $K = G^\theta$, so $H \subset U \subset K$. Now $\Sigma^+ = \Sigma_k^+ \cup \Sigma_m^+$ (disjoint) where $\Sigma_k^+$ consists of the compact positive roots ($\overline{\mathfrak{g}}^\phi \subset \mathfrak{k}_{\mathbb{C}}$) and $\Sigma_m^+$ consists of the noncompact positive roots ($\overline{\mathfrak{g}}^\phi \not\subset \mathfrak{k}_{\mathbb{C}}$). If $\lambda \in L'$ we have

$$q(\lambda) = |\{\phi \in \Sigma_k^+ : \langle\lambda,\phi\rangle < 0\}| + |\{\phi \in \Sigma_m^+ : \langle \lambda, \phi\rangle > 0\}|. \tag{7.2.2}$$

Recall the statement of Theorem 3.5.9. The main result of §7 is

THEOREM 7.2.3. *Let $[\mu] \in \widehat{U}$, say $[\mu] = [\chi \otimes \mu^0]$ as in (7.1.3). Let $\lambda$ be the highest weight of $\mu^0$ for the positive $\mathfrak{h}_{\mathbb{C}}$–root system $\Sigma^+ \cap \Phi^r$ of $\mathfrak{u}_{\mathbb{C}}$. Then $\lambda \in L$ and $[\mu] \in \widehat{U}_\zeta$ where $\zeta \in \widehat{Z}$ coincides with $e^\lambda$ on $Z \cap G^0$ and $[\chi] \in \widehat{Z_G(G^0)}_\zeta$. Assume $\lambda + \rho \in L'$. Then $H_2^{0,q}(Y;\mathbb{E}_\mu) = 0$ whenever $q \neq q(\lambda+\rho)$, and the natural action of $G$ on $H_2^{0,q(\lambda+\rho)}(Y;\mathbb{E}_\mu)$ is the $\zeta$–discrete series class $[\pi_{\chi,\lambda+\rho}]$.*

Theorem 7.2.3 gives a number of explicit geometric realizations of the relative discrete series representations of $G$. The case where $G$ is a connected semisimple Lie group with finite center and $U = H$ is due to W. Schmid ([**40**], [**41**]); to some extent we follow his ideas. The case where $G$ is a connected semisimple Lie group with finite center and $Y = G(x)$ is a hermitian symmetric space was proved by M. S. Narasimhan and K. Okamoto [**35**]. Some results for groups with possibly infinite center were proved by Harish–Chandra [**15**] and J. A. Tirao [**46**]. Also, W. Schmid (unpublished) and R. Parthasarathy ([**36**], [**37**]) obtained realizations on spaces of square integrable harmonic spinors. Finally, R. Hotta [**30**] realized discrete series representations of connected semisimple groups of finite center on certain eigenspaces of the Casimir operator.

We carry out the proof of Theorem 7.2.3 in §§7.3 through 7.7.

**7.3.** We reduce Theorem 7.2.3 to the case $G = G^\dagger$.

Choose a system $\{g_1, \cdots, g_r\}$ of coset representatives of $G$ modulo $G^\dagger$. According to Lemma 7.2.1, the topological components of $Y = G(x)$ are the $Y_i = G^\dagger(g_i x)$. Let ${}^i\pi_\mu^q$ denote the representation of $G^\dagger$ on

$$H_2^{0,q}(Y;\mathbb{E}_\mu|_{Y_i}) = \{\omega \in H_2^{0,q}(Y;\mathbb{E}_\mu) \ : \ \omega \text{ is supported in } Y_i\}.$$

Evidently $H_2^{0,q}(Y;\mathbb{E}_\mu) = H_2^{0,q}(Y;\mathbb{E}_\mu|_{Y_1}) \oplus \cdots \oplus H_2^{0,q}(Y;\mathbb{E}_\mu|_{Y_r})$ as orthogonal direct sum. Thus $\pi_\mu^{0,q} = {}^1\pi_\mu^{0,q} \oplus \cdots \oplus {}^r\pi_\mu^{0,q}$. Also, $\pi_\mu^q(g_i)$ sends $H_2^{0,q}(Y;\mathbb{E}_\mu|_{Y_j})$ to $H_2^{0,q}(Y;\mathbb{E}_\mu|_{Y_k})$ where $g_i Y_k = Y_j$, i.e., where $g_i^{-1} g_j \in g_k G^\dagger$. In summary,

LEMMA 7.3.1. *$\pi_\mu^q = \mathrm{Ind}_{G^\dagger}^G({}^i\pi_\mu^{0,q})$ for $1 \leqq i \leqq r$.*

We know from Theorem 3.5.7 that $\widehat{G}_{disc}$ consists of the classes $[\pi] = [\mathrm{Ind}_{G^\dagger}^G(\pi^\dagger)]$ where $[\pi^\dagger] \in \widehat{G^\dagger}_{disc}$. Further, $\Theta_\pi$ is supported in $G^\dagger$, where it coincides with $\Theta_{\pi_{G^\dagger}}$.

Now Lemma 7.3.1 tells us that, if Theorem 7.2.3 holds for $G^\dagger$ with each of the $\mathbb{E}_{\mu|_{Y_i}}$. Then Theorem 7.2.3 is valid for $G$ with $\mathbb{E}_\mu$. In summary,

LEMMA 7.3.2. *In the proof of* Theorem 7.2.3 *we may assume* $G = G^\dagger$.

**7.4.** We reduce Theorem 7.2.3 to the case where $G$ is connected. Using Lemma 7.3.2 we assume $G = G^\dagger$. Thus $G = Z_G(G^0)G^0$ and $Y = G(x)$ is connected. Recall $[\mu] = [\chi \otimes \mu^0]$ with $[\chi] \in \widehat{Z_G(G^0)}$ and $[\mu^0] \in \widehat{U^0}$, so $E_\mu = E_\chi \otimes E_{\mu^0}$. Now $[\mu^0]$ specifies a $G^0$–homogeneous holomorphic vector bundle $\mathbb{E}_{\mu^0} \to Y$. Let $\pi^q_{\mu^0}$ denote the representation of $G^0$ on $H^{0,q}_2(Y; \mathbb{E}_{\mu^0})$.

LEMMA 7.4.1. $\pi^q_\mu = \chi \otimes \pi^q_{\mu^0}$ *for all* $q \geqq 0$.

PROOF. $Z_G(G^0)$ acts trivially on $X$, so it acts trivially on the bundle of ordinary $(0,q)$–forms over the orbit $Y \subset X$. Thus $Z_G(G^0)$ acts on $L^{0,q}_2(Y; \mathbb{E}_\mu)$ as a type I primary representation $\omega_\chi$. In particular $\pi^q_\mu|_{Z_G(G^0)}$ is a multiple of $\chi$. But $\mu|_{U^0} = (\dim \chi)\mu^0$ so $\pi^q_\mu|_{G^0} = (\dim \chi)\pi^q_{\mu^0}$. We conclude $\pi^q_\mu = \chi \otimes \pi^q_{\mu^0}$. □

We know from Proposition 3.5.2 that $\widehat{G^\dagger}_{disc}$ consists of the $[\chi \otimes \pi^0]$ where $[\chi] \in \widehat{Z_G(G^0)}$ and $[\pi^0] \in \widehat{G^0}_{disc}$ agree on $Z_{G^0}$. The distribution character $\Theta_{\chi\otimes\pi^0} = (\text{trace } \chi)\Theta_{\pi^0}$. If Theorem 7.2.3 holds for $G^0$ with $\mathbb{E}_{\mu^0}$ now Lemma 7.4.1 ensures the result for $G$ with $\mathbb{E}_\mu$. In summary

LEMMA 7.4.2. *In the proof of* Theorem 7.2.3 *we may assume $G$ is connected.*

**7.5.** We reduce Theorem 7.2.3 to the case where $G = G^0$ and $U = H^0$.

Choose a Borel subgroup $B \subset Q$ of $\overline{G}_{\mathbb{C}}$. Denote $X' = \overline{G}_{\mathbb{C}}/B$ and consider the $G$-equivariant projection $r : X' \to X$ defined by $r(\bar{g}B) = \bar{g}Q$. Now choose a base point $x' \in r^{-1}(x)$ defined by $\mathfrak{b}_{x'} = \mathfrak{q}_\emptyset$ relative to $(\mathfrak{h}_{\mathbb{C}}, \Pi)$. Since $\tau\phi = -\phi$ for every $\mathfrak{h}$-root, the isotropy subalgebra of $\mathfrak{g}$ at $x'$ is just $\mathfrak{h}$. Now $Y' = G(x')$ is open in $X'$ and $H = \{g \in G : g(x') = x'\}$, and $r : Y' \to Y$ is $G$-equivariant and holomorphic.

Following Lemma 7.4.2, we assume $G$ connected, so $U$ and $H$ are connected. Now $\lambda$ is the highest weight of $\mu$ and $e^\lambda \in \widehat{H}$ specifies

$$(7.5.1) \qquad \begin{aligned} &\mathbb{L}_\lambda \to Y' : \ G\text{–homogeneous holomorphic line bundle.} \\ &\pi^q_\lambda : \ \text{representation of } G \text{ on } H^{0,q}_2(Y; \mathbb{L}_\lambda). \end{aligned}$$

LEMMA 7.5.2. $[\pi^q_\lambda] = [\pi^q_\mu]$.

PROOF. This is a Leray spectral sequence argument. Let $\mathcal{O}(Y'; \mathbb{L}_\lambda)$ denote the sheaf of germs of holomorphic sections of $\mathbb{L}_\lambda \to Y'$. Each integer $s \geqq 0$ gives a sheaf $\mathcal{R}^s(Y; \mathbb{L}_\lambda) \to Y$, associated to the presheaf that assigns the sheaf cohomology group $H^s(Y' \cap r^{-1}D; \mathcal{O}(Y'; \mathbb{L}_\lambda))$ to an open set $D \subset Y$. Since $r : Y' \to Y$ is a holomorphic fiber bundle, $\mathcal{R}^s(\mathbb{L}_\lambda)$ is the sheaf of germs of holomorphic sections of the holomorphic vector bundle over $Y$ whose fiber at $y \in Y$ is $H^s(Y' \cap r^{-1}(y); \mathcal{O}(\mathbb{L}_\lambda))$.

Recall our Borel-Weil Theorem from Proposition 1.1.9 with $q_0 = 0$, and apply it to $Y' \cap r^{-1}(y) = U(x') \cong U/H$. That says $H^0(Y' \cap r^{-1}(y); \mathcal{O}(\mathbb{L}_\lambda)) = E_\mu$ as $U$-module and $H^s(Y' \cap r^{-1}(y); \mathcal{O}(\mathbb{L}_\lambda)) = 0$ for $s > 0$. Now $\mathcal{R}^0(\mathbb{L}_\lambda) = \mathcal{O}(\mathbb{E}_\mu)$ and $H^s(Y; \mathbb{L}_\lambda) = 0$ for $s > 0$.

Our analysis of the direct image sheaves $\mathcal{R}^s(\mathbb{L}_\lambda)$ shows that the Leray spectral sequence collapses for $r : Y' \to Y$, so each $H^q(Y'; \mathcal{O}(\mathbb{L}_\lambda)) = H^q(Y; \mathcal{O}(\mathbb{E}_\mu))$ as $G$-modules. More to the point, we carry the spectral sequence over from sheaf cohomology to Dolbeault cohomology and use the Andreotti–Vesentini theory ((7.1.9) and (7.1.10)) to restrict considerations to square integrable forms. Then the resultant square integrable Leray spectral sequence collapses and we conclude that each $H_2^{0,q}(Y';\mathbb{L}_\lambda) = H_2^{0,q}(Y;\mathbb{E}_\mu)$ as $G$-modules. □

As immediate consequence of Lemmas 7.4.2 and 7.5.2 we have

LEMMA 7.5.3. *In the proof of* Theorem 7.2.3 *we may assume that $G$ is connected, that $Q$ is a Borel subgroup of $\overline{G}_{\mathbb{C}}$ and that $U = H$.*

**7.6.** Next, we prove the formula $\sum_{q\geqq 0}\Theta^{disc}_{\pi^q_\mu} = (-1)^{|\Sigma^+|+q(\lambda+\rho)}\Theta_{\pi_{\chi,\lambda+\rho}}$. By Lemma 7.5.3 we may assume that $G$ is connected, that $Q$ is a Borel subgroup of $\overline{G}_{\mathbb{C}}$ and that $U = H$. $K/Z$ is the maximal compact subgroup of $G/Z$ that contains the compact Cartan subgroup $H/Z$. If $[\pi] \in \widehat{G}_\zeta$ Lemma 3.2.1, and an argument [**11**, §5] of Harish–Chandra say that

(7.6.1)
$$\begin{aligned}
&\pi|_K = \sum_{\widehat{K}_\zeta} m_\kappa \cdot \kappa \text{ where } 0 \leqq m_\kappa \leqq n_G(\dim \kappa)\\
&(\pi|_K)(f) = \int_K f(k)\pi(k)dk,\ f \in C_c^\infty(K), \text{ is of trace class, and}\\
&T_\pi : C_c^\infty(K) \to \mathbb{C} \text{ defined by } f \mapsto \text{trace } (\pi|_K)(f) \text{ is a distribution on } K.
\end{aligned}$$

Harish–Chandra's argument [**16**, §12] now shows that $T_\pi|_{K\cap G'}$ is a real analytic function on $K \cap G'$ and that $T_\pi|_{K\cap G'} = \Theta_\pi|_{K\cap G'}$ .

Recall the Cartan involution $\theta$ of $G$ with fixed point set $K$. Fix a nondegenerate invariant bilinear form $\langle\ ,\ \rangle$ on $\mathfrak{g}_{\mathbb{C}}$ that restricts to the Killing form on the derived algebra and is negative definite on $\mathfrak{k} = (\mathfrak{k} \cap [\mathfrak{g},\mathfrak{g}]) \oplus \mathfrak{c}$. That gives us a positive definite $\mathrm{Ad}(K)$-invariant hermitian inner product $(u, v) = -\langle u, \theta\tau v\rangle$ on $\mathfrak{g}_{\mathbb{C}}$ where $\tau$ is complex conjugation of $\mathfrak{g}_{\mathbb{C}}$ over $\mathfrak{g}$ .

Consider the nilpotent algebra $\mathfrak{n} = \sum_{\phi\in\Sigma^+} \overline{\mathfrak{g}}^{-\phi} = \mathfrak{q}_x^u \subset \mathfrak{g}_{\mathbb{C}}$ . Denote

(7.6.2) $$\Lambda(\mathrm{Ad}^*) = \sum_{j\geqq 0} \Lambda^j(\mathrm{Ad}^*) : \text{ representation of } \mathfrak{q}_x \text{ on } \Lambda\mathfrak{n}^* = \sum_{j\geqq 0}\Lambda^j\mathfrak{n}^* .$$

The inner product $(\ ,\ )$ gives $\mathfrak{n}$, thus $\mathfrak{n}^*$, thus also $\Lambda\mathfrak{n}^*$, a Hilbert space structure; and $\mathrm{Ad}^*(\mathfrak{h})$ acts by skew-hermitian transformations.

Fix $[\pi] \in \widehat{G}_\zeta$ and write $\mathcal{H}_\pi$ for its representation space. Let $\mathcal{H}_\pi^0$ denote the space of $K$-finite vectors in $\mathcal{H}_\pi$ . It is dense and consists of analytic vectors, by (7.6.1). Now $\mathfrak{h}$ acts on $\mathcal{H}_\pi^0 \otimes \Lambda\mathfrak{n}^*$ by $\pi\otimes\mathrm{Ad}^*$ by skew–hermitian transformations. Let $\{y_1, \cdots, y_n\}$ be a basis of $\mathfrak{n}$, let $\{\omega^j\}$ be the dual basis of $\mathfrak{n}^*$, and $e(\omega^j) : \Lambda\mathfrak{n}^* \to \Lambda\mathfrak{n}^*$ the exterior product. Then $\delta := \sum\{(\pi(\omega_j) \otimes e(\omega^j)) + \frac{1}{2}1 \otimes (e(\omega_j)\,\mathrm{Ad}^*(y_j))\}$ is the coboundary operator $\mathcal{H}_\pi^0 \to \mathcal{H}_\pi^0$ of the Lie algebra cohomology for the action of $\mathfrak{h}$. It has formal adjoint $\delta^* = \sum\{(-\pi(\tau y_j) \otimes i(\omega^j)) + (\frac{1}{2}1 \otimes \mathrm{Ad}^*(y_j)^* i(\omega^j))\}$ where $i(\omega^j)$ denotes interior product. Now $\delta + \delta^*$ is a densely defined symmetric operator on $\mathcal{H}_\pi \otimes \Lambda\mathfrak{n}^*$.

Choose a basis $\{z_i\}$ of $\mathfrak{k}_{\mathbb{C}}$ that is orthonormal relative to $(\ ,\ )$. Then $\Omega_K = \sum z_i z_i \in \mathcal{U}(\mathfrak{k})$ is independent of choice of the basis $\{z_i\}$. In particular $\Omega_K$ is a linear combination, positive coefficients, of the Casimir operators of the simple ideals of $\mathfrak{k}$ plus the Laplacian on the center of $\mathfrak{k}$. Thus (7.6.1) $\pi(\Omega_K)$ is symmetric non-negative on $\mathcal{H}_\pi^0$ and has a unique self adjoint extension $\pi(\Omega_K)$ to $\mathcal{H}_\pi$ . Further $\mathcal{H}_\pi$ is the discrete direct sum of the (all non–negative) eigenspaces of $\pi(\Omega_K)$. As

$$\{[\kappa] \in \widehat{K}_\zeta : \kappa(\text{ Casimir element of } \mathcal{U}(\mathfrak{k})) \leqq c\}$$

is finite for every real $c$, (7.6.1) also says that the sum of the eigenspaces of $\pi(\Omega_K)$ for eigenvalues $\leqq c$ has finite dimension. Thus $(1+\pi(\Omega_K))^{-1}$ is a self adjoint compact operator on $\mathcal{H}_\pi$. With this preparation, Wilfried Schmid's arguments [**40**, §3]. are valid in our situation. We state the result.

LEMMA 7.6.3. *The closure of $\delta+\delta^*$ from the domain $\mathcal{H}_\pi^0 \otimes \Lambda\mathfrak{n}^*$ is the unique self adjoint extension of $\delta+\delta^*$ on $\mathcal{H}_\pi \otimes \Lambda\mathfrak{n}^*$. Each*

$$\mathcal{H}^q(\pi): \text{ kernel of } \delta+\delta^* \text{ on } \mathcal{H}_\pi \otimes \Lambda^q\mathfrak{n}^*$$

*is a finite dimensional $H$-module. Define $f_\pi = \sum(-1)^q(\text{character of } H \text{ on } \mathcal{H}^q(\pi))$. Let $\Delta$ and $\rho$ be as in (7.2.1) and $n=\dim_{\mathbb{C}}\mathfrak{n}=|\Sigma^+|$. Then*

$$f_\pi|_{H\cap G'} = (-1)^n\Delta e^\rho \cdot T_\pi|_{H\cap G'}.$$

Let $d\pi$ denote Plancherel measure on $\widehat{G}_\zeta$, so $L_2(G/Z,\zeta)=\int_{\widehat{G}_\zeta}\mathcal{H}_\pi\widehat{\otimes}\mathcal{H}_\pi^*\,d\pi$. We have the unitary $G$-module structure $L_2^{0,q}(Y;\mathbb{L}_\lambda)=\int_{\widehat{G}_\zeta}\mathcal{H}_\pi\widehat{\otimes}\{\mathcal{H}_\pi^*\otimes\Lambda^q\mathfrak{n}^*\otimes L_\lambda\}^H d\pi$ where $L_\lambda$ is the representation space of $e^\lambda$, where $H$ acts on $\mathcal{H}_\pi^*\otimes\Lambda^q\mathfrak{n}^*\otimes L_\lambda$ by $\pi^*\otimes\mathrm{Ad}^*\otimes e^\lambda$, and where $(\cdot)^H$ denotes the fixed points of $H$ there. Now $\overline{\partial}: A^{0,q}(Y;\mathbb{L}_\lambda)\to A^{0,q+1}(Y;\mathbb{L}_\lambda)$ and its formal adjoint $\overline{\partial}^*$ act by

$$\overline{\partial}(f\cdot\omega^J\cdot\ell)=\textstyle\sum_{1\le k\le n}(y_k(f)\cdot e(\omega^k)\omega^J\cdot\ell)+\frac{1}{2}\sum_{1\le k\le n}(f\cdot e(\omega^k)\,\mathrm{Ad}^*(y_k)\omega^J\cdot\ell)$$

and

$$\overline{\partial}^*(f\cdot\omega^I\cdot\ell)=-\textstyle\sum_{1\le k\le n}(\tau(y_k)f\cdot i(\omega^k)\omega^I\cdot\ell)+\frac{1}{2}\sum_{1\le k\le n}(f\cdot\mathrm{Ad}^*(y_k)^*i(\omega^k)\omega^I\cdot\ell)$$

where $I$ and $J$ are multi-indices and $\ell\in L_\lambda$. These correspond to the formulae for $\delta$ and $\delta^*$. The argument of [**40**, Lemmas 5 and 6] shows that $[\pi]\mapsto\{\mathcal{H}^q(\pi^*)\otimes L_\lambda\}^H$ is a measurable assignment of Hilbert spaces on $\widehat{G}_\zeta$, and that

$$\mathcal{H}_2^{0,q}(Y;\mathbb{L}_\lambda)=\int_{\widehat{G}_\zeta}\mathcal{H}_\pi\otimes\{\mathcal{H}^q(\pi^*)\otimes L_\lambda\}^H d\pi, \text{ unitary } G\text{–module; i.e.,}$$

$$\pi_\lambda^q=\int_{\widehat{G}_\zeta}\dim(\mathcal{H}^q(\pi^*)\otimes L_\lambda)^H\cdot\pi\,d\pi, \text{ which has discrete part} \tag{7.6.4}$$

$${}^0\pi_\lambda^q=\textstyle\sum_{\widehat{G}_{\zeta-disc}}\dim(\mathcal{H}^q(\pi^*)\otimes L_\lambda)^H\cdot\pi.$$

The equation for ${}^0\pi_\lambda^q$ is summation over the discrete set $W_G\backslash\{\nu\in L' : e^{\nu-\rho}|_Z=\zeta\}$.

Write $\Theta_{\pi_\lambda^q}^{disc}$ for the formal sum of the characters of the irreducible (in this case, $\zeta$-discrete) subrepresentations of $\pi_\lambda^q$. Define

$$\begin{aligned}F_\lambda &=\textstyle\sum_{q\geqq 0}(-1)^q\dim(\mathcal{H}^q(\pi^*)\otimes L_\lambda)^H\cdot\Theta_\pi\\ &=\textstyle\sum_{\pi\in\widehat{G}_{\zeta\text{-disc}}}\Big(\sum_{q\geqq 0}(-1)^q\dim(\mathcal{H}^q(\pi^*)\otimes L_\lambda)^H\cdot\Theta_\pi\\ &=\textstyle\sum_{\widehat{G}_{\zeta-disc}}(\text{coefficient of } e^{-\lambda} \text{ in } f_{\pi^*})\,\Theta_\pi\end{aligned}$$

because $\dim(\mathcal{H}^q(\pi^*)\otimes L_\lambda)^H$ is the multiplicity of $e^{-\lambda}$ for $H$ on $\mathcal{H}^q(\pi^*)$. Let $\nu\in L'$ with $e^{\nu-\rho}|_Z=\zeta$, so $[\pi_\nu]\in\widehat{G}_{\zeta-disc}$. By (7.6.1),

$$\Delta\,T_{\pi_\nu^*}|_{H\cap G'}=\Delta\,T_{\pi_{-\nu}}|_{H\cap G'}=(-1)^{q(\nu)}\sum_{w\in W_G}\det(w)e^{-w(\nu)}.$$

Thus Lemma 7.6.4 says $f_{\pi_\nu^*}=(-1)^{n+q(\nu)}\sum_{w\in W_G}\det(w)e^{-w(\nu)}$. In particular the coefficient of $e^{-\lambda}$ in $f_{\pi_\nu^*}$ is 0 if $\lambda+\rho\notin W_G(\nu)$, is $(-1)^{n+q(\nu)}\det(w)$ if $w(\nu)=\lambda+\rho$

for some $w \in W_G$. Now $F_\lambda = (-1)^{n+q(\nu)}\Theta_{\pi_{\lambda+\rho}}$. In view of Lemma 7.5.3, we have just proved the alternating sum formula

(7.6.5) $$\sum_{q\geqq 0}\Theta^{disc}_{\pi^q_\mu} = (-1)^{|\Sigma^+|+q(\lambda+\rho)}\Theta_{\pi_{\chi,\lambda+\rho}}.$$

This is a key step in the proof of Theorem 7.2.3.

**7.7.** We complete the proof of Theorem 7.2.3. The crux of the matter is the vanishing statement,

$$\text{if } \lambda+\rho \in L' \text{ then } H^{0,q}_2(Y;\mathbb{E}_\mu) = 0 \text{ for } q \neq q(\lambda+\rho)$$

combined with the alternating sum formula (7.6.5).

The vanishing statement was proved by Griffiths and Schmid [**8**, Theorem 7.8] for the case where $G$ is a connected semisimple Lie group with finite center, $Q$ is a Borel subgroup of $\overline{G}_{\mathbb{C}}$, and $\lambda+\rho$ is "sufficiently" far from the walls of the Weyl chamber that contains it. Then the requirement of "sufficiently" far from the wall was eliminated by Schmid [**41**] using methods not available earlier. Both proofs go through without change in our case.

Now we have $H^{0,q}_2(Y;\mathbb{L}_\lambda) = 0$ for $q \neq q(\lambda+\rho)$. Using the alternating sum formula (7.6.5) and linear independence of the $\Theta_\pi$ for $[\pi] \in \widehat{G}_{\zeta-disc}$, we see that

(7.7.1) $$[\pi_{\lambda+\rho}] \text{ is the discrete part } [{}^0\pi^{q(\lambda+\rho)}_\lambda] \text{ of } [\pi^{q(\lambda+\rho)}_\lambda].$$

Corollary 5.1.5 applied to $\overline{\zeta}$, with (7.6.1), tells us that

$$\{[\pi] \in \widehat{G}_\zeta \setminus \widehat{G}_{\zeta-disc} \mid T_\pi \neq 0\} \text{ has Plancherel measure zero in } \widehat{G}_\zeta.$$

Lemma 7.6.3 says $f_{\pi^*} = 0$ for almost all $[\pi] \in \widehat{G}_\zeta \setminus \widehat{G}_{\zeta-disc}$. If $q \neq q(\lambda+\rho)$ now (7.6.4) and $H^{0,q}_2(Y;\mathbb{E}_\mu) = 0$ force $(\mathcal{H}^q(\pi^*)\otimes L_\lambda)^H = 0$, so $e^{-\lambda}$ has multiplicity 0 in the representation of $H$ on $\mathcal{H}^q(\pi^*)$. If $f_{\pi^*} = 0$ then also $e^{-\lambda}$ has multiplicity 0 in the representation of $H$ on $\mathcal{H}^{q(\lambda+\rho)}(\pi^*)$, so $(\mathcal{H}^{q(\lambda+\rho)}(\pi^*)\otimes L_\lambda)^H = 0$. In summary,

(7.7.2) $$(\mathcal{H}^{q(\lambda+\rho)}(\pi^*)\otimes L_\lambda)^H = 0 \text{ for almost all } [\pi] \in \widehat{G}_\zeta \setminus \widehat{G}_{\zeta-disc}.$$

The measure $(\dim(\mathcal{H}^{q(\lambda+\rho)}(\pi^*)\otimes L_\lambda)^H d\pi$ on $\widehat{G}_\zeta$ is concentrated on $\widehat{G}_{\zeta-disc}$ by (7.7.2). Now (7.6.4) says that $[\pi^{q(\lambda+\rho)}_\lambda] = [{}^0\pi^{q(\lambda+\rho)}_\lambda]$, so $[\pi^{q(\lambda+\rho)}_\lambda] = [\pi_{\lambda+\rho}]$ by (7.7.1). That completes the proof of Theorem 7.2.3.

## 8. Measurable Orbits and Nondegenerate Series

Let $G$ be a reductive Lie group from the class $\widetilde{\mathcal{H}}$ of general real reductive Lie groups defined in §3.1. If $\zeta \in \widehat{Z}$ then Theorem 5.1.1 shows that Plancherel measure on $\widehat{G}_\zeta$ is supported by the constituents of $H$–series classes that transform by $\zeta$, as $H$ runs over the conjugacy classes of Cartan subgroups of $G$. Here we work out geometric realizations for all these $H$–series classes. Our method is a reduction to the special case of the relative discrete series ($H/Z$ compact) that we studied in §7.

Fix a Cartan subgroup $H = T \times A$ in $G$ and an associated cuspidal parabolic subgroup $P = MAN$ of $G$. We work over measurable orbits $Y = G(x) \subset X = \overline{G}_{\mathbb{C}}/Q$ such that (i) the $G$–normalizer $N_{[x]}$ of the holomorphic arc $S_{[x]}$ is open in $P$ and (ii) $U = \{m \in M : m(x) = x\}$ is compact modulo $Z$. In §8.1 we first check that $G$ has isotropy subgroup $UAN$ at $x$, $U = Z_M(M^0)U^0$, and that $N_{[x]} = M^\dagger AN$. If $[\mu] \in \widehat{U}$ and $\sigma \in \mathfrak{a}^*$ we show that the $G$–homogeneous complex vector bundle

$$p : \mathbb{E}_{\mu,\sigma} \to G/UAN = Y \text{ associated to } \mu \otimes e^{\rho_{\mathfrak{a}}+i\sigma}$$

is holomorphic over each holomorphic arc component of $Y$, in an essentially unique manner.

Let $K$ be the fixed point set of a Cartan involution that preserves $H$. Since $\mu$ is unitary we get a $K$–invariant hermitian metric on $\mathbb{E}_{\mu,\sigma}$. Since $U/Z$ is compact we get a $K$–invariant assignment of hermitian metrics on the holomorphic arc components of $Y$. This results in Hilbert spaces $\mathcal{H}_2^{p,q}(Y;\mathbb{E}_{\mu,\sigma})$ of "square integrable partially harmonic $(p,q)$–forms" on $Y$ with values in $\mathbb{E}_{\mu,\sigma}$: measurable $\omega$ such that (i) $\omega|_{S_{[kx]}}$ is a harmonic $(p,q)$–form on $S_{[kx]}$ with values in $\mathbb{E}_{\mu,\sigma}|_{S_{[kx]}}$ and $L_2$ norm $||\omega|_{S_{[kx]}}|| < \infty$ for almost all $k \in K$ and (ii) $\int_{K/Z} ||\omega|_{S_{[kx]}}||^2 d(kZ) < \infty$. We end §8.1 by showing that the natural action of $G$ on $\mathcal{H}_2^{p,q}(Y;\mathbb{E}_{\mu,\sigma})$ is a unitary representation.

The representation of $G$ on $\mathcal{H}_2^{0,q}(Y;\mathbb{E}_{\mu,\sigma})$ is denoted $\pi^q_{\mu,\sigma}$. Let $\eta^q_\mu$ denote the representation of $M^\dagger$ on $\mathcal{H}_2^{0,q}(S_{[x]};\mathbb{E}_{\mu,\sigma}|_{S_{[x]}})$; we studied these in Section 7. Now we have a representation

$$\eta^q_{\mu,\sigma}(man) = e^{i\sigma}(a)\eta^q_\mu(m) \text{ of } N_{[x]} = M^\dagger AN.$$

In §8.2 we prove $[\pi^q_{\mu,\sigma}] = [\operatorname{Ind}_{N_{[x]}}^G(\eta^q_{\mu,\sigma})]$.

Our main result is Theorem 8.3.2. Split $[\mu] = [\chi \otimes \mu^0]$ where $[\chi] \in \widehat{Z_M(M^0)}$ and $[\mu^0] \in \widehat{U^0}$, where $[\mu^0]$ has highest weight $\nu$ such that $\nu + \rho_{\mathfrak{t}}$ is $\mathfrak{m}$–regular. Then the $H$–series constituents of $\pi^q_{\mu,\sigma}$ are just its irreducible subrepresentations. Their sum ${}^H\pi^q_{\mu,\sigma}$ has distribution character $\Theta^H_{\pi^q_{\mu,\sigma}}$. Further

$$\sum\nolimits_{q \geqq 0} (-1)^q \Theta^H_{\pi^q_{\mu,\sigma}} = (-1)^{|\Sigma_{\mathfrak{t}}^+| + q_M(\nu+\rho_{\mathfrak{t}})} \Theta_{\pi_{\chi,\nu+\rho_{\mathfrak{t}},\sigma}}.$$

Also, if $q \neq q_M(\nu+\rho_{\mathfrak{t}})$ then $\mathcal{H}_2^{0,q}(S_{[x]};\mathbb{E}_{\mu,\sigma}) = 0$. This combines with the alternating sum formula and some consequences of the Plancherel Theorem, yielding

$$[\pi^{q_M(\nu+\rho_{\mathfrak{t}})}_{\mu,\sigma}] = [\pi_{\chi,\nu+\rho_{\mathfrak{t}},\sigma}], \ H\text{–series class.}$$

The proof is a matter of applying the results from Section 7 to every holomorphic arc component of $Y$ and combining those results by means of the induced representation theorem of §8.2.

**8.1.** $G$ is a general real reductive Lie group from our class $\widetilde{\mathcal{H}}$ defined in §3.1. As noted at the end of §6.2, $\overline{G} = G/Z_G(G^0)$ is a linear semisimple group with complexification $\overline{G}_{\mathbb{C}} = \operatorname{Int}(\mathfrak{g}_{\mathbb{C}})$, and $G$ acts on the complex flag manifolds of $\overline{G}_{\mathbb{C}}$.

For the remainder of Section 8 we fix a Cartan subgroup $H = T \times A$ of $G$ and an associated cuspidal parabolic subgroups $P = MAN$. In order to realize the $H$–series of $G$ we work with a complex flag manifold $X = \overline{G}_{\mathbb{C}}/Q$ and a measurable $G$–orbit $Y = G(x) \subset X$ such that the $G$–normalizers of the holomorphic arc components of $Y$ in $X$ have the property

(8.1.1) $$N_{[x]} = \{g \in G : gS_{[x]} = S_{[x]} \text{ has Lie algebra } \mathfrak{p}.$$

Since the orbit $Y = G(x)$ is measurable, it is partially complex and of flag type. Thus $S_{[x]}$ is an open $M^0$–orbit on the smaller flag manifold $\overline{M}_{\mathbb{C}}(x)$ where $\overline{M} = M/Z_G(G^0)$ and $AN$ acts trivially on $S_{[x]}$. The isotropy group of $G$ at $x$ is $UAN$ where $T \subset U \subset M$. We require that

(8.1.2) $$U/Z_G(G^0) = \{m \in M : m(x) = x\}/Z_G(G^0) \text{ is compact.}$$

The $G$–orbits discussed studied in Theorem 6.5.5 form the special case in which the orbit is integrable We obtain a number of examples of that class from the construction in the paragraph after the proof of Theorem 6.5.5.

LEMMA 8.1.3. *Suppose that* $G(x) \subset X$ *is a measurable orbit, that* $N_{[x]} = \{g \in G : gS_{[x]} = S_{[x]}$ *has Lie algebra* $\mathfrak{p}$, *and that the isotropy group of* $G$ *at* $x$ *is* $UAN$ *with* $U/Z_G(G^0)$ *compact. Then the open orbit* $M(x) \subset \overline{M}_{\mathbb{C}}(x)$ *is measurable and integrable. Further* $U = Z_M(M^0)U^0$, $U \cap M^0 = U^0$, $UM^0 = M^\dagger$, *and* $M/M^\dagger$ *generates the topological components of* $M(x)$. *Finally,* $N_{[x]} = M^\dagger AN$, *and* $G/M^\dagger G^0$ *enumerates the topological components of* $Y = G(x)$.

PROOF. The open orbit $M(x) \subset \overline{M}_{\mathbb{C}}(x) = \overline{M}_{\mathbb{C}}/(Q \cap \overline{M}_{\mathbb{C}})$ satisfies (7.1.1). Applying Lemma 7.1.2 to it, we get the first two assertions. For the third, $N_{[x]} = UN^0_{[x]} = UM^0AN = M^\dagger AN$, and the $G$–normalizer of $G^0(x)$ is $UG^0 = UM^0G^0 = M^\dagger G^0$. □

REMARK 8.1.4. $G^\dagger \subset M^\dagger G^0$ in general, but one can have $G^\dagger \neq M^\dagger G^0$. For example let $G = SL(2,R) \cup \left(\begin{smallmatrix} 1 & 0 \\ 0 & -1 \end{smallmatrix}\right) SL(2,R)$ and $\mathfrak{h} = \{\left(\begin{smallmatrix} a & 0 \\ 0 & -a \end{smallmatrix}\right) : a \text{ real}\}$. Then $M^\dagger = M = \{\pm\left(\begin{smallmatrix} 1 & 0 \\ 0 & 1 \end{smallmatrix}\right), \left(\pm\begin{smallmatrix} 1 & 0 \\ 0 & -1 \end{smallmatrix}\right)\}$ so $M^\dagger G^0 = G \neq G^\dagger = G^0$.

Fix $[\mu] \in \widehat{U}$ and $\sigma \in \mathfrak{a}^*$, so $[\mu \otimes e^{i\sigma}] \in \widehat{U \times A}$. As usual, $\Sigma_{\mathfrak{a}}^+$ is the positive $\mathfrak{a}$–root system on $\mathfrak{g}$ such that $\mathfrak{n}$ is the sum of the negative $\mathfrak{a}$–root spaces, and $\rho_{\mathfrak{a}} = \frac{1}{2}\sum_{\phi \in Sigma_{\mathfrak{a}}^+}(\dim \mathfrak{g}^\phi)\phi$, so $\mathfrak{a}$ acts on $\mathfrak{n}$ with trace $-2\rho_{\mathfrak{a}}$. Now $UAN$ acts on the representation space $V_\mu$ of $\mu$ by

$$\gamma_{\mu,\sigma}(uan) = e^{\rho_{\mathfrak{a}}+i\sigma}(a)\mu(u).$$

That specifies the associated $G$–homogeneous complex vector bundle

$$p : \mathbb{E}_{\mu,\sigma} \to G/UAN = G(x) = Y. \tag{8.1.5}$$

LEMMA 8.1.6. *There is a unique assignment of complex structures to the parts* $p^{-1}S_{[gx]}$ *of* $\mathbb{E}_{\mu,\sigma}$ *over the holomorphic arc components of* $Y$, *such that each restriction* $\mathbb{E}_{\mu,\sigma}|_{S_{[gx]}} \to S_{[gx]}$ *is an* $N_{[gx]}$*–homogeneous holomorphic vector bundle. The assignment is a* $G$*–equivariant real analytic tangent space distribution on* $\mathbb{E}_{\mu,\sigma}$.

PROOF. Lemma 7.1.4 says that $p^{-1}S_{[gx]}$ has a unique complex structure for which $\mathbb{E}_{\mu,\sigma}|_{S_{[gx]}} \to S_{[gx]}$ is an $\mathrm{Ad}(g)M^\dagger$–homogeneous holomorphic vector bundle. Each $\mathrm{Ad}(g)(an)$ is trivial on $S_{[gx]} = gS_{[x]}$ and multiplies all fibers of $\mathbb{E}_{\mu,\sigma}|_{S_{[gx]}}$ by the same scalar $e^{\rho_{\mathfrak{a}}+i\sigma}(a)$. Now the complex structure on $p^{-1}S_{[gx]}$ is invariant by the action of $\mathrm{Ad}(g)N_{[x]} = N_{[gx]}$, so $\mathbb{E}_{\mu,\sigma}|_{S_{[gx]}} \to S_{[gx]}$ is an $N_{[gx]}$–homogeneous holomorphic vector bundle. Finally, the assignment of complex structures to the $p^{-1}S_{[gx]}$ is $G$–invariant by uniqueness, thus also real analytic. □

If $z \in Y = G(x)$ we have the holomorphic tangent space $T_z$ to $S_{[z]}$ at $z$. Evidently $\{T_z\}_{z \in Y}$ is a $G$–invariant complex tangent space distribution on $Y$, so it is real analytic. Thus $\mathbb{T} := \bigcup_{z \in Y} T_z$ is a $G$–homogeneous real analytic sub–bundle of the complexified tangent bundle of $Y$. Given non-negative integers $p$ and $q$, the space of *partially smooth* $(p,q)$*–forms* on $Y$ with values in $\mathbb{E}_{\mu,\sigma}$ is

$$\begin{aligned} A^{p,q}(Y;\mathbb{E}_{\mu,\sigma}) &: \text{measurable sections } \alpha \text{ of } \mathbb{E}_{\mu,\sigma} \otimes \Lambda^p\mathbb{T}^* \otimes \Lambda^q\overline{\mathbb{T}}^* \\ &\quad \text{where } \alpha \text{ is } C^\infty \text{ on each holomorphic arc component of } Y. \end{aligned} \tag{8.1.7}$$

If $\alpha \in A^{p,q}(Y;\mathbb{E}_{\mu,\sigma})$ and $z \in Y$ then $\alpha|_{S_{[z]}}$ is a smooth $(p,q)$–form on $S_{[z]}$ with values in $\mathbb{E}_{\mu,\sigma}|_{S_{[z]}}$, in the ordinary sense. The $\overline{\partial}$ operator of $X$ specifies operators $\overline{\partial} : A^{p,q}(Y;\mathbb{E}_{\mu,\sigma}) \to A^{p,q+1}(Y;\mathbb{E}_{\mu,\sigma})$.

We need hermitian metrics for the harmonic theory. Let $\theta$ be a Cartan involution of $G$ with $\theta(H) = H$ and denote $K = \{g \in G \mid \theta(g) = g\}$ as usual. Then $K \cap N_{[x]} = K \cap M^\dagger$ can be assumed to contain $U$, and we have an $M^\dagger$–invariant hermitian metric on the complex manifold $S_{[x]}$. Every holomorphic arc component of $G(x)$ is an $S_{[kx]}$, $k \in K$. Give $S_{[kx]}$ the hermitian metric such that the $k : S_{[x]} \to S_{[kx]}$ are hermitian isometries. In other words, we have a $K$–invariant hermitian metric on the fibers of the bundle $\mathbb{T} \to Y$. Similarly the unitary structure of $E_\mu$ specifies an $M^\dagger$–invariant hermitian metric on the fibers of $\mathbb{E}_{\mu,\sigma} \to Y$. Now we have $K$–invariant hermitian metrics on the fibers of the bundles $\mathbb{E}_{\mu,\sigma} \otimes \Lambda^p\mathbb{T}^* \otimes \Lambda^q\overline{\mathbb{T}^*} \to Y$. As in (7.1.5), that specifies Hodge–Kodaira operators

$$(8.1.8) \qquad A^{p,q}(Y;\mathbb{E}_{\mu,\sigma}) \overset{\#}{\to} A^{n-p,n-q}(Y;\mathbb{E}^*_{\mu,\sigma}) \overset{\widetilde{\#}}{\to} A^{p,q}(Y;\mathbb{E}_{\mu,\sigma})$$

where $n = \dim_{\mathbb{C}} S_{[x]}$. It also specifies a pre Hilbert space

$$(8.1.9) \quad A_2^{p,q}(Y;\mathbb{E}_{\mu,\sigma}) = \left\{ \alpha \in A^{p,q}(Y;\mathbb{E}_{\mu,\sigma}) \,\middle|\, \int_{K/Z} \left( \int_{S_{[kx]}} \alpha\bar{\wedge}\#\alpha \right) d(kZ) < \infty \right\}$$

whose inner product is $\langle \alpha, \beta \rangle = \int_{K/Z} \left( \int_{S_{[kx]}} \alpha\bar{\wedge}\#\beta \right) d(kZ)$.

We define *square integrable partially-$(p,q)$-form* on $Y$ with values in $\mathbb{E}_{\mu,\sigma}$ to mean an element of

$$(8.1.10) \qquad L_2^{p,q}(Y;\mathbb{E}_{\mu,\sigma}) : \text{ Hilbert space completion of } A_2^{p,q}(Y;\mathbb{E}_{\mu,\sigma}).$$

$\overline{\partial}$ is densely defined on $L_2^{p,q}(\mathbb{E}_{\mu,\sigma})$ with formal adjoint $\overline{\partial}^* = -\widetilde{\#}\overline{\partial}\#$; this follows from the corresponding standard fact (7.1.8) over each holomorphic arc component. The analogue of the Hodge–Kodaira–Laplacian is

$$(8.1.11) \qquad \square = (\overline{\partial} + \overline{\partial}^*)^2 = \overline{\partial}\overline{\partial}^* + \overline{\partial}^*\overline{\partial},$$

which is elliptic and essentially self adjoint over every holomorphic arc component. Now $\square$ is essentially self adjoint on $L_2^{p,q}(Y;\mathbb{E}_{\mu,\sigma})$ from the domain consisting of $C^\infty$ forms with support compact modulo $Z$. We write $\square$ for the closure, which is the unique self–adjoint extension on $L_2^{p,q}(Y;\mathbb{E}_{\mu,\sigma})$. The kernel

$$(8.1.12) \qquad H_2^{p,q}(Y;\mathbb{E}_{\mu,\sigma}) = \{\omega \in L_2^{p,q}(Y;\mathbb{E}_{\mu,\sigma}) \mid \square\omega = 0\}$$

is the space of *square integrable partially harmonic* $(p,q)$*–forms* on $Y$ with values in $\mathbb{E}_{\mu,\sigma}$. $H_2^{p,q}(Y;\mathbb{E}_{\mu,\sigma})$ is the subspace of $A_2^{p,q}(Y;\mathbb{E}_{\mu,\sigma})$ consisting of all elements $\omega$ such that $\omega|_{S_{[kx]}}$ is harmonic a.e. in $K/Z$. It is a closed subspace of $L_2^{p,q}(Y;\mathbb{E}_{\mu,\sigma})$ and there is an orthogonal direct sum decomposition

$$(8.1.13) \quad L_2^{p,q}(Y;\mathbb{E}_{\mu,\sigma}) = c\ell\, \overline{\partial} L_2^{p,q-1}(Y;\mathbb{E}_{\mu,\sigma}) \oplus \overline{\partial}^* L_2^{p,q+1}(Y;\mathbb{E}_{\mu,\sigma}) \oplus H_2^{p,q}(Y;\mathbb{E}_{\mu,\sigma})$$

obtained by applying (7.1.10) to each holomorphic arc component.

LEMMA 8.1.14. *The action* $[\widetilde{\pi}_{\mu,\sigma}^{p,q}\alpha](z) = g(\alpha(g^{-1}z))$ *of* $G$ *on* $L_2^{p,q}(Y;\mathbb{E}_{\mu,\sigma})$ *is a unitary representation.*

PROOF. $\mathbb{E}_{\mu,\sigma} \otimes \Lambda^p \mathbb{T}^* \otimes \Lambda^q \overline{\mathbb{T}}^*$ has fiber $E_\mu^{p,q} := E_\mu \otimes \Lambda^p T_x^* \otimes \Lambda^q \overline{T}_x^*$ over $x$. If $\mu^{p,q}$ denotes the representation of $U$ on $E_\mu^{p,q}$ then $UAN$ acts on $E_\mu^{p,q}$ by

$$\gamma_{\mu,\sigma}^{p,q}(uan) = e^{\rho_{\mathfrak{a}}+i\sigma}(a)\mu^{p,q}(u) = e^{\rho_{\mathfrak{a}}}(a) \cdot {}'\gamma_{\mu,\sigma}^{p,q}(uan)$$

where ${}'\gamma_{\mu,\sigma}^{p,q} = \mu^{p,q} \otimes e^{i\sigma}$ is unitary. Since $e^{\rho_{\mathfrak{a}}}(a)$ is the square root of the determinant of $uan$ on the real tangent space $\mathfrak{g}/(\mathfrak{u}+\mathfrak{a}+\mathfrak{n})$ to $Y$ at $x$, now $\widetilde{\pi}_{\mu,\sigma}^{p,q}$ is the unitarily induced representation $\mathrm{Ind}_{UAN}^G({}'\gamma_{\mu,\sigma}^{p,q})$. □

The representation $\widetilde{\pi}_{\mu,\sigma}^{p,q}$ commutes with $\overline{\partial}$, hence also with $\overline{\partial}^*$, so $H_2^{p,q}(Y;\mathbb{E}_{\mu,\sigma})$ is a closed $G$–invariant subspace of $L_2^{p,q}(Y;\mathbb{E}_{\mu,\sigma})$. Thus we have

(8.1.15) $\pi_{\mu,\sigma}^{p,q}$ : unitary representation of $G$ on $H_2^{p,q}(Y;\mathbb{E}_{\mu,\sigma})$ and $\pi_{\mu,\sigma}^q = \pi_{\mu,\sigma}^{0,q}$

The program of Section 8 is to represent the various $H$–series of unitary representation classes of $G$ by the various $\pi_{\mu,\sigma}^q$.

**8.2.** We set up $\pi_{\mu,\sigma}^q$ as an induced representation from $N_{[x]} = M^\dagger AN$. Write $\mathbb{E}_\mu = \mathbb{E}_{\mu,\sigma}|_{S_{[x]}} \to S_{[x]}$. It is the $M^\dagger$–homogeneous hermitian holomorphic vector bundle defined by $[\mu] \in \widehat{U}$ as in Lemma 7.1.4. That gives us the unitary representations $\eta_\mu^q$ of $M^\dagger$ on $H_2^{0,q}(S_{[x]};\mathbb{E}_\mu)$. The formula $\eta_{\mu,\sigma}^q(man) = e^{i\sigma}(a)\eta_\mu^q(m)$ defines a unitary representation of $N_{[x]} = M^\dagger AN$ on $H_2^{0,q}(S_{[x]};\mathbb{E}_\mu)$.

THEOREM 8.2.1. $[\pi_{\mu,\sigma}^q] = [\mathrm{Ind}_{N_{[x]}}^G(\eta_{\mu,\sigma}^q)]$.

PROOF. Let $\widetilde{\pi} = \widetilde{\pi}_{\mu,\sigma}^{0,q}$, the representation of $G$ on $L_2^{p,q}(Y;\mathbb{E}_{\mu,\sigma})$. Let ${}'\gamma$ denote the representation of $UAN$ on $E_\mu \otimes \Lambda^q(\overline{T}_x^*)$; it is the ${}'\gamma_{\mu,\sigma}^{0,q} \otimes e^{i\sigma}$ of the proof of Lemma 8.1.14. That lemma was proved (if $p=0$) by showing $[\widetilde{\pi}] = [\mathrm{Ind}_{UAN}^G({}'\gamma)]$.

Let $\widetilde{\eta}$ denote the representation of $M^\dagger$ on $L_2^{0,q}(S_{[x]};\mathbb{E}_\mu)$, and ${}'\eta$ the representation of $M^\dagger AN$ by ${}'\eta(man) = e^{i\sigma}(a)\widetilde{\eta}(m)$. Then $[\widetilde{\eta}] = [\mathrm{Ind}_U^{M^\dagger}(\mu^{0,q})]$, and so $[{}'\eta] = [\mathrm{Ind}_{UAN}^{M^\dagger AN}(\gamma)]$.

Induction by stages now says that $\widetilde{\pi}$ is unitarily equivalent to $\mathrm{Ind}_{M^\dagger AN}^G({}'\eta)$. We need the equivalence. Let $f$ be in the representation space of $\mathrm{Ind}_{M^\dagger AN}^G({}'\eta)$. In other words $f : G \to L_2^{0,q}(S_{[x]};\mathbb{E}_\mu)$ is Borel measurable, $f$ transforms by $f(gman) = e^{-\rho_{\mathfrak{a}}}(a) \cdot {}'\eta(man)^{-1}f(g)$ for $g \in G$ and $man \in M^\dagger AN$, and we have global norms $\int_{K/Z} ||f(k)||^2 d(kZ) < \infty$. For almost all $g \in G$ we may view $f(g) \in L_2^{0,q}(S_{[x]};\mathbb{E}_\mu)$ as a Borel–measurable function $M^\dagger AN \to E_\mu^q = E_\mu \otimes \Lambda^q(\overline{T}_x^*)$ such that

$$f(g)(puan) = {}'\gamma(uan)^{-1}[f(g)(p)] \text{ for } p \in M^\dagger AN\,, uan \in UAN$$

$$\text{and } \int_{M^\dagger/U} ||f(g)(m)||^2 d(mU) < \infty.$$

Now define

(8.2.2) $$F = \Gamma(f) : G \to E_\mu^q = E_\mu \otimes \Lambda^q(\overline{T}_x^*) \text{ by } F(g) = f(g)(1).$$

Then $F$ is Borel measurable. Use ${}'\eta = \mathrm{Ind}_{UAN}^{M^\dagger AN}({}'\gamma)$ to compute

$$\begin{aligned} F(guan) &= f(guan)(1) = [e^{-\rho_{\mathfrak{a}}}(a) \cdot {}'\eta(uan)^{-1}f(g)](1) \\ &= e^{-\rho_{\mathfrak{a}}}(a) \cdot {}'\gamma(uan)^{-1}[f(g)(1)] = e^{-\rho_{\mathfrak{a}}}(a) \cdot {}'\gamma(uan)^{-1}F(g) \end{aligned}$$

and

$$\int_{K/Z}\Big\{\int_{M^\dagger/U} ||F(km)||^2\, d(mU)\Big\}d(kZ) = \int_{K/Z}\Big\{\int_{M^\dagger/U} ||f(k)(m)||^2\, d(mU)\Big\}d(kZ) < \infty.$$

Thus $f \mapsto \Gamma(f) = F$ is the desired equivalence $\widetilde{\pi} \simeq \mathrm{Ind}\,^G_{M^\dagger AN}('\eta)$.

In the construction just above, $f$ is in the representation space of $\mathrm{Ind}\,^g_{N[x]}(\eta^q_{\mu,\sigma})$ precisely when almost every $f(g)$ is annihilated by the Hodge–Kodaira–Laplace operator of $\mathbb{E}_\mu \to S_{[x]}$. That is equivalent to $\Box\Gamma(f) = 0$. Thus the equivalence $\Gamma$ of (8.2.2) restricts to the equivalence asserted in Theorem 8.2.1. $\Box$

**8.3.** We now come to the geometric realization of the various $H$–series of unitary representations of $G$. Note that these are the standard induced representations. They are unitary, in fact tempered.

We construct a particular positive $\mathfrak{h}_\mathbb{C}$–root system $\Sigma^+$. The choice $P = MAN$ us a choice $\Sigma^+_\mathfrak{a}$ of positive $\mathfrak{a}$–root system on $\mathfrak{g}$. Let $\overline{M} = M/Z_G(G^0)$, and choose a simple $\mathfrak{t}_\mathbb{C}$–root system $\Pi_\mathfrak{t}$ on $\mathfrak{m}_\mathbb{C}$ such that the parabolic subalgebra $\mathfrak{q} \cap \overline{\mathfrak{m}}_\mathbb{C}$ of $\overline{\mathfrak{m}}_\mathbb{C}$ is specified by $\Pi_\mathfrak{t}$ and a subset $\Phi_\mathfrak{t}$. $\Sigma^+$ will be the positive $\mathfrak{h}_\mathbb{C}$–root system on $\mathfrak{g}_\mathbb{C}$ determined by $\Sigma^+_\mathfrak{a}$ and $\Sigma^+_\mathfrak{t}$, and $\Pi$ is the simple root system for $\Sigma^+$. From Proposition 6.5.5 the measurable orbit $G(x)$ is integrable exactly when $\mathfrak{q} = \mathfrak{q}_\Phi$ with $\Phi = \Phi_\mathfrak{t} \cup (\Pi \setminus \Pi_\mathfrak{t})$. As we had done before for $\mathfrak{g}$ let

$$\rho_\mathfrak{t} = \tfrac{1}{2}\sum_{\phi\in\Sigma^+_\mathfrak{t}} \phi\,,\ \Delta_{M,T} = \prod_{\mathfrak{t}^+}(e^{\phi/2} - e^{-\phi/2})\,,\ \varpi_\mathfrak{t}(\nu) = \prod_{\mathfrak{t}^+}\langle \nu, \phi\rangle.$$

Replacing $G$ be a $\mathbb{Z}_2$ extension if necessary, Lemma 4.4.3 ensures that $e^{\rho_\mathfrak{t}}$ and $\Delta_{M,T}$ are well defined on $T$. As we did before for $(G, H)$ denote

(8.3.1) $\quad L_\mathfrak{t} = \{\nu \in i\mathfrak{t}^* \mid e^\nu \text{ is well defined on } T^0\}$ and $L''_\mathfrak{t} = \{\nu \in L_\mathfrak{t} \mid \varpi_\mathfrak{t}(\nu) \neq 0\}$.

Then $\rho_\mathfrak{t} \in L''_\mathfrak{t}$. If $\nu + \rho_\mathfrak{t} \in L''_\mathfrak{t}$ then $\Sigma^+_\mathfrak{t}$ specifies $q_M(\nu + \rho_\mathfrak{t}) \in \mathbb{Z}$ as in (7.2.2).

Since $U = Z_M(M^0)U^0$ and $U \cap M^0 = U^0$, $\widehat{U}$ consists of all $[\chi \otimes \mu^0]$ with $\chi \in \widehat{Z_M(M^0)}$ consistent with $[\mu^0] \in \widehat{U^0}$.

We review some aspects of our basic setup and then come to our main result.

$G$ is a general real reductive Lie group as defined in §3.1, $Q$ is a parabolic subgroup of $\overline{G}_\mathbb{C}$, and $Y = G(x) \subset X = \overline{G}_\mathbb{C}$ is a measurable integrable orbit partially complex orbit of flag type as described in §6.5. $H = T \times A \in \mathrm{Car}(G)$, $P = MAN$ is an associated cuspidal parabolic subgroup of $G$, and we suppose that $U = \{m \in M \mid m(x) = x\}$ is compact modulo $Z_G(G^0)$. $S_{[x]}$ is the holomorphic arc component of $Y$ through $x$ and its $G$–normalizer is $N_{[x]} = M^\dagger AN$ denotes the $G$–normalizer of the holomorphic arc component $S_{[x]}$.

Recall that $\pi^q_{\mu,\sigma}$ denotes the unitary representation of $G$ on $H^{0,q}_2(Y; \mathbb{E}_{\mu,\sigma})$ and that ${}^H\pi^q_{\mu,\sigma}$ denotes the sum of its irreducible subrepresentations.

THEOREM 8.3.2. *Let $[\mu] \in \widehat{U}$, say $[\mu] = [\chi \otimes \mu^0]$ as above. Let $\nu$ be the highest weight of $\mu^0$ in the $\mathfrak{t}_\mathbb{C}$–root system $\Sigma^+_\mathfrak{t} \cap \Phi^r_\mathfrak{t}$ of $\mathfrak{u}_\mathbb{C}$. Then $\nu \in L_\mathfrak{t}$ and $\mu \in \widehat{U}_\zeta$ where $\zeta \in \widehat{Z}$ agrees with $e^\nu$ on $Z \cap M^0$ and where $[\chi] \in \widehat{Z_M(M^0)}_\zeta$. Let $\sigma \in \mathfrak{a}^*$ and suppose that $\nu + \rho_\mathfrak{t} \in L''_\mathfrak{t}$.*

1. *If $q \neq q(\nu + \rho_\mathfrak{t})$ then $H^{0,q}_2(Y; E_{\mu,\sigma}) = 0$, so $\pi^q_{\mu,\sigma}$ does not occur*
2. *If $q = q(\nu + \rho_\mathfrak{t})$ then $G$ acts on $H^{0,q}_2(Y; E_{\mu,\sigma})$ by the $H$–series representation $\pi_{\chi,\nu+\rho_\mathfrak{t},\sigma}$. Every $H$–series representation is obtained in this way.*

PROOF. Theorem 7.2.3 says that the representation $\eta_\mu^q$ of $M^\dagger$ on $H_2^{0,q}(S_{[x]};\mathbb{E}_\mu)$ is trivial if $q \neq q_M(\nu+\rho_{\mathfrak{t}})$, and if $q = q_M(\nu+\rho_{\mathfrak{t}})$ it is equivalent to the relative discrete series representation $\eta_{\chi,\nu+\rho_{\mathfrak{t}}}$ of $M^\dagger$. Then the representation of $M$ on $H_2^{0,q}(S_{[x]};\mathbb{E}_\mu)$ is the relative discrete series representation (which we temporarily denote $\eta'_{\chi,\nu+\rho_{\mathfrak{t}}}$) $\mathrm{Ind}_{M^\dagger}^M(\eta_{\chi,\nu+\rho_{\mathfrak{t}}})$ of $M$. Finally, if $\sigma \in \mathfrak{a}^*$ then $G$ acts on $H_2^{0,q}(Y;E_{\mu,\sigma})$ by the $H$–series representation $\mathrm{Ind}_{MAN}^G\big(\mathrm{Ind}_{M^\dagger AN}^{MAN}(\eta_{\chi,\nu+\rho_{\mathfrak{t}}}\otimes e^{i\sigma})\big) = \pi_{\chi,\nu+\rho_{\mathfrak{t}},\sigma}$ by Theorem 8.2.1. □

**8.4.** Theorem 8.3.2 gives explicit geometric realizations for the standard tempered representations, i.e., for the various $H$–series classes of unitary representations of $G$. Theorem 7.2.3 is the special case of the relative discrete series. In view of the Plancherel Theorem 5.1.1 we now have, for every $\zeta \in \widehat{Z}$, explicit geometric realizations for a subset of $\widehat{G}_\zeta$ that supports Plancherel measure there.

## References


[1] J. D. Adams, M. van Leeuwen, P. E. Trapa & D. A. Vogan, Unitary representations of real reductive groups, to appear. {arXiv: 1212.2192v4}

[2] Aldo Andreotti and Edoardo Vesentini, *Carleman estimates for the Laplace-Beltrami equation on complex manifolds*, Inst. Hautes Études Sci. Publ. Math. **25** (1965), 81–130. MR0175148

[3] Raoul Bott, *Homogeneous vector bundles*, Ann. of Math. (2) **66** (1957), 203–248. MR0089473

[4] François Bruhat, *Sur les représentations induites des groupes de Lie* (French), Bull. Soc. Math. France **84** (1956), 97–205. MR0084713

[5] William Casselman and Dragan Miličić, *Asymptotic behavior of matrix coefficients of admissible representations*, Duke Math. J. **49** (1982), no. 4, 869–930. MR683007

[6] Jacques Dixmier, *Les C*-algèbres et leurs représentations* (French), Cahiers Scientifiques, Fasc. XXIX, Gauthier-Villars & Cie, Éditeur-Imprimeur, Paris, 1964. MR0171173

[7] Roger Godement, *Sur les relations d'orthogonalité de V. Bargmann. II. Démonstration générale* (French), C. R. Acad. Sci. Paris **225** (1947), 657–659. MR0021945

[8] Phillip Griffiths and Wilfried Schmid, *Locally homogeneous complex manifolds*, Acta Math. **123** (1969), 253–302. MR0259958

[9] Harish-Chandra, *Representations of a semisimple Lie group on a Banach space. I*, Trans. Amer. Math. Soc. **75** (1953), 185–243. MR0056610

[10] Harish-Chandra, *Representations of semisimple Lie groups. II*, Trans. Amer. Math. Soc. **76** (1954), 26–65. MR0058604

[11] Harish-Chandra, *Representations of semisimple Lie groups. III*, Trans. Amer. Math. Soc. **76** (1954), 234–253. MR0062747

[12] Harish-Chandra, *The Plancherel formula for complex semisimple Lie groups*, Trans. Amer. Math. Soc. **76** (1954), 485–528. MR0063376

[13] Harish-Chandra, *Representations of semisimple Lie groups. IV*, Amer. J. Math. **77** (1955), 743–777. MR0072427

[14] Harish-Chandra, *Representations of semisimple Lie groups. V*, Amer. J. Math. **78** (1956), 1–41. MR0082055

[15] Harish-Chandra, *Representations of semisimple Lie groups. VI. Integrable and square-integrable representations*, Amer. J. Math. **78** (1956), 564–628. MR0082056

[16] Harish-Chandra, *The characters of semisimple Lie groups*, Trans. Amer. Math. Soc. **83** (1956), 98–163. MR0080875

[17] Harish-Chandra, *Invariant eigendistributions on a semisimple Lie group*, Trans. Amer. Math. Soc. **119** (1965), 457–508. MR0180631

[18] Harish-Chandra, *Discrete series for semisimple Lie groups. I. Construction of invariant eigendistributions*, Acta Math. **113** (1965), 241–318. MR0219665

[19] Harish-Chandra, *Discrete series for semisimple Lie groups. I. Construction of invariant eigendistributions*, Acta Math. **113** (1965), 241–318. MR0219665

[20] Harish-Chandra, *Two theorems on semi-simple Lie groups*, Ann. of Math. (2) **83** (1966), 74–128. MR0194556

[21] Harish-Chandra, *Harmonic analysis on semisimple Lie groups*, Bull. Amer. Math. Soc. **76** (1970), 529–551. MR0257282

[22] Harish-Chandra, *On the theory of the Eisenstein integral*, Conference on Harmonic Analysis (Univ. Maryland, College Park, Md., 1971), Springer, Berlin, 1972, pp. 123–149. Lecture Notes in Math., Vol. 266. MR0399355

[23] Harish-Chandra, *Harmonic analysis on real reductive groups. I. The theory of the constant term*, J. Functional Analysis **19** (1975), 104–204. MR0399356

[24] Harish-Chandra, *Harmonic analysis on real reductive groups. II. Wavepackets in the Schwartz space*, Invent. Math. **36** (1976), 1–55. MR0439993

[25] Harish-Chandra, *Harmonic analysis on real reductive groups. III. The Maass-Selberg relations and the Plancherel formula*, Ann. of Math. (2) **104** (1976), no. 1, 117–201. MR0439994

[26] Henryk Hecht, Dragan Miličić, Wilfried Schmid, and Joseph A. Wolf, *Localization and standard modules for real semisimple Lie groups. I. The duality theorem*, Invent. Math. **90** (1987), no. 2, 297–332. MR910203

[27] Rebecca A. Herb and Joseph A. Wolf, *The Plancherel theorem for general semisimple groups*, Compositio Math. **57** (1986), no. 3, 271–355. MR829325

[28] Rebecca A. Herb and Joseph A. Wolf, *Rapidly decreasing functions on general semisimple groups*, Compositio Math. **58** (1986), no. 1, 73–110. MR834048

[29] Takeshi Hirai, *The characters of some induced representations of semi-simple Lie groups*, J. Math. Kyoto Univ. **8** (1968), 313–363. MR0238994

[30] Ryoshi Hotta, *On a realization of the discrete series for semisimple Lie groups*, J. Math. Soc. Japan **23** (1971), 384–407. MR0306405

[31] A. W. Knapp and Gregg J. Zuckerman, *Classification of irreducible tempered representations of semisimple groups*, Ann. of Math. (2) **116** (1982), no. 2, 389–455. MR672840

[32] A. W. Knapp and Gregg J. Zuckerman, *Classification of irreducible tempered representations of semisimple groups. II*, Ann. of Math. (2) **116** (1982), no. 3, 457–501. MR678478

[33] R. P. Langlands, On the classification of irreducible representations of real algebraic groups, mimeographed notes, Institute for Advanced Study, 1973. Also in "Representation Theory and Harmonic Analysis on Semisimple Lie Groups", Mathematical Surveys and Monographs **31**, Amer. Math. Soc. 1989, 101–170.

[34] Ronald L. Lipsman, *On the characters and equivalence of continuous series representations.*, J. Math. Soc. Japan **23** (1971), 452–480. MR0289718

[35] M. S. Narasimhan and K. Okamoto, *An analogue of the Borel-Weil-Bott theorem for hermitian symmetric pairs of non-compact type*, Ann. of Math. (2) **91** (1970), 486–511. MR0274657

[36] R. Parthasarathy, *A note on the vanishing of certain "$L^2$-cohomologies"*, J. Math. Soc. Japan **23** (1971), 676–691. MR0333076

[37] R. Parthasarathy, *Dirac operator and the discrete series*, Ann. of Math. (2) **96** (1972), 1–30. MR0318398

[38] Marc A. Rieffel, *Square-integrable representations of Hilbert algebras*, J. Functional Analysis **3** (1969), 265–300. MR0244780

[39] L. Preiss Rothschild, *Orbits in a real reductive Lie algebra*, Trans. Amer. Math. Soc. **168** (1972), 403–421. MR0349778

[40] Wilfried Schmid, *On a conjecture of Langlands*, Ann. of Math. (2) **93** (1971), 1–42. MR0286942

[41] Wilfried Schmid, *$L^2$-cohomology and the discrete series*, Ann. of Math. (2) **103** (1976), no. 2, 375–394. MR0396856

[42] Wilfried Schmid and Joseph A. Wolf, *Globalizations of Harish-Chandra modules*, Bull. Amer. Math. Soc. (N.S.) **17** (1987), no. 1, 117–120. MR888885

[43] Wilfried Schmid and Joseph A. Wolf, *Geometric quantization and derived functor modules for semisimple Lie groups*, J. Funct. Anal. **90** (1990), no. 1, 48–112. MR1047577

[44] Birgit Speh and David A. Vogan Jr., *Reducibility of generalized principal series representations*, Acta Math. **145** (1980), no. 3-4, 227–299. MR590291

[45] Juan A. Tirao and Joseph A. Wolf, *Homogeneous holomorphic vector bundles*, Indiana Univ. Math. J. **20** (1970/1971), 15–31. MR0263110

[46] Juan A. Tirao, *Square integrable representations of semisimple Lie groups*, Trans. Amer. Math. Soc. **190** (1974), 57–75. MR0338270

[47] David A. Vogan, *Irreducible characters of semisimple Lie groups. III. Proof of Kazhdan-Lusztig conjecture in the integral case*, Invent. Math. **71** (1983), no. 2, 381–417. MR689650

[48] David A. Vogan Jr., *The Kazhdan-Lusztig conjecture for real reductive groups*, Representation theory of reductive groups (Park City, Utah, 1982), Progr. Math., vol. 40, Birkhäuser Boston, Boston, MA, 1983, pp. 223–264. MR733817

[49] Joseph A. Wolf, *Spaces of constant curvature*, 6th ed., AMS Chelsea Publishing, Providence, RI, 2011. MR2742530

[50] Joseph A. Wolf, *The action of a real semisimple group on a complex flag manifold. I. Orbit structure and holomorphic arc components*, Bull. Amer. Math. Soc. **75** (1969), 1121–1237. MR0251246

[51] Joseph A. Wolf, *The action of a real semisimple Lie group on a complex flag manifold. II. Unitary representations on partially holomorphic cohomology spaces*, American Mathematical Society, Providence, R.I., 1974. Memoirs of the American Mathematical Society, No. 138. MR0393350

[52] Joseph A. Wolf, *Partially harmonic spinors and representations of reductive Lie groups*, J. Functional Analysis **15** (1974), 117–154. MR0393351

[53] Hon-Wai Wong, *Dolbeault cohomological realization of Zuckerman modules associated with finite rank representations*, J. Funct. Anal. **129** (1995), no. 2, 428–454. MR1327186

[54] Hon-Wai Wong, *Cohomological induction in various categories and the maximal globalization conjecture*, Duke Math. J. **96** (1999), no. 1, 1–27. MR1663911

Department of Mathematics, University of California, Berkeley, California 94720–3840

*Email address*: jawolf@math.berkeley.edu

## Selected Published Titles in This Series

714 **Jens Gerlach Christensen, Susanna Dann, and Matthew Dawson, Editors,** Representation Theory and Harmonic Analysis on Symmetric Spaces, 2018

708 **Christian Ausoni, Kathryn Hess, Brenda Johnson, Ieke Moerdijk, and Jérôme Scherer, Editors,** An Alpine Bouquet of Algebraic Topology, 2018

707 **Nitya Kitchloo, Mona Merling, Jack Morava, Emily Riehl, and W. Stephen Wilson, Editors,** New Directions in Homotopy Theory, 2018

706 **Yeonhyang Kim, Sivaram K. Narayan, Gabriel Picioroaga, and Eric S. Weber, Editors,** Frames and Harmonic Analysis, 2018

705 **Graham J. Leuschke, Frauke Bleher, Ralf Schiffler, and Dan Zacharia, Editors,** Representations of Algebras, 2018

704 **Alain Escassut, Cristina Perez-Garcia, and Khodr Shamseddine, Editors,** Advances in Ultrametric Analysis, 2018

703 **Andreas Malmendier and Tony Shaska, Editors,** Higher Genus Curves in Mathematical Physics and Arithmetic Geometry, 2018

702 **Mark Grant, Gregory Lupton, and Lucile Vandembroucq, Editors,** Topological Complexity and Related Topics, 2018

701 **Joan-Carles Lario and V. Kumar Murty, Editors,** Number Theory Related to Modular Curves, 2018

700 **Alexandre Girouard, Dmitry Jakobson, Michael Levitin, Nilima Nigam, Iosif Polterovich, and Frédéric Rochon, Editors,** Geometric and Computational Spectral Theory, 2017

699 **Mark L. Agranovsky, Matania Ben-Artzi, Catherine Bénéteau, Lavi Karp, Dmitry Khavinson, Simeon Reich, David Shoikhet, Gilbert Weinstein, and Lawrence Zalcman, Editors,** Complex Analysis and Dynamical Systems VII, 2017

698 **Alexander M. Blokh, Leonid A. Bunimovich, Paul H. Jung, Lex G. Oversteegen, and Yakov G. Sinai, Editors,** Dynamical Systems, Ergodic Theory, and Probability: in Memory of Kolya Chernov, 2017

697 **Fabrizio Broglia, Françoise Delon, Max Dickmann, Danielle Gondard-Cozette, and Victoria Ann Powers, Editors,** Ordered Algebraic Structures and Related Topics, 2017

696 **Ara S. Basmajian, Yair N. Minsky, and Alan W. Reid, Editors,** In the Tradition of Ahlfors–Bers, VII, 2017

695 **Katrina Barron, Elizabeth Jurisich, Antun Milas, and Kailash Misra, Editors,** Lie Algebras, Vertex Operator Algebras, and Related Topics, 2017

694 **Manjul Bhargava, Robert Guralnick, Gerhard Hiss, Klaus Lux, and Pham Huu Tiep, Editors,** Finite Simple Groups: Thirty Years of the Atlas and Beyond, 2017

693 **Michael Cwikel and Mario Milman, Editors,** Functional Analysis, Harmonic Analysis, and Image Processing, 2017

692 **Anatole Katok, Yakov Pesin, and Federico Rodriguez Hertz, Editors,** Modern Theory of Dynamical Systems, 2017

691 **Farrell Brumley, Maria Paula Gómez Aparicio, and Alberto Mínguez, Editors,** Around Langlands Correspondences, 2017

690 **Andrés Eduardo Caicedo, James Cummings, Peter Koellner, and Paul B. Larson, Editors,** Foundations of Mathematics, 2017

689 **Erica Flapan, Allison Henrich, Aaron Kaestner, and Sam Nelson, Editors,** Knots, Links, Spatial Graphs, and Algebraic Invariants, 2017

688 **Jeffrey Bergen, Stefan Catoiu, and William Chin, Editors,** Groups, Rings, Group Rings, and Hopf Algebras, 2017

For a complete list of titles in this series, visit the
AMS Bookstore at **www.ams.org/bookstore/conmseries/**.